Israel Gohberg
Seymour Goldberg
Marinus A. Kaashoek

Classes of Linear Operators Vol. I

Springer Basel AG

Authors' addresses:

I. Gohberg
School of Mathematical Sciences
Raymond and Beverly Sackler
Faculty of Exact Sciences
Tel-Aviv University
69978 Tel-Aviv
Israel

S. Goldberg
Department of Mathematics
University of Maryland
College Park, MD 20742
USA

M. A. Kaashoek
Faculteit Wiskunde en Informatica
Vrije Universiteit
De Boelelaan 1081
1081 HV Amsterdam
The Netherlands

Deutsche Bibliothek Cataloging-in-Publication Data

Gochberg, Izrail':
Classes of linear operators / Israel Gohberg ; Seymour
Goldberg ; Marinus A. Kaashoek. – Basel ; Boston ; Berlin :
Birkhäuser.
NE: Goldberg, Seymour:; Kaashoek, Marinus A.:
Vol. 1 (1990)
  (Operator theory ; Vol. 49)

ISBN 978-3-0348-7511-0   ISBN 978-3-0348-7509-7 (eBook)
DOI 10.1007/978-3-0348-7509-7

NE: GT

Originally published by Birkhäuser Verlag Basel in 1990.

PREFACE

After the book "Basic Operator Theory" by Gohberg-Goldberg was published, we, that is the present authors, intended to continue with another book which would show the readers the large variety of classes of operators and the important role they play in applications. The book was planned to be of modest size, but due to the profusion of results in this area of analysis, the number of topics grew larger than expected. Consequently, we decided to divide the material into two volumes — the first volume being presented now.

During the past years, courses and seminars were given at our respective institutions based on parts of the texts. These were well received by the audience and enabled us to make appropriate choices for the topics and presentation for the two volumes. We would like to thank G.J. Groenewald, A.B. Kuijper and A.C.M. Ran of the Vrije Universiteit at Amsterdam, who provided us with lists of remarks and corrections.

We are now aware that the Basic Operator Theory book should be revised so that it may suitably fit in with our present volumes. This revision is planned to be the last step of an induction and not the first.

We gratefully acknowledge the support from the mathematics departments of Tel Aviv University, the University of Maryland at College Park, and the Vrije Universiteit at Amsterdam, which enabled us to visit and confer with each other. We also thank the Nathan and Lillian Silver Chair in Mathematical Analysis and Operator Theory for its financial assistance.

March 15, 1990

The authors

TABLE OF CONTENTS

VOLUME I

VOLUME II

PART V: TRIANGULAR REPRESENTATIONS

PART VI: TOEPLITZ OPERATORS

INTRODUCTION

These two volumes constitute texts for graduate courses in linear operator theory. The reader is assumed to have a knowledge of both complex analysis and the first elements of operator theory. The texts are intended to concisely present a variety of classes of linear operators, each with its own character, theory, techniques and tools. For each of the classes, various differential and integral operators motivate or illustrate the main results. Although each class is treated separately and the first impression may be that of many different theories, interconnections appear frequently and unexpectedly. The result is a beautiful, unified and powerful theory.

The classes we have chosen are representatives of the principal important classes of operators, and we believe that these illustrate the richness of operator theory, both in its theoretical developments and in its applications. Because we wanted the books to be of reasonable size, we were selective in the classes we chose and restricted our attention to the main features of the corresponding theories. However, these theories have been updated and enhanced by new developments, many of which appear here for the first time in an operator-theory text.

The books present a wide panorama of modern operator theory. They are not encyclopedic in nature and do not delve too deeply into one particular area. In our opinion it is this combination that will make the books attractive to readers who know basic operator theory.

The exposition is self-contained and has been simplified and polished in an effort to make advanced topics accessible to a wide audience of students and researchers in mathematics, science and engineering.

The classes encompass compact operators, various subclasses of compact operators (such as trace class and Hilbert-Schmidt operators), Fredholm operators (bounded and unbounded), Wiener-Hopf and Toeplitz operators, selfadjoint operators (bounded and unbounded), and integral and differential operators on finite and infinite intervals. The two volumes also contain an introduction to the theory of Banach algebras with applications to algebras of Toeplitz operators, the first elements of the theory of operator semigroups with applications to initial value problems, the theory of triangular representation, the method of factorization for general operators and for matrix functions, an introduction to the theory of characteristic operator functions for contractions. Also included are recent developments concerning extension and completion problems for operator matrices and matrix functions.

The first volume is divided into Parts I–IV.

Part I discusses the elements of spectral theory that apply to arbitrary bounded operators. The topics discussed include spectral decomposition theorems, Riesz projections, functional calculus and eigenvalues of finite type. Analytic equivalence and an analysis of linear operator pencils are elements that appear here for the first time in a text book. Also included is the spectral theorem for bounded selfadjoint operators, which is presented in an unconventional form as a further refinement of the Riesz spectral projection theory.

Part II presents different classes of compact operators including trace class operators and Hilbert-Schmidt operators. Trace and determinant are introduced as natural generalizations of their matrix counter parts. This part also contains theorems about the growth of the resolvent and completeness of eigenvectors and generalized eigenvectors. Integral operators with semi-separable kernels, which arise in problems of networks and systems, are also treated here.

Part III describes Wiener-Hopf integral operators and begins with an introduction to the theory of Fredholm operators. Here the main results are index and factorization theorems. As a novelty this part contains a complete treatment of Wiener-Hopf integral equations with a rational matrix symbol based on connections with mathematical systems theory. In the latter framework a finite dimensional analogue of the transport equation is treated.

Part IV treats unbounded linear operators. Several results about bounded operators are extended to this class. Many examples of unbounded operators arising from ordinary and partial differential equations are given. This part also contains an introduction to the theory of strongly continuous semigroups with applications to initial value problems and transport theory.

The second volume is divided into Parts V–IX. The titles are as follows. Part V: Triangular Representations; Part VI: Contractive Operators; Part VII: Toeplitz Operators; Part VIII: Banach Algebras and Algebras of Operators; Part IX: Extension and Completion Problems.

PART I
GENERAL SPECTRAL THEORY

This part is devoted to elements of spectral theory that can be applied to arbitrary bounded operators regardless of the class they belong to. Three main topics are discussed, namely separation of spectra and functional calculus (Chapters I and IV), isolated eigenvalues of finite type which behave like eigenvalues of matrices (Chapter II), and analytic equivalence of operators for the case when the spectral parameter is linear as well as for nonlinear dependence (Chapter III). The spectral theory for bounded selfadjoint operators, which is the main topic of Chapter V, is presented here as a further refinement of the spectral separation theorems of F. Riesz.

CHAPTER I

RIESZ PROJECTIONS AND FUNCTIONAL CALCULUS

This chapter is concerned with the part of the spectral theory which is applicable to all bounded linear operators. It contains theorems on decomposition of operators corresponding to separated parts of the spectrum, the general version of the functional calculus and applications to operator and differential equations and to stability problems.

To make clear the approach followed in this chapter let us first recall the case when the operator is a compact selfadjoint operator A acting on a Hilbert space H. In that case H decomposes into an orthogonal sum of eigenspaces, namely

$$H = \operatorname{Ker} A \oplus \operatorname{Ker}(\lambda_1 - A) \oplus \operatorname{Ker}(\lambda_2 - A) \oplus \cdots$$

where $\lambda_1, \lambda_2, \ldots$ is the sequence of distinct non-zero eigenvalues of A. Without the selfadjointness condition such a decomposition of the space does not hold true. This is already clear in the finite dimensional case. For example,

$$A = \begin{bmatrix} 0 & 1 \\ 0 & 0 \end{bmatrix}$$

on $H = \mathbf{C}^2$ has only one eigenvalue, namely, $\lambda = 0$, and the corresponding eigenspace is different from H.

The finite dimensional case provides a hint for the type of decomposition one may be looking for in the non-selfadjoint case. Assume H is finite dimensional, and let $\lambda_1, \ldots, \lambda_r$ be the different eigenvalues of A. In H there exists a basis such that the matrix J_A of A with respect to this basis has Jordan normal form, that is, J_A appears as a block diagonal matrix such that the blocks on the diagonal are elementary Jordan blocks of the form

$$(1) \qquad T = \begin{bmatrix} \lambda_j & 1 & & \\ & \lambda_j & \ddots & \\ & & \ddots & 1 \\ & & & \lambda_j \end{bmatrix}.$$

Let M_ν be the space spanned by the basis vectors that correspond to the elementary Jordan blocks in J_A with λ_ν on the main diagonal. Then

$$(2) \qquad H = M_1 \oplus \cdots \oplus M_r,$$

each space M_ν is invariant under A and the restriction of A to M_ν has a single eigenvalue, namely λ_ν.

To find the space M_ν one does not have to know the Jordan basis. There is a direct way. In fact,

$$(3) \qquad M_\nu = \operatorname{Im}\left[\frac{1}{2\pi i} \int_{\Gamma_\nu} (\lambda - A)^{-1} d\lambda\right],$$

where Γ_ν is a contour around λ_ν separating λ_ν from the other eigenvalues. To see this, note that for the $k \times k$ matrix T given by (1)

$$(\lambda - T)^{-1} = \begin{bmatrix} (\lambda - \lambda_j)^{-1} & (\lambda - \lambda_j)^{-2} & \cdots & (\lambda - \lambda_j)^{-k} \\ & (\lambda - \lambda_j)^{-1} & & \vdots \\ & & \ddots & (\lambda - \lambda_j)^{-2} \\ & & & (\lambda - \lambda_j)^{-1} \end{bmatrix},$$

and hence $\frac{1}{2\pi i} \int_{\Gamma_\nu} (\lambda - T)^{-1} d\lambda$ is equal to the $k \times k$ identity matrix if $j = \nu$ and equal to the zero matrix otherwise. It follows that

$$\left[\frac{1}{2\pi i} \int_{\Gamma_\nu} (\lambda - A)^{-1} d\lambda \right] x = \begin{cases} x & \text{if } x \in M_\nu, \\ 0 & \text{if } x \in M_j, j \neq \nu. \end{cases}$$

In particular (3) holds.

In formula (3) the finite dimensionality of the space does not play a role anymore. In fact, as we shall see in this chapter, formula (3) can also be used in the infinite dimensional case to obtain spectral decompositions of the space similar to the one given in (2). To carry out this program requires the use of methods of contour integration and of complex analysis for vector and operator valued functions. The first section of this chapter is devoted to the latter topics.

I.1 PRELIMINARIES ABOUT OPERATORS AND OPERATOR-VALUED FUNCTIONS

We begin with a few words about terminology and notation. All linear spaces in this book are vector spaces over $\mathbf{C}$ (the field of complex numbers). Unless stated otherwise, an operator is a bounded linear operator acting between Banach or Hilbert spaces. The identity on a linear space X is denoted by I_X or just by I. In expressions like $\lambda I - A$ we shall often omit the symbol I and write $\lambda - A$. The word *subspace* denotes a closed linear manifold in a Banach or Hilbert space. Given Banach spaces X and Y the symbol $\mathcal{L}(X, Y)$ stands for the Banach space of all bounded linear operators from X into Y (endowed with the operator norm). We shall write $\mathcal{L}(X)$ instead of $\mathcal{L}(X, X)$.

Next, we recall (see [GG]) some basic facts from the spectral theory of a bounded linear operator. Let $A: X \to X$ be a bounded linear operator acting on a Banach space X. By definition the *resolvent set* $\rho(A)$ of A is the set of all complex numbers λ such that for each $y \in X$ the equation $\lambda x - Ax = y$ has a unique solution $x \in X$. Equivalently, $\lambda \in \rho(A)$ if and only if $\lambda - A$ is an invertible operator, that is, there is a bounded linear operator $R(\lambda)$ on X such that

$$(1) \qquad\qquad R(\lambda)(\lambda - A) = (\lambda - A)R(\lambda) = I.$$

The complement of $\rho(A)$ in $\mathbf{C}$ is called the *spectrum* of A and is denoted by $\sigma(A)$. It is well-known (see [GG], Theorem X.6.1) that $\sigma(A)$ is a bounded closed subset of $\mathbf{C}$. The operator $R(\lambda)$ appearing in (1) will be denoted by $(\lambda - A)^{-1}$ and the operator function $(\cdot - A)^{-1}$ will be called the *resolvent* of A.

In what follows some basic theorems of complex analysis are extended to vector and operator valued functions. We start with the definition of contour integrals of the form:

$$(2) \qquad \frac{1}{2\pi i} \int_{\Gamma} g(\lambda)d\lambda,$$

where the integrand is a function with values in some Banach space.

First, let us make clear what kind of contours are used in (2). We call Γ a *Cauchy contour* if Γ is the oriented boundary of a bounded Cauchy domain in **C**. By definition, a *Cauchy domain* is a disjoint union in **C** of a finite number of non-empty open connected sets $\Delta_1, \ldots, \Delta_r$, say, such that $\overline{\Delta}_i \cap \overline{\Delta}_j = \emptyset$ $(i \neq j)$ and for each j the boundary of Δ_j consists of a finite number of non-intersecting closed rectifiable Jordan curves which are oriented in such a way that Δ_j belongs to the inner domains of the curves. If σ is a compact subset of a (nonempty) open set $\Omega \subset \mathbf{C}$, then one can always find a Cauchy contour Γ in Ω such that σ belongs to the inner domain of Γ. To see this, construct in the complex plane a grid of congruent hexagons of diameter less one third of the distance between σ and $\mathbf{C}\backslash\Omega$, and let Δ be the interior of the union of all closed hexagons of the grid which have a non-empty intersection with σ. Then the boundary of Δ is a Cauchy contour of the desired type.

Let Γ be a Cauchy contour, and let $g: \Gamma \to Z$ be a continuous function on Γ with values in the Banach space Z. Then (as in complex function theory) the integral (2) is defined as a Stieltjes integral, but now its convergence has to be understood in the norm of Z. Thus the value of (2) is a vector in Z which appears as the limit (in the norm of Z) of the corresponding Stieltjes sum. From this definition it is clear that

$$(3) \qquad F\left(\frac{1}{2\pi i} \int_{\Gamma} g(\lambda)d\lambda\right) = \frac{1}{2\pi i} \int_{\Gamma} F(g(\lambda))d\lambda$$

for any continuous linear functional F on Z. Note that the integrand of the second integral in (3) is just a scalar-valued function. Often the integral in (2) can be computed if g has additional analyticity properties.

Let Ω be a non-empty open set in **C**, and let Z be a Banach space. The function $g: \Omega \to Z$ is said to be *analytic at* $\lambda_0 \in \Omega$ if in some neighbourhood $\mathcal{U}$ of λ_0 in Ω the function g can be represented as the sum of a power series in $\lambda - \lambda_0$, i.e.,

$$(4) \qquad g(\lambda) = \sum_{n=0}^{\infty} (\lambda - \lambda_0)^n g_n, \qquad \lambda \in \mathcal{U}.$$

Here $g_0, g_1, \ldots$ are vectors in Z (which do not depend on λ) and the series in (4) converges in the norm of Z. If g is analytic at each point of Ω, then g is called *analytic on* Ω. For such a function the Cauchy integral formula holds true. Indeed, assume $g: \Omega \to Z$ is analytic on Ω and let Γ be a Cauchy contour such that Γ and its inner domain are in Ω. Then

$$(5) \qquad g(\lambda_0) = \frac{1}{2\pi i} \int_\Gamma \frac{1}{\lambda - \lambda_0} g(\lambda) d\lambda$$

for any point λ_0 inside Γ.

To prove (5), let y be the vector in Z defined by the right hand side of (5). Take an arbitrary continuous linear functional F on Z. Note that $F \circ g$ is a scalar-valued analytic function. Hence, by the usual Cauchy integral formula,

$$F(g(\lambda_0)) = \frac{1}{2\pi i} \int_\Gamma \frac{1}{\lambda - \lambda_0} F(g(\lambda)) d\lambda.$$

Now use (3) to conclude that $F(g(\lambda_0)) = F(y)$. Since F is an arbitrary continuous linear functional on Z, the Hahn-Banach theorem implies that $g(\lambda_0) = y$, which proves (5).

Let Ω be a non-empty open set in $\mathbf{C}$, and let $g: \Omega \to Z$ have values in the Banach space Z. The function g is called *differentiable* on Ω if for each λ_0 in Ω the *derivative*

$$g'(\lambda_0) := \lim_{\lambda \to \lambda_0} \frac{1}{\lambda - \lambda_0} \big(g(\lambda) - g(\lambda_0)\big)$$

exists in the norm of Z. Obviously, analyticity of g implies that g is differentiable. The converse statement is also true. For scalar functions this is a well-known fact from complex function theory. To prove it for vector functions, assume that g is differentiable on Ω, and let λ_0 be an arbitrary point of Ω. Choose a circle Γ with centre at λ_0 and with radius r in such a way that Γ and its inner domain are in Ω. We shall show that

$$(6) \qquad g(\mu) = \frac{1}{2\pi i} \int_\Gamma \frac{1}{\lambda - \mu} g(\lambda) d\lambda, \qquad |\mu - \lambda_0| < r.$$

Note that differentiability of g implies that g is continuous, and so the right hand side of (6) is well-defined. Let F be an arbitrary continuous linear functional on Z. Then the scalar function $F \circ g$ is differentiable on Ω, and hence analytic on Ω. So (5) holds for $F \circ g$ instead of g. Since F is an arbitrary continuous linear functional on Z, we can apply (3) and the Hahn-Banach theorem to conclude that (6) holds. From (6) and

$$\frac{1}{\lambda - \mu} = \sum_{n=0}^\infty \frac{1}{(\lambda - \lambda_0)^{n+1}} (\mu - \lambda_0)^n$$

for $|\mu - \lambda_0| < r$ and $\lambda \in \Gamma$, it follows that

$$(7) \qquad g(\mu) = \sum_{n=0}^\infty (\mu - \lambda_0)^n \left(\frac{1}{2\pi i} \int_\Gamma \frac{1}{(\lambda - \lambda_0)^{n+1}} g(\lambda) d\lambda \right),$$

which proves the analyticity of g.

Let us specify some of the preceding results for the case when Z is the Banach space $\mathcal{L}(X,Y)$. Let X and Y be Banach spaces, and let $g:\Gamma \to \mathcal{L}(X,Y)$ be a continuous function. Then the value of the integral (2) is a bounded linear operator from X into Y, and for each $x \in X$ we have

$$(8) \qquad \left(\frac{1}{2\pi i}\int_\Gamma g(\lambda)d\lambda\right)x = \frac{1}{2\pi i}\int_\Gamma g(\lambda)x d\lambda.$$

Furthermore, if $A: X_1 \to X$ and $B: Y \to Y_1$ are bounded linear operators acting between Banach spaces, then

$$(9) \qquad B\left(\frac{1}{2\pi i}\int_\Gamma g(\lambda)d\lambda\right)A = \frac{1}{2\pi i}\int_\Gamma Bg(\lambda)A d\lambda.$$

Let $g: \Omega \to \mathcal{L}(X,Y)$ be analytic on Ω, and assume that for a point λ_0 in Ω the operator $g(\lambda_0)$ is invertible. Then there exists an open neighbourhood $\mathcal{U}$ of λ_0 in Ω such that $g(\lambda)$ is invertible for $\lambda \in \mathcal{U}$ and $g(\cdot)^{-1}$ is analytic on $\mathcal{U}$. The last statement follows from the formula

$$(10) \qquad (g(\lambda)^{-1})' = -g(\lambda)^{-1}g(\lambda)'g(\lambda)^{-1}, \qquad \lambda \in \mathcal{U},$$

and the fact that differentiability is the same as analyticity.

Let $A: X \to X$ be a bounded linear operator on the Banach space X. The resolvent $R(\cdot) = (\cdot - A)^{-1}$ is one of the main operator functions which we have to study. From what we proved in the previous paragraph (see also [GG], Theorem X.8.1) it is clear that the resolvent $R(\cdot)$ is analytic on the open set $\rho(A)$. We also know (see [GG], Section X.6) that

$$R(\lambda) = (\lambda - A)^{-1} = \sum_{n=0}^{\infty} \frac{1}{\lambda^{n+1}}A^n, \qquad |\lambda| > \|A\|.$$

From these facts it follows that $\sigma(A)$ is a non-empty set whenever $X \neq (0)$. Indeed, assume that $\sigma(A) = \emptyset$. Take a vector x in X and a continuous linear functional F on X. Then the function $f(\cdot) = F((\cdot - A)^{-1}x)$ is a bounded entire function. Hence, by Liouville's theorem $f(\lambda)$ is constant. Since $(\lambda - A)^{-1} \to 0$ if $\lambda \to \infty$, it follows that $f(\lambda) \equiv 0$ on $\mathbb{C}$. This holds for any continuous linear functional F on X. So, by the Hahn-Banach theorem, $(\lambda - A)^{-1}x = 0$ for all λ. By the way, the above argument also shows that the Liouville theorem carries over to entire functions that have their values in a Banach space.

I.2 SPECTRAL DECOMPOSITION AND RIESZ PROJECTION

In this section A is a bounded linear operator on a Banach space X. If N is a subspace of X invariant under A, then $A|N$ denotes the *restriction* of A to N, which has to be considered as an operator from N into N.

Assume that the spectrum of A is the disjoint union of two non-empty closed subsets σ and τ. We want to show that to this decomposition of the spectrum there corresponds a direct sum decomposition of the space, $X = M \oplus L$, such that M and L are A-invariant subspaces of X, the spectrum of the restriction $A|M$ is precisely equal to σ and that of $A|L$ to τ. To prove that such a spectral decomposition exists we study (cf. formula (3) in the introduction to this chapter) the operator

$$(1) \qquad P = \frac{1}{2\pi i} \int_{\Gamma} (\lambda - A)^{-1} d\lambda.$$

A set σ is called an *isolated part* of $\sigma(A)$ if both σ and $\tau := \sigma(A)\backslash\sigma$ are closed subsets of $\sigma(A)$. Given an isolated part σ of $\sigma(A)$ we define P_σ to be the bounded linear operator on X given by the right hand side (1), where we assume that Γ is a Cauchy contour (in the resolvent set of A) around σ separating σ from $\tau = \sigma(A)\backslash\sigma$. By the latter we mean that σ belongs to the inner domain of Γ and τ to the outer domain of Γ. The existence of such contours has been proved in the previous section, where one also finds the definition of the integral (1). Since $(\lambda - A)^{-1}$ is an analytic operator function (in λ) on the resolvent set of A, a standard argument of complex function theory shows that the definition of P_σ does not depend on the particular choice of the contour Γ. The operator P_σ is called the *Riesz projection* of A corresponding to the isolated part σ. The use of the word projection is justified by the next lemma.

LEMMA 2.1. *The operator P_σ is a projection, i.e., $P_\sigma^2 = P_\sigma$.*

PROOF. Let Γ_1 and Γ_2 be Cauchy contours around σ separating σ from $\tau = \sigma(A)\backslash\sigma$. Assume that Γ_1 is in the inner domain of Γ_2. Then

$$P_\sigma^2 = \left(\frac{1}{2\pi i} \int_{\Gamma_1} (\lambda - A)^{-1} d\lambda\right)\left(\frac{1}{2\pi i} \int_{\Gamma_2} (\mu - A)^{-1} d\mu\right)$$

$$= \left(\frac{1}{2\pi i}\right)^2 \int_{\Gamma_1}\int_{\Gamma_2} (\lambda - A)^{-1}(\mu - A)^{-1} d\mu d\lambda.$$

Now use the so-called *resolvent equation*,

$$(2) \qquad (\lambda - A)^{-1} - (\mu - A)^{-1} = (\mu - \lambda)(\lambda - A)^{-1}(\mu - A)^{-1}, \qquad \lambda, \mu \in \rho(A),$$

and write $P_\sigma^2 = Q - R$, where

$$Q = \left(\frac{1}{2\pi i}\right)^2 \int_{\Gamma_1}\int_{\Gamma_2} \frac{1}{\mu - \lambda}(\lambda - A)^{-1} d\mu d\lambda$$

$$= \frac{1}{2\pi i} \int_{\Gamma_1} (\lambda - A)^{-1}\left(\frac{1}{2\pi i}\int_{\Gamma_2} \frac{1}{\mu - \lambda} I d\mu\right) d\lambda$$

$$= \frac{1}{2\pi i} \int_{\Gamma_1} (\lambda - A)^{-1} d\lambda = P_\sigma,$$

and

$$R = \left(\frac{1}{2\pi i}\right)^2 \int\limits_{\Gamma_1}\int\limits_{\Gamma_2} \frac{1}{\mu - \lambda}(\mu - A)^{-1}d\mu d\lambda$$

$$= \left(\frac{1}{2\pi i}\right)^2 \int\limits_{\Gamma_2}\int\limits_{\Gamma_1} \frac{1}{\mu - \lambda}(\mu - A)^{-1}d\lambda d\mu$$

$$= \frac{1}{2\pi i}\int\limits_{\Gamma_2} (\mu - A)^{-1}\left(\frac{1}{2\pi i}\int\limits_{\Gamma_1} \frac{1}{\mu - \lambda}Id\lambda\right)d\mu = 0.$$

Here we used

$$\int\limits_{\Gamma_2} \frac{d\mu}{\mu - \lambda} = 2\pi i \quad (\lambda \in \Gamma_1), \qquad \int\limits_{\Gamma_1} \frac{d\lambda}{\mu - \lambda} = 0 \quad (\mu \in \Gamma_2),$$

and these identities hold true, because Γ_1 is in the inner domain of Γ_2. Furthermore, the interchange of the integrals in the computation for R is justified by the fact that the integrand is a continuous operator function on $\Gamma_1 \times \Gamma_2$.

  THEOREM 2.2. *Let σ be an isolated part of $\sigma(A)$, and put $M = \operatorname{Im} P_\sigma$ and $L = \operatorname{Ker} P_\sigma$. Then $X = M \oplus L$, the spaces M and L are A-invariant subspaces and*

$$(3) \qquad\qquad \sigma(A|M) = \sigma, \qquad \sigma(A|L) = \sigma(A)\backslash\sigma.$$

  PROOF. Since P_σ is a projection, it is clear that M and L are closed subspaces and $X = M \oplus L$. From $A(\lambda - A)^{-1} = (\lambda - A)^{-1}A$ for $\lambda \in \rho(A)$, it follows that $AP_\sigma = P_\sigma A$, which implies that M and L are invariant under A. It remains to prove (3).

  Let Γ be a Cauchy contour around σ separating σ from $\tau := \sigma(A)\backslash\sigma$. For $\mu \notin \Gamma$ put

$$S(\mu) = \frac{1}{2\pi i}\int\limits_{\Gamma} \frac{1}{\mu - \lambda}(\lambda - A)^{-1}d\lambda.$$

Since P_σ commutes with A, we know that P_σ commutes with the resolvent $(\lambda - A)^{-1}$, and hence P_σ commutes with $S(\mu)$. Thus the spaces M and L are invariant under $S(\mu)$. One computes that

$$S(\mu)(A - \mu) = (A - \mu)S(\mu)$$

$$= \frac{1}{2\pi i}\int\limits_{\Gamma} \frac{1}{\mu - \lambda}(A - \mu)(\lambda - A)^{-1}d\lambda$$

$$= \frac{1}{2\pi i}\int\limits_{\Gamma} \frac{-1}{\mu - \lambda}Id\lambda - \frac{1}{2\pi i}\int\limits_{\Gamma} (\lambda - A)^{-1}d\lambda$$

$$= \begin{cases} I - P_\sigma & \text{for } \mu \text{ inside } \Gamma, \\ -P_\sigma I & \text{for } \mu \text{ outside } \Gamma. \end{cases}$$

Take $\mu \notin \sigma$. Without loss of generality we assume that Γ has been chosen such that μ is outside Γ. But then the above computation shows that

$$(A - \mu)S(\mu)x = S(\mu)(A - \mu)x = -x, \qquad x \in M.$$

Since $S(\mu)M \subset M$, it follows that $A - \mu$ maps M in a one-one way onto M and $(\mu - A|M)^{-1} = S(\mu)|M$. Thus $\mu \in \rho(A|M)$, and we may conclude that $\sigma(A|M) \subset \sigma$. In a similar way one shows that $\sigma(A|L) \subset \tau$.

Finally, take $\lambda \notin \sigma(A|M) \cup \sigma(A|L)$. Then $\lambda - A$ maps M (resp., L) in a one-one way onto M (resp., L). It follows that $\lambda \in \rho(A)$. So

$$\sigma(A) \subset \sigma(A|M) \cup \sigma(A|L) \subset \sigma \cup \tau = \sigma(A),$$

and (3) is proved. $\square$

When applied to $\sigma = \sigma(A)$, Theorem 2.2 yields

$$(4) \qquad\qquad\qquad P_{\sigma(A)} = I.$$

Indeed, put $\sigma = \sigma(A)$, and let M and L be as in Theorem 2.2. Then $\sigma(A|L) = \sigma(A)\backslash\sigma = \emptyset$. This can happen only when $L = (0)$. Thus $M = X$ and (4) is proved.

COROLLARY 2.3. *Assume $\sigma(A)$ is a disjoint union of two closed subsets σ and τ. Then*

$$(5) \qquad\qquad\qquad P_\sigma + P_\tau = I, \qquad P_\sigma \cdot P_\tau = 0.$$

PROOF. Choose Cauchy contours Γ, Γ_1 and Γ_2 as indicated by the following picture:

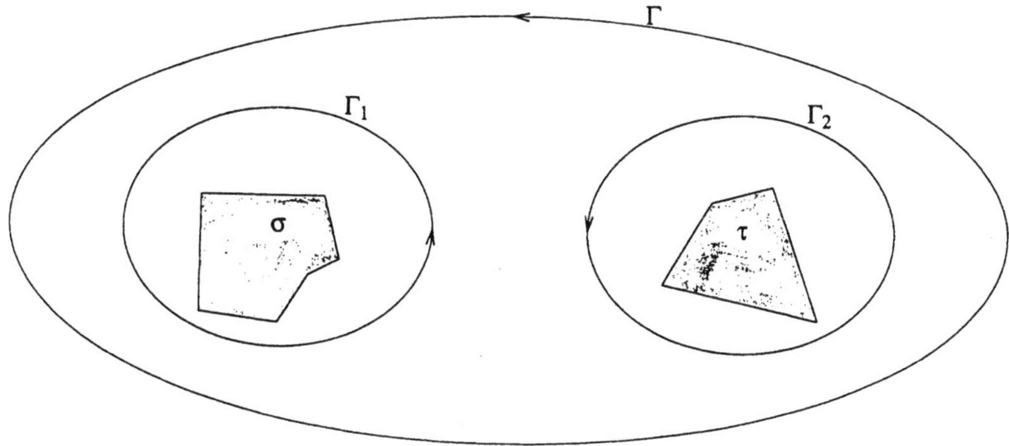

A standard argument from complex function theory shows that

$$(6) \qquad \frac{1}{2\pi i}\int_{\Gamma_1}(\lambda - A)^{-1}d\lambda + \frac{1}{2\pi i}\int_{\Gamma_2}(\lambda - A)^{-1}d\lambda = \frac{1}{2\pi i}\int_{\Gamma}(\lambda - A)^{-1}d\lambda.$$

The left hand side of (6) is equal to $P_\sigma + P_\tau$. By formula (4) the right hand side of (6) is I. Thus the first identity in (5) is proved. Using this, we have $P_\sigma P_\tau = P_\sigma(I - P_\sigma) = 0$. $\square$

The next proposition shows that the identities in (3) determine the spaces M and L uniquely.

PROPOSITION 2.4. *Assume X is a direct sum of A-invariant subspaces M and L. Then $\sigma(A) = \sigma(A|M) \cup \sigma(A|L)$, and if $\sigma(A|M) \cap \sigma(A|L) = \emptyset$, then*

$$(7) \qquad M = \operatorname{Im} P_{\sigma(A|M)}, \qquad L = \operatorname{Ker} P_{\sigma(A|M)}.$$

PROOF. Since M and L are invariant under A, the operator $\lambda - A$ has the following 2×2 operator matrix representation relative to the direct sum $X = M \oplus L$:

$$\lambda - A = \begin{bmatrix} \lambda - A_1 & 0 \\ 0 & \lambda - A_2 \end{bmatrix},$$

where $A_1 = A|M$ and $A_2 = A|L$. But this implies that $\lambda \in \rho(A)$ if and only if $\lambda \in \rho(A_1) \cap \rho(A_2)$, and in that case

$$(8) \qquad (\lambda - A)^{-1} = \begin{bmatrix} (\lambda - A_1)^{-1} & 0 \\ 0 & (\lambda - A_2)^{-1} \end{bmatrix}.$$

Thus $\rho(A) = \rho(A_1) \cap \rho(A_2)$, and $\sigma(A) = \sigma(A_1) \cup \sigma(A_2)$.

Next, assume $\sigma(A_1) \cap \sigma(A_2) = \emptyset$. Take a Cauchy contour Γ around $\sigma(A_1)$ which separates $\sigma(A_1)$ from $\sigma A_2)$, and integrate (8) over Γ. Because of (4) and (5),

$$\frac{1}{2\pi i} \int_\Gamma (\lambda - A_1)^{-1} d\lambda = I_M, \qquad \frac{1}{2\pi i} \int_\Gamma (\lambda - A_2)^{-1} d\lambda = 0.$$

Thus

$$P_{\sigma(A_1)} = \frac{1}{2\pi i} \int_\Gamma (\lambda - A)^{-1} d\lambda = \begin{bmatrix} I_M & 0 \\ 0 & 0 \end{bmatrix},$$

which proves (7). $\square$

PROPOSITION 2.5. *Assume X is a Hilbert space, and let σ be an isolated part of $\sigma(A)$. Then $\bar{\sigma} = \{\bar{\lambda} \mid \lambda \in \sigma\}$ is an isolated part of $\sigma(A^*)$ and*

$$(9) \qquad \left(P_\sigma(A)\right)^* = P_{\bar{\sigma}}(A^*).$$

PROOF. The first statement about σ is trivial. Let Γ be a Cauchy contour around σ with separates σ from $\sigma(A)\backslash\sigma$. Let $\alpha: J \to \mathbb{C}$ be a parametrization of Γ. Denote by $\bar{\Gamma}$ the curve parametrized by the function $\overline{\alpha(t)}$, $t \in J$, and let $-\bar{\Gamma}$ be the same curve with the opposite orientation. Then $-\bar{\Gamma}$ is a Cauchy contour around $\bar{\sigma}$ which separates

$\bar{\sigma}$ from $\sigma(A^*)\backslash\bar{\sigma}$. Now

$$
\begin{aligned}
(P_\sigma(A))^* &= \left(\frac{1}{2\pi i}\int_\Gamma (\lambda - A)^{-1}d\lambda\right)^* \\
&= \lim\left\{\frac{1}{2\pi i}\sum_j (\alpha(t_j) - \alpha(t_{j-1}))(\alpha(s_j) - A)^{-1}\right\}^* \\
&= \lim \frac{-1}{2\pi i}\sum_j \overline{(\alpha(t_j)} - \overline{\alpha(t_{j-1})})(\overline{\alpha(s_j)} - A^*)^{-1} \\
&= \frac{-1}{2\pi i}\int_{\overline{\Gamma}} (\lambda - A^*)^{-1}d\lambda = P_{\bar{\sigma}}(A^*). \quad \Box
\end{aligned}
$$

I.3 FUNCTIONAL CALCULUS

Let $A: X \to X$ be a bounded linear operator acting on the Banach space X. Note that

$$
\begin{aligned}
A^n(\lambda - A)^{-1} &= A^{n-1}(A - \lambda + \lambda)(\lambda - A)^{-1} \\
&= \lambda A^{n-1}(\lambda - A)^{-1} - A^{n-1}.
\end{aligned}
$$

Proceeding in this way one finds that

$$
(1) \qquad A^n(\lambda - A)^{-1} = \lambda^n(\lambda - A)^{-1} - \sum_{j=0}^{n-1}\lambda^{n-1-j}A^j.
$$

Now, let Γ be a Cauchy contour around the spectrum $\sigma(A)$. Note that the integral over Γ of the last term in (1) is the zero operator. It follows that

$$
\begin{aligned}
A^n &= A^n\left(\frac{1}{2\pi i}\int_\Gamma (\lambda - A)^{-1}d\lambda\right) \\
&= \frac{1}{2\pi i}\int_\Gamma A^n(\lambda - A)^{-1}d\lambda = \frac{1}{2\pi i}\int_\Gamma \lambda^n(\lambda - A)^{-1}d\lambda.
\end{aligned}
$$

Let $p(\lambda) = \sum_{n=0}^r \alpha_n\lambda^n$ be a complex polynomial. The preceding calculations imply

$$
(2) \qquad p(A) := \sum_{n=0}^r \alpha_n A^n = \frac{1}{2\pi i}\int_\Gamma p(\lambda)(\lambda - A)^{-1}d\lambda.
$$

The latter expression for $p(A)$ is the starting point of a new definition.

An open set Ω in $\mathbb{C}$ is called an *open neighbourhood of a set σ* if $\sigma \subset \Omega$. By $\mathcal{F}(A)$ we denote the family of all complex-valued functions f that are analytic on some open neighbourhood (which may depend on f) of $\sigma(A)$. All complex polynomials belong to $\mathcal{F}(A)$. Given $f \in \mathcal{F}(A)$ we define

$$f(A) := \frac{1}{2\pi i} \int_{\partial \Delta} f(\lambda)(\lambda - A)^{-1} d\lambda.$$

Here Δ is a bounded Cauchy domain in the open neighbourhood Ω of $\sigma(A)$ on which f is defined and $\sigma(A) \subset \Delta \subset \overline{\Delta} \subset \Omega$. Note that $f(A)$ is a bounded linear operator on X. A standard argument of complex function theory shows that its definition does not depend on the particular choice of the Cauchy domain Δ. In other words, if f and g belong to $\mathcal{F}(A)$ and $f(\lambda) = g(\lambda)$ on some open neighbourhood of $\sigma(A)$, then $f(A) = g(A)$.

In this setting the Riesz projections appear by using functions which take only the values 1 and 0. More precisely, if σ is an isolated part of $\sigma(A)$, then the corresponding Riesz projection P_σ is equal to $h(A)$, where h is a function which takes the value 1 on an open neighbourhood of σ and the value 0 on an open neighbourhood of the complement $\sigma(A) \backslash \sigma$. In particular, $f(A) = I$ (resp., $f(A) = 0$) whenever f is equal to 1 (resp., 0) on an open neighbourhood of $\sigma(A)$.

THEOREM 3.1. *Given f, g in $\mathcal{F}(A)$, then*

(i) $(f + g)(A) = f(A) + g(A)$,

(ii) $(\alpha f)(A) = \alpha f(A)$, $\alpha \in \mathbf{C}$,

(iii) $(fg)(A) = f(A)g(A)$.

PROOF. Let Ω be the intersection of the domains of f and g. On Ω the functions $f + g$, αf and fg are defined by

$$(f + g)(\lambda) = f(\lambda) + g(\lambda), \qquad (\alpha f)(\lambda) = \alpha f(\lambda),$$
$$(fg)(\lambda) = f(\lambda) g(\lambda), \qquad \lambda \in \Omega.$$

Obviously, $f + g$, αf and fg are in $\mathcal{F}(A)$. Since $f(\lambda)$ appears linearly in the definition of $f(A)$, the statements (i) and (ii) are easy to check. We prove (iii).

Let Δ_1 and Δ_2 be bounded Cauchy domains such that

$$\sigma(A) \subset \Delta_1 \subset \overline{\Delta}_1 \subset \Delta_2 \subset \overline{\Delta}_2 \subset \Omega,$$

and for $\nu = 1, 2$ let Γ_ν be the boundary of Δ_ν. Then

$$f(A)g(A) = \left(\frac{1}{2\pi i}\right)^2 \left(\int_{\Gamma_1} f(\lambda)(\lambda - A)^{-1} d\lambda\right)\left(\int_{\Gamma_2} g(\mu)(\mu - A)^{-1} d\mu\right)$$

$$= \left(\frac{1}{2\pi i}\right)^2 \int_{\Gamma_1} \int_{\Gamma_2} f(\lambda)g(\mu)(\lambda - A)^{-1}(\mu - A)^{-1} d\mu d\lambda$$

$$= \left(\frac{1}{2\pi i}\right)^2 \int_{\Gamma_1} \int_{\Gamma_2} f(\lambda)g(\mu)(\mu - \lambda)^{-1}\{(\lambda - A)^{-1} - (\mu - A)^{-1}\} d\mu d\lambda$$

$$= \gamma_1 - \gamma_2,$$

where

$$\gamma_1 = \left(\frac{1}{2\pi i}\right)^2 \int_{\Gamma_1}\int_{\Gamma_2} f(\lambda)g(\mu)(\mu - \lambda)^{-1}(\lambda - A)^{-1}d\mu d\lambda$$

$$= \frac{1}{2\pi i}\int_{\Gamma_1}\left(\frac{1}{2\pi i}\int_{\Gamma_2}(\mu - \lambda)^{-1}g(\mu)d\mu\right)f(\lambda)(\lambda - A)^{-1}d\lambda$$

$$= \frac{1}{2\pi i}\int_{\Gamma_1} f(\lambda)g(\lambda)(\lambda - A)^{-1}d\lambda = (fg)(A),$$

and

$$\gamma_2 = \left(\frac{1}{2\pi i}\right)^2 \int_{\Gamma_1}\int_{\Gamma_2} f(\lambda)g(\mu)(\mu - \lambda)^{-1}(\mu - A)^{-1}d\mu d\lambda$$

$$= \left(\frac{1}{2\pi i}\right)^2 \int_{\Gamma_2}\int_{\Gamma_1} f(\lambda)g(\mu)(\mu - \lambda)^{-1}(\mu - A)^{-1}d\lambda d\mu$$

$$= \frac{1}{2\pi i}\int_{\Gamma_2}\left(\frac{1}{2\pi i}\int_{\Gamma_1}(\mu - \lambda)^{-1}f(\lambda)d\lambda\right)g(\mu)(\mu - A)^{-1}d\mu$$

$$= 0.$$

To get the last equality we use the fact that all points μ on Γ_2 do not belong to $\overline{\Delta}_1$. Hence for $\mu \in \Gamma_2$ the function $(\mu - \cdot)^{-1}f(\cdot)$ is analytic on an open neighbourhood of $\overline{\Delta}_1$, and thus, by Cauchy's theorem, $\int_{\Gamma_1}(\mu - \lambda)^{-1}f(\lambda)d\lambda = 0$. Note that in the above computations the change in the order of the integrals is allowed because of the continuity of the function

$$(\lambda, \mu) \mapsto f(\lambda)g(\mu)(\mu - \lambda)^{-1}(\mu - A)^{-1}$$

on $\Gamma_1 \times \Gamma_2$. $\quad\square$

THEOREM 3.2. *Let $f(\lambda) = \sum_{\nu=0}^{\infty}\alpha_\nu\lambda^\nu$ for λ in some open neighbourhood Ω of $\sigma(A)$. Then $f \in \mathcal{F}(A)$ and*

$$(3) \qquad\qquad f(A) = \sum_{\nu=0}^{\infty}\alpha_\nu A^\nu,$$

where the latter series converges in the operator norm.

PROOF. Let Γ be the boundary of a bounded Cauchy domain Δ such that $\sigma(A) \subset \Delta \subset \overline{\Delta} \subset \Omega$. Let ρ be the radius of convergence of the series $\sum_{\nu=0}^{\infty}\alpha_\nu\lambda^\nu$. Since the latter series converges for $\lambda \in \Omega$, it follows that the open set Ω is a subset of the open disc $\{\lambda \in \mathbb{C} \,|\, |\lambda| < \rho\}$. This implies that f is analytic on Ω, and hence $f \in \mathcal{F}(A)$. Furthermore, we see that the series $\sum_{\nu=0}^{\infty}\alpha_\nu\lambda^\nu$ converges uniformly on the compact set $\overline{\Delta}$ and hence also on Γ. For $n \geq 0$ put $s_n(\lambda) = \sum_{\nu=0}^{n}\alpha_\nu\lambda^\nu$, and let $\ell(\Gamma)$ denote the length of the contour Γ. Then

$$\left\| f(A) - \sum_{\nu=0}^{n} \alpha_\nu A^\nu \right\| = \left\| f(A) - s_n(A) \right\|$$

$$= \left\| \frac{1}{2\pi i} \int_\Gamma (f(\lambda) - s_n(\lambda))(\lambda - A)^{-1} d\lambda \right\|$$

$$\leq \frac{1}{2\pi} \ell(\Gamma) \left(\max_{\lambda \in \Gamma} |f(\lambda) - s_n(\lambda)| \right) \max_{\lambda \in \Gamma} \|(\lambda - A)^{-1}\|$$

$$\longrightarrow 0 \qquad (n \to \infty),$$

which proves (3). $\quad\square$

The next theorem is known as the *spectral mapping theorem*.

THEOREM 3.3. *If* $f \in \mathcal{F}(A)$, *then*

$$(4) \qquad \sigma\big(f(A)\big) = \{f(\lambda) \mid \lambda \in \sigma(A)\} = f[\sigma(A)].$$

PROOF. Let f be analytic on the open neighbourhood Ω of $\sigma(A)$. Take $\lambda_0 \in \sigma(A)$, and define a function g on Ω by

$$g(z) = (z - \lambda_0)^{-1}\big(f(z) - f(\lambda_0)\big), \qquad \lambda_0 \neq z \in \Omega,$$

and $g(\lambda_0) = f'(\lambda_0)$, where f' denotes the derivative of f. Obviously, $g \in \mathcal{F}(A)$ and

$$(5) \qquad f(z) - f(\lambda_0) = (z - \lambda_0)g(z) = g(z)(z - \lambda_0), \qquad z \in \Omega.$$

Now apply Theorem 3.1. It follows from (5) that

$$f(A) - f(\lambda_0)I = (A - \lambda_0 I)g(A) = g(A)(A - \lambda_0 I).$$

Since $\lambda_0 \in \sigma(A)$, the operator $A - \lambda_0 I$ is not invertible. But then the same conclusion holds true for $f(A) - f(\lambda_0)I$, which implies that $f(\lambda_0) \in \sigma\big(f(A)\big)$. We have proved that $f[\sigma(A)] \subset \sigma\big(f(A)\big)$.

To prove the reverse inclusion, take $\beta \in \mathbb{C}\backslash f[\sigma(A)]$. Then there exists an open neighbourhood Ω_0 of $\sigma(A)$ such $\Omega_0 \subset \Omega$ and $f(z) - \beta \neq 0$ for $z \in \Omega_0$. Put $h(z) = (f(z) - \beta)^{-1}$ for $z \in \Omega_0$. Then $h \in \mathcal{F}(A)$ and

$$h(z)\big(f(z) - \beta\big) = 1 = \big(f(z) - \beta\big)h(z), \qquad z \in \Omega_0,$$

and we can apply Theorem 3.1 to show that

$$h(A)\big(f(A) - \beta I\big) = I = \big(f(A) - \beta I\big)h(A).$$

Thus $f(A) - \beta I$ is invertible, and hence $\beta \in \mathbb{C}\backslash\sigma\big(f(A)\big)$. We see that $f[\sigma(A)] \supset \sigma\big(f(A)\big)$, and the theorem is proved. $\quad\square$

I.4 AN OPERATOR EQUATION

Let X and Y be Banach spaces, and let $A\colon Y \to Y$, $B\colon X \to X$ and $C\colon X \to Y$ be given operators. Consider the operator equation

$$(1) \qquad\qquad AZ - ZB = C.$$

In (1) the unknown is an operator Z from X into Y. Note that equation (1) is solvable if and only if there exists an operator $Z\colon X \to Y$ such that the following operator matrix identity holds true:

$$(2) \qquad \begin{bmatrix} I_Y & Z \\ 0 & I_X \end{bmatrix} \begin{bmatrix} A & C \\ 0 & B \end{bmatrix} \begin{bmatrix} I_Y & -Z \\ 0 & I_X \end{bmatrix} = \begin{bmatrix} A & 0 \\ 0 & B \end{bmatrix}.$$

The various factors in (2) act as bounded linear operators on $Y \oplus X$.

From (2) one sees that the problem to solve (1) has to do with similarity of operators. Indeed, the first and third factors in the left hand side of (2) are invertible and

$$\begin{bmatrix} I_Y & Z \\ 0 & I_X \end{bmatrix}^{-1} = \begin{bmatrix} I_Y & -Z \\ 0 & I_X \end{bmatrix}.$$

Thus the operators

$$(3) \qquad \begin{bmatrix} A & C \\ 0 & B \end{bmatrix}, \qquad \begin{bmatrix} A & 0 \\ 0 & B \end{bmatrix}$$

are similar whenever equation (1) is solvable. If the spaces X and Y are finite dimensional, then the converse statement also holds true (see Gohberg-Lancaster-Rodman [1], page 342). In the infinite dimensional case it is unknown whether similarity of the operators in (3) implies that (1) is solvable. Another application connected with equation (1) will be given in Section I.6.

THEOREM 4.1. *Assume that the spectra of the operators $A\colon Y \to Y$ and $B\colon X \to X$ are disjoint. Then the operator equation (1) has a unique solution $Z\colon X \to Y$ which is given by*

$$(4) \qquad \begin{aligned} Z &= \frac{1}{2\pi i} \int_{\Gamma_A} (\lambda - A)^{-1} C (\lambda - B)^{-1} d\lambda \\ &= \frac{-1}{2\pi i} \int_{\Gamma_B} (\lambda - A)^{-1} C (\lambda - B)^{-1} d\lambda, \end{aligned}$$

where Γ_A and Γ_B are Cauchy contours around $\sigma(A)$ and $\sigma(B)$, respectively, which separate $\sigma(A)$ from $\sigma(B)$.

PROOF. The fact that $\sigma(A)$ and $\sigma(B)$ are disjoint compact sets allows us to choose Cauchy contours Γ_A and Γ_B as indicated. First, let us assume that $Z\colon X \to Y$ is a solution of (1). Then

$$Z(\lambda - B) - (\lambda - A)Z = C, \qquad \lambda \in \mathbb{C},$$

and hence

$$(5) \qquad (\lambda - A)^{-1}Z - Z(\lambda - B)^{-1} = (\lambda - A)^{-1}C(\lambda - B)^{-1}$$

for each $\lambda \in \rho(A) \cap \rho(B)$. In particular, the identity (5) holds true for $\lambda \in \Gamma_A$. Next, use that $\sigma(A)$ is inside Γ_A and $\sigma(B)$ is outside Γ_A, and integrate (5) over Γ_A. We see that Z is given by the first identity in (4). If we replace in the above argument Γ_A by Γ_B, then we obtain the second identity in (4).

It remains to show that (1) is solvable. To do this, let Z be the bounded linear operator from X into Y defined by the first identity in (4). Then

$$
\begin{aligned}
AZ &= \frac{1}{2\pi i} \int_{\Gamma_A} A(\lambda - A)^{-1}C(\lambda - B)^{-1}d\lambda \\
&= \frac{1}{2\pi i} \int_{\Gamma_A} (A - \lambda + \lambda)(\lambda - A)^{-1}C(\lambda - B)^{-1}d\lambda \\
&= \frac{1}{2\pi i} \int_{\Gamma_A} \lambda(\lambda - A)^{-1}C(\lambda - B)^{-1}d\lambda - \frac{1}{2\pi i} \int_{\Gamma_A} C(\lambda - B)^{-1}d\lambda.
\end{aligned}
$$

Note that the second integral in the last identity is equal to the zero operator. Thus

$$
\begin{aligned}
AZ &= \frac{1}{2\pi i} \int_{\Gamma_A} (\lambda - A)^{-1}C(\lambda - B)^{-1}(\lambda - B + B)d\lambda \\
&= \frac{1}{2\pi i} \int_{\Gamma_A} (\lambda - A)^{-1}Cd\lambda + \frac{1}{2\pi i} \int_{\Gamma_A} (\lambda - A)^{-1}C(\lambda - B)^{-1}Bd\lambda \\
&= C + ZB.
\end{aligned}
$$

It follows that Z is a solution of (1). $\square$

Consider the Banach space $\mathcal{L}(X, Y)$ of all bounded linear operators from X into Y, and define $J: \mathcal{L}(X, Y) \to \mathcal{L}(X, Y)$ by $J(Z) = AZ - ZB$, where A and B are as in the beginning of this section. Obviously, J is a bounded linear operator. Theorem 4.1 tells us that J is invertible if $\sigma(A) \cap \sigma(B) = \emptyset$. Note that

$$(J - \lambda I)(Z) = (A - \lambda)Z - ZB, \qquad \lambda \in \mathbb{C}.$$

It follows that $J - \lambda I$ is invertible, whenever $\sigma(A - \lambda) \cap \sigma(B) = \emptyset$, and hence

$$(6) \qquad \sigma(J) \subset \{\alpha - \beta \mid \alpha \in \sigma(A), \beta \in \sigma(B)\}.$$

D.C. Kleinecke (see Introduction of Lumer-Rosenblum [1]) has shown that the spectrum of J is equal to the right hand side of (6).

I.5 THE DIFFERENTIAL EQUATION $y' = Ay$

Let A be a bounded linear operator on the Banach space X. Consider the equation

$$(1) \qquad y'(t) = Ay(t), \qquad 0 \le t < \infty.$$

A function $y \colon [0, \infty) \to X$ is said to be a *solution* of (1) if y is continuous on $[0, \infty)$ and

$$\lim_{h \to 0} \left\| Ay(t) - \frac{1}{h}(y(t+h) - y(t)) \right\| = 0$$

for each $t > 0$. Of course one expects the solutions of (1) to be of the form $y(t) = e^{tA}x$, where x is some vector in X. By definition

$$(2) \qquad e^{tA} = \frac{1}{2\pi i} \int_{\Gamma} e^{t\lambda}(\lambda - A)^{-1} d\lambda,$$

where Γ is a Cauchy contour around $\sigma(A)$. Hence, according to Theorem 3.2, the operator e^{tA} is also given by

$$(3) \qquad e^{tA} = \sum_{\nu=0}^{\infty} \frac{1}{\nu!} t^{\nu} A^{\nu}.$$

LEMMA 5.1. *The function* $t \mapsto e^{tA}$ *from* $\mathbf{R}$ *into* $\mathcal{L}(X)$ *is differentiable and*

$$(4) \qquad \frac{d}{dt}(e^{tA}) = Ae^{tA}.$$

PROOF. Take a fixed $t \in \mathbf{R}$, and choose $h \ne 0$ in $\mathbf{R}$ such that $|h| \|A\| \le \frac{1}{2}$. From the functional calculus it follows that $e^{(t+h)A} = e^{hA}e^{tA}$. Thus

$$
\begin{aligned}
\left\| \frac{1}{h}\{e^{(t+h)A} - e^{tA}\} - Ae^{tA} \right\|
&\le \left\| \frac{1}{h}\{e^{hA} - I\} - A \right\| \cdot \|e^{tA}\| \\
&= \left\| \sum_{\nu=2}^{\infty} \frac{1}{\nu!} h^{\nu-1} A^{\nu} \right\| \cdot \|e^{tA}\| \\
&\le |h| \cdot \|e^{tA}\| \left(\sum_{\nu=2}^{\infty} |h|^{\nu-2} \|A\|^{\nu} \right) \\
&\le 2|h| \cdot \|e^{tA}\| \cdot \|A\|^2 \to 0 \qquad (h \to 0),
\end{aligned}
$$

and (4) is proved. $\square$

THEOREM 5.2. *The initial value problem*

$$(5) \qquad \begin{cases} y'(t) = Ay(t), & 0 \le t < \infty, \\ y(0) = x, \end{cases}$$

has precisely one solution, namely $y(t) = e^{tA}x$.

PROOF. Put $y(t) = e^{tA}x$. From the functional calculus it follows that $e^{0A} = I$, and hence $y(0) = x$. But then we can apply Lemma 5.1 to show that y is a solution of (5).

Assume $w: [0, \infty) \to X$ is a second solution of (5). Consider $g(t) = e^{-tA}w(t)$ for $t \geq 0$. Then

$$(6) \qquad \frac{d}{dt}g(t) = e^{-tA}(-A)w(t) + e^{-tA}Aw(t) = 0, \qquad t > 0.$$

Let F be an arbitrary continuous linear functional on X. Then $h = F \circ g$ is a continuous scalar function on $[0, \infty)$ and according to (6) the derivative $h'(t) = 0$ for $t > 0$. It follows that h is constant on $[0, \infty)$, and thus

$$F\big(e^{-tA}w(t)\big) = F(x), \qquad t \geq 0.$$

An application of the Hahn-Banach theorem yields that $e^{-tA}w(t) = x$ for $t \geq 0$. By the functional calculus $(e^{hA})^{-1} = e^{-hA}$ for any $h \in \mathbf{R}$. Hence $w(t) = e^{tA}x$, $t \geq 0$. $\square$

A solution y of (1) is called *asymptotically stable* if $y(t) \to 0$ if $t \to \infty$. If every solution of (1) is asymptotically stable, then the equation (1) is said to be *asymptotically stable*. (Since $y(t) \equiv 0$ is a solution of (1), the vector $x = 0$ is an equilibrium point of the differential equation (1). In this terminology asymptotic stability of (1) is equivalent to the requirement that $x = 0$ is a so-called asymptotically stable equilibrium point for (1).)

THEOREM 5.3. *If $\sigma(A)$ lies in the open half plane $\Re\lambda < 0$, then equation (1) is asymptotically stable. Conversely, if (1) is asymptotically stable, then $\sigma(A)$ lies in the closed half plane $\Re\lambda \leq 0$. If, in addition, X is finite dimensional, then asymptotic stability of (1) is equivalent to the statement that $\sigma(A)$ belongs to the open half plane $\Re\lambda < 0$.*

To prove Theorem 5.3 we shall make use of the following lemma.

LEMMA 5.4. *If $\sigma(A)$ belongs to the open half plane $\Re\lambda < \gamma$, then there exists a constant C (depending on A and γ) such that*

$$(7) \qquad \|e^{tA}\| \leq Ce^{\gamma t}, \qquad t \geq 0.$$

Conversely, from (7) it follows that $\sigma(A)$ belongs to the closed half plane $\Re\lambda \leq \gamma$.

PROOF. Assume $\sigma(A)$ belongs to the open half plane $\Re\lambda < \gamma$. Then we may choose a Cauchy contour Γ around $\sigma(A)$ such that Γ also belongs to $\Re\lambda < \gamma$. Let $\ell(\Gamma)$ be the length of Γ. Note that $|e^{t\lambda}| \leq e^{t\gamma}$ for $\lambda \in \Gamma$ and $t \geq 0$. Thus

$$\|e^{tA}\| = \frac{1}{2\pi}\left\|\int_\Gamma e^{t\lambda}(\lambda - A)^{-1}d\lambda\right\|$$

$$\leq \left(\frac{1}{2\pi}\ell(\Gamma)\max_{\lambda \in \Gamma}\|(\lambda - A)^{-1}\|\right)e^{\gamma t}, \qquad t \geq 0,$$

which proves (7).

To prove the converse statement, assume that (7) holds true. Then for $t \geq 0$ the spectrum of e^{tA} belongs to the closed disc with center at 0 and radius $Ce^{\gamma t}$. According to the spectral mapping theorem (Theorem 3.3) this implies that

$$e^{t\Re\lambda} = |e^{\lambda t}| \leq Ce^{\gamma t}, \qquad \lambda \in \sigma(A).$$

This holds for each $t \geq 0$. Now take a fixed $\lambda \in \sigma(A)$. Then $\exp\{t(\Re\lambda - \gamma)\} \leq C$ for $t \geq 0$, which implies that $\Re\lambda \leq \gamma$. Thus $\Re\sigma(A) \leq \gamma$. $\square$

PROOF OF THEOREM 5.3. Assume $\Re\sigma(A) < 0$. Since $\sigma(A)$ is compact, there exists $\varepsilon > 0$ such that $\sigma(A)$ belongs to the open half plane $\Re\lambda < -\varepsilon$. By Lemma 5.4 this implies that

$$\|e^{tA}\| \leq Ce^{-\varepsilon t}, \qquad t \geq 0$$

for some constant C. So for each $x \in X$ we may conclude that $e^{tA}x \to 0$ if $t \to \infty$. Hence (1) is asymptotically stable.

Next, assume that (1) is asymptotically stable. Take $x \in X$. According to our assumption $e^{tA}x \to 0$ if $t \to \infty$. In particular, there exists a constant C_x (depending on x) such that $\|e^{tA}x\| \leq C_x < \infty$ for all $t \geq 0$. But then we can apply the principle of uniform boundedness to show that there exists a constant C such that $\|e^{tA}\| \leq C < \infty$ for $t \geq 0$. This implies (Lemma 5.4) that $\Re\sigma(A) \leq 0$.

Finally, assume (1) is asymptotically stable and $\dim X < \infty$. We want to show that $\Re\sigma(A) < 0$. We already know that $\Re\sigma(A) \leq 0$. Suppose $\lambda = ib \in \sigma(A)$ with $b \in \mathbf{R}$. Since X is finite dimensional, λ is an eigenvalue of A. Let $x \neq 0$ be a corresponding eigenvector. Then $e^{tA}x = e^{itb}x$. It follows that

$$\|x\| = \|e^{itb}x\| = \|e^{tA}x\| \to 0, \qquad t \to \infty.$$

Thus $x = 0$, which is a contradiction. $\square$

The next theorem is another useful application of Lemma 5.4.

THEOREM 5.5. *Let $A\colon Y \to Y$, $B\colon X \to X$ and $C\colon X \to Y$ be operators acting between Banach spaces, and assume that the spectra of A and B are in the open half plane $\Re\lambda < 0$. Then the operator equation*

$$(8) \qquad\qquad AZ + ZB = C$$

has a unique solution Z in $\mathcal{L}(X,Y)$, namely

$$(9) \qquad\qquad Z = -\int_0^\infty e^{tA}Ce^{tB}\,dt.$$

PROOF. From the condition on $\sigma(A)$ and $\sigma(B)$ it follows that $\sigma(A) \cap \sigma(-B) = \emptyset$. According to Theorem 4.1 this implies that the equation

$$AZ + ZB = AZ - Z(-B) = C$$

has a unique solution Z in $\mathcal{L}(X,Y)$. We have to prove that Z is given by (9).

The integral in (9) is considered as an improper Riemann-Stieltjes integral. Note that the integrand is continuous. The fact that $\sigma(A)$ and $\sigma(B)$ are in the open half plane $\Re\lambda < 0$ implies (see the first part of the proof of Theorem 5.3) that there exist positive constants ε and M such that

$$\|e^{tA}\| \leq Me^{-\varepsilon t}, \qquad \|e^{tB}\| \leq Me^{-\varepsilon t} \qquad (t \geq 0).$$

It follows that the integral in (9) is well-defined. Next, observe that

$$\frac{d}{dt}(e^{tA}Ce^{tB}) = Ae^{tA}Ce^{tB} + e^{tA}Ce^{tB}B.$$

Hence

$$-C = \lim_{\tau \to \infty}(e^{\tau A}Ce^{\tau B} - C)$$

$$= \lim_{\tau \to \infty}\left\{ A\left(\int_0^\tau e^{tA}Ce^{tB}dt\right) + \left(\int_0^\tau e^{tA}Ce^{tB}dt\right)B\right\}$$

$$= -(AZ + ZB),$$

where Z is the right hand side of (9). $\square$

I.6 LYAPUNOV'S THEOREM

A bounded linear operator A on a Hilbert space H is said to be *strictly positive* if A is selfadjoint and there exists a constant $\delta > 0$ such that

$$(1) \qquad\qquad \langle Ax,x\rangle \geq \delta\langle x,x\rangle, \qquad x \in H.$$

Here $\langle\cdot,\cdot\rangle$ is the inner product on H. If $-A$ is strictly positive, then A is called *strictly negative*. The next theorem is known as *Lyapunov's theorem*.

THEOREM 6.1. *The spectrum of an operator A on a Hilbert space H belongs to the open half plane $\Re\lambda < 0$ if and only if there exists a strictly positive operator Z on H such that $ZA + A^*Z$ is strictly negative.*

Here A^* stands for the adjoint of A. The proof of Theorem 6.1 will be based on the following lemma.

LEMMA 6.2. *Let A be an operator on a Hilbert space. If $A + A^*$ is strictly negative, then $\sigma(A)$ belongs to the open half plane $\Re\lambda < 0$.*

PROOF. Choose $\delta > 0$ such that $\langle(A+A^*)x,x\rangle \leq -\delta\langle x,x\rangle$ for $x \in H$. Take a fixed $x \in H$ and consider the function $\varphi(t) = \|e^{tA}x\|^2$. Note that

$$\frac{d}{dt}\varphi(t) = \frac{d}{dt}\langle e^{tA}x, e^{tA}x\rangle$$

$$= \frac{d}{dt}\langle e^{tA^*}e^{tA}x, x\rangle$$

$$= \langle e^{tA^*}(A^* + A)e^{tA}x, x\rangle$$

$$= \langle(A^* + A)e^{tA}x, e^{tA}x\rangle \leq -\delta\varphi(t).$$

Thus, $(e^{\delta t}\varphi(t))' = \delta e^{\delta t}\varphi(t) + e^{\delta t}\varphi'(t) \leq 0$, which implies that $e^{\delta t}\varphi(t) \leq \varphi(0)$ for $t \geq 0$. Thus

$$\|e^{tA}x\| \leq \exp\left(-\frac{1}{2}\delta t\right)\|x\|, \qquad t \geq 0.$$

We conclude that $\|e^{tA}\| \leq \exp(-\frac{1}{2}\delta t)$ for $t \geq 0$. According to Lemma 5.4 this shows that $\Re\sigma(A) < 0$. $\square$

PROOF OF THEOREM 6.1. Assume there exists a strictly positive operator Z on H such that $ZA + A^*Z$ is strictly negative. Choose $\delta > 0$ in such a way that (1) holds with A replaced by Z. On H we introduce a new inner product $[\cdot,\cdot]$ by setting

$$[x,y] = \langle Zx, y\rangle \qquad (x, y \in H).$$

Let $\|\!|\cdot|\!\|$ be the corresponding norm. Note that

$$(2) \qquad \delta\|x\|^2 \leq \|\!|x|\!\|^2 \leq \|Z\|\|x\|^2, x \in H.$$

Thus the original norm $\|\cdot\|$ and the new norm $\|\!|\cdot|\!\|$ are equivalent. In particular, H endowed with $[\cdot,\cdot]$ is again a Hilbert space. It follows that A is a bounded linear operator on H endowed with the new $\|\!|\cdot|\!\|$. By $A^\#$ we denote the adjoint of A with respect to the inner product $[\cdot,\cdot]$. For x and y in H we have

$$[Ax,y] = \langle ZAx, y\rangle = \langle x, A^*Zy\rangle$$
$$[x, A^\#y] = \langle Zx, A^\#y\rangle = \langle x, ZA^\#y\rangle.$$

We conclude that $ZA^\# = A^*Z$. Thus

$$[(A + A^\#)x, x] = \langle ZAx, x\rangle + \langle ZA^\#x, x\rangle$$
$$= \langle(ZA + A^*Z)x, x\rangle.$$

According to our hypotheses $ZA + A^*Z$ is strictly negative. Now use that $\|\cdot\|$ and $\|\!|\cdot|\!\|$ are equivalent norms. It follows that $A + A^\#$ is also strictly negative. But then we can apply Lemma 6.2 to show that $\Re\sigma(A) < 0$.

To prove the converse, assume that $\Re\sigma(A) < 0$. Then $\sigma(A)$ and $\sigma(A^*)$ are both in the open half plane $\Re\lambda < 0$, and Theorem 5.5 tells us that the equation $ZA + A^*Z = -2I$ has a unique solution, namely

$$(3) \qquad Z := 2\int_0^\infty e^{tA^*}e^{tA}dt.$$

Since $-2I$ is strictly negative, it suffices to show that Z (defined by (3)) is strictly positive. From $(e^{tA})^* = e^{tA^*}$ it is clear that $e^{tA^*}e^{tA}$ is selfadjoint, and hence

$$(4) \qquad \langle Zx, x\rangle = 2\int_0^\infty \langle e^{tA^*}e^{tA}x, x\rangle dt = 2\int_0^\infty \|e^{tA}x\|^2 dt.$$

For $t \geq 0$ we have

$$\|x\| = \|e^{-tA}e^{tA}x\| \leq \|e^{-tA}\|\|e^{tA}x\|$$
$$\leq e^{t\|A\|}\|e^{tA}x\|,$$

and thus

$$\|e^{tA}x\| \geq e^{-t\|A\|}\|x\| \qquad (x \in H, t \geq 0).$$

By inserting the latter inequality in (4) one sees that

$$\langle Zx, x \rangle \geq 2 \int_0^\infty e^{-2t\|A\|}\|x\|^2 = \left(\frac{1}{\|A\|}\right)\|x\|^2.$$

Thus Z is strictly positive. $\square$

Let A be an operator on a Hilbert space H with spectrum in $\Re\lambda < 0$, and let $y\colon [0, \infty) \to H$ be a solution of the differential equation

$$y'(t) = Ay(t), \qquad 0 \leq t < \infty.$$

Then we know (see the previous section) that

$$\|y(t)\| \leq Ce^{-\varepsilon t}, \qquad t \geq 0,$$

for some positive constants C and ε. Lyapunov's theorem allows us to say a bit more about the behaviour of the function y, namely, there exists $\varepsilon > 0$ and an equivalent norm $\|\| \cdot \|\|$ on H such that $e^{\varepsilon t}\|\|y(t)\|\|$ is decreasing as a function of t. In fact, $\varepsilon > 0$ and $\|\| \cdot \|\|$ may be taken as follows:

$$\|\|x\|\| := \langle Zx, x \rangle^{1/2}, \qquad \langle (ZA + A^*Z)x, x \rangle \leq -\varepsilon\|x\|^2 \quad (x \in H),$$

where Z is a strictly positive operator with the property described in Theorem 6.1. The proof of Lemma 6.2 and the first part of the proof of Theorem 6.1 show that with this choice of ε and $\|\| \cdot \|\|$ the function $e^{\varepsilon t}\|\|y(t)\|\|$ is decreasing indeed.

CHAPTER II
EIGENVALUES OF FINITE TYPE

The results of the previous chapter are developed further for the case when the operator has isolated eigenvalues with properties of the same type as eigenvalues of finite matrices. The problem of completeness of eigenvectors and generalized eigenvectors appears in a natural way. The results are applied to compact operators. This chapter also contains limit theorems for spectra and the infinite dimensional version of Schur's lemma about triangular forms.

II.1 DEFINITION AND MAIN PROPERTIES

In this section A is a bounded linear operator acting on a Banach space X. Let σ be an isolated part of $\sigma(A)$. We are interested in conditions which guarantee that the corresponding Riesz projection P_σ has finite rank. For this purpose we need the following definition. A point $\lambda_0 \in \sigma(A)$ is called an *eigenvalue of finite type* if the space X admits a direct sum decomposition, $X = M \oplus L$, with the following properties:

(E_1) M and L are A-invariant subspaces,

(E_2) $\dim M < \infty$,

(E_3) $\sigma(A|M) = \{\lambda_0\}$, $\lambda_0 \notin \sigma(A|L)$.

Since the spectrum of an operator acting on a finite dimensional space consists of eigenvalues only, conditions (E_2) and (E_3) imply that λ_0 is an eigenvalue of $A|M$ and hence also of A.

An operator may have eigenvalues which are not of finite type. For example, let $S: \ell_2 \to \ell_2$ be the backwards shift of ℓ_2, i.e., $S(\alpha_1, \alpha_2, \alpha_3, \ldots) = (\alpha_2, \alpha_3, \alpha_4, \ldots)$. Then each λ in the open unit disc is an eigenvalue of S. Indeed,

$$S(1, \lambda, \lambda^2, \ldots) = \lambda(1, \lambda, \lambda^2, \ldots), \qquad |\lambda| < 1.$$

But none of these eigenvalues are of finite type, because (as follows from the next theorem) eigenvalues of finite type are isolated points in the spectrum.

THEOREM 1.1. *A point λ_0 in $\sigma(A)$ is an eigenvalue of finite type if and only if λ_0 is an isolated point in $\sigma(A)$ and the corresponding Riesz projection $P_{\{\lambda_0\}}$ has finite rank.*

PROOF. Note that λ_0 is an isolated point of $\sigma(A)$ means that the set $\{\lambda_0\}$ is an isolated part of $\sigma(A)$. Now let λ_0 be such a point, and assume that the corresponding Riesz projection $P_{\{\lambda_0\}}$ has finite rank. Put $M = \operatorname{Im} P_{\{\lambda_0\}}$ and $L = \operatorname{Ker} P_{\{\lambda_0\}}$. Then Theorem I.2.2 tells us that $X = M \oplus L$ and the properties (E_1), (E_2) and (E_3) hold. Thus λ_0 is an eigenvalue of finite type.

Conversely, assume λ_0 is an eigenvalue of A of finite type. So X admits a direct sum decomposition, $X = M \oplus L$, with the properties (E_1), (E_2) and (E_3). Apply

Proposition I.2.4. It follows that $\sigma(A) = \{\lambda_0\} \cup \sigma(A|L)$. Since $\lambda_0 \notin \sigma(A|L)$, we conclude that λ_0 is an isolated point of $\sigma(A)$ and $M = \operatorname{Im} P_{\{\lambda_0\}}$. But $\dim M < \infty$. So $P_{\{\lambda_0\}}$ has finite rank. $\square$

Let λ_0 be an eigenvalue of A of finite type, and let $X = M \oplus L$ be a direct sum decomposition with the properties (E_1), (E_2) and (E_3). From Proposition I.2.4 it follows that M and L are uniquely determined. In fact, $M = \operatorname{Im} P_{\{\lambda_0\}}$ and $L = \operatorname{Ker} P_{\{\lambda_0\}}$. The dimension of the space M will be called the *algebraic multiplicity* of the eigenvalue λ_0 and is denoted by $m(\lambda_0; A)$. In other words,

$$(1) \qquad\qquad m(\lambda_0; A) = \operatorname{rank} P_{\{\lambda_0\}}.$$

The *geometric multiplicity* of λ_0 as an eigenvalue of A is, by definition, equal to $\dim \operatorname{Ker}(\lambda_0 - A)$.

COROLLARY 1.2. *Let σ be an isolated part of $\sigma(A)$. Then the corresponding Riesz projection P_σ has finite rank if and only if σ consists of a finite number of eigenvalues of A of finite type. Furthermore, in that case*

$$(2) \qquad\qquad \operatorname{rank} P_\sigma = \sum_{\lambda \in \sigma} m(\lambda; A).$$

PROOF. The space $M := \operatorname{Im} P_\sigma$ is an A-invariant subspace of X and $\sigma(A|M) = \sigma$. Assume $\operatorname{rank} P_\sigma$ is finite. Then we know from Linear Algebra that $\sigma(A|M)$ consists of a finite number of eigenvalues, say $\lambda_1, \ldots, \lambda_r$, and M admits a direct sum decomposition $M = M_1 \oplus M_2 \oplus \cdots \oplus M_r$ such that $AM_j \subset M_j$ and $\sigma(A|M_j) = \{\lambda_j\}$ for $j = 1, \ldots, r$. Now put $L_j = M_1 \oplus \cdots \oplus M_{j-1} \oplus \cdots \oplus M_{j+1} \oplus M_r \oplus \operatorname{Ker} P_\sigma$. Then $X = M_j \oplus L_j$ and the conditions (E_1), (E_2) and (E_3) are fulfilled. It follows that σ consists of a finite number of eigenvalues of finite type.

To prove the converse, assume $\sigma = \{\lambda_1, \ldots, \lambda_r\}$, where $\lambda_1, \ldots, \lambda_r$ are (different) eigenvalues of A of finite type. Then (cf. Corollary I.2.3)

$$\operatorname{Im} P_\sigma = \operatorname{Im} P_{\{\lambda_1\}} \oplus \cdots \oplus \operatorname{Im} P_{\{\lambda_r\}}.$$

Hence $\operatorname{rank} P_\sigma = \sum_{j=1}^{r} \operatorname{rank} P_{\{\lambda_j\}} = \sum_{j=1}^{r} m(\lambda_j; A) < \infty$. $\square$

The next theorem describes the behaviour of the resolvent in a neighbourhood of an eigenvalue of finite type.

THEOREM 1.3. *Let λ_0 be an eigenvalue of A of finite type. Then the resolvent $(\lambda - A)^{-1}$ admits at λ_0 an expansion of the following type:*

$$(3) \qquad\qquad (\lambda - A)^{-1} = \sum_{\nu = -q}^{\infty} (\lambda - \lambda_0)^\nu B_\nu$$

where $B_{-1}, \ldots, B_{-q}$ are operators of finite rank. Here $q \leq m(\lambda_0; A)$ and the series in (3) converges in the operator norm for all λ in some punctured disc $0 < |\lambda - \lambda_0| < \varepsilon$.

PROOF. Put $M = \operatorname{Im} P_{\{\lambda_0\}}$ and $L = \operatorname{Ker} P_{\{\lambda_0\}}$. Then $X = M \oplus L$ and relative to this direct sum decomposition $\lambda - A$ admits the following 2×2 operator matrix representation:

$$\lambda - A = \begin{bmatrix} \lambda - A_1 & 0 \\ 0 & \lambda - A_2 \end{bmatrix},$$

where $A_1 = A|M$ and $A_2 = A|L$. Since λ_0 is an isolated point of $\sigma(A)$, it follows that for some $\delta > 0$

$$(4) \qquad (\lambda - A)^{-1} = \begin{bmatrix} (\lambda - A_1)^{-1} & 0 \\ 0 & (\lambda - A_2)^{-1} \end{bmatrix}, \qquad 0 < |\lambda - \lambda_0| < \delta.$$

Note that A_1 acts on a finite dimensional space and A_1 has precisely one point in its spectrum, namely λ_0. So M has a basis such that the matrix of A_1 relative to this basis has a Jordan normal form with λ_0 on the main diagonal. Consider the following single $m \times m$ Jordan block:

$$J = \begin{bmatrix} \lambda_0 & 1 & & \\ & \lambda_0 & 1 & \\ & & \ddots & 1 \\ & & & \ddots & \lambda_0 \end{bmatrix}.$$

As is well-known

$$(\lambda - J)^{-1} = \begin{bmatrix} (\lambda - \lambda_0)^{-1} & (\lambda - \lambda_0)^{-2} & \cdots & (\lambda - \lambda_0)^{-m} \\ & (\lambda - \lambda_0)^{-1} & \cdots & (\lambda - \lambda_0)^{-m+1} \\ & & \ddots & \vdots \\ & & & (\lambda - \lambda_0)^{-1} \end{bmatrix}.$$

But then we may conclude that

$$(5) \qquad (\lambda - A_1)^{-1} = \sum_{j=1}^{q} (\lambda - \lambda_0)^{-j}(A_1 - \lambda_0)^{j-1}, \qquad \lambda \neq \lambda_0,$$

where $q \leq \dim M = m(\lambda_0; A)$. Next, recall that $\lambda_0 \notin \sigma(A|L)$. Thus $\lambda_0 - A_2$ is invertible, and hence

$$(6) \qquad (\lambda - A_2)^{-1} = \sum_{\nu=0}^{\infty} -(\lambda - \lambda_0)^{\nu}(A_2 - \lambda_0)^{-\nu-1}$$

for $|\lambda - \lambda_0| < \|(\lambda_0 - A_2)^{-1}\|^{-1}$. From (4), (5) and (6) it is now clear that (3) holds provided we define

$$B_\nu = \begin{bmatrix} (A_1 - \lambda_0)^{-\nu-1} & 0 \\ 0 & 0 \end{bmatrix}, \qquad \nu = -1, \ldots, -q,$$

$$B_\nu = \begin{bmatrix} 0 & 0 \\ 0 & -(A_2 - \lambda_0)^{-\nu-1} \end{bmatrix}, \qquad \nu = 0, 1, 2, \ldots.$$

Note that the operators $B_{-1}, \ldots, B_{-q}$ are of finite rank. $\square$

Let λ_0 be an isolated point in $\sigma(A)$. Then (with arguments similar to the ones used in the proof of Theorem 1.3) it can be shown that the resolvent $(\lambda - A)^{-1}$ admits the following expansion:

$$(7) \qquad (\lambda - A)^{-1} = \sum_{\nu=-\infty}^{\infty} (\lambda - \lambda_0)^{\nu} B_{\nu}$$

in some punctured disc $0 < |\lambda - \lambda_0| < \varepsilon$. Note that in this expansion the coefficient B_{-1} is precisely the Riesz projection $P_{\{\lambda_0\}}$. Thus if (7) holds and rank $B_{-1} < \infty$, then λ_0 is an eigenvalue of finite type. Furthermore, in that case $B_{\nu} = 0$ for $\nu < -m(\lambda_0; A)$.

II.2 JORDAN CHAINS

Let $A : X \to X$ be an operator acting on the Banach space X, and let λ_0 be an eigenvalue of A. An ordered set $\{x_0, x_1, \ldots, x_{r-1}\}$ in X is called a *Jordan chain* of A at λ_0 if $x_0 \neq 0$ and

$$(1) \qquad Ax_0 = \lambda_0 x_0, \qquad Ax_j = \lambda_0 x_j + x_{j-1} \qquad (j = 1, \ldots, r-1).$$

Note that the first vector in a Jordan chain is an eigenvector of A. The elements $x_1, \ldots, x_{r-1}$ in (1) are called *generalized eigenvectors* of A.

Jordan chains may be characterized in the following way. Let $x_0, \ldots, x_{r-1}$ be vectors in X. Then $\{x_0, \ldots, x_{r-1}\}$ is a Jordan chain of A at λ_0 if and only if the vectors $x_0, \ldots, x_{r-1}$ are linearly independent, the space $M_0 = \mathrm{span}\{x_0, \ldots, x_{r-1}\}$ is invariant under A, and the matrix $A|M_0$ relative to the basis $x_0, \ldots, x_{r-1}$ is a single Jordan block with λ_0 on the main diagonal, i.e., with respect to $x_0, \ldots, x_{r-1}$

$$\mathrm{matrix}(A|M_0) = \begin{bmatrix} \lambda_0 & 1 & & & \\ & \lambda_0 & 1 & & \\ & & & \ddots & \\ & & & \ddots & 1 \\ & & & & \lambda_0 \end{bmatrix}.$$

PROPOSITION 2.1. *If λ_0 is an eigenvalue of finite type, then $\mathrm{Im}\, P_{\{\lambda_0\}}$ has a basis of eigenvectors and generalized eigenvectors.*

PROOF. The space $M = \mathrm{Im}\, P_{\{\lambda_0\}}$ is finite dimensional and A-invariant. Thus M has a basis such that relative to this basis the matrix of $A|M$ has a Jordan normal form. It follows that the basis elements are eigenvectors or generalized eigenvectors of A. $\square$

PROPOSITION 2.2. *The vectors $x_0, \ldots, x_{r-1}$ form a Jordan chain of A at λ_0 if and only if $x_0 \neq 0$ and*

$$(2) \qquad y(t) = e^{\lambda_0 t} \left(\sum_{\nu=0}^{r-1} \frac{1}{\nu!} t^{\nu} x_{r-1-\nu} \right)$$

satisfies the differential equation $y'(t) = Ay(t)$, $-\infty < t < \infty$.

PROOF. Assume $x_0, \ldots, x_{r-1}$ form a Jordan chain for A at λ_0. Then for λ in the resolvent set of A

$$(3) \qquad (\lambda - A)^{-1} x_j = \sum_{\nu=0}^{j} \frac{1}{(\lambda - \lambda_0)^{\nu+1}} x_{j-\nu}, \qquad j = 0, \ldots, r-1.$$

Take a Cauchy contour Γ around the spectrum of A. One computes that

$$e^{tA} x_{r-1} = \frac{1}{2\pi i} \int_\Gamma e^{t\lambda} (\lambda - A)^{-1} x_{r-1} d\lambda$$

$$= e^{t\lambda_0} \sum_{n=0}^{\infty} \frac{1}{n!} t^n \left(\frac{1}{2\pi i} \int_\Gamma (\lambda - \lambda_0)^n (\lambda - A)^{-1} x_{r-1} d\lambda \right)$$

$$= e^{t\lambda_0} \sum_{n=0}^{r-1} \frac{1}{n!} t^n x_{r-1-n} = y(t).$$

According to Lemma I.5.1 this implies that $y(\cdot)$ is a solution of $y'(t) = Ay(t)$, $-\infty < t < \infty$.

Next, assume that the function $y(\cdot)$ given by (2) satisfies the differential equation $y'(t) = Ay(t)$, $-\infty < t < \infty$. Then

$$(A - \lambda_0) y(t) = y'(t) - \lambda_0 y(t)$$

$$= e^{\lambda_0 t} \sum_{j=0}^{r-2} \frac{1}{j!} t^j x_{r-2-j}.$$

It follows that $(A - \lambda_0) x_0 = 0$ and $(A - \lambda_0) x_j = x_{j-1}$ for $j = 1, \ldots, r-1$. Thus $x_0, \ldots, x_{r-1}$ is a Jordan chain. $\square$

Solutions of $y'(t) = Ay(t)$, $-\infty < t < \infty$, of the form (2) are called *elementary solutions*. Note that the elementary solutions are precisely those solutions for which the initial value at 0 is an eigenvector or generalized eigenvector. If the initial value is a linear combination of eigenvectors and generalized eigenvectors (corresponding to possibly different eigenvalues), then obviously the solution is a linear combination of elementary solutions. Thus, if the linear span of all eigenvectors and generalized eigenvectors of A is dense in the space X, then any solution of $y'(t) = Ay(t)$, $-\infty < t < \infty$, can be approximated uniformly on finite intervals by linear combinations of elementary solutions. From these observations the importance of the completeness of eigenvectors and generalized eigenvectors is apparent. Completeness is also important if one wants to calculate $f(A)$ for functions f different from e^t. For compact operators the *problem of completeness* (i.e., under what conditions is the linear span of eigenvectors and generalized eigenvectors dense) is one of the main topics of Part II of the present book.

Let λ_0 be an eigenvalue of A of finite type, and let $\{x_0, x_1, \ldots, x_{r-1}\}$ be a Jordan chain of A at λ_0. Then (3) holds. Take a Cauchy contour Γ around λ_0 which separates λ_0 from $\sigma(A)\backslash\{\lambda_0\}$. By integrating (3) over Γ one sees that $P_{\{\lambda_0\}}x_j = x_j$ for $j = 0,\ldots,r-1$. It follows (cf. Proposition 2.1) that $\operatorname{Im} P_{\{\lambda_0\}}$ is precisely the space spanned by all eigenvectors and generalized eigenvectors of A corresponding to λ_0. For this reason $\operatorname{Im} P_{\{\lambda_0\}}$ is often called the *generalized eigenspace* of A corresponding to the eigenvalue λ_0. Note that the generalized eigenspace of A corresponding to λ_0 is equal to $\operatorname{Ker}(\lambda_0 - A)^r$ for any $r \geq m(\lambda_0; A)$.

II.3 EIGENVALUES OF COMPACT OPERATORS

THEOREM 3.1. *Let A be a compact operator on a Banach space, and let σ be an isolated part of $\sigma(A)$. If 0 does not belong to σ, then the corresponding Riesz projection P_σ has finite rank.*

PROOF. Let Γ be a Cauchy contour around σ which separates σ from $\sigma(A)\backslash\sigma$. Since $0 \notin \sigma$, we may assume without loss of generality that 0 belongs to the outer domain of Γ. It follows that

$$\begin{aligned}
P_\sigma &= \frac{1}{2\pi i}\int_\Gamma (\lambda - A)^{-1}d\lambda \\
&= \frac{1}{2\pi i}\int_\Gamma \frac{1}{\lambda}(\lambda - A + A)(\lambda - A)^{-1}d\lambda \\
&= \frac{1}{2\pi i}\int_\Gamma \frac{1}{\lambda}Id\lambda + \frac{1}{2\pi i}\int_\Gamma \frac{1}{\lambda}A(\lambda - A)^{-1}d\lambda \\
&= A\left(\frac{1}{2\pi i}\int_\Gamma \frac{1}{\lambda}(\lambda - A)^{-1}d\lambda\right).
\end{aligned}$$

Thus P_σ is compact. Since P_σ is a projection, this implies that $\operatorname{rank} P_\sigma$ is finite. □

COROLLARY 3.2. *Any non-zero point in the spectrum of a compact operator is an eigenvalue of finite type.*

PROOF. Let λ_0 be a non-zero point in the spectrum of the compact operator A. From [GG], Section XI.6 we already know that λ_0 is an isolated point in $\sigma(A)$. (Let us remark that in [GG] the latter statement is proved under a certain additional restriction on A; a general proof will be given in Section XI.8 of the chapter on Fredholm operators.) According to Theorem 3.1 the operator $P_{\{\lambda_0\}}$ has finite rank. Thus by Theorem 1.1 the point λ_0 is an eigenvalue of finite type. □

In the remaining part of this section A is a compact operator acting on a separable Hilbert space H. We know now that the non-zero part of the spectrum of A is a finite or countable set of eigenvalues of finite type. This set can only accumulate at 0. In what follows

$$(1) \qquad\qquad \lambda_1(A), \lambda_2(A), \lambda_3(A), \ldots$$

denotes the (finite or infinite) sequence of non-zero eigenvalues of A with the following convention:

(i)   *each non-zero eigenvalue of A appears in the sequence (1) as many times as the value of its algebraic multiplicity,*

(ii) *the eigenvalues are ordered according to decreasing absolute value, i.e., $|\lambda_1(A)| \geq |\lambda_2(A)| \geq |\lambda_3(A)| \geq \cdots$.*

The following lemma is known as the *lemma of Schur*.

LEMMA 3.3. *Let A be a compact operator on a Hilbert space H, and let E_A denote the smallest closed linear manifold of H containing all the eigenvectors and generalized eigenvectors of A corresponding to non-zero eigenvalues. In E_A there exists an orthonormal basis $\varphi_1, \varphi_2, \ldots$ such that for $j = 1, 2, \ldots$*

$$(2) \qquad A\varphi_j = a_{1j}\varphi_1 + \cdots + a_{jj}\varphi_j, \qquad a_{jj} = \lambda_j(A).$$

PROOF. Let $\mu_1, \mu_2, \ldots$ be the different non-zero eigenvalues of A. Assume $|\mu_1| \geq |\mu_2| \geq \cdots$, and let m_j be the algebraic multiplicity of μ_j. Obviously,

$$E_A = \overline{\operatorname{span}\{\operatorname{Im} P_{\{\mu_j\}} \mid j = 1, 2, \ldots\}}.$$

In $\operatorname{Im} P_{\{\mu_j\}}$ we choose a basis $\omega_{j1}, \ldots, \omega_{jm_j}$ such that the matrix of $A|\operatorname{Im} P_{\{\mu_j\}}$ relative to this basis has a Jordan normal form. This is done for each j. Consider the set

$$V = \{\omega_{11}, \ldots, \omega_{1m_1}, \omega_{21}, \ldots, \omega_{2m_2}, \omega_{31}, \ldots\}.$$

Then V is linearly independent set of vectors and $E_A = \overline{\operatorname{span} V}$. Now apply the Gram-Schmidt orthonormalization procedure to V. The resulting orthonormal system $\varphi_1, \varphi_2, \ldots$ has the desired properties. $\square$

In the special case that $E_A = H$ the lemma of Schur shows that H has an orthonormal basis $\varphi_1, \varphi_2, \ldots$ such that the matrix of A with respect to this basis is an upper triangular matrix with non-zero diagonal elements.

In general we may write $H = E_A \oplus E_A^\perp$. Let us consider the 2×2 operator matrix of A relative to the decomposition $H = E_A \oplus E_A^\perp$. Since E_A is invariant under A, the element in the left lower corner is the zero operator. Thus

$$(3) \qquad A = \begin{bmatrix} A_{11} & A_{12} \\ 0 & A_{22} \end{bmatrix} : E_A \oplus E_A^\perp \to E_A \oplus E_A^\perp.$$

LEMMA 3.4. *The operator A_{22} in (3) is a compact operator and $\sigma(A_{22})$ has no non-zero elements.*

PROOF. Let Q be the orthogonal projection of H onto $E_A^\perp$. Then $A_{22}x = QAx$ for each $x \in E_A^\perp$. Since A is compact, it follows that the same is true for A_{22}. Assume that μ is a non-zero element in $\sigma(A_{22})$. Then $\bar\mu$ is a non-zero element in $\sigma(A_{22}^*)$, and hence $\bar\mu$ is an eigenvalue of A_{22}^*. Let $x_0 \neq 0$ in $E_A^\perp$ be a corresponding eigenvector. From $AE_A \subset E_A$ it follows that $A^*E_A^\perp \subset E_A^\perp$, and hence $A_{22}^* = A^* \mid E_A^\perp$. But then $A^*x_0 = \bar\mu x_0$, and we conclude that

$$(4) \qquad 0 \neq x_0 \in E_A^\perp \cap \operatorname{Im} P_{\{\bar\mu\}}(A^*).$$

On the other hand $\operatorname{Im} P_{\{\mu\}}(A) \subset E_A$. So we can apply Proposition I.2.5 to show that

$$E_A^{\perp} \subset \left(\operatorname{Im} P_{\{\mu\}}(A)\right)^{\perp} = \operatorname{Ker} P_{\{\mu\}}(A)^*$$
$$= \operatorname{Ker} P_{\{\overline{\mu}\}}(A^*),$$

which contradicts (4). Thus $\sigma(A_{22})$ has no non-zero elements. $\square$

A compact operator whose spectrum consists of the zero element only is called a *Volterra operator*. To study such operators one cannot use the functional calculus of Section I.3, but one needs more advanced methods. In Part II we shall develop some of these more advanced methods. Note that the system of eigenvectors and generalized eigenvectors of A is complete if, for example, $\dim E_A^{\perp} < \infty$ or the Volterra operator A_{22} in (3) is the zero operator.

II.4 CONTINUITY OF SPECTRA AND EIGENVALUES

Let $A: X \to X$ be a bounded linear operator acting on the Banach space X. In this section we describe what happens to the spectrum or parts of the spectrum of A if the operator A is subjected to a small perturbation.

THEOREM 4.1. *Let Ω be an open neighbourhood of $\sigma(A)$. Then there exists $\varepsilon > 0$ such that $\sigma(B) \subset \Omega$ for any operator B on X with $\|A - B\| < \varepsilon$.*

PROOF. Choose a Cauchy contour Γ in Ω around $\sigma(A)$. Put

$$(1) \qquad \gamma = \min\{\|(\lambda - A)^{-1}\|^{-1} \mid \lambda \in \Gamma\}.$$

Assume that $\|A - B\| \le \frac{1}{2}\gamma$. Then

$$(2) \qquad \|(\lambda - A) - (\lambda - B)\| \le \frac{1}{2}\|(\lambda - A)^{-1}\|^{-1}, \qquad \lambda \in \Gamma.$$

Since $\lambda - A$ is invertible for $\lambda \in \Gamma$, we can apply Corollary II.8.2 in [GG] to show that $\sigma(B) \cap \Gamma = \emptyset$ and

$$\|(\lambda - A)^{-1} - (\lambda - B)^{-1}\| \le \frac{\|(\lambda - A)^{-1}\|^2 \|A - B\|}{1 - \|(\lambda - A)^{-1}\|\|A - B\|}$$
$$\le 2\|(\lambda - A)^{-1}\|^2 \|A - B\|, \qquad \lambda \in \Gamma.$$

Let P be the Riesz projection corresponding to the part of $\sigma(B)$ inside Γ. Then

$$\|I - P\| = \left\|\frac{1}{2\pi i} \int_{\Gamma} [(\lambda - A)^{-1} - (\lambda - B)^{-1}]d\lambda\right\|$$
$$\le \frac{1}{2\pi} \int_{\Gamma} \|(\lambda - A)^{-1} - (\lambda - B)^{-1}\|d\lambda$$
$$\le C\|A - B\|,$$

where

$$(3) \qquad C = \frac{1}{\pi} \int_{\Gamma} \|(\lambda - A)^{-1}\|^2 d\lambda < \infty.$$

Now, put $\varepsilon = \min\{\frac{1}{2}\gamma, (C+1)^{-1}\}$ and take $\|A - B\| < \varepsilon$. Then $\|I - P\| < 1$. But $I - P$ is a projection, and thus $I - P = 0$. Note that $I - P$ is the Riesz projection corresponding to the part of $\sigma(B)$ outside Γ. Since $I - P = 0$, it follows that $\sigma(B)$ is inside Γ, and hence $\sigma(B) \subset \Omega$. $\square$

The next theorem concerns the behaviour of the eigenvalues of finite type under small perturbations.

THEOREM 4.2. *Let σ be a finite set of eigenvalues of A of finite type, and let Γ be a contour around σ which separates σ from $\sigma(A)\backslash\sigma$. Then there exists $\varepsilon > 0$ such that for any operator B on X with $\|A - B\| < \varepsilon$ the following holds true: $\sigma(B) \cap \Gamma = \emptyset$, the part of $\sigma(B)$ inside Γ is a finite set of eigenvalues of finite type and*

$$(4) \qquad \sum_{\lambda \text{ inside } \Gamma} m(\lambda; B) = \sum_{\lambda \text{ inside } \Gamma} m(\lambda; A).$$

To prove Theorem 4.2 we use the following lemma (due to B. Sz-Nagy [1,2]).

LEMMA 4.3. *Let P and Q be projections of the Banach space X. If $\|P - Q\| < 1$, then*

$$(5) \qquad X = \operatorname{Ker} P \oplus \operatorname{Im} Q, \qquad X = \operatorname{Im} P \oplus \operatorname{Ker} Q.$$

Furthermore, P and Q have the same rank.

PROOF. Our assumption implies that the operators $S := I - P + Q$ and $T := I - Q + P$ are invertible. Take $x \in X$, and put $y = S^{-1}x$. Then $x = Sy = (I - P)y + Qy \in \operatorname{Ker} P + \operatorname{Im} Q$. Furthermore, if $z \in \operatorname{Ker} P \cap \operatorname{Im} Q$, then $Tz = z - Qz + Pz = 0$, and hence $z = 0$. This proves the first identity in (5). By interchanging the roles of P and Q one obtains the second identity in (5). $\square$

PROOF OF THEOREM 4.2. Put $\varepsilon = \min\{\frac{1}{2}\gamma, (C+1)^{-1}\}$, where the constants γ and C are defined as in (1) and (3), respectively. Since $\varepsilon \leq \frac{1}{2}\gamma$, one can use the same arguments as in the proof of Theorem 4.1 to show that $\sigma(B) \cap \Gamma = \emptyset$. Let P (resp. Q) be the Riesz projection corresponding to the part of $\sigma(A)$ (resp. $\sigma(B)$) inside Γ. Then $P = P_\sigma$ has finite rank (Corollary 1.2) and (see the proof of Theorem 4.1)

$$\|P - Q\| \leq C\|A - B\| < 1.$$

Now apply Lemma 4.3. We obtain $\operatorname{rank} Q = \operatorname{rank} P < \infty$. But then we can use Corollary 1.2 to finish the proof. $\square$

COROLLARY 4.4. *The limit in the operator norm of a sequence of Volterra operators is again a Volterra operator.*

PROOF. Let $A_1, A_2, \ldots$ be a sequence of Volterra operators on the Banach space X, and assume that the sequence converges in the operator norm to the bounded linear operator A. Thus $\|A_n - A\| \to 0$ for $n \to \infty$. Then A is compact (cf. [GG], Theorem II.14.3). We want to show that $\sigma(A)$ consists of the zero element only. Suppose not, and let $0 \neq \lambda \in \sigma(A)$. Then λ is an eigenvalue of finite type for A. Choose a circle in the resolvent set of A with centre at λ and radius $r < |\lambda|$ such that λ is the only point of $\sigma(A)$ inside Γ. Now apply Theorem 4.2. So for n sufficiently large the operator A_n must have an eigenvalue of finite type inside Γ. Since A_n is Volterra and 0 is not inside Γ, this is impossible. Thus A is Volterra. $\square$

If we drop the compactness condition on the operators in Corollary 4.4, then the statement does not remain true. More precisely, one can construct (see Rickart [1], page 282) a sequence $A_1, A_2, \ldots$ of bounded linear operators on a Hilbert space H, which converges in the operator norm to an operator A, such that $\sigma(A_n) = \{0\}$ for $n = 1, 2, \ldots$ and $\sigma(A)$ contains non-zero elements.

COROLLARY 4.5. *A Volterra operator on a Hilbert space is the limit in the operator norm of a sequence of finite rank Volterra operators.*

PROOF. Let A be a Volterra operator on a Hilbert space H. Thus A is compact and A has no non-zero elements in the spectrum. Since A is compact and acts on a Hilbert space, there exists (see [GG], Theorem VIII.4.2) a sequence $A_1, A_2, \ldots$ of finite rank operators such that $\|A_n - A\| \to 0$ $(n \to \infty)$. Let E_n be the closed linear manifold in H containing all the eigenvectors and generalized eigenvectors of A_n corresponding to non-zero eigenvalues. Obviously, $\dim E_n < \infty$. Relative to the decomposition $H = E_n \oplus E_n^\perp$ the operator A_n admits a 2×2 operator matrix representation, namely

$$A_n = \begin{bmatrix} A_{11}^{(n)} & A_{12}^{(n)} \\ 0 & A_{22}^{(n)} \end{bmatrix}.$$

From Lemma 3.3 we know that the space E_n has an orthonormal basis $\varphi_1^{(n)}, \ldots, \varphi_{r_n}^{(n)}$ such that the matrix of $A_{11}^{(n)}$ relative to this basis has the following upper triangular form:

$$\begin{bmatrix} \lambda_1(A_n) & * & \cdots & * \\ & \lambda_2(A_n) & \cdots & * \\ & & \ddots & \vdots \\ & & & \lambda_{r_n}(A_n) \end{bmatrix}.$$

Define D_n on E_n by setting $D_n \varphi_j^{(n)} = \lambda_j(A_n)\varphi_j^{(n)}$, and consider

$$V_n = \begin{bmatrix} A_{11}^{(n)} - D_n & A_{12}^{(n)} \\ 0 & A_{22}^{(n)} \end{bmatrix}.$$

Since $\dim E_n < \infty$, the operator $A_n - V_n$ has finite rank, and hence $V_n = A_n - (A_n - V_n)$ is a finite rank operator. According to Lemma 3.4 the operator $A_{22}^{(n)}$ has no non-zero eigenvalues. By construction the same is true for $A_{11}^{(n)} - D_n$. Hence V_n has no non-zero eigenvalues, and thus V_n is a finite rank Volterra operator. It remains to show that

$$(6) \qquad \|A_n - V_n\| = \|D_n\| = \max_j |\lambda_j(A)| \to 0 \qquad (n \to \infty).$$

To prove (6), take $\Omega = \{\lambda \in \mathbf{C} \,|\, |\lambda| < \delta\}$, where δ is some positive number. Note that $\sigma(A) = \{0\} \subset \Omega$. According to Theorem 4.1 this implies that there exists an integer N such that $\sigma(A_n) \subset \Omega$ for $n \geq N$. So for each j we have $|\lambda_j(A_n)| < \delta$ whenever $n \geq N$. This proves (6). $\square$

CHAPTER III

ANALYTIC EQUIVALENCE

In this chapter we study a concept of equivalence which allows one to compare spectra and spectral properties of different operators with the aim to find their similarities. Local and global aspects of the equivalence concept are studied separately. Equivalence is also used for the analysis of operators which depend analytically (and not necessarily linearly) on the spectral parameter.

III.1 A FIRST EXAMPLE

The type of equivalence studied in this chapter has its origins in the theory of differential equations. Consider the n-th order equation:

$$(1) \qquad \varphi^{(n)}(t) + A_{n-1}\varphi^{(n-1)}(t) + \cdots + A_0\varphi(t) = 0,$$

where $A_0,\ldots,A_{n-1}$ are operators acting on a Banach space X and φ is an X-valued function on $\mathbf{R}$. The usual way to deal with the equation (1) is to replace it by the following linear system:

$$\begin{cases} \dfrac{d}{dt}\varphi_1(t) - \varphi_2(t) = 0, \\[2mm] \qquad\vdots \\[2mm] \dfrac{d}{dt}\varphi_{n-1}(t) - \varphi_n(t) = 0, \\[2mm] \dfrac{d}{dt}\varphi_n(t) + A_0\varphi_1(t) + \cdots + A_{n-1}\varphi_n(t) = 0, \end{cases}$$

which can be written in the form

$$(2) \qquad \frac{d}{dt}\Phi(t) = C\Phi(t),$$

with

$$(3) \qquad C = \begin{bmatrix} 0 & I_X & 0 & & 0 \\ 0 & 0 & I_X & & 0 \\ \vdots & \vdots & \vdots & \ddots & I_X \\ -A_0 & -A_1 & -A_2 & \cdots & -A_{n-1} \end{bmatrix}, \qquad \Phi = \begin{bmatrix} \varphi_1 \\ \varphi_2 \\ \vdots \\ \varphi_n \end{bmatrix}.$$

The operator C acts as a bounded linear operator on the space X^n (the direct sum of n copies of X) and Φ is an X^n-valued function. Equations (1) and (2) are equivalent in the following sense: If φ is a solution of (1), then Φ with $\varphi_j = \varphi^{(j-1)}$, $j = 1,\ldots,n$, is a solution of (2), and, conversely, if Φ is a solution of (2), then $\varphi = \varphi_1$ is a solution of (1). It follows (cf. Section I.5) that the general solution of (1) is given by

$$\varphi(t) = Qe^{tC}\eta,$$

where η is an arbitrary vector in X^n and Q assigns to a vector in X^n its first X-component, i.e.,

$$Q = [I \quad 0 \quad \cdots \quad 0] : X^n \to X.$$

The equivalence between (1) and its linearized version (2) can be made more explicit in the following way. Let $L(\lambda)$ be the monic operator polynomial

$$L(\lambda) = \lambda^n I + \lambda^{n-1} A_{n-1} + \cdots + A_0.$$

Then the following identity holds true

$$(4) \qquad \begin{bmatrix} L(\lambda) & & & \\ & I_X & & \\ & & \ddots & \\ & & & I_X \end{bmatrix} = F(\lambda)(\lambda I_{X^n} - C)E(\lambda), \qquad \lambda \in \mathbb{C}.$$

Here C is given by the first equality in (3) and

$$E(\lambda) = \begin{bmatrix} I_X & & & \\ \lambda I_X & I_X & & \\ \vdots & \vdots & \ddots & \\ \lambda^{n-1} I_X & \lambda^{n-2} I_X & \cdots & I_X \end{bmatrix},$$

$$F(\lambda) = \begin{bmatrix} B_{n-1}(\lambda) & B_{n-2}(\lambda) & \cdots & & B_0(\lambda) \\ -I_X & 0 & \cdots & & 0 \\ & -I_X & \cdots & & 0 \\ & & & & \\ & & \cdots & -I_X & 0 \end{bmatrix},$$

with $B_0(\lambda) = I_X$ and $B_{\nu+1}(\lambda) = \lambda B_\nu(\lambda) + A_{n-1-\nu}$ for $\nu = 0, 1, \ldots, n-2$. From the lower triangular form of $E(\lambda)$ it is evident that $E(\lambda)$ is an invertible operator on X^n. A cyclic permutation of the columns transforms $F(\lambda)$ into an upper triangular operator matrix with I_X or $-I_X$ on the main diagonal. Hence also $F(\lambda)$ is an invertible operator on X^n. Furthermore, $E(\lambda)$, $F(\lambda)$ and their inverses $E(\lambda)^{-1}$, $F(\lambda)^{-1}$ are polynomials in λ. In particular, all these operators depend analytically on the variable λ. It is this type of analytic equivalence which forms the main topic of this chapter.

III.2 GLOBAL EQUIVALENCE

Let Ω be an open set in $\mathbb{C}$, and for each λ in Ω let $T(\lambda) : X_1 \to Y_1$ and $S(\lambda) : X_2 \to Y_2$ be bounded linear operators which act between (complex) Banach spaces.

The operator functions $T(\cdot)$ and $S(\cdot)$ are called (*globally*) *equivalent on* Ω if there exist operator functions $E\colon \Omega \to \mathcal{L}(X_1, X_2)$ and $F\colon \Omega \to \mathcal{L}(Y_2, Y_1)$, which are analytic on Ω, such that

$$(1.a) \qquad\qquad T(\lambda) = F(\lambda)S(\lambda)E(\lambda), \qquad \lambda \in \Omega,$$

and, in addition, $E(\lambda)$ and $F(\lambda)$ are invertible for each $\lambda \in \Omega$. In that case also

$$(1.b) \qquad\qquad S(\lambda) = F(\lambda)^{-1}T(\lambda)E(\lambda)^{-1}, \qquad \lambda \in \Omega$$

and the operator functions $E(\cdot)^{-1}$ and $F(\cdot)^{-1}$ are again analytic on Ω. Formula (4) in the previous section provides a first example of global equivalence. The next example describes another context in which this notion appears.

Let $A\colon X \to Y$ and $B\colon Y \to X$ be operators acting between Banach spaces. Then the operator functions $T(\lambda) = (\lambda I_Y - AB) \oplus I_X$ and $S(\lambda) = (\lambda I_X - BA) \oplus I_Y$ are globally equivalent on $C\backslash\{0\}$. In fact,

$$(2) \qquad \begin{bmatrix} \lambda I_Y - AB & 0 \\ 0 & I_X \end{bmatrix} = F(\lambda) \begin{bmatrix} \lambda I_X - BA & 0 \\ 0 & I_Y \end{bmatrix} E(\lambda), \qquad \lambda \neq 0,$$

where

$$E(\lambda) = \begin{bmatrix} -\lambda^{-1}B & \lambda^{-1}I_X \\ \lambda I_Y - AB & A \end{bmatrix}, \qquad E(\lambda)^{-1} = \begin{bmatrix} -A & \lambda^{-1}I_Y \\ \lambda I_X - BA & \lambda^{-1}B \end{bmatrix},$$

$$F(\lambda) = \begin{bmatrix} -A & I_Y - \lambda^{-1}AB \\ I_X & \lambda^{-1}B \end{bmatrix}, \qquad F(\lambda)^{-1} = \begin{bmatrix} -\lambda^{-1}B & I_X - \lambda^{-1}BA \\ I_Y & A \end{bmatrix}.$$

From the equivalence in (2) it follows that $\lambda \neq 0$ belongs to $\sigma(AB)$ if and only if λ belongs to $\sigma(BA)$. In other words:

$$(3) \qquad\qquad \sigma(AB)\backslash\{0\} = \sigma(BA)\backslash\{0\}.$$

Later (e.g., in the next section and in Corollary VII.6.2) we shall see that the equivalence in (2) implies that several other spectral characteristics of AB and BA are the same.

Given an operator function $T\colon \Omega \to \mathcal{L}(X, Y)$ and a Banach space Z, we call the operator function

$$\begin{bmatrix} T(\cdot) & 0 \\ 0 & I_Z \end{bmatrix} \colon \Omega \to \mathcal{L}(X \oplus Z, Y \oplus Z)$$

the Z-*extension* of $T(\cdot)$. According to formula (4) in the previous section a suitable extension of a monic operator polynomial is equivalent on C to a linear function. The next theorem gives another example of linearization by extension and equivalence.

THEOREM 2.1. *Let* Γ *be a Cauchy contour around* 0 *in* C, *and let* $T(\cdot)$ *be an operator function, which is analytic on the inner domain* Ω *of* Γ, *continuous on* $\Omega \cup \Gamma$

and whose values are operators on the Banach space X. Denote by $C(\Gamma, X)$ the Banach space of all X-valued continuous functions on Γ endowed with the supremum norm. Let

$$Z = \left\{ f \in C(\Gamma, X) \,\Big|\, \int_{\Gamma} \frac{1}{\zeta} f(\zeta) d\zeta = 0 \right\},$$

and define A on $C(\Gamma, X)$ by setting

$$(4) \qquad (Af)(z) = zf(z) - \frac{1}{2\pi i} \int_{\Gamma} [I - T(\zeta)] f(\zeta) d\zeta, \qquad z \in \Gamma.$$

Then the Z-extension of $T(\lambda)$ is equivalent on Ω to $\lambda - A$.

PROOF. Let $\tau: X \to C(\Gamma, X)$ be the canonical embedding, i.e., $\tau(x)(z) = x$ for each $z \in \Gamma$ and $x \in X$. Furthermore, define $\omega: C(\Gamma, X) \to X$ by

$$\omega f = \frac{1}{2\pi i} \int_{\Gamma} \frac{1}{\zeta} f(\zeta) d\zeta.$$

Obviously, $\omega \tau = I_X$ and $P := \tau \omega$ is a projection of $C(\Gamma, X)$ with $\operatorname{Ker} P = Z$. Let $J: X \oplus Z \to C(\Gamma, X)$ be given by $J(x, g) = \tau x + g$. Then J is invertible and $J^{-1} f = \big(\omega f, (I - P) f \big)$.

Next, consider on $C(\Gamma, X)$ the following two auxiliary operators:

$$(Vf)(z) = zf(z), \qquad (Mf)(z) = T(z)f(z), \qquad z \in \Gamma.$$

The set Ω is in the resolvent set of V. In fact, for each $\lambda \in \Omega$ we have

$$((\lambda - V)^{-1} f)(z) = (\lambda - z)^{-1} f(z), \qquad z \in \Gamma.$$

Since M commutes with V, the operator M also commutes with $(\lambda - V)^{-1}$ for each $\lambda \in \Omega$. For $\lambda \in \Omega$ we define:

$$B(\lambda) = I + PV(\lambda - V)^{-1} - PV(\lambda - V)^{-1} M,$$
$$E(\lambda) = (\lambda - V)^{-1} J,$$
$$F(\lambda) = J^{-1} \big(I - PB(\lambda)(I - P) \big).$$

Obviously, $E(\lambda): X \oplus Z \to C(\Gamma, X)$ and $F(\lambda): C(\Gamma, X) \to X \oplus Z$ are invertible operators which depend analytically on the variable λ in Ω. Note that $A = V - PV + PVM$. It follows that for $\lambda \in \Omega$

$$\begin{aligned}
F(\lambda)(\lambda - A)E(\lambda) &= F(\lambda)[I + PV(\lambda - V)^{-1} - PVM(\lambda - V)^{-1}]J \\
&= J^{-1}[I - PB(\lambda)(I - P)]B(\lambda)J \\
&= J^{-1}[B(\lambda) - PB(\lambda)(I - P)B(\lambda)]J \\
&= J^{-1}[B(\lambda) - PB(\lambda)(I - P)]J \\
&= J^{-1}PB(\lambda)PJ + J^{-1}(I - P)J.
\end{aligned}$$

Now, use the Cauchy integral formula to show that $B(\lambda)\tau = \tau T(\lambda)$. We conclude that

$$F(\lambda)(\lambda - A)E(\lambda) = \begin{bmatrix} T(\lambda) & 0 \\ 0 & I_Z \end{bmatrix}, \qquad \lambda \in \Omega,$$

and the theorem is proved. $\square$

It can be shown (see Gohberg-Kaashoek-Lay [2]) that the spectrum of the operator A defined by (4) is given by

$$\sigma(A) = \Gamma \cup \{\lambda \in \Omega \mid T(\lambda) \text{ not invertible}\}.$$

The next theorem shows that for an operator function of the form $\lambda - A$ the procedure of linearization by extension and equivalence does not simplify further the operator A and leads to operators that are similar to A.

THEOREM 2.2. *Let A_1 and A_2 be operators acting on the Banach spaces X_1 and X_2, respectively, and suppose that for some Banach spaces Z_1 and Z_2 the extensions $(\lambda - A_1) \oplus I_{Z_1}$ and $(\lambda - A_2) \oplus I_{Z_2}$ are equivalent on some open set Ω containing $\sigma(A_1) \cup \sigma(A_2)$. Then A_1 and A_2 are similar. In fact, if the equivalence is given by*

$$(5) \qquad \begin{bmatrix} \lambda - A_1 & 0 \\ 0 & I_{Z_1} \end{bmatrix} = F(\lambda) \begin{bmatrix} \lambda - A_2 & 0 \\ 0 & I_{Z_2} \end{bmatrix} E(\lambda), \qquad \lambda \in \Omega,$$

then $SA_1 = A_2 S$, where $S: X_1 \to X_2$ is an invertible operator defined by

$$(6) \qquad S = \frac{1}{2\pi i} \int_\Gamma (\lambda - A_2)^{-1} \pi_2 F(\lambda)^{-1} \tau_1 d\lambda,$$

and its inverse is equal to

$$(7) \qquad S^{-1} = \frac{1}{2\pi i} \int_\Gamma \pi_1 E(\lambda)^{-1} \tau_2 (\lambda - A_2)^{-1} d\lambda.$$

Here Γ is the boundary of some bounded Cauchy domain Δ such that $(\sigma(A_1) \cup \sigma(A_2)) \subset \Delta \subset \overline{\Delta} \subset \Omega$; for $i = 1, 2$ the map $\pi_i: X_i \oplus Z_i \to X_i$ is the projection of $X_i \oplus Z_i$ onto X_i and $\tau_i: X_i \to X_i \oplus Z_i$ is the natural embedding of X_i into $X_i \oplus Z_i$.

PROOF. From the equivalence (5) it follows that the integrands in (6) and (7) satisfy the following identities:

$$(8) \qquad (\lambda - A_2)^{-1} \pi_2 F(\lambda)^{-1} \tau_1 = \pi_2 E(\lambda) \tau_1 (\lambda - A_1)^{-1}, \qquad \lambda \in \Omega,$$

$$(9) \qquad \pi_1 E(\lambda)^{-1} \tau_2 (\lambda - A_2)^{-1} = (\lambda - A_1)^{-1} \pi_1 F(\lambda) \tau_2, \qquad \lambda \in \Omega.$$

Let the contour Γ be as in the theorem. Since the integrals

$$\frac{1}{2\pi i} \int_\Gamma \pi_2 F(\lambda)^{-1} \tau_1 d\lambda, \qquad \frac{1}{2\pi i} \int_\Gamma \pi_2 E(\lambda) \tau_1 d\lambda$$

are equal to the zero operator, one computes that

$$A_2 S = \frac{1}{2\pi i} \int_\Gamma (A_2 - \lambda + \lambda)(\lambda - A_2)^{-1} \pi_2 F(\lambda)^{-1} \tau_1 d\lambda$$

$$= \frac{1}{2\pi i} \int_\Gamma \lambda(\lambda - A_2)^{-1} \pi_2 F(\lambda)^{-1} \tau_1 d\lambda$$

$$= \frac{1}{2\pi i} \int_\Gamma \lambda \pi_2 E(\lambda) \tau_1 (\lambda - A_1)^{-1} d\lambda$$

$$= \frac{1}{2\pi i} \int_\Gamma \pi_2 E(\lambda) \tau_1 (\lambda - A_1 + A_1)(\lambda - A_1)^{-1} d\lambda$$

$$= \left(\frac{1}{2\pi i} \int_\Gamma \pi_2 E(\lambda) \tau_1 (\lambda - A_1)^{-1} d\lambda \right) A_1 = S A_1.$$

Let T be the operator defined by the right hand side of (7). It remains to show that TS and ST are identity operators. To do this we shall use the resolvent equation (cf. formula (2) in Section I.2) and the fact that the operators

$$(10) \qquad \pi_1 E(\lambda)^{-1} \tau_2 (\lambda - A_2)^{-1} \pi_2 F(\lambda)^{-1} \tau_1 - (\lambda - A_1)^{-1}$$

$$(11) \qquad \pi_2 E(\lambda) \tau_1 (\lambda - A_1)^{-1} \pi_1 F(\lambda) \tau_2 - (\lambda - A_2)^{-1}$$

depend analytically on λ in Ω. To see that the function in (10) is analytic on Ω we employ the equivalence (5). First we take the inverses of the left hand side and the right hand side of (5). Next we multiply both sides from the left by τ_1 and from the right by π_1. This shows that

$$(\lambda - A_1)^{-1} = \pi_1 E(\lambda)^{-1} \begin{bmatrix} (\lambda - A_2)^{-1} & 0 \\ 0 & I_{Z_2} \end{bmatrix} F(\lambda)^{-1} \tau_1,$$

for $\lambda \in \Omega \backslash \{\sigma(A_1) \cup \sigma(A_2)\}$. It follows that for such λ the expression in (10) is equal to

$$\pi_1 E(\lambda)^{-1} \begin{bmatrix} 0 & 0 \\ 0 & I_{Z_2} \end{bmatrix} F(\lambda)^{-1} \tau_1.$$

Since $E(\cdot)^{-1}$ and $F(\cdot)^{-1}$ are analytic on Ω, we see that (10) extends to a function which is analytic on Ω. Now, let Γ_1 and Γ_2 be contours with the same properties as Γ, and assume that Γ_1 is in the inner domain of Γ_2. Then, by (8) and (9),

$$ST = \left(\frac{1}{2\pi i} \right)^2 \int_{\Gamma_1} \int_{\Gamma_2} \pi_2 E(\lambda) \tau_1 (\lambda - A_1)^{-1} (\mu - A_1)^{-1} \pi_1 F(\mu) \tau_2 d\mu d\lambda$$

$$= \left(\frac{1}{2\pi i} \right)^2 \int_{\Gamma_1} \int_{\Gamma_2} \pi_2 E(\lambda) \tau_1 (\mu - \lambda)^{-1} [(\lambda - A_1)^{-1} - (\mu - A_1)^{-1}] \pi_1 F(\mu) \tau_2 d\mu d\lambda$$

$$= A - B,$$

where

$$A = \left(\frac{1}{2\pi i}\right)^2 \int_{\Gamma_1}\int_{\Gamma_2} \pi_2 E(\lambda)\tau_1(\mu - \lambda)^{-1}(\lambda - A_1)^{-1}\pi_1 F(\mu)\tau_2 \, d\mu \, d\lambda$$

$$= \frac{1}{2\pi i}\int_{\Gamma_1} \pi_2 E(\lambda)\tau_1(\lambda - A_1)^{-1}\left(\frac{1}{2\pi i}\int_{\Gamma_2}(\mu - \lambda)^{-1}\pi_1 F(\mu)\tau_2 \, d\mu\right) d\lambda$$

$$= \frac{1}{2\pi i}\int_{\Gamma_1} \pi_2 E(\lambda)\tau_1(\lambda - A_1)^{-1}\pi_1 F(\lambda)\tau_2 \, d\lambda$$

$$= \frac{1}{2\pi i}\int_{\Gamma_1}(\lambda - A_2)^{-1} d\lambda = I_{X_2};$$

$$B = \left(\frac{1}{2\pi i}\right)^2 \int_{\Gamma_1}\int_{\Gamma_2} \pi_2 E(\lambda)\tau_1(\mu - \lambda)^{-1}(\mu - A_1)^{-1}\pi_1 F(\mu)\tau_2 \, d\mu \, d\lambda$$

$$= \frac{1}{2\pi i}\int_{\Gamma_2}\left(\frac{1}{2\pi i}\int_{\Gamma_1}(\mu - \lambda)^{-1}\pi_2 E(\lambda)\tau_1 \, d\lambda\right)(\mu - A_1)^{-1}\pi_1 F(\mu)\tau_2 \, d\mu$$

$$= 0.$$

The last equality follows from the fact that μ is not in the inner domain of Γ_1 and hence $(\mu - \cdot)^{-1}\pi_2 E(\cdot)\tau_1$ is analytic on Γ_1 and its inner domain. We have now proved that $ST = I_{X_2}$. In a similar way, using the analyticity of (10) instead of (11), one obtains $TS = I_{X_1}$. $\square$

For linear functions $\lambda - A_1$ and $\lambda - A_2$ global equivalence on $\mathbf{C}$ means just that A_1 and A_2 are similar. This follows from the next corollary.

COROLLARY 2.3. *Two operators A_1 and A_2 are similar if and only if $\lambda - A_1$ and $\lambda - A_2$ are equivalent on some open set containing $\sigma(A_1)$ and $\sigma(A_2)$.*

PROOF. If A_1 and A_2 are similar, then, obviously, $\lambda - A_1$ and $\lambda - A_2$ are equivalent on $\mathbf{C}$. Theorem 2.2 gives the reverse implication. $\square$

III.3 LOCAL EQUIVALENCE

Let Ω be an open set in $\mathbf{C}$, and let $T(\cdot)$ and $S(\cdot)$ be operator functions defined on Ω. Given λ_0 in Ω we say that $T(\cdot)$ and $S(\cdot)$ are *(locally) equivalent* at λ_0 if there exists an open neighbourhood $\mathcal{U}$ of λ_0 in Ω such that

$$T(\lambda) = F(\lambda)S(\lambda)E(\lambda), \qquad \lambda \in \mathcal{U},$$

where $E(\lambda)$ and $F(\lambda)$ are invertible operators which depend analytically on λ in $\mathcal{U}$. In other words, using the terminology of the previous section, the operator functions $T(\cdot)$ and $S(\cdot)$ are equivalent at λ_0 if they are globally equivalent on an open neighbourhood of λ_0. In what follows we shall be concerned mainly with the case when $T(\lambda) = \lambda - A_1$ and $S(\lambda) = \lambda - A_2$ with A_1 and A_2 bounded linear operators on a Banach space X.

THEOREM 3.1. *Let $\lambda - A_1$ and $\lambda - A_2$ be equivalent at the point λ_0, and assume that λ_0 is an eigenvalue of finite type for A_1. Then λ_0 is an eigenvalue of finite type for A_2 and the restrictions $A_1|\operatorname{Im} P_{\{\lambda_0\}}(A_1)$ and $A_2|\operatorname{Im} P_{\{\lambda_0\}}(A_2)$ are similar.*

PROOF. For some open neighbourhood $\mathcal{U}$ of λ_0 we have

$$(1) \qquad \lambda - A_2 = F(\lambda)(\lambda - A_1)E(\lambda), \qquad \lambda \in \mathcal{U}.$$

Here $E(\lambda)$ and $F(\lambda)$ are invertible operators on the Banach space X and $E(\cdot)$ and $F(\cdot)$ are analytic on $\mathcal{U}$. According to our hypothesis λ_0 is an isolated point in $\sigma(A_1)$. From (1) it follows that λ_0 is also an isolated point in $\sigma(A_2)$. For $\nu = 1, 2$ let P_ν be the Riesz projection of A_ν corresponding to the part $\{\lambda_0\}$. Write A_ν as a 2×2 operator matrix relative to the decomposition $X = \operatorname{Im} P_\nu \oplus \operatorname{Ker} P_\nu$:

$$A_\nu = \begin{bmatrix} A_{\nu 1} & 0 \\ 0 & A_{\nu 2} \end{bmatrix}, \qquad \nu = 1, 2.$$

It remains to show that A_{11} and A_{21} are similar.

Note that $\lambda_0 - A_{12}$ and $\lambda_0 - A_{22}$ are invertible. So without loss of generality (replace $\mathcal{U}$ by a smaller neighbourhood if necessary) we may assume that $\lambda - A_{12}$ and $\lambda - A_{22}$ are invertible for each λ in $\mathcal{U}$. Define

$$E_0(\lambda) = \begin{bmatrix} I_{\operatorname{Im} P_1} & 0 \\ 0 & \lambda - A_{12} \end{bmatrix} E(\lambda), \qquad \lambda \in \mathcal{U},$$

$$F_0(\lambda) = \begin{bmatrix} I_{\operatorname{Im} P_2} & 0 \\ 0 & (\lambda - A_{22})^{-1} \end{bmatrix} F(\lambda), \qquad \lambda \in \mathcal{U}.$$

Then $E_0(\lambda)$ and $F_0(\lambda)$ are invertible for each $\lambda \in \mathcal{U}$ and as functions $E_0(\cdot)$ and $F_0(\cdot)$ are analytic on $\mathcal{U}$. Furthermore,

$$\begin{bmatrix} \lambda - A_{21} & 0 \\ 0 & I_{\operatorname{Ker} P_2} \end{bmatrix} = F_0(\lambda) \begin{bmatrix} \lambda - A_{11} & 0 \\ 0 & I_{\operatorname{Ker} P_1} \end{bmatrix} E_0(\lambda)$$

for $\lambda \in \mathcal{U}$. Now, recall that $\sigma(A_{11}) = \sigma(A_{21}) = \{\lambda_0\}$. Thus $(\lambda - A_{11}) \oplus I_{\operatorname{Ker} P_1}$ and $(\lambda - A_{21}) \oplus I_{\operatorname{Ker} P_2}$ are (globally) equivalent on the open set $\mathcal{U}$ containing $\sigma(A_{11})$ and $\sigma(A_{21})$. But then we can apply Theorem 2.2 to show that A_{11} and A_{21} are similar. $\square$

THEOREM 3.2. *Let A_1 and A_2 be compact operators, and assume that $\lambda - A_1$ and $\lambda - A_2$ are equivalent at each point of C. Then A_1 and A_2 are similar.*

PROOF. For some open neighbourhood $\mathcal{U}$ of 0 we have

$$(2) \qquad \lambda - A_2 = F(\lambda)(\lambda - A_1)E(\lambda), \qquad \lambda \in \mathcal{U}.$$

Here $E(\lambda)$ and $F(\lambda)$ are invertible operators on the Banach space X and $E(\cdot)$ and $F(\cdot)$ are analytic on $\mathcal{U}$. Let σ be the part of the spectrum of A_1 outside $\mathcal{U}$. We know that σ consists of a finite number of eigenvalues of finite type (Section II.3). Since $\lambda - A_1$ and $\lambda - A_2$ are equivalent at each point of C, the operators A_1 and A_2 have the same spectrum. It follows that the part of $\sigma(A_2)$ outside $\mathcal{U}$ coincides with σ. For $\nu = 1,2$ let P_ν be the Riesz projection of A_ν corresponding to σ. Write A_ν as a 2×2 operator matrix relative to the decomposition $X = \operatorname{Ker} P_\nu \oplus \operatorname{Im} P_\nu$:

$$A_\nu = \begin{bmatrix} A_{\nu 1} & 0 \\ 0 & A_{\nu 2} \end{bmatrix}, \qquad \nu = 1,2.$$

From Theorem 3.1 we know that A_{12} and A_{22} are similar. It remains to prove that A_{11} and A_{21} are similar.

Note that $\sigma(A_{12}) = \sigma(A_{22}) = \sigma$. Thus $\lambda - A_{12}$ and $\lambda - A_{22}$ are invertible for each λ in $\mathcal{U}$. From (2) it follows that

$$\begin{bmatrix} \lambda - A_{21} & 0 \\ 0 & I_{\operatorname{Im} P_2} \end{bmatrix} = F_0(\lambda) \begin{bmatrix} \lambda - A_{11} & 0 \\ 0 & I_{\operatorname{Im} P_1} \end{bmatrix} E_0(\lambda), \qquad \lambda \in \mathcal{U},$$

where

$$E_0(\lambda) = \begin{bmatrix} I_{\operatorname{Ker} P_1} & 0 \\ 0 & \lambda - A_{12} \end{bmatrix} E(\lambda), \qquad \lambda \in \mathcal{U},$$

$$F_0(\lambda) = \begin{bmatrix} I_{\operatorname{Ker} P_2} & 0 \\ 0 & (\lambda - A_{22})^{-1} \end{bmatrix} F(\lambda), \qquad \lambda \in \mathcal{U}.$$

So $(\lambda - A_{21}) \oplus I_{\operatorname{Im} P_2}$ and $(\lambda - A_{11}) \oplus I_{\operatorname{Im} P_1}$ are globally equivalent on $\mathcal{U}$, and hence we can apply Theorem 2.2 to show that A_{11} and A_{21} are similar. $\square$

Note that in Theorem 3.2 it is not necessary to assume that both A_1 and A_2 are compact. In fact it suffices to assume that one of the operators is compact, and then the similarity implies that the other is also compact.

The question whether or not Theorem 3.2 holds for arbitrary (not necessarily compact) bounded linear operators is an unsolved problem.

If two operator functions $T(\cdot)$ and $S(\cdot)$ are globally equivalent on an open set Ω, then, obviously, $T(\cdot)$ and $S(\cdot)$ are (locally) equivalent at each point of Ω. For certain special classes of operator functions the converse statement is also true, however, in general, local equivalence at each point of Ω does not imply global equivalence on Ω (see Gohberg-Kaashoek-Lay [2], Leiterer [1], and Apostol [1] for further information). Note that the problem mentioned in the preceding paragraph can be phrased as follows: If $\lambda - A_1$ and $\lambda - A_2$ are locally equivalent at each point of $\mathbf{C}$, does it follow that $\lambda - A_1$ and $\lambda - A_2$ are globally equivalent on $\mathbf{C}$?

III.4 MATRICIAL COUPLING AND EQUIVALENCE

In this section we present a general method to obtain equivalence. This method is based on the notion of matricial coupling of operators which is defined as follows. Two operators $T: X_1 \to Z_1$ and $S: Z_2 \to X_2$, acting between Banach spaces, are said to be *matricially coupled* if they are related in the following way:

$$(1) \qquad \begin{bmatrix} T & A_{12} \\ A_{21} & A_{22} \end{bmatrix}^{-1} = \begin{bmatrix} B_{11} & B_{12} \\ B_{21} & S \end{bmatrix}.$$

More precisely, this means that one can construct an invertible 2×2 operator matrix

$$(2) \qquad \begin{bmatrix} A_{11} & A_{12} \\ A_{21} & A_{22} \end{bmatrix} : X_1 \oplus X_2 \to Z_1 \oplus Z_2,$$

with $A_{11} = T$, such that its inverse is given by

$$(3) \qquad \begin{bmatrix} B_{11} & B_{12} \\ B_{21} & B_{22} \end{bmatrix} : Z_1 \oplus Z_2 \to X_1 \oplus X_2,$$

where $B_{22} = S$. The 2×2 operator matrices appearing in (2) and (3) are called the *coupling matrices* and we shall refer to (1) as the *coupling relation*.

To give an example of matricially coupled operators, let $A : X \to Y$ and $B : Y \to X$ be operators acting between Banach spaces. Then the operators $\lambda I_Y - AB$ and $\lambda I_X - BA$ are matricially coupled for each non-zero λ. In fact,

$$(4) \qquad \begin{bmatrix} \lambda I_Y - AB & A \\ -\lambda^{-1}B & \lambda^{-1}I_X \end{bmatrix}^{-1} = \begin{bmatrix} \lambda^{-1}I_Y & -A \\ \lambda^{-1}B & \lambda I_X - BA \end{bmatrix}, \qquad \lambda \neq 0.$$

The next theorem shows that a matricial coupling of T and S is of particular interest if one of the operators is more simple than the other.

THEOREM 4.1. *Assume* $T : X_1 \to Z_1$ *and* $S : Z_2 \to X_2$ *are matricially coupled operators, and let the coupling relation be given by*

$$(5) \qquad \begin{bmatrix} T & A_{12} \\ A_{21} & A_{22} \end{bmatrix}^{-1} = \begin{bmatrix} B_{11} & B_{12} \\ B_{21} & S \end{bmatrix}.$$

Then

$$(6) \qquad \begin{bmatrix} T & 0 \\ 0 & I_{X_2} \end{bmatrix} = F \begin{bmatrix} S & 0 \\ 0 & I_{Z_1} \end{bmatrix} E,$$

where E and F are invertible 2×2 operator matrices

$$(7) \qquad E = \begin{bmatrix} A_{21} & A_{22} \\ T & A_{12} \end{bmatrix}, \qquad F = \begin{bmatrix} -A_{12} & TB_{11} \\ I_{X_2} & B_{21} \end{bmatrix},$$

$$(8) \qquad E^{-1} = \begin{bmatrix} B_{12} & B_{11} \\ S & B_{21} \end{bmatrix}, \qquad F^{-1} = \begin{bmatrix} -B_{21} & SA_{22} \\ I_{Z_1} & A_{12} \end{bmatrix}.$$

PROOF. By direct computation, using (5). $\quad\square$

Theorem 4.1 leads to a global equivalence theorem if the entries in (5) depend analytically on a parameter λ. Let Ω be an open set in $\mathbb{C}$, and let

$$(9) \qquad T(\cdot) : \Omega \to \mathcal{L}(X_1, Z_1), \qquad S(\cdot) \in \mathcal{L}(Z_2, X_2)$$

be analytic operator-valued functions. We say that $T(\cdot)$ and $S(\cdot)$ are *analytically matricially coupled* on Ω if $T(\cdot)$ and $S(\cdot)$ are related in the following way:

$$(10) \qquad \begin{bmatrix} T(\lambda) & A_{12}(\lambda) \\ A_{21}(\lambda) & A_{22}(\lambda) \end{bmatrix}^{-1} = \begin{bmatrix} B_{11}(\lambda) & B_{12}(\lambda) \\ B_{21}(\lambda) & S(\lambda) \end{bmatrix}, \qquad \lambda \in \Omega,$$

where the operators $A_{12}(\lambda)$, $A_{21}(\lambda)$, $A_{22}(\lambda)$ (and hence also the operators $B_{11}(\lambda)$, $B_{12}(\lambda)$, $B_{21}(\lambda)$) depend analytically on the variable λ. Formula (4) implies that the operator functions $\lambda I_Y - AB$ and $\lambda I_X - BA$ are analytically matricially coupled on $\mathbb{C}\backslash\{0\}$.

THEOREM 4.2. *Assume that the analytic operator functions $T(\cdot)$ and $S(\cdot)$ in (9) are analytically matricially coupled. Then the X_2-extension of $T(\cdot)$ is globally equivalent on Ω to the Z_1-extension of $S(\cdot)$.*

PROOF. Take $\lambda \in \Omega$. According to our hypothesis $T(\lambda)$ and $S(\lambda)$ are matricially coupled. So we may apply Theorem 4.1. Since the entries of the coupling matrices in (10) depend analytically on the parameter λ, the same is true for the entries of the operator matrices E and F appearing in (7) and (8). Thus

$$(11) \qquad \begin{bmatrix} T(\lambda) & 0 \\ 0 & I_{X_2} \end{bmatrix} = F(\lambda) \begin{bmatrix} S(\lambda) & 0 \\ 0 & I_{Z_1} \end{bmatrix} E(\lambda), \qquad \lambda \in \Omega,$$

where $E(\cdot)$ and $F(\cdot)$ are analytic and for each $\lambda \in \Omega$ the operators $E(\lambda)$ and $F(\lambda)$ are invertible. This gives the desired result. $\square$

The conditions of Theorem 4.2 are symmetric with respect to A_{ij} and B_{ij} appearing in the coupling relation (10), but the equivalence relation in (11) is not. In fact, under the conditions of Theorem 4.2 one can also prove another equivalence relation which shows that the Z_2-extension of $T(\cdot)$ is globally equivalent on Ω to the X_1-extension of $S(\cdot)$.

Note that the equivalence in formula (2) of Section III.2 may be obtained as a corollary of Theorem 4.1 and formula (4).

Theorem 4.1 allows us to compare the invertibility properties of matricially coupled operators. The following corollary will be useful later (e.g., in Chapters IX and XIII).

COROLLARY 4.3. *Let T and S be matricially coupled operators, and let the coupling relation be given by*

$$(12) \qquad \begin{bmatrix} T & A_{12} \\ A_{21} & A_{22} \end{bmatrix}^{-1} = \begin{bmatrix} B_{11} & B_{12} \\ B_{21} & S \end{bmatrix}.$$

Then

$$(13) \qquad \operatorname{Ker} T = \{B_{12}z \mid z \in \operatorname{Ker} S\}, \qquad \operatorname{Ker} S = \{A_{21}x \mid x \in \operatorname{Ker} T\},$$

$$(14) \qquad \operatorname{Im} T = \{z \in Z_1 \mid B_{21}z \in \operatorname{Im} S\}, \qquad \operatorname{Im} S = \{x \in X_2 \mid A_{12}x \in \operatorname{Im} T\},$$

$$(15) \qquad \dim \operatorname{Ker} T = \dim \operatorname{Ker} S, \qquad \operatorname{codim} \operatorname{Im} T = \operatorname{codim} \operatorname{Im} S.$$

Furthermore, the operator T is invertible if and only if S is invertible, and in that case

$$(16) \qquad T^{-1} = B_{11} - B_{12} S^{-1} B_{21}, \qquad S^{-1} = A_{22} - A_{21} T^{-1} A_{12}.$$

PROOF. Since T and S are matricially coupled, we may apply Theorem 4.1. The relation (6) implies

$$\operatorname{Ker} \begin{bmatrix} T & 0 \\ 0 & I_{X_2} \end{bmatrix} = E^{-1} \operatorname{Ker} \begin{bmatrix} S & 0 \\ 0 & I_{Z_1} \end{bmatrix},$$

which yields the first identity in (13). From (6) it also follows that $y \in \operatorname{Im} T$ if and only if

$$\begin{bmatrix} y \\ 0 \end{bmatrix} \in F \operatorname{Im} \begin{bmatrix} S & 0 \\ 0 & I_{Z_1} \end{bmatrix} = \left\{ \begin{bmatrix} -A_{12} S z_2 + T B_{11} z_1 \\ S z_2 + B_{21} z_1 \end{bmatrix} \,\middle|\, z_1 \in Z_1, z_2 \in Z_2 \right\}.$$

The coupling relation (12) implies that $A_{12} B_{21} + T B_{11} = I_{Z_1}$. Thus $y \in \operatorname{Im} T$ if and only if $y = z_1$ with $B_{21} z_1 \in \operatorname{Im} S$, which proves the first identity in (14). The second identities in (13) and (14) follow from the first by interchanging the roles of T and S. Formula (15) and the statement about invertibility are direct consequences of formula (6). To get (16) note that

$$(17) \qquad \begin{bmatrix} T^{-1} & 0 \\ 0 & I_{X_2} \end{bmatrix} = E^{-1} \begin{bmatrix} S^{-1} & 0 \\ 0 & I_{Z_1} \end{bmatrix} F^{-1}.$$

Now compute the $(1,1)$-entry of the 2×2 operator matrix defined by the right hand side of (17). This gives the first identity in (16). The second is obtained by interchanging the roles of T and S. $\square$

To illustrate the results of this section we show that the usual method of reducing the inversion of an operator $I - K$, with K finite rank, to that of a matrix (see [GG], Theorem II.7.1) can be understood and made more precise in the present context. Assume that $K \colon X \to X$ is given by

$$K x = \sum_{j=1}^{n} \varphi_j(x) y_j, \qquad x \in X,$$

where $y_1, \dots, y_n$ are given vectors in the Banach space X and $\varphi_1, \dots, \varphi_n$ are continuous linear functionals on X. Define $A \colon X \to \mathbf{C}^n$ and $B \colon \mathbf{C}^n \to X$ by

$$A x = \begin{bmatrix} \varphi_1(x) \\ \vdots \\ \varphi_n(x) \end{bmatrix}, \qquad B \begin{bmatrix} \alpha_1 \\ \vdots \\ \alpha_n \end{bmatrix} = \sum_{j=1}^{n} \alpha_j y_j.$$

Note that $G = AB$ acts on $\mathbf{C}^n$ and its matrix with respect to the standard basis of $\mathbf{C}^n$ is given by

$$\operatorname{mat}(G) = \big(\varphi_i(y_j) \big)_{i,j=1}^{n}.$$

Since $K = BA$, the operator functions $\lambda I_X - K$ and $\lambda I_n - G$ are analytically matricially coupled on $\mathbb{C}\backslash\{0\}$ (cf. formula (4)). Here I_n is the identity on $\mathbb{C}^n$. It follows (Corollary 4.3) that for $\lambda \neq 0$

$$(\lambda I_X - K)^{-1} = \frac{1}{\lambda}I_X + \frac{1}{\lambda}B(\lambda I_n - G)^{-1}A$$

whenever $\det(\lambda I_n - G) \neq 0$. Furthermore, the non-zero eigenvalues of K and G are the same, the corresponding multiplicities are equal and the relationship between the Jordan chains of K and G (which is not obvious) becomes transparent.

CHAPTER IV
LINEAR OPERATOR PENCILS

In this chapter we study linear pencils $\lambda G - A$, where G and A are bounded linear operators acting between complex Banach spaces X and Y. The simplest example is the case when $X = Y$ and $G = I$, the identity operator on X. This case was already considered in the previous chapters. If G is invertible, then spectral problems concerning the pencil $\lambda G - A$ are easily reduced to those of $\lambda I - G^{-1}A$. In this chapter we consider the more general case when G is not invertible, but we assume that the pencil is regular, i.e., for some $\lambda_0 \in \mathbb{C}$ the operator $\lambda_0 G - A$ is invertible. The latter property allows us to write

$$(1) \qquad \lambda G - A = (\lambda - \lambda_0)(\lambda_0 G - A)[(\lambda - \lambda_0)^{-1}I + (\lambda_0 G - A)^{-1}G].$$

From this expression it is already clear that the Riesz theory developed in Chapters I and II may be extended to regular pencils. In what follows we make this extension in a more preferable and direct way without using the identity (1). The results are applied to difference equations and parallel those of Chapter I.

IV.1 SPECTRAL DECOMPOSITION

By a *linear operator pencil* acting between X and Y (or on X if $X = Y$) we shall mean a linear operator polynomial $\lambda G - A$, where $G: X \to Y$ and $A: X \to Y$ are bounded linear operators acting between complex Banach spaces and λ is a complex variable. Often the words linear and operator will be omitted, and we shall just use the term pencil.

The *spectrum* of the pencil $\lambda G - A$ will be denoted by $\sigma(G, A)$ and is, by definition, the subset of the extended complex plane $\mathbb{C}_\infty = \mathbb{C} \cup \{\infty\}$ determined by the following properties. The point $\infty \in \sigma(G, A)$ if and only if G is not invertible, and $\sigma(G, A) \cap \mathbb{C}$ consists of all those $\lambda \in \mathbb{C}$ for which $\lambda G - A$ is not invertible. As for the case when $X = Y$ and $G = I$ one proves that $\sigma(G, A)$ is nonempty whenever $X \neq (0)$ or $Y \neq (0)$ (see the end of Section I.1). With respect to the usual topology on $\mathbb{C}_\infty$ (see [C], page 8) the spectrum $\sigma(G, A)$ is compact. Its complement (in $\mathbb{C}_\infty$) is the *resolvent set* of $\lambda G - A$, which is denoted by $\rho(G, A)$. Note that $\infty \in \rho(G, A)$ if and only if G is invertible. It may happen that $\rho(G, A)$ is empty. E.g., take $X = Y = \mathbb{C}^2$ and

$$G = A = \begin{bmatrix} 1 & 0 \\ 0 & 0 \end{bmatrix}.$$

The pencil $\lambda G - A$ is said to be *regular* if $\rho(G, A) \neq \emptyset$. Given a nonempty subset Δ of $\mathbb{C}_\infty$, we say that $\lambda G - A$ is Δ-*regular* if $\Delta \subset \rho(G, A)$. We shall study Γ-regular pencils, where Γ is a Cauchy contour (see Section I.1). Recall that the *inner domain* Δ_+ of a Cauchy contour Γ consists of all points inside Γ. By definition the *outer domain* Δ_- of Γ is the set $\Delta_- = \mathbb{C}_\infty \backslash (\Delta_+ \cup \Gamma)$. Note that $\infty \in \Delta_-$. The next theorem may be viewed as the analogue of Lemma I.2.1 and Theorem I.2.2.

THEOREM 1.1. *Let Γ be a Cauchy contour with Δ_+ and Δ_- as inner and outer domain, respectively, and let $\lambda G - A$ be a Γ-regular linear operator pencil acting between X and Y. Put*

$$(1) \qquad P = \frac{1}{2\pi i} \int_\Gamma (\lambda G - A)^{-1} G \, d\lambda \colon X \to X,$$

$$(2) \qquad Q = \frac{1}{2\pi i} \int_\Gamma G (\lambda G - A)^{-1} \, d\lambda \colon Y \to Y.$$

Then P and Q are projections on X and Y, respectively, and relative to the decompositions $X = \operatorname{Ker} P \oplus \operatorname{Im} P$ and $Y = \operatorname{Ker} Q \oplus \operatorname{Im} Q$ the following partitioning holds true:

$$(3) \qquad \lambda G - A = \begin{bmatrix} \lambda G_1 - A_1 & 0 \\ 0 & \lambda G_2 - A_2 \end{bmatrix} \colon \operatorname{Ker} P \oplus \operatorname{Im} P \to \operatorname{Ker} Q \oplus \operatorname{Im} Q,$$

where $\lambda G_1 - A_1$ and $\lambda G_2 - A_2$ are Γ-regular pencils such that

$$(4) \qquad \sigma(G_1, A_1) = \Delta_- \cap \sigma(G, A), \qquad \sigma(G_2, A_2) = \Delta_+ \cap \sigma(G, A).$$

PROOF. We have to modify the arguments which have been used to derive the properties of the Riesz projections (see the proofs of Lemma I.2.1 and Theorem I.2.2). Only the main differences will be explained. First, note that for a linear operator pencil a *generalized resolvent equation* holds true, namely

$$(5) \qquad (\lambda G - A)^{-1} - (\mu G - A)^{-1} = (\mu - \lambda)(\lambda G - A)^{-1} G (\mu G - A)^{-1},$$

where λ and μ are in $\rho(G, A)$. Introduce the following auxiliary operator

$$K = \frac{1}{2\pi i} \int_\Gamma (\lambda G - A)^{-1} \, d\lambda \colon Y \to X.$$

Obviously,

$$(6) \qquad P = KG, \qquad Q = GK.$$

Using the resolvent identity (5), the usual contour integration arguments show that $KGK = K$ (cf., the proof of Lemma I.2.1), and hence the identities in (6) imply that P and Q are projections. We also have

$$(7) \qquad GP = QG, \qquad AP = QA, \qquad K = KQ = PK.$$

The first identity in (7) is obvious, the third follows from (6) and the fact that $K = KGK$, and the second identity in (7) is an easy consequence of the following formula:

$$(8) \qquad A(\lambda G - A)^{-1} G = G(\lambda G - A)^{-1} A, \qquad \lambda \in \rho(G, A).$$

Formula (7) allows us to partition the operators G, A and K in the following way:

$$(9) \qquad G = \begin{bmatrix} G_1 & 0 \\ 0 & G_2 \end{bmatrix} : \operatorname{Ker} P \oplus \operatorname{Im} P \to \operatorname{Ker} Q \oplus \operatorname{Im} Q,$$

$$(10) \qquad A = \begin{bmatrix} A_1 & 0 \\ 0 & A_2 \end{bmatrix} : \operatorname{Ker} P \oplus \operatorname{Im} P \to \operatorname{Ker} Q \oplus \operatorname{Im} Q,$$

$$(11) \qquad K = \begin{bmatrix} 0 & 0 \\ 0 & L \end{bmatrix} : \operatorname{Ker} Q \oplus \operatorname{Im} Q \to \operatorname{Ker} P \oplus \operatorname{Im} P.$$

From (9) and (10) we conclude that (3) holds. Since $\lambda G - A$ is invertible for $\lambda \in \Gamma$, it follows that the same holds true for $\lambda G_1 - A_1$ and $\lambda G_2 - A_2$. Thus $\lambda G_1 - A_1$ and $\lambda G_2 - A_2$ are Γ-regular. The identities in (6) imply that G_2 is invertible and $G_2^{-1} = L$. In particular, $\infty \notin \sigma(G_2, A_2)$. Next consider

$$T(\lambda) = \frac{1}{2\pi i} \int_{\Gamma} (\lambda - \zeta)^{-1}(\zeta G - A)^{-1} d\zeta : Y \to X.$$

The operator $T(\lambda)$ is well-defined and bounded for $\lambda \notin \Gamma$. One checks that

$$(12) \qquad T(\lambda)(\lambda G - A) = \begin{cases} P - I & \text{for} \quad \lambda \in \Delta_+, \\ P & \text{for} \quad \lambda \in \Delta_- \backslash \{\infty\}; \end{cases}$$

$$(13) \qquad (\lambda G - A)T(\lambda) = \begin{cases} Q - I & \text{for} \quad \lambda \in \Delta_+, \\ Q & \text{for} \quad \lambda \in \Delta_- \backslash \{\infty\}. \end{cases}$$

From the generalized resolvent equation (5) and contour integration arguments it follows that $T(\lambda)Q = PT(\lambda)$, $\lambda \in \Gamma$, and hence $T(\lambda)$ partitions as follows:

$$T(\lambda) = \begin{bmatrix} T_1(\lambda) & 0 \\ 0 & T_2(\lambda) \end{bmatrix} : \operatorname{Ker} Q \oplus \operatorname{Im} Q \to \operatorname{Ker} P \oplus \operatorname{Im} P, \qquad \lambda \notin \Gamma.$$

But then we can use (12) and (13) to conclude that

$$(14) \qquad (\lambda G_1 - A_1)^{-1} = -T_1(\lambda), \qquad \lambda \in \Delta_+,$$

$$(15) \qquad (\lambda G_2 - A_2)^{-1} = T_2(\lambda), \qquad \lambda \in \Delta_- \backslash \{\infty\}.$$

We already know that Γ belongs to $\rho(G_1, A_1)$ and $\rho(G_2, A_2)$. Thus (14) implies that $\Delta_+ \cup \Gamma$ is a subset of $\rho(G_1, A_1)$, and hence $\sigma(G_1, A_1) \subset \Delta_-$. Since $\infty \in \rho(G_2, A_2)$, we also get that $\Delta_- \cup \Gamma$ is contained in $\rho(G_2, A_2)$. Thus $\sigma(G_2, A_2) \subset \Delta_+$. Obviously, $\sigma(G, A) = \sigma(G_1, A_1) \cup \sigma(G_2, A_2)$, and hence (4) holds true. $\quad \square$

If in Theorem 1.1 the spectrum $\sigma(G, A)$ lies inside Γ, then the projections P and Q are the identity operators on X and Y, respectively. To see this, note that according to (4) the inclusion $\sigma(G, A) \subset \Delta_+$ implies that $\sigma(G_1, A_1) = \emptyset$. But if the latter holds, then $\operatorname{Ker} P$ and $\operatorname{Ker} Q$ must consist of the zero element only, and hence P and Q are identity operators. In a similar way one shows that $\sigma(G, A) \subset \Delta_-$ implies that P and Q are both zero.

From the remarks made in the previous paragraph it also follows that the projections P and Q appearing in Theorem 1.1 are uniquely determined by the spectral properties of the pencils $\lambda G_1 - A_1$ and $\lambda G_2 - A_2$. We make this more explicit in the next corollary which concerns the case when 0 is inside Γ.

COROLLARY 1.2. *Let Γ be a Cauchy contour with Δ_+ and Δ_- as inner and outer domain, respectively, and let $\lambda G - A$ be a Γ-regular linear operator pencil acting between the spaces X and Y. Assume that 0 is inside Γ. Then there exists a projection Q on Y and an invertible operator $E: Y \to X$ such that relative to the decomposition $Y = \operatorname{Ker} Q \oplus \operatorname{Im} Q$ the following partitioning holds true:*

$$(16) \qquad (\lambda G - A)E = \begin{bmatrix} \lambda \Omega_1 - I_1 & 0 \\ 0 & \lambda I_2 - \Omega_2 \end{bmatrix} : \operatorname{Ker} Q \oplus \operatorname{Im} Q \to \operatorname{Ker} Q \oplus \operatorname{Im} Q,$$

where I_1 (resp. I_2) denotes the identity operator on $\operatorname{Ker} Q$ (resp. $\operatorname{Im} Q$), the pencil $\lambda \Omega_1 - I_1$ is $(\Delta_+ \cup \Gamma)$-regular and $\lambda I_2 - \Omega_2$ is $(\Delta_- \cup \Gamma)$-regular. Furthermore, Q and E (and hence also the operators Ω_1 and Ω_2) are uniquely determined. In fact

$$(17) \qquad Q = \frac{1}{2\pi i} \int_\Gamma G(\zeta G - A)^{-1} d\zeta,$$

$$(18) \qquad E = \frac{1}{2\pi i} \int_\Gamma (1 - \zeta^{-1})(\zeta G - A)^{-1} d\zeta,$$

$$(19) \qquad \Omega = \begin{bmatrix} \Omega_1 & 0 \\ 0 & \Omega_2 \end{bmatrix} = \frac{1}{2\pi i} \int_\Gamma (\zeta - \zeta^{-1})G(\zeta G - A)^{-1} d\zeta.$$

PROOF. Let P and Q be given by (1) and (2). Then the partitioning (3) holds true and the spectra of the pencils $\lambda G_1 - A_1$ and $\lambda G_2 - A_2$ are given by (4). Since $0 \in \Delta_+$, the operator A_1 is invertible. We also know that G_2 is invertible (because $\infty \in \Delta_-$). Now, put

$$(20) \qquad E = \begin{bmatrix} A_1^{-1} & 0 \\ 0 & G_2^{-1} \end{bmatrix} : \operatorname{Ker} Q \oplus \operatorname{Im} Q \to \operatorname{Ker} P \oplus \operatorname{Im} P,$$

$$(21) \qquad \Omega_1 = G_1 A_1^{-1}, \qquad \Omega_2 = A_2 G_2^{-1}.$$

With Q, E, Ω_1 and Ω_2 defined in this way, (16) holds true and the pencils $\lambda\Omega_1 - I_1$ and $\lambda I_2 - \Omega_2$ have the desired regularity properties.

Next we prove the uniqueness of Q and E. So let us assume that for some projection Q on Y and some invertible operator $E: Y \to X$ the identity (16) holds true, with $\lambda\Omega_1 - I_1$, and $\lambda I_2 - \Omega_2$ regular on $\Delta_+ \cup \Gamma$ and $\Delta_- \cup \Gamma$, respectively. Formula (16) implies that

$$(22) \qquad GE = \begin{bmatrix} \Omega_1 & 0 \\ 0 & I_2 \end{bmatrix}; \qquad AE = \begin{bmatrix} I_1 & 0 \\ 0 & \Omega_2 \end{bmatrix}.$$

Thus

$$\frac{1}{2\pi i}\int_\Gamma G(\zeta G - A)^{-1}d\zeta = \frac{1}{2\pi i}\int_\Gamma GE[(\zeta G - A)E]^{-1}d\zeta$$

$$= \frac{1}{2\pi i}\int_\Gamma \begin{bmatrix} \Omega_1(\zeta\Omega_1 - I_1)^{-1} & 0 \\ 0 & (\zeta I_2 - \Omega_2)^{-1} \end{bmatrix} d\zeta$$

$$= \begin{bmatrix} 0 & 0 \\ 0 & I_2 \end{bmatrix} = Q.$$

Here we used the regularity properties of the pencils $\lambda\Omega_1 - I_1$ and $\lambda I_2 - \Omega_2$. It follows that Q is given by (17), and so Q is uniquely determined. From

$$(23) \qquad (\lambda G - A)^{-1} = E\begin{bmatrix} (\lambda\Omega_1 - I_1)^{-1} & 0 \\ 0 & (\lambda I_2 - \Omega_2)^{-1} \end{bmatrix}, \qquad \lambda \in \Gamma,$$

and the properties of the pencils $\lambda\Omega_1 - I_1$ and $\lambda I_2 - \Omega_2$ we may conclude that

$$(24) \qquad \frac{1}{2\pi i}\int_\Gamma (\zeta G - A)^{-1}d\zeta = E\begin{bmatrix} 0 & 0 \\ 0 & I_2 \end{bmatrix} = EQ,$$

$$(25) \qquad \frac{1}{2\pi i}\int_\Gamma -\zeta^{-1}(\zeta G - A)^{-1}d\zeta = E\begin{bmatrix} I_1 & 0 \\ 0 & 0 \end{bmatrix} = E(I - Q).$$

To prove the last formula one uses that $0 \in \Delta_+$ and $\infty \in \Delta_-$, and thus for some $\varepsilon > 0$

$$-\zeta^{-1}(\zeta\Omega_1 - I_1)^{-1} = \sum_{\nu=0}^\infty \zeta^{\nu-1}\Omega_1^\nu, \qquad 0 < |\zeta| < \varepsilon,$$

$$-\zeta^{-1}(\zeta I_2 - \Omega_2)^{-1} = \sum_{\nu=0}^\infty -\zeta^{-\nu-2}\Omega_2^\nu, \qquad 0 < |\zeta^{-1}| < \varepsilon.$$

Since $E = EQ + E(I - Q)$, we obtain that E is given by (18). In particular, E is uniquely determined.

To prove (19) we first use (23) and the regularity properties of $\lambda\Omega_1 - I_1$ and $\lambda I_2 - \Omega_2$ to show that

$$(26) \qquad \frac{1}{2\pi i} \int_\Gamma \zeta(\zeta G - A)^{-1} d\zeta = E \begin{bmatrix} 0 & 0 \\ 0 & \Omega_2 \end{bmatrix}.$$

Formulas (25) and (26), together with the first identity in (22), yield

$$\frac{1}{2\pi i} \int_\Gamma (\zeta - \zeta^{-1}) G(\zeta G - A)^{-1} d\zeta = GE \begin{bmatrix} 0 & 0 \\ 0 & \Omega_2 \end{bmatrix} + GE \begin{bmatrix} I_1 & 0 \\ 0 & 0 \end{bmatrix} = \Omega. \quad \square$$

Let $\lambda G - A$ and Γ be as in Corollary 1.2. We shall call the 2×2 operator matrix in (16) the Γ-*spectral decomposition* of the pencil $\lambda G - A$, and we shall refer to the operator Ω in (19) as the *associated operator* corresponding to $\lambda G - A$ and Γ. For the projection Q and the operator E in Corollary 1.2 we shall use the words *separating projection* and *right equivalence operator*, respectively. For later purposes we list the following identities (cf. (22)):

$$(27) \qquad AE(I - Q) = I - Q, \qquad AEQ = \Omega Q,$$

$$(28) \qquad GE(I - Q) = \Omega(I - Q), \qquad GEQ = Q.$$

Later we shall apply Corollary 1.2 for the case when Γ is equal to the unit circle T. Note that in that case the regularity properties of the pencils $\lambda\Omega_1 - I_1$ and $\lambda I_2 - \Omega_2$ are just equivalent to the requirement that the spectrum of Ω lies in the open unit circle.

IV.2   A SECOND OPERATOR EQUATION

In this section we consider the operator equation

$$(1) \qquad A_1 Z G_2 - G_1 Z A_2 = C.$$

Here $A_1, G_1 \colon X_1 \to Y_1$ and $A_2, G_2 \colon X_2 \to Y_2$ are given operators acting between Banach spaces. The problem is for a given $C \in \mathcal{L}(X_2, Y_1)$ to find $Z \in \mathcal{L}(Y_2, X_1)$ such that (1) holds. The operator equation considered in Section I.4 is a special case of (1). In fact, to get equation (I.4.1) from (1) we have to take $X_1 = Y_1$, $X_2 = Y_2$ and G_1, G_2 should be the identity operators on X_1 and X_2, respectively. The next theorem is the analogue of Theorem I.4.1.

THEOREM 2.1. *If the spectra of the pencils $\lambda G_1 - A_1$ and $\lambda G_2 - A_2$ are disjoint, then for any right hand $C \in \mathcal{L}(X_2, Y_1)$ equation (1) has a unique solution $Z \in \mathcal{L}(Y_2, X_1)$. More precisely, if Γ is a Cauchy contour such that $\sigma(G_1, A_1)$ is in the inner domain of Γ, and $\sigma(G_2, A_2)$ is in the outer domain of Γ, then*

$$(2) \qquad Z = \frac{1}{2\pi i} \int_\Gamma (\lambda G_1 - A_1)^{-1} C(\lambda G_2 - A_2)^{-1} d\lambda.$$

PROOF. Since $\sigma(G_1, A_1) \cap \sigma(G_2, A_2) = \emptyset$, the point ∞ cannot be in both spectra. So without loss of generality we may assume that $\infty \notin \sigma(G_1, A_1)$. Then $\sigma(G_1, A_1)$ is a compact subset of $\mathbf{C}$ which lies in the open set $V = \mathbf{C} \backslash \sigma(G_2, A_2)$. Choose a bounded Cauchy domain Δ such that $\sigma(G_1, A_1) \subset \Delta \subset \overline{\Delta} \subset V$, and let Γ be the oriented boundary of Δ. Then Γ is a Cauchy contour, $\sigma(G_1, A_1)$ is in the inner domain of Γ and $\sigma(G_2, A_2)$ is in the outer domain of Γ. So it suffices to prove the second part of the theorem, i.e., we have to show that (2) gives the unique solution of (1). Recall (see the remark after the proof of Theorem 1.1) that from the location of the spectra it follows that

$$(3) \qquad \frac{1}{2\pi i} \int_\Gamma G_1 (\lambda G_1 - A_1)^{-1} d\lambda = I, \qquad \frac{1}{2\pi i} \int_\Gamma (\lambda G_2 - A_2)^{-1} G_2 d\lambda = 0,$$

where now I stands for the identity operator on Y_1.

Let Z be given by (2). Then $Z \in \mathcal{L}(Y_2, X_1)$ and because of (3)

$$
\begin{aligned}
A_1 Z G_2 &= \frac{1}{2\pi i} \int_\Gamma A_1 (\lambda G_1 - A_1)^{-1} C (\lambda G_2 - A_2)^{-1} G_2 d\lambda \\
&= \frac{1}{2\pi i} \int_\Gamma -C(\lambda G_2 - A_2)^{-1} G_2 d\lambda \\
&\qquad + \frac{1}{2\pi i} \int_\Gamma \lambda G_1 (\lambda G_1 - A_1)^{-1} C (\lambda G_2 - A_2)^{-1} G_2 d\lambda \\
&= \frac{1}{2\pi i} \int_\Gamma G_1 (\lambda G_1 - A_1)^{-1} C (\lambda G_2 - A_2)^{-1} (\lambda G_2 - A_2 + A_2) d\lambda \\
&= \frac{1}{2\pi i} \int_\Gamma G_1 (\lambda G_1 - A_1)^{-1} C d\lambda + G_1 Z A_2 \\
&= C + G_1 Z A_2.
\end{aligned}
$$

Hence Z is a solution of (1).

Conversely, if Z is a solution of (1). Then

$$
\begin{aligned}
C &= A_1 Z G_2 - \lambda G_1 Z G_2 + \lambda G_1 Z G_2 - G_1 Z A_2 \\
&= -(\lambda G_1 - A_1) Z G_2 + G_1 Z (\lambda G_2 - A_2).
\end{aligned}
$$

It follows that

$$
\begin{aligned}
&\frac{1}{2\pi i} \int_\Gamma (\lambda G_1 - A_1)^{-1} C (\lambda G_2 - A_2)^{-1} d\lambda \\
&\qquad = \frac{-1}{2\pi i} \int_\Gamma Z G_2 (\lambda G_2 - A_2)^{-1} d\lambda + \frac{1}{2\pi i} \int_\Gamma (\lambda G_1 - A_1)^{-1} G_1 Z d\lambda \\
&\qquad = Z.
\end{aligned}
$$

Here we used the analogue of (3) with the order of G_ν and $(\lambda G_\nu - A_\nu)^{-1}$ interchanged ($\nu = 1, 2$). We have now proved that equation (1) is uniquely solvable and its unique solution is given by (2). $\square$

COROLLARY 2.2. *Let $T_1 \colon X_1 \to X_1$ and $T_2 \colon X_2 \to X_2$ be bounded linear operators acting between Banach spaces and with spectra in the open unit disc. Then for any $C \in \mathcal{L}(X_2, X_1)$ the equation*

$$(4) \qquad\qquad T_1 Z T_2 - Z = C$$

has a unique solution $Z \in \mathcal{L}(X_2, X_1)$, namely

$$(5) \qquad\qquad Z = -\sum_{\nu=0}^{\infty} T_1^\nu C T_2^\nu.$$

PROOF. We apply Theorem 2.1 with $\lambda G_1 - A_1 = \lambda I_1 - T_1$ and $\lambda G_2 - A_2 = \lambda T_2 - I_2$. Here I_ν denotes the identity operator on X_ν ($\nu = 1, 2$). Obviously, $\sigma(G_1, A_1) = \sigma(T_1)$ and

$$\sigma(G_2, A_2) = \{\lambda^{-1} \mid \lambda \in \sigma(T_2)\},$$

where we use the convention that $0^{-1} = \infty$. Since $\sigma(T_1)$ and $\sigma(T_2)$ are in the open unit disc, the spectra $\sigma(G_1, A_1)$ and $\sigma(G_2, A_2)$ are disjoint, and Theorem 2.1 implies that (4) has a unique solution which is given by

$$(6) \qquad\qquad Z = \frac{1}{2\pi i} \int_{\mathsf{T}} (\lambda I_1 - T_1)^{-1} C (\lambda T_2 - I_2)^{-1} d\lambda.$$

Here T is the unit circle endowed with the counter clockwise orientation. But

$$(\lambda I_1 - T_1)^{-1} = \sum_{\nu=1}^{\infty} \frac{1}{\lambda^\nu} T_1^{\nu-1}, \qquad \lambda \in \mathsf{T},$$

$$(\lambda T_2 - I_2)^{-1} = -\sum_{\nu=0}^{\infty} \lambda^\nu T_2^\nu, \qquad \lambda \in \mathsf{T},$$

because $\sigma(T_1)$ and $\sigma(T_2)$ are in the open unit disc. It follows that the right hand side of (6) is equal to the right hand side of (5). $\square$

IV.3 HOMOGENEOUS DIFFERENCE EQUATION

In this section we consider the difference equation:

$$(1) \qquad\qquad \begin{cases} G x_{n+1} = A x_n, & n = 0, 1, 2, \ldots, \\ x_0 = y. \end{cases}$$

Here G and A are bounded linear operators acting on a Banach space X and y is a given vector in X. We shall assume that the pencil $\lambda G - A$ is regular. If $G = I$, the identity operator on X, or, more generally, if G is invertible, then (1) is always solvable and the general solution of (1) is given by

$$(2) \qquad x_n = (G^{-1}A)^n y, \qquad n = 0, 1, 2, \dots .$$

If G is not invertible, the situation is different and in that case it may happen that (1) is never solvable for $y \neq 0$. For example, let G be the zero operator and A the identity operator on X. Then the pencil $\lambda G - A$ is regular, but (1) is only solvable for $y = 0$, and in the latter case the solution is the trivial one, namely $x_n = 0$ $(n = 0, 1, 2, \dots)$. The next theorem is an analogue of Theorem I.5.3.

THEOREM 3.1. *Let $\lambda G - A$ be a regular linear operator pencil acting on X. If the spectrum $\sigma(G, A)$ belongs to the open unit disc, then for each y in X equation (1) is solvable and the (unique) solution $x_0, x_1, x_2, \dots$ converges to zero. Conversely, if for each y in X equation (1) is solvable and its solution $x_0, x_1, x_2, \dots$ converges to zero, then $\sigma(G, A)$ belongs to the closed unit disc. If, in addition, $\dim X < \infty$, then $\sigma(G, A)$ belongs to the open unit disc.*

PROOF. Let $\mathbf{D}$ denote the open unit disc, and assume that $\sigma(G, A) \subset \mathbf{D}$. In particular, $\infty \notin \sigma(G, A)$, and thus G is invertible. But then equation (1) is solvable for each $y \in X$ and given y the general solution of (1) is described by (2). Note that $\sigma(G^{-1}A) = \sigma(G, A)$. Since $\sigma(G^{-1}A)$ is compact and lies in $\mathbf{D}$, there exists $0 < \rho < 1$ such that

$$\sigma(G^{-1}A) \subset \{\lambda \in \mathbf{C} \,|\, |\lambda| < \rho\}.$$

Let Γ be the circle with center o and radius ρ, and let the orientation on Γ be counter clockwise. Then, by the operational calculus,

$$\|(G^{-1}A)^n\| = \left\| \frac{1}{2\pi i} \int_{\Gamma} \lambda^n (\lambda - G^{-1}A)^{-1} d\lambda \right\|$$

$$\leq \rho^{n+1} \left(\max_{\lambda \in \Gamma} \|(\lambda G - A)^{-1} G\| \right)$$

$$\longrightarrow 0 \qquad (n \to \infty).$$

Thus (2) implies that for each $y \in X$ the solution of (1) tends to zero if $n \to \infty$.

Next, assume that for each $y \in X$ equation (1) is solvable and that the solutions tend to zero if $n \to \infty$. We want to show that $\sigma(G, A) \subset \overline{\mathbf{D}}$. Take $y \in X$. Since (1) is solvable, there exists $x \in X$ such that $Gx = Ay$. This holds for each y. Hence $\operatorname{Im} A \subset \operatorname{Im} G$, and thus $\operatorname{Im}(\lambda G - A) \subset \operatorname{Im} G$. Now use that $\lambda G - A$ is regular. So for some $\lambda_0 \in \mathbf{D}$ the operator $\lambda_0 G - A$ is invertible, and hence $\operatorname{Im} G = X$. Our hypotheses also imply that $\operatorname{Ker} G = (0)$. Assume not, i.e., $\operatorname{Ker} G \neq (0)$. Then we can construct a solution $x_0, x_1, x_2, \dots$ of (1) such that $\|x_j\| \geq j$ for $j = 0, 1, 2, \dots$. Indeed, assume we have constructed $x_0, \dots, x_{n-1}$ such that $Gx_{j+1} = Ax_j$ for $j = 0, \dots, n-1$ and $\|x_j\| \geq j$ for $j = 0, \dots, n$. Since $\operatorname{Im} G = X$, there exists $x \in X$ such that $Gx = Ax_n$. Choose $z \in \operatorname{Ker} G$ such that $\|x + z\| \geq n + 1$, and define $x_{n+1} = x + z$. Proceeding in this way yields the desired solution. But the existence of such a solution contradicts our hypotheses. So $\operatorname{Ker} G = (0)$. We have now proved that G is invertible.

Since G is invertible, the solution of (1) is given by (2). So our hypotheses imply that for each $y \in X$

$$\lim_{n \to \infty} (G^{-1}A)^n y = 0.$$

But then we can use the principle of uniform boundedness to show that

$$(3) \qquad\qquad M := \sup_n \|(G^{-1}A)^n\| < \infty.$$

Now take $\lambda \in \sigma(G, A)$. Then $\lambda \in \sigma(G^{-1}A)$ and, by the spectral mapping theorem (Theorem I.3.3), the point λ^n belongs to $\sigma\big((G^{-1}A)^n\big)$. According to (3), this implies $|\lambda^n| \leq M$ for $n = 0, 1, 2, \ldots$, which is possible only if $|\lambda| \leq 1$. This proves that $\sigma(G, A) \subset \overline{\mathbf{D}}$.

Finally, assume additionally that $\dim X < \infty$. Take $\lambda_0 \in \sigma(G, A)$, and suppose that $|\lambda_0| = 1$. Then λ_0 is an eigenvalue of $G^{-1}A$, i.e., there exists $y \neq 0$ such that $G^{-1}Ay = \lambda_0 y$. But this implies that

$$\|(G^{-1}A)^n y\| = \|\lambda_0^n y\| = \|y\| \neq 0$$

for each n, and hence $(G^{-1}A)^n y$ does not tend to zero if $n \to \infty$. This is a contradiction, and it follows that $\sigma(G, A) \subset \mathbf{D}$. $\square$

Let A be a bounded linear operator on the Banach space X. The difference equation

$$(4) \qquad\qquad x_{n+1} = Ax_n, \qquad n = 0, 1, 2, \ldots$$

is said to be *asymptotically stable* if every solution $x_0, x_1, \ldots$ of (4) converges to zero if $n \to \infty$. Since (4) is solvable for each initial value $x_0 = y$ in X, Theorem 3.1 yields the following corollary.

COROLLARY 3.2. *If $\sigma(A)$ lies in the open unit disc $\mathbf{D}$, then equation (4) is asymptotically stable. Conversely, if (4) is asymptotically stable, then $\sigma(A)$ lies in the closed unit disc. If, in addition, X is finite dimensional, then asymptotic stability of (4) is equivalent to $\sigma(A) \subset \mathbf{D}$.*

CHAPTER V
SPECTRAL THEORY FOR BOUNDED SELFADJOINT OPERATORS

A compact selfadjoint operator A acting on a Hilbert space can be represented in the form

$$(1) \qquad A = \sum_j \lambda_j \Delta E_j,$$

where $\{\lambda_j\}$ is the set of non-zero eigenvalues of A, the operator ΔE_j is the orthogonal projection onto the eigenspace $\mathrm{Ker}(\lambda_j - A)$ and the series converges in the operator norm. The aim of this chapter is to obtain an analogous representation for an arbitrary bounded selfadjoint operator. The first step is to rewrite the right hand side of (1) as a Stieltjes integral,

$$(2) \qquad A = \int \lambda\, dE(\lambda),$$

with the operator-valued integrator given by $E(\lambda) = \sum_{\lambda_j \leq \lambda} \Delta E_j$. We shall see that the representation (2) also holds for a non-compact selfadjoint operator. In the latter case $E(\lambda)$ is the orthogonal projection onto the maximal A-invariant subspace M such that the spectrum of $A|M$ is contained in $(-\infty, \lambda]$. The construction of such spectral subspaces is given in Section 2 of this chapter. Their general properties are discussed in the first section. In the third section bounded resolutions of the identity are introduced and integrals of the type appearing in (2) are defined. The representation (2) is established in Section 4. In Section 5 spectrum and resolvent are described in terms of the resolution of the identity. In Section 6 the functional calculus is used to construct the square root of a non-negative operator and, as an application, the polar decomposition is obtained. The last section concerns the spectral theorem for unitary operators.

V.1 SPECTRAL SUBSPACES

Let A be a bounded linear operator acting on a complex Banach space X, and let σ be a closed subset of $\mathbf{C}$. An A-invariant subspace M of X is called the *spectral subspace* of A associated with σ if M has the following two properties:

(i) $\sigma(A|M) \subset \sigma$,

(ii) if N is any A-invariant subspace of X such that $\sigma(A|N) \subset \sigma$, then $N \subset M$.

In other words, M is the spectral subspace of A associated with σ if and only if M is the largest A-invariant subspace M such that $\sigma(A|M) \subset \sigma$. Note that the properties (i) and (ii) determine M uniquely. If σ is an isolated part of $\sigma(A)$, then the spectral theory of Section I.2 shows that the range of the Riesz projection $P_\sigma(A)$ is the spectral subspace of A associated with σ. Indeed, in that case $\sigma\big(A|\operatorname{Im} P_\sigma(A)\big) \subset \sigma$, because of Theorem I.2.2, and if N is an A-invariant subspace such that $\sigma(A|N) \subset \sigma$, then

$$P_\sigma(A)x = P_\sigma(A|N)x = x, \qquad x \in N,$$

which shows that $N \subset \operatorname{Im} P_\sigma(A)$.

PROPOSITION 1.1. *Let M be an A-invariant subspace of X, and let σ be a closed subset of $\sigma(A)$ such that $\mathbb{C}\backslash\sigma$ is connected. Then M is the spectral subspace of A associated with σ if and only if M is the largest A-invariant subspace of X such that $(\lambda - A)^{-1}x$ has an analytic continuation to $\mathbb{C}\backslash\sigma$ for each $x \in M$.*

PROOF. Let N be any A-invariant subspace such that $(\lambda - A)^{-1}x$ has an analytic continuation to $\mathbb{C}\backslash\sigma$ for each $x \in N$. We shall prove that $\sigma(A|N) \subset \sigma$.

Take $x \in N$. By our hypotheses on N there exists an X-valued function $u(\cdot)$, defined and analytic on $\mathbb{C}\backslash\sigma$, such that

$$(1) \qquad u(\lambda) = (\lambda - A)^{-1}x, \qquad \lambda \in \rho(A).$$

It follows that $(\lambda - A)u(\lambda) = x$ for $\lambda \in \rho(A)$. Since $\mathbb{C}\backslash\sigma$ is an open connected set which contains $\rho(A)$ as a subset, we conclude (by analytic continuation) that

$$(2) \qquad (\lambda - A)u(\lambda) = x, \qquad \lambda \in \mathbb{C}\backslash\sigma.$$

We claim that $u(\lambda) \in N$ for each $\lambda \in \mathbb{C}\backslash\sigma$. Assume not, then $u(\lambda_0) \notin N$ for some $\lambda_0 \notin \sigma$. The space N is closed. So, by the Hahn-Banach theorem, there exists a continuous linear functional g on X which annihilates N such that $g\big(u(\lambda_0)\big) \neq 0$. We have $(\lambda - A)^{-1}x \in N$ for $|\lambda| > \|A\|$, because N is A-invariant. Thus (1) implies that $g\big(u(\lambda)\big) = 0$ for $|\lambda| > \|A\|$. Since $g\big(u(\cdot)\big)$ is analytic on $\mathbb{C}\backslash\sigma$, we see that $g\big(u(\lambda)\big) = 0$ for each $\lambda \in \mathbb{C}\backslash\sigma$, including $\lambda = \lambda_0$, which is a contradiction. Thus $u(\lambda) \in N$ for all $\lambda \in \mathbb{C}\backslash\sigma$. But then (2) shows that $(\lambda - A)N = N$ for $\lambda \in \mathbb{C}\backslash\sigma$.

To prove that $\sigma(A|N) \subset \sigma$, it remains to show that $\operatorname{Ker}\big(\lambda - (A|N)\big) = \{0\}$ for each $\lambda \in \mathbb{C}\backslash\sigma$. Again take $x \in N$, and let $u(\cdot)$ be as in the previous paragraph. Assume that $Ax = \lambda_0 x$, where $\lambda_0 \in \mathbb{C}\backslash\sigma$. We want to show that $x = 0$. If $\lambda_0 \in \rho(A)$, then this is automatically true. Therefore assume $\lambda_0 \notin \rho(A)$. But then it follows that

$$(\lambda - A)^{-1}x = (\lambda - \lambda_0)^{-1}x, \qquad \lambda \in \rho(A).$$

Thus $u(\lambda) = (\lambda - \lambda_0)^{-1}x$ for $\lambda \in \rho(A)$. By analytic continuation, the latter identity also holds for $\lambda \in \mathbb{C}\backslash\sigma$, $\lambda \neq \lambda_0$. But $u(\cdot)$ is also analytic at λ_0. This can only happen when $x = 0$. Thus $\operatorname{Ker}\big(\lambda_0 - (A|N)\big) = \{0\}$, and we have shown that $\sigma(A|N) \subset \sigma$.

Now, assume that M is the spectral subspace of A associated with σ. According to our hypotheses on M, the resolvent set $\rho(A)$ is contained in $\rho(A|M)$. Thus for $\lambda \in \rho(A)$ we have

$$(3) \qquad (\lambda - A)^{-1}x = [\lambda - (A|M)]^{-1}x, \qquad x \in M.$$

Since $\sigma(A|M) \subset \sigma$, it follows that $(\lambda - A)^{-1}x$ has an analytic continuation to $\mathbb{C}\backslash\sigma$ for each $x \in M$. Let N be another A-invariant subspace with this property. Then, by the result of the first three paragraphs of the proof, $\sigma(A|N) \subset \sigma$. But this implies that $N \subset M$, because of property (ii) in the definition of a spectral subspace. Thus M is the largest A-invariant subspace with the property that $(\lambda - A)^{-1}x$ has an analytic continuation to $\mathbb{C}\backslash\sigma$ for each $x \in M$.

To prove the converse implication, assume that M is as in the last sentence of the previous paragraph. Then, as we have seen above, $\sigma(A|M) \subset \sigma$. Let N be another A-invariant subspace such that $\sigma(A|N) \subset \sigma$. Then (3) holds with M replaced by N, and thus $(\lambda - A)^{-1}x$ has an analytic continuation to $\mathbb{C}\backslash\sigma$ for each $x \in N$. But M is the largest A-invariant subspace with this property. So $N \subset M$, and we have proved that M is the spectral subspace of A associated with σ. $\square$

Let M be an A-invariant subspace, and assume that M is the spectral subspace of A associated with the set σ. Of course, one would like to have

$$(4) \qquad \sigma(A|M) = \sigma \cap \sigma(A).$$

If σ is an isolated part of $\sigma(A)$, then this equality holds true (see Theorem I.2.2). However, in general it fails. In Lyubich-Macaev [1] an example is given of an operator A such that $\sigma(A) = [0,1]$ and $\sigma(A) = \sigma(A|L)$ for any A-invariant subspace $L \neq \{0\}$. Thus, if σ is a proper closed subinterval of $[0,1]$, then for the latter operator the spectral subspace M associated with σ consists of the zero element only and (4) does not hold. This example shows that one has to be careful with the notion of a spectral subspace. In this context it should also be mentioned that it may happen that a bounded linear operator on a Banach space has no non-trivial (i.e., containing non-zero vectors but not equal to the whole space) invariant subspace at all (see Beauzamy [1], [2], Enflo [1], and Read [1], [2] for examples). For an arbitrary Hilbert space operator the existence of a non-trivial invariant subspace is still an open problem. In the next section we shall construct spectral subspaces for selfadjoint operators, and we shall see (in Section 5) that they satisfy the identity (4) up to a natural modification.

V.2 SPECTRAL SUBSPACES FOR SELFADJOINT OPERATORS

In what follows H is a complex Hilbert space. The inner product on H will be denoted by $\langle \cdot, \cdot \rangle$. As usual, $\mathcal{L}(H)$ stands for the set of all (bounded linear) operators on H. Recall (see [GG], Section II.12) that an operator $A \in \mathcal{L}(H)$ is called *selfadjoint* if $A = A^*$, that is, $\langle Ax, y \rangle = \langle x, Ay \rangle$ for each x and y in H. For a selfadjoint operator A the numbers $\langle Ax, x \rangle$ are real, and hence we may define

$$(1) \qquad m(A) := \inf_{\|x\|=1} \langle Ax, x \rangle, \qquad M(A) := \sup_{\|x\|=1} \langle Ax, x \rangle.$$

THEOREM 2.1. *If $A \in \mathcal{L}(H)$ is selfadjoint, then its spectrum $\sigma(A)$ is real, $\sigma(A) \subset [m(A), M(a)]$ and*

$$(2) \qquad \|(\lambda - A)^{-1}\| \leq |\Im\lambda|^{-1}, \qquad \Im\lambda \neq 0.$$

Furthermore, $[m(A), M(A)]$ is the smallest closed interval containing $\sigma(A)$.

PROOF. Let $\lambda = a + ib$ with a, b real and $b \neq 0$. Since $a - A$ is selfadjoint,

$$(3) \qquad \|(\lambda - A)x\|^2 = \|(a - A)x\|^2 + |b|^2\|x\|^2 \geq |b|^2\|x\|^2$$

for each $x \in H$. Therefore $\lambda - A$ is injective. From (3) it also follows that $\mathrm{Im}(\lambda - A)$ is closed. Indeed, assume that $(\lambda - A)x_n \to y$ for $n \to \infty$. Then (3) implies that (x_n) is a Cauchy-sequence. Since H is a Hilbert space, it follows that $x_0 = \lim_{n \to \infty} x_n$ exists, and, from the continuity of A, we may conclude that $y = (\lambda - A)x_0 \in \text{Im}(\lambda - A)$. Thus $\text{Im}(\lambda - A)$ is closed. But then

$$\text{Im}(\lambda - A) = \overline{\text{Im}(\lambda - A)} = \text{Ker}(\lambda - A)^\perp = \{0\}^\perp = H,$$

and $\lambda - A$ is bijective. This proves that $\lambda \notin \sigma(A)$. Replacing x in formula (3) by $(\lambda - A)^{-1}y$, yields (2).

We know now that $\sigma(A)$ is real. Take $\lambda \in \mathbf{R}$ and $\lambda > M(A)$. Then

$$(4) \qquad \|(\lambda - A)x\|\|x\| \geq \langle (\lambda - A)x, x \rangle \geq (\lambda - M(A))\|x\|^2.$$

Since $\lambda - M(A) > 0$, formula (4) implies that $\lambda - A$ is injective and $\text{Im}(\lambda - A)$ is closed. It follows (as in the preceding paragraph) that $\lambda - A$ is bijective. Thus $\lambda \notin \sigma(A)$. In a similar way one shows that $\lambda < m(A)$ implies that $\lambda \notin \sigma(A)$, and therefore $\sigma(A) \subset [m(A), M(A)]$.

To prove the last part of the theorem, let $\sigma(A) \subset [m, M]$. We have to show that $m \leq m(A)$ and $M \geq M(A)$. Assume that $m(A) < m$. Consider the operator $T := A - m(A)I$. From our hypothesis it follows that T is invertible and $\langle Tx, x \rangle \geq 0$. By applying the Cauchy-Schwarz inequality to the (possibly nondefinite) inner product $[x, y] := \langle Tx, y \rangle$, one sees that

$$|\langle Tx, y \rangle|^2 \leq \langle Tx, x \rangle \langle Ty, y \rangle, \qquad x, y \in H.$$

By taking $y = Tx$ the following inequality is obtained:

$$(5) \qquad \|Tx\|^4 \leq \langle Tx, x \rangle \langle T^2 x, Tx \rangle, \qquad x \in H.$$

From the definition of $m(A)$ we know that there exists a sequence $x_1, x_2, \ldots$ in H, $\|x_n\| = 1$ $(n = 1, 2, \ldots)$, such that $\langle Tx_n, x_n \rangle \to 0$ if $n \to \infty$. But then the inequality (5) implies that $\|Tx_n\| \to 0$ if $n \to \infty$, which contradicts the invertibility of T. Indeed,

$$\|Tx_n\| \geq \|T^{-1}\|^{-1}\|x_n\| = \|T^{-1}\|^{-1} > 0, \qquad n \geq 1.$$

Hence we must have $m \leq m(A)$. In a similar way one shows that $M(A) \leq M$. $\square$

Let $A \in \mathcal{L}(H)$ be selfadjoint. From Theorem 2.1 we know that $\lambda - A$ is invertible for $\lambda \notin [m(A), M(A)]$. For later purposes we note the following two inequalities:

$$(6a) \qquad \|(\lambda - A)^{-1}x\| \leq |\lambda - M(A)|^{-1}, \qquad \Re\lambda > M(A),$$

$$(6b) \qquad \|(\lambda - A)^{-1}x\| \leq |\lambda - m(A)|^{-1}, \qquad \Re\lambda < m(A).$$

To prove (6a) write $\lambda = a + ib$ with a, b real and $a > M(A)$. Take $x \in H$. From (4) we know that

$$\|(a - A)x\| \geq (a - M(A))\|x\|.$$

Inserting this inequality in (3) yields

$$\|(\lambda - A)x\|^2 \geq \{|a - M(A)|^2 + |b|^2\}\|x\|^2$$
$$\geq |\lambda - M(A)|^2\|x\|^2,$$

which proves (6a). In a similar way one proves (6b).

THEOREM 2.2. *Let $A \in \mathcal{L}(H)$ be selfadjoint, and take $s < t$ in $\mathbf{R}$. Then the spectral subspace of A associated with $[s,t]$ exists and is equal to the space*

$$(7) \qquad \mathrm{Ker}(s - A) \oplus \overline{\mathrm{Im}\,\Omega_{s,t}} \oplus \mathrm{Ker}(t - A).$$

Here

$$(8) \qquad \Omega_{s,t} = \frac{1}{2\pi i} \int\limits_{\Gamma_{s,t}} (t - \zeta)(s - \zeta)(\zeta - A)^{-1}d\zeta$$

and $\Gamma_{s,t}$ is the boundary of the rectangle with vertices $s \pm i$ and $t \pm i$.

To prove Theorem 2.2 we develop for operators of the type (8) a calculus which resembles the functional calculus developed in Sections I.2 and I.3. This will be done in the next two lemmas.

LEMMA 2.3. *For $s < t$ in $\mathbf{R}$ the operator $\Omega_{s,t}$ in (8) is a well-defined selfadjoint operator, $\Omega_{s,t}B = B\Omega_{s,t}$ for any $B \in \mathcal{L}(H)$ that commutes with A, and*

(i) $\Omega_{s,t} = (t - A)(s - A)$ *if* $\sigma(A) \subset [s,t]$,

(ii) $\Omega_{s,t} = 0$ *if* $\sigma(A) \cap (s,t) = \emptyset$.

PROOF. The contour $\Gamma_{s,t}$ appearing in formula (8) is assumed to be positively oriented, i.e., the open rectangle with boundary $\Gamma_{s,t}$ belongs to the inner domain of $\Gamma_{s,t}$. Note that $\Gamma_{s,t}$ intersects $\sigma(A)$ in at most two points, namely s and t. If both s and t do not belong to $\sigma(A)$, then $\Gamma_{s,t}$ has an empty intersection with $\sigma(A)$ and, according to the functional calculus of Section I.3,

$$(9) \qquad \Omega_{s,t} = (s - A)(t - A)P_\sigma(A),$$

where $P_\sigma(A)$ is the Riesz projection of A corresponding to $\sigma = \sigma(A) \cap [s,t]$. From (9) it is natural to expect that the statements (i) and (ii) hold true.

If s and/or t belong to $\sigma(A)$, then (8) has to be understood as:

$$(10) \qquad \Omega_{s,t} := \lim_{\varepsilon \downarrow 0} \frac{1}{2\pi i} \int\limits_{\Gamma_{s,t}^\varepsilon} (t - \zeta)(s - \zeta)(\zeta - A)^{-1}d\zeta,$$

where $\Gamma_{s,t}^\varepsilon$ is the part of $\Gamma_{s,t}$ outside the open discs $|s - \zeta| < \varepsilon$ and $|t - \zeta| < \varepsilon$. By Theorem 2.1 the function

$$(11) \qquad \zeta \mapsto \|(t - \zeta)(s - \zeta)(\zeta - A)^{-1}\|$$

is bounded on $\Gamma_{s,t}\backslash\{s,t\}$, and hence the limit in (10) exists in the norm of $\mathcal{L}(H)$. It follows that $\Omega_{s,t}$ is a well-defined bounded linear operator. Since A is selfadjoint and $\Gamma_{s,t}$ is symmetric with respect to the real line, the arguments used in the proof of Proposition I.2.5 show that $\Omega_{s,t}^* = \Omega_{s,t}$.

If $B \in \mathcal{L}(H)$ commutes with A, then B commutes with the integrand in (8), and hence the convergence of the integral in the norm of $\mathcal{L}(H)$ implies that B commutes with $\Omega_{s,t}$.

To prove (i), assume that $\sigma(A) \subset [s,t]$. Take $0 < \varepsilon \leq 1$, and let $\Lambda_{s,t}^\varepsilon$ be the closed contour which is the union of the curve $\Gamma_{s,t}^\varepsilon$ and the curves

$$\left\{ t + \varepsilon e^{i\varphi} \ \Big| \ -\frac{1}{2}\pi \leq \varphi \leq \frac{1}{2}\pi \right\},$$

$$\left\{ s + \varepsilon e^{i\varphi} \ \Big| \ \frac{1}{2}\pi \leq \varphi \leq \frac{3}{2}\pi \right\}.$$

The orientation on $\Lambda_{s,t}^\varepsilon$ is given so that $[s,t]$ is in the inner domain of $\Lambda_{s,t}^\varepsilon$. Since $\sigma(A) \subset [s,t]$, we know from Section I.3 that

$$(12) \qquad \frac{1}{2\pi i} \int_{\Lambda_{s,t}^\varepsilon} (t-\zeta)(s-\zeta)(\zeta-A)^{-1}d\zeta = (t-A)(s-A).$$

From $\sigma(A) \subset [s,t]$ it also follows that $s \leq m(A)$ and $t \geq M(A)$. But then we can use the inequalities (2), (6a) and (6b) to show that there exists a constant C, not depending on ε, such that

$$\|(t-\zeta)(s-\zeta)(\zeta-A)^{-1}\| \leq C, \qquad \zeta \in \Lambda_{s,t}^\varepsilon\backslash\Gamma_{s,t}.$$

Since the length of the curve $\Lambda_{s,t}^\varepsilon\backslash\Gamma_{s,t}$ tends to zero if $\varepsilon \downarrow 0$, we conclude that the left hand side of (12) tends to $\Omega_{s,t}$ in the norm of $\mathcal{L}(H)$ if $\varepsilon \downarrow 0$. The right hand side of (12) is independent of ε, and thus (i) holds true.

Next, we prove (ii). Assume $\sigma(A) \cap (s,t) = \emptyset$. First we consider the case when $\sigma(A) \subset (-\infty, s]$. Take $\varepsilon > 0$, and let $\Delta_{s,t}^\varepsilon$ be the contour which consists of the part of $\Gamma_{s,t}$ outside the open disc $|s - \zeta| < \varepsilon$ and the curve

$$\left\{ s + \varepsilon e^{i\varphi} \ \Big| \ -\frac{1}{2}\pi \leq \varphi \leq \frac{\pi}{2}\varphi \right\}.$$

For $\varepsilon > 0$ sufficiently small $\Delta_{s,t}^\varepsilon$ is a well-defined closed contour and, by our assumption on $\sigma(A)$,

$$(13) \qquad \frac{1}{2\pi i} \int_{\Delta_{s,t}^\varepsilon} (t-\zeta)(s-\zeta)(\zeta-A)^{-1}d\zeta = 0.$$

From $\sigma(A) \subset (-\infty, s]$ it also follows that $s \leq M(A)$, and hence we can use the inequalities (2) and (6a) to show that there exists a constant C, independent of ε, such that

$$\|(t-\zeta)(s-\zeta)(\zeta-A)^{-1}\| \leq C, \qquad \zeta \in \Delta_{s,t}^\varepsilon\backslash\Gamma_{s,t}.$$

As in the proof of (i), it follows that

$$\Omega_{s,t} = \lim_{\varepsilon \downarrow 0}\left(\frac{1}{2\pi i}\int\limits_{\Delta_{s,t}^{\varepsilon}} (t-\zeta)(s-\zeta)(\zeta - A)^{-1}d\zeta\right),$$

and hence $\Omega_{s,t} = 0$. In a similar way one shows that $\Omega_{s,t} = 0$ if $\sigma(A) \subset [t,\infty)$.

For the general case consider the set $\sigma = \sigma(A)\cap(-\infty, s]$. By our assumption on $\sigma(A)$, the set σ is an isolated part of $\sigma(A)$. Let P be the corresponding Riesz projection. Put $M = \operatorname{Im} P$ and $L = \operatorname{Ker} P$. The M and L are invariant under A. Furthermore, the operators $A|M$ and $A|L$ are selfadjoint. Now

$$(14a) \qquad \Omega_{s,t}x = \frac{1}{2\pi i}\int\limits_{\Gamma_{s,t}} (t-\zeta)(s-\zeta)[\zeta - (A|M)]^{-1}xd\zeta, \qquad x \in M,$$

$$(14b) \qquad \Omega_{s,t}x = \frac{1}{2\pi i}\int\limits_{\Gamma_{s,t}} (t-\zeta)(s-\zeta)[\zeta - (A|L)]^{-1}xd\zeta, \qquad x \in L.$$

From Section I.2 we know that $\sigma(A|M) \subset (-\infty, s]$ and $\sigma(A|L) \subset [t,\infty)$. Thus the integrals in (14a) and (14b) are zero by what has been proved so far. Since $H = M \oplus L$, it follows that $\Omega_{s,t} = 0$. $\square$

For $s < t$ in $\mathbf{R}$ let $\mathcal{F}_{s,t}$ denote the set of all complex-valued functions that are continuous on the set

$$(15) \qquad \Xi(s,t) = \mathbf{C}\backslash\{(-\infty, s)\cup(t,\infty)\}$$

and analytic on the interior of $\Xi(s,t)$. For $g \in \mathcal{F}_{s,t}$ we define

$$(16) \qquad \Omega_{s,t}(g) = \frac{1}{2\pi i}\int\limits_{\Gamma_{s,t}} g(\zeta)(t-\zeta)(s-\zeta)(\zeta - A)^{-1}d\zeta.$$

The integral in (16) has to be understood in the same way as the integral in (8). Note that g is continuous on $\Gamma_{s,t}$. Since $(t-\zeta)(s-\zeta)(\zeta-A)^{-1}$ is bounded in the operator norm on $\Gamma_{s,t}\backslash\{s,t\}$, it follows that the same holds true for the integrand in (16), and hence the (improper) integral in (16) converges in the norm of $\mathcal{L}(H)$. Obviously, $\Omega_{s,t} = \Omega_{s,t}(1)$, where 1 stands for the function which is identically equal to one. As for $\Omega_{s,t}$ (see Lemma 2.3) one can prove that $\Omega_{s,t}(g)B = B\Omega_{s,t}(g)$ for any operator B that commutes with A. Note that $\Omega_{s,t}(g)$ is selfadjoint if g is symmetric with respect to the real line, i.e., if

$$\overline{g(\zeta)} = g(\overline{\zeta}), \qquad \zeta \in \Xi(s,t).$$

(Here the bar denotes the operation of complex conjugation.) To prove the last statement one uses the symmetry in $\Gamma_{s,t}$ and arguments of the type used in the proof of Proposition I.2.5.

To derive properties of $\Omega_{s,t}(g)$ it is sometimes convenient to replace $\Gamma_{s,t}$ by other curves. Therefore we mention that in (16) (as well as in (8)) the curve $\Gamma_{s,t}$ may be replaced by any simple closed rectifiable oriented Jordan curve with the following properties. The curve Γ is symmetric with respect to the real line, the open interval (s,t) belongs to the inner domain of Γ and there exists $\rho > 0$ such that the part of Γ in the disc $|s - \zeta| < \rho$ (resp. $|t - \zeta| < \rho$) lies in the sector $s + re^{i\varphi}$ (resp. $t + e^{i\varphi}$) with $0 \le r < \rho$ and $\left|\frac{1}{2}\pi \pm \varphi\right| \le \frac{1}{4}\pi$. By Theorem 2.1 this non-tangential behaviour of Γ guarantees that the integrand in (16) is again bounded on $\Gamma\backslash\{s,t\}$ in the operator norm, and hence the integral in (16) is also well-defined for Γ instead of $\Gamma_{s,t}$. We shall refer to a curve Γ with the above properties as an (s,t)-*admissible curve*. The usual argument of complex function theory shows that the value of the integral in (16) does not change if $\Gamma_{s,t}$ is replaced by an (s,t)-admissible curve.

LEMMA 2.4. *Let $A \in \mathcal{L}(H)$ be selfadjoint, and let $g \in \mathcal{F}_{s,t}$. Then*

$$(17) \qquad \Omega_{s,t}(g)^2 = (t - A)(s - A)\Omega_{s,t}(g^2),$$

and for any complex polynomial p

$$(18) \qquad p(A)\Omega_{s,t}(g) = \Omega_{s,t}(pg).$$

PROOF. First we shall prove (18). To do this, it suffices to consider the case when $p(\zeta) = \zeta^n$. By formula (1) in Section I.3,

$$A^n\Omega_{s,t}(g) = \Omega_{s,t}(\zeta^n g) - \frac{1}{2\pi i}\int_{\Gamma_{s,t}} g(\zeta)(t - \zeta)(s - \zeta)\left(\sum_{j=0}^{n-1}\zeta^{n-1-j}A^j\right)d\zeta.$$

By analyticity the latter integral is zero, and so (18) holds true.

The proof of (17) will be based on the following identity:

$$(19) \qquad \Omega_{s,t}(g)\Omega_{\alpha,\beta}(g) = (t - A)(s - A)\Omega_{\alpha,\beta}(g^2), \qquad s < \alpha < \beta < t.$$

To prove (19) we apply the same type of arguments as in the proof of Theorem I.3.1(iii). First we replace the curve $\Gamma_{s,t}$ by a curve Γ which is (s,t)-admissible and contains $\Gamma_{\alpha,\beta}$ in its inner domain. It follows that

$$
\begin{aligned}
\Omega_{s,t}(g)\Omega_{\alpha,\beta}(g) &= \left(\frac{1}{2\pi i}\right)^2 \int_{\Gamma_{s,t}}\int_{\Gamma_{\alpha,\beta}} g(\zeta)g(\lambda)(t - \zeta)(s - \zeta)(\beta - \lambda)(\alpha - \lambda) \\
&\qquad \cdot (\zeta - A)^{-1}(\lambda - A)^{-1}d\lambda d\zeta \\
&= \left(\frac{1}{2\pi i}\right)^2 \int_{\Gamma}\int_{\Gamma_{\alpha,\beta}} g(\zeta)g(\lambda)(t - \zeta)(s - \zeta)(\beta - \lambda)(\alpha - \lambda) \\
&\qquad \cdot (\lambda - \zeta)^{-1}\{(\zeta - A)^{-1} - (\lambda - A)^{-1}\}d\lambda d\zeta \\
&= \eta_1 - \eta_2,
\end{aligned}
$$

where

$$\eta_1 = \left(\frac{1}{2\pi i}\right)^2 \int_\Gamma g(\zeta)(t-\zeta)(s-\zeta)(\zeta-A)^{-1}$$
$$\cdot \left(\int_{\Gamma_{\alpha,\beta}} \frac{g(\lambda)(\beta-\lambda)(\alpha-\lambda)}{\lambda-\zeta}d\lambda\right)d\zeta,$$
$$= 0,$$

because the curve Γ is in the outer domain of $\Gamma_{\alpha,\beta}$, and

$$\eta_2 = \left(\frac{1}{2\pi i}\right)^2 \int_{\Gamma_{\alpha,\beta}} g(\lambda)(\beta-\lambda)(\alpha-\lambda)(\lambda-A)^{-1}$$
$$\cdot \left(\int_\Gamma \frac{g(\zeta)(t-\zeta)(s-\zeta)}{\lambda-\zeta}d\zeta\right)d\lambda$$
$$= \frac{-1}{2\pi i} \int_{\Gamma_{\alpha,\beta}} g(\lambda)^2 (t-\lambda)(s-\lambda)(\beta-\lambda)(\alpha-\lambda)(\lambda-A)^{-1}d\lambda$$
$$= -(t-A)(s-A)\Omega_{\alpha,\beta}(g^2).$$

Here we used that $\Gamma_{\alpha,\beta}$ is in the inner domain of Γ, and we applied formula (18) with $p(\zeta) = (t-\zeta)(s-\zeta)$ and with g^2 in place of g. The change in the order of integration is justified by the fact that the integrand is integrable on $\Gamma \times \Gamma_{\alpha,\beta}$. We have now proved (19).

To derive (17) from (19) it suffices to show that for $g \in \mathcal{F}_{s,t}$

$$(20) \qquad \Omega_{\alpha,\beta}(g) \to \Omega_{s,t}(g), \qquad \beta \uparrow t, \ \alpha \downarrow s,$$

with convergence in the norm of $\mathcal{L}(H)$. Take $s \le \alpha < \beta \le t$, and let $\varepsilon > 0$ be given. For $0 < \delta \le 1$ let $\Gamma_{\alpha,\beta}(\delta)$ be the oriented boundary of the rectangle with vertices $\alpha \pm i\delta$ and $\beta \pm i\delta$. The orientation is such that the interval (α,β) belongs to the inner domain of $\Gamma_{\alpha,\beta}(\delta)$. We write $V_{\alpha,\beta}(\delta)$ for the vertical parts of $\Gamma_{\alpha,\beta}(\delta)$ and $H_{\alpha,\beta}(\delta)$ for the horizontal parts. Of course, $\Gamma_{\alpha,\beta}(\delta)$ is an (α,β)-admissible curve. By Theorem 2.1,

$$\|g(\zeta)(\beta-\zeta)(\alpha-\zeta)(\zeta-A)^{-1}\| \le (t-s+1)\sup\{|g(\zeta)| \mid s \le \Re\zeta \le t, \ |\Im\zeta| \le \delta\}.$$

Since the length of $V_{\alpha,\beta}(\delta)$ is 4δ, it follows that we may choose $\delta \in (0,1]$ in such a way that

$$(21) \qquad \left\|\frac{1}{2\pi i} \int_{V_{\alpha,\beta}(\delta)} g(\zeta)(\beta-\zeta)(\alpha-\zeta)(\zeta-A)^{-1}d\zeta\right\| < \frac{1}{4}\varepsilon$$

for each $(\alpha,\beta) \subset (s,t)$. Fix $\delta \in (0,1]$ such that (21) holds. Since the set $H_{s,t}(\delta)$ is compact, we can find a constant $C \ge 0$ so that for each $\zeta \in H_{s,t}(\delta)$

$$(22) \qquad \|g(\zeta)(t-\zeta)(s-\zeta)(\zeta-A)^{-1}\| \le C,$$

$$(23) \qquad \|g(\zeta)(\beta - \zeta)(\alpha - \zeta)(\zeta - A)^{-1} - g(\zeta)(t - \zeta)(s - \zeta)(\zeta - A)^{-1}\|$$
$$\leq C(|\alpha\beta - st| + |\alpha - s| + |\beta - t|).$$

The length of the curve $H_{\alpha,\beta}(\delta)$ is less than or equal to $2(t-s)$. Thus, by using inequality (23), we see that there exists $\rho_0 > 0$ such that $|\alpha - s| + |\beta - t| < \rho_0$ implies that

$$(24) \qquad \left\| \frac{1}{2\pi i} \int\limits_{H_{\alpha,\beta}(\delta)} g(\zeta)(\beta - \zeta)(\alpha - \zeta)(\zeta - A)^{-1} d\zeta \right.$$
$$\left. - \frac{1}{2\pi i} \int\limits_{H_{\alpha,\beta}(\delta)} g(\zeta)(t - \zeta)(s - \zeta)(\zeta - A)^{-1} d\zeta \right\| < \frac{1}{4}\varepsilon.$$

Write $\Lambda_{\alpha,\beta}(\delta)$ for $H_{s,t}(\delta)\backslash H_{\alpha,\beta}(\delta)$. Since the length of $\Lambda_{\alpha,\beta}(\delta)$ is equal to $2(|\alpha - s| + |\beta - t|)$, inequality (22) shows that there exists $0 < \rho < \rho_0$ such that

$$(25) \qquad \left\| \frac{1}{2\pi i} \int\limits_{\Lambda_{\alpha,\beta}(\delta)} g(\zeta)(t - \zeta)(s - \zeta)(\zeta - A)^{-1} d\zeta \right\| < \frac{1}{4}\varepsilon,$$

whenever $|\alpha - s| + |\beta - t| < \rho$. From (21), (24) and (25), the statement (20) is clear. $\square$

PROOF OF THEOREM 2.2. Assume $Ax = tx$. Then $(\zeta - A)^{-1}x = (\zeta - t)^{-1}x$, and hence $\Omega_{s,t}x = 0$. Since $\Omega_{s,t}$ is selfadjoint, it follows that

$$(26a) \qquad \operatorname{Ker}(t - A) \subset \operatorname{Ker}\Omega_{s,t} = (\overline{\operatorname{Im}\Omega_{s,t}})^{\perp}.$$

Here (as well as in (7)) the bar means that one has to take the closure of the corresponding set. In a similar way one proves that

$$(26b) \qquad \operatorname{Ker}(s - A) \subset \operatorname{Ker}\Omega_{s,t} = (\overline{\operatorname{Im}\Omega_{s,t}})^{\perp}.$$

We also know that eigenvectors of A corresponding to different eigenvalues are orthogonal, because A is selfadjoint. Thus $\operatorname{Ker}(s - A) \subset \operatorname{Ker}(t - A)^{\perp}$ and $\operatorname{Ker}(t - A) \subset \operatorname{Ker}(s - A)^{\perp}$. This together with (26a) and (26b) implies that the set given by (7) is a well-defined closed linear manifold of H. In what follows it will be denoted by $L(s,t)$. Since $A\Omega_{s,t} = \Omega_{s,t}A$, by Lemma 2.3, we know that $\operatorname{Im}\Omega_{s,t}$ is invariant under A. But then, by the continuity of A, the same is true for $\overline{\operatorname{Im}\Omega_{s,t}}$. Also $\operatorname{Ker}(s - A)$ and $\operatorname{Ker}(t - A)$ are invariant under A. So we conclude that $L(s,t)$ is an A-invariant subspace.

We proceed by showing that

$$(27) \qquad s\|x\|^2 \leq \langle Ax, x \rangle \leq t\|x\|^2, \qquad x \in L(s,t).$$

To achieve this, let g_0 be the analytic continuation of the function

$$(28) \qquad \sqrt{t - \lambda}, \qquad -\infty < \lambda < t,$$

to the region $\mathbf{C}\backslash[t, \infty)$. Put $g_0(t) = 0$. Note that

$$(29) \qquad g_0(\zeta)^2 = t - \zeta, \qquad \zeta \in \Xi(s,t).$$

Here $\Xi(s,t)$ is as in (15). We conclude that $g_0 \in \mathcal{F}_{s,t}$, and, by Lemma 2.4,

$$
\begin{aligned}
\Omega_{s,t}(g_0)^2 &= (t-A)(s-A)\Omega_{s,t}(g_0^2) \\
&= \frac{1}{2\pi i} \int_{\Gamma_{s,t}} (t-\zeta)^3 (s-\zeta)^2 (\zeta - A)^{-1} d\zeta \\
&= (t-A)\Omega_{s,t}^2.
\end{aligned}
$$

From the definition of g_0 it follows that g_0 is symmetric with respect to the real line. Hence $\Omega_{s,t}(g_0)$ is selfadjoint. Also $\Omega_{s,t}$ is selfadjoint. Thus for each $x \in H$ we have

$$
\begin{aligned}
\langle (t-A)\Omega_{s,t}x, \Omega_{s,t}x \rangle &= \langle (t-A)\Omega_{s,t}^2 x, x \rangle \\
&= \langle \Omega_{s,t}(g_0)^2 x, x \rangle = \|\Omega_{s,t}(g_0)x\|^2 \geq 0.
\end{aligned}
$$

By continuity it follows that $\langle (t-A)y, y \rangle \geq 0$ for each $y \in \overline{\operatorname{Im} \Omega_{s,t}}$. Now take $x \in L(s,t)$, and write $x = x_1 + y + x_2$, where $x_1 \in \operatorname{Ker}(s-A)$, $y \in \overline{\operatorname{Im} \Omega_{s,t}}$ and $x_2 \in \operatorname{Ker}(t-A)$. Since the elements x_1, y and x_2 are mutually orthogonal, we see that

$$
\begin{aligned}
\langle (t-A)x, x \rangle &= \langle (t-A)x_1, x_1 \rangle + \langle (t-A)y, y \rangle + \langle (t-A)x_2, x_2 \rangle \\
&= (t-s)\|x_1\|^2 + \langle (t-A)y, y \rangle \geq 0,
\end{aligned}
$$

and the second inequality in (27) is proved. To establish the first inequality in (27) one repeats the above arguments with the function (28) replaced by

$$
(30) \qquad\qquad \sqrt{\lambda - s}, \qquad s < \lambda < \infty.
$$

Since $A|L(s,t)$ is selfadjoint, the inequalities in (27) imply that $\sigma\big(A|L(s,t)\big)$ is contained in $[s,t]$ (cf. Theorem 2.1). We want to show that $L(s,t)$ is the largest A-invariant subspace with this property. Thus, let N be another A-invariant subspace such that $\sigma(A|N) \subset [s,t]$. We have to prove that $N \subset L(s,t)$. Take $x \in N$. Note that $N^\perp$ is also invariant under A, because A is selfadjoint. It follows (cf. formula (8) in Section I.2) that

$$
(\zeta - A)^{-1}x = [\zeta - (A|N)]^{-1}x, \qquad \zeta \in \rho(A).
$$

Now apply Lemma 2.3(i) to $A|N$ in place of A. Since $A|N$ is selfadjoint, this yields

$$
\begin{aligned}
\Omega_{s,t}x &= \frac{1}{2\pi i} \int_{\Gamma_{s,t}} (t-\zeta)(s-\zeta)[\zeta - (A|N)]^{-1}x \, d\zeta \\
&= [t - (A|N)][s - (A|N)]x \\
&= (t-A)(s-A)x,
\end{aligned}
$$

and we have proved that $(t-A)(s-A)N \subset \operatorname{Im} \Omega_{s,t}$. It follows that

$$
\begin{aligned}
N &= [N \cap \operatorname{Ker}(t-A)] + [N \cap \operatorname{Ker}(s-A)] + \overline{(t-A)(s-A)N} \\
&\subset \operatorname{Ker}(t-A) + \operatorname{Ker}(s-A) + \overline{\operatorname{Im} \Omega_{s,t}} \\
&= L(s,t),
\end{aligned}
$$

which completes the proof of Theorem 2.2. □

COROLLARY 2.5. *Let $A \in \mathcal{L}(H)$ be selfadjoint and $t \in \mathbf{R}$. Then the spectral subspace of A associated with $(-\infty, t]$ exists and is equal to the space*

$$(31) \qquad L := \mathrm{Ker}(t - A) \oplus \overline{\mathrm{Im}\left(\frac{1}{2\pi i} \int_{\Gamma_t} (t - \zeta)(\zeta - A)^{-1} d\zeta \right)}.$$

Here Γ_t is the boundary of the rectangle with vertices $t \pm i$ and $t - 1 - |t - m(A)| \pm i$. Furthermore,

$$(32a) \qquad \langle (A - t)x, x \rangle \leq 0, \qquad x \in L,$$

$$(32b) \qquad \langle (A - t)x, x \rangle \geq 0, \qquad x \in L^{\perp}.$$

PROOF. Put $r := t - 1 - |t - m(A)|$. By definition, $r < \min(t, m(A))$, and hence $r - A$ is invertible. Let $L(r, t)$ be the spectral subspace of A associated with $[r, t]$. From Theorem 2.2 we know that $L(r, t)$ exists. Obviously, $\sigma(A|L(r, t))$ is contained in $(-\infty, t]$. Let N be another A-invariant subspace such that $\sigma(A|N) \subset (-\infty, t]$. Since $N^{\perp}$ is also invariant under A, we have (see Proposition I.2.4)

$$\sigma(A|N) \subset \sigma(A) \subset [m(A), M(A)].$$

It follows that $\sigma(A|N) \subset [r, t]$, because $r < m(A)$. But then $N \subset L(r, t)$, and $L(r, t)$ is also the spectral subspace of A associated with $(-\infty, t]$.

Next, we show that $L(r, t) = L$, where L is given by (31). Let us assume that Γ_t is oriented in such a way that the open rectangle with boundary Γ_t belongs to the inner domain of Γ_t, and put

$$(33) \qquad \Omega_t = \frac{1}{2\pi i} \int_{\Gamma_t} (t - \zeta)(\zeta - A)^{-1} d\zeta.$$

The operator Ω_t is well-defined for the same reason as the operator $\Omega_{s,t}$ in (8) is well-defined. Note that for $s = r$ the curve Γ_t is equal to the curve $\Gamma_{s,t}$ appearing in Theorem 2.2. It follows that

$$\Omega_t(r - A) = (r - A)\Omega_t = \Omega_{r,t},$$

where $\Omega_{r,t}$ is defined by (8) with s replaced by r. Since $r - A$ is invertible, we see that $\mathrm{Im}\,\Omega_{r,t}$ and $\mathrm{Im}\,\Omega_t$ coincide, and hence

$$L(r, t) = \mathrm{Ker}(r - A) \oplus \overline{\mathrm{Im}\,\Omega_{r,t}} \oplus \mathrm{Ker}(t - A)$$
$$= \overline{\mathrm{Im}\,\Omega_t} \oplus \mathrm{Ker}(t - A) = L,$$

which proves formula (31).

Since $L = L(r, t)$, formula (32a) is just a corollary of the second inequality in (27) (applied to $L(r, t)$ instead of $L(s, t)$). To prove (32b), choose $u > \max(t, M(A))$. We shall prove that $L^\perp = \overline{\operatorname{Im} \Omega_{t,u}}$. As soon as this identity has been established, we can apply to first inequality in (27) (with (s, t) replaced by (t, u)) to show that (32b) holds true.

We know that $A|L$ is a selfadjoint operator which has its spectrum in $(-\infty, t]$. Thus Lemma 2.3(ii) applied to $A|L$ shows that for each $x \in L$

$$\Omega_{t,u} x = \frac{1}{2\pi i} \int_{\Gamma_{t,u}} (u - \zeta)(t - \zeta)[\zeta - (A|L)]^{-1} x \, d\zeta = 0.$$

In particular, $\overline{\operatorname{Im} \Omega_{t,r}} \subset \operatorname{Ker} \Omega_{t,u}$, and hence

$$\overline{\operatorname{Im} \Omega_{t,u}} = (\operatorname{Ker} \Omega_{t,u})^\perp \subset (\overline{\operatorname{Im} \Omega_{t,r}})^\perp.$$

We already know (see formula (26a)) that $\overline{\operatorname{Im} \Omega_{t,u}} \subset \operatorname{Ker}(t - A)^\perp$, and so

$$\overline{\operatorname{Im} \Omega_{t,u}} \subset \{\operatorname{Ker}(t - A) + \overline{\operatorname{Im} \Omega_{r,t}}\}^\perp$$
$$= \{\operatorname{Ker}(t - A) + \overline{\operatorname{Im} \Omega_t}\}^\perp = L^\perp.$$

To prove the reverse inclusion, note that (by formula (18))

$$\Omega_{r,t}(u - A) + \Omega_{t,u}(r - A) = (u - A)\Omega_{r,t} + (r - A)\Omega_{t,u}$$
$$= \frac{1}{2\pi i} \int_{\Gamma_{r,t}} (u - \zeta)(t - \zeta)(r - \zeta)(\zeta - A)^{-1} d\zeta$$
$$+ \frac{1}{2\pi i} \int_{\Gamma_{t,u}} (r - \zeta)(u - \zeta)(t - \zeta)(\zeta - A)^{-1} d\zeta$$
$$= \frac{1}{2\pi i} \int_{\Gamma_{r,u}} (u - \zeta)(t - \zeta)(r - \zeta)(\zeta - A)^{-1} d\zeta.$$

Since $\sigma(A) \subset [m(A), M(A)] \subset (r, u)$, the last integral can be computed by using the functional calculus of Section I.3. It follows that

$$(34) \qquad \Omega_{r,t}(u - A) + \Omega_{t,u}(r - A) = (u - A)(t - A)(r - A).$$

Thus if $x \in L^\perp$ and $x \perp \overline{\operatorname{Im} \Omega_{t,u}}$, then x is orthogonal to the range of the left hand side of (34), which implies that x is orthogonal to the range of the right hand side of (34). But then x must be orthogonal to $\operatorname{Im}(t - A)$ (because $u - A$ and $r - A$ are invertible). However $\operatorname{Ker}(t - A) \subset L$ and $x \perp L$. So $x = 0$, and thus $\overline{\operatorname{Im} \Omega_{t,u}} = L^\perp$. $\square$

V.3 RESOLUTION OF THE IDENTITY

A family $\{E(t)\}_{t\in\mathbf{R}}$ of orthogonal projections on the Hilbert space H is called
a (*bounded*) *resolution of the identity* supported by the compact interval $[m, M]$ if

(C$_1$) $\operatorname{Im} E(s) \subset \operatorname{Im} E(t)$ whenever $s \leq t$,

(C$_2$) $\operatorname{Im} E(s) = \cap\{\operatorname{Im} E(t) \mid t > s\}$,

(C$_3$) $E(t) = 0$ if $t < m$,

(C$_4$) $E(t) = I$ if $t > M$.

Later, in Chapter XVI, we shall also need resolutions of the identity that are not bounded,
but all resolutions in the present chapter will be bounded, and therefore in what follows
we omit the word bounded.

Note that condition (C$_1$) is equivalent to the requirement that

$$E(s) = E(t)E(s) = E(s)E(t), \qquad s \leq t,$$

and hence the projections in a resolution of the identity commute with one another.
Condition (C$_2$) means that the resolution is required to be continuous from the right.
To be more precise, the following proposition holds.

PROPOSITION 3.1. *Let $\{E(t)\}_{t\in\mathbf{R}}$ be a resolution of the identity on H.
Then*

$$E(\lambda)x = \lim_{t\downarrow\lambda} E(t)x, \qquad x \in H.$$

PROOF. Take a fixed $x \in H$. Let $\lambda < s < t$. The operator $\Delta E := E(t) - E(s)$
is an orthogonal projection, because of condition (C$_1$). Furthermore, $E(s)\Delta E = 0$, and
thus

$$\langle(\Delta E)x, E(s)x\rangle = \langle E(s)(\Delta E)x, x\rangle = 0,$$

which shows that $(\Delta E)x \perp E(s)x$. So the Pythagorean equality gives:

$$(1) \qquad \|E(t)x - E(s)x\|^2 = \|E(t)x\|^2 - \|E(s)x\|^2, \qquad s < t.$$

It follows that the function $\|E(\cdot)x\|$ is a monotonely increasing nonnegative function, and
thus $\lim_{t\downarrow\lambda}\|E(t)\|$ exists. Since H is a complete metric space, the inequality (1) and the
Cauchy criterion for convergence imply that

$$(2) \qquad z := \lim_{t\downarrow\lambda} E(t)x$$

exists. We have to show that $z = E(\lambda)x$.

Take $\mu > \lambda$. Then, by condition (C$_1$),

$$E(\mu)z = \lim_{t\downarrow\lambda} E(\mu)E(t)x = \lim_{\substack{t\to\lambda \\ \lambda < t \leq \mu}} E(\mu)E(t)x$$

$$= \lim_{\substack{t\to\lambda \\ \lambda < t \leq \mu}} E(t)x = z.$$

This shows that $z \in \operatorname{Im} E(\mu)$ for each $\mu > \lambda$. But then we can apply condition (C_2) to show that $z \in \operatorname{Im} E(\lambda)$. Thus (use again condition (C_1))

$$z = E(\lambda)z = \lim_{t \downarrow \lambda} E(\lambda)E(t)x$$

$$= \lim_{t \downarrow \lambda} E(\lambda)x = E(\lambda)x. \quad \square$$

Together with Theorem 2.2 the next theorem shows how one may construct resolutions of the identity.

THEOREM 3.2. *Let $A \in \mathcal{L}(H)$ be selfadjoint, and for $t \in \mathbf{R}$ let $E(t)$ be the orthogonal projection of H onto the spectral subspace of A associated with $(-\infty, t]$. Then $\{E(t)\}_{t \in \mathbf{R}}$ is a resolution of the identity supported by the interval $[m(A), M(A)]$. Furthermore, if $B \in \mathcal{L}(H)$ commutes with A, then B commutes with each $E(t)$.*

PROOF. Let $L(t)$ be the spectral subspace of A associated with $(-\infty, t]$. From Corollary 2.5 we know that $L(t)$ is well-defined and given by formula (31) in Section 2. Take $s < t$. We know that $L(s)$ is A-invariant and $\sigma(A|L(s))$ is contained in $(-\infty, s]$. Hence, also $\sigma(A|L(s)) \subset (-\infty, t]$. But $L(t)$ is the largest A-invariant subspace N with $\sigma(A|N) \subset (-\infty, t]$. Thus $L(s) \subset L(t)$, and (C_1) is established. Since $\sigma(A) \subset [m(A), M(A)]$, the functional calculus of Section I.3 shows that $L(t) = 0$ for $t < m(A)$ and

$$L(t) = \operatorname{Ker}(t - A) \oplus \overline{\operatorname{Im}(t - A)} = H, \qquad t > M(A).$$

Thus conditions (C_3) and (C_4) are fulfilled.

Next we check condition (C_2). Write $L(s + 0)$ for the space $\cap\{L(t) \mid t > s\}$. Obviously, $L(s) \subset L(s + 0)$. The subspace $L(s + 0)$ is invariant under A, because $L(t)$ is invariant under A for each $t > s$. Take $t > s$, and let $N = L(t) \cap L(s + 0)^{\perp}$. Then N is also invariant under A, and so (cf. Proposition I.2.4)

$$\sigma\big(A|L(s + 0)\big) \subset \sigma\big(A|L(t)\big) \subset (-\infty, t].$$

It follows that

$$\sigma\big(A|L(s + 0)\big) \subset \bigcap_{t > s}(-\infty, t] = (-\infty, s].$$

But $L(s)$ is the largest A-invariant subspace L with $\sigma(A|L) \subset (-\infty, s]$. Hence $L(s + 0) \subset L(s)$, and (C_2) is satisfied.

To prove the last part of the theorem, assume that $B \in \mathcal{L}(H)$ commutes with A. Let $\Omega_{s,t}$ be the operator defined by formula (8) in Section 2. By Lemma 2.3 the operator B commutes with $\Omega_{s,t}$ for any $s < t$ in $\mathbf{R}$. Now fix $t \in \mathbf{R}$. From the proof of Corollary 2.5 we know that

$$L(t) = \operatorname{Ker}(t - A) \oplus \overline{\operatorname{Im} \Omega_{r,t}}, \qquad L(t)^{\perp} = \overline{\operatorname{Im} \Omega_{t,u}},$$

provided that $r < \min\big(t, m(A)\big)$ and $u > \max\big(t, M(A)\big)$. Since B commutes with A and with the operators $\Omega_{r,t}$ and $\Omega_{t,u}$, the spaces $L(t)$ and $L(t)^{\perp}$ are invariant under B. But this is equivalent to the statement that B commutes with the orthogonal projection on $L(t)$. $\square$

We shall refer to the resolution defined in Theorem 3.2 as the *resolution of the identity for the selfadjoint operator A*. In the next section we shall see that a selfadjoint operator may be reconstructed from its resolution. As a first step in this direction we show here how a resolution of the identity can be used to build new operators using a Stieltjes type of integral.

Let $\{E(t)\}_{t \in \mathbf{R}}$ be a resolution of the identity supported by the interval $[m, M]$. Choose $\alpha < m$ and $\beta > M$, and let f be a complex-valued continuous function on $[\alpha, \beta]$. Let P be a partition, $\alpha = \lambda_0 < \lambda_1 < \cdots < \lambda_n = \beta$, of the interval $[\alpha, \beta]$, and let $\tau = \{t_1, \ldots, t_n\}$ be a set of points such that $\lambda_{j-1} \leq t_j \leq \lambda_j$ for $j = 1, \ldots, n$. As usual the width of the partition P (i.e., the maximal length of the subintervals $[\lambda_{j-1}, \lambda_j]$) is denoted by $\nu(P)$. Consider the Stieltjes type sum:

$$(3) \qquad S_\tau(f; P) = \sum_{j=1}^{n} f(t_j)\big(E(\lambda_j) - E(\lambda_{j-1})\big),$$

and define

$$(4) \qquad \int_\alpha^\beta f(\lambda)\,dE(\lambda) := \lim_{\nu(P) \to 0} S_\tau(f; P).$$

Note that $S_\tau(f; P)$ is a bounded linear operator on H. A standard argument from the theory of Riemann-Stieltjes integration shows that the limit in the right hand side of (4) exists in the norm of $\mathcal{L}(H)$. Indeed, take $\varepsilon > 0$, and let P and Q be partitions of $[\alpha, \beta]$ with $\nu(P) < \delta$ and $\nu(Q) < \delta$, where $\delta > 0$ has been chosen in such a way that $|f(t) - f(s)| < \varepsilon/2$ whenever $|t - s| < \delta$. Consider Stieltjes type sums for P and Q:

$$S_\tau(f; P) = \sum_{i=1}^{n} f(t_i)\{E(\lambda_i) - E(\lambda_{i-1})\},$$

$$S_\sigma(f; Q) = \sum_{j=1}^{r} f(s_j)\{E(\mu_j) - E(\mu_{j-1})\}.$$

First, assume that Q is finer than P, and put $t'_j = t_i$ if $[\mu_{j-1}, \mu_j] \subset [x_{i-1}, x_i]$. It follows that $|t'_j - s_j| < \delta$, and hence $|f(t'_j) - f(s_j)| < \varepsilon/2$ for $j = 1, \ldots, r$. Now use that for each $x \in H$ the vectors $(E(\mu_j) - E(\mu_{j-1}))x$, $j = 1, \ldots, r$, are mutually orthogonal. So

$$\|(S_\tau(f; P) - S_\sigma(f; Q))x\|^2 = \left\| \sum_{j=1}^{r} (f(t'_j) - f(s_j))(E(\mu_j) - E(\mu_{j-1}))x \right\|^2$$

$$= \sum_{j=1}^{r} |f(t'_j) - f(s_j)|^2 \big\|(E(\mu_j) - E(\mu_{j-1}))x\big\|^2$$

$$\leq (\varepsilon/2)^2 \left(\sum_{j=1}^{r} \left\| (E(\mu_j) - E(\mu_{j-1}))x \right\|^2 \right)$$

$$= (\varepsilon/2)^2 \left\| (E(\beta) - E(\alpha))x \right\|^2$$

$$= (\varepsilon/2)^2 \|x\|^2,$$

and hence $\|S_\tau(f;P) - S_\sigma(f;Q)\| < \varepsilon/2$. But then, always (without the assumption that Q is finer than P), we have $\|S_\tau(f;P) - S_\sigma(f;Q)\| < \varepsilon$. Since $\mathcal{L}(H)$, endowed with the operator norm, is a Banach space, we may conclude that the limit in (4) exists in $\mathcal{L}(H)$.

The value of the integral in (4) depends only on the values of f on $[m, M]$. To see this, choose the partition $P = \{\alpha = \lambda_0 < \lambda_1 < \cdots < \lambda_n = \beta\}$ in such a way that m and M belong to P. So $m = \lambda_k$ and $M = \lambda_\ell$ for some k and ℓ. It follows that

$$(5) \qquad S_\tau(f;P) = \sum_{j=k+1}^{\ell} f(t_j)\{E(\lambda_j) - E(\lambda_{j-1})\} + R_\tau(f;P),$$

where

$$R_\tau(f;P) = \left(\sum_{j=1}^{k} + \sum_{j=\ell+1}^{n} \right) [f(t_j)\{E(\lambda_j) - E(\lambda_{j-1})\}]$$

$$= f(t_k)E(m) + f(t_{\ell+1})(I - E(M))$$

$$\to f(m)E(m) + f(M)(I - E(M)), \qquad \nu(P) \to 0,$$

because of the continuity of f. Since the first term in the right hand side of (5) only involves values of f on $[m, M]$, we conclude that indeed the integral in (4) is uniquely determined by the values of f on $[m, M]$.

For a complex-valued continuous function f on $[m, M]$ we define

$$(6) \qquad \int_{m-0}^{M+0} f(\lambda)dE(\lambda) := \int_{\alpha}^{\beta} \widetilde{f}(\lambda)dE(\lambda),$$

where $[\alpha, \beta]$ is any interval with $\alpha < m$ and $\beta > M$ and $\widetilde{f}$ is an arbitrary continuous extension of f to $[\alpha, \beta]$. The remark made in the previous paragraph shows that the integral in the left hand side of (6) is well-defined.

THEOREM 3.3. *Let $\{E(t)\}_{t\in\mathbf{R}}$ be a resolution of the identity on H supported by the interval $[m, M]$, and let $C([m, M])$ be the linear space of complex-valued continuous function on $[m, M]$. Then the map*

$$\mathcal{J} : C([m, M]) \to \mathcal{L}(H), \qquad \mathcal{J}(f) = \int_{m-0}^{M+0} f(\lambda)dE(\lambda)$$

has the following properties:

(i) $\mathcal{J}$ *is linear,*

(ii) $\mathcal{J}(fg) = \mathcal{J}(f)\mathcal{J}(g)$,

(iii) $\mathcal{J}(f)^* = \mathcal{J}(\overline{f})$,

(iv) $\mathcal{J}(e) = I$, *where* $e(t) = 1$ *for* $m \leq t \leq M$,

(v) $\|\mathcal{J}f\| \leq \max\{|f(\lambda)| \mid m \leq \lambda \leq M\}$.

Furthermore, if $T \in \mathcal{L}(H)$ *commutes with* $E(t)$ *for each* $t \in \mathbf{R}$*, then* $\mathcal{J}(f)$ *commutes with* T.

PROOF. Choose $\alpha < m$ and $\beta > M$, and let P be a partition of $[\alpha, \beta]$. Each $f \in C([m, M])$ is extended to a continuous function on $[\alpha, \beta]$, also denoted by f, by setting

$$f(t) = \begin{cases} f(m) & \text{for} \quad \alpha \leq t \leq m, \\ f(M) & \text{for} \quad M \leq t \leq \beta. \end{cases}$$

Then

$$(7) \qquad \mathcal{J}(f) = \lim_{\nu(P) \to 0} S_\tau(f; P),$$

with convergence in the norm of $\mathcal{L}(H)$. Hence it suffices to prove the theorem for $S_\tau(f; P)$ instead of $\mathcal{J}(f)$.

Obviously, $S_\tau(f; P)$ is linear in f, and so (i) is proved. Set $P = \{\alpha = \lambda_0 < \lambda_1 < \cdots < \lambda_n = \beta\}$. Note that

$$\big(E(\lambda_j) - E(\lambda_{j-1})\big)\big(E(\lambda_i) - E(\lambda_{i-1})\big)$$
$$= \delta_{ij}\big(E(\lambda_j) - E(\lambda_{j-1})\big), \qquad i, j = 1, \ldots, n,$$

where δ_{ij} is the Kronecker delta. This implies that $S_\tau(fg; P) = S_\tau(f, P)S_\tau(g; P)$, and hence (ii) is proved. Since

$$\big(E(\lambda_j) - E(\lambda_{j-1})\big)^* = E(\lambda_j) - E(\lambda_{j-1})$$

for $i = 1, \ldots, n$, we have $S_\tau(f; P)^* = S_\tau(\overline{f}, P)$, which proves (iii). Statement (iv) follows from

$$S_\tau(e; P) = \sum_{j=1}^{n} \big(E(\lambda_j) - E(\lambda_{j-1})\big) = I.$$

To prove (v), recall that for each x in H the vectors $\big(E(\lambda_j) - E(\lambda_{j-1})\big)x$, $j = 1, \ldots, n,$

are mutually orthogonal, and hence

$$\|S_\tau(f;P)x\|^2 = \sum_{j=1}^{n} \|f(t_j)\{E(\lambda_j) - E(\lambda_{j-1})\}x\|^2$$

$$= \sum_{j=1}^{n} |f(t_j)|^2 \|\{E(\lambda_j) - E(\lambda_{j-1})\}x\|^2$$

$$\leq \gamma^2 \sum_{j=1}^{n} \|\{E(\lambda_j) - E(\lambda_{j-1})\}x\|^2$$

$$= \gamma^2 \|\{E(\beta) - E(\alpha)\}x\|^2 = \gamma^2\|x\|^2,$$

where

$$\gamma = \max_{\alpha \leq t \leq \beta} |f(t)| = \max_{m \leq t \leq M} |f(t)|.$$

The above calculations show that $\|S_\tau(f;P)\| \leq \gamma$, which proves (v).

Finally, assume that $T \in \mathcal{L}(H)$ commutes with $E(t)$ for each $t \in \mathbf{R}$. Then $TS_\tau(f;P) = S_\tau(f;P)T$, and (7) implies that $\mathcal{J}(f)$ commutes with T. $\square$

V.4 THE SPECTRAL THEOREM

The first theorem of this section is called the *spectral theorem for a bounded selfadjoint operator.*

THEOREM 4.1. *Let $A \in \mathcal{L}(H)$ be selfadjoint, and let $\{E(t)\}_{t \in \mathbf{R}}$ be the resolution of the identity for A. Then*

$$(1) \qquad\qquad A = \int_{m(A)-0}^{M(A)+0} \lambda \, dE(\lambda).$$

PROOF. By definition (see the previous section) $E(t)$ is the orthogonal projection of H onto the spectral subspace of A associated with $(-\infty, t]$. Take $s < t$, and put $L_0 = \mathrm{Im}\{E(t) - E(s)\}$. Obviously, $L_0 = \mathrm{Im}\,E(t) \cap \mathrm{Ker}\,E(s)$. We know that $\mathrm{Im}\,E(t)$ and $\mathrm{Im}\,E(s)$ are invariant under A. Since A is selfadjoint, also $\mathrm{Ker}\,E(s)$ is invariant under A, and hence L_0 is invariant under A. By formulas (32a) and (32b) in Section 2,

$$(2) \qquad\qquad s\|x\|^2 \leq \langle Ax, x \rangle \leq t\|x\|^2, \qquad x \in L_0.$$

We are now ready to prove (1). Choose $\alpha < m(A)$ and $\beta > M(A)$. Let $P = \{\alpha = \lambda_0 < \lambda_1 < \cdots < \lambda_n = \beta\}$ be a partition of $[\alpha, \beta]$. From the result of the previous paragraph we know that

$$0 \leq \langle (A - \lambda_{j-1})y, y \rangle \leq (\lambda_j - \lambda_{j-1})\|y\|^2$$

for each $y \in \mathrm{Im}\{E(\lambda_j) - E(\lambda_{j-1})\}$, and hence

$$(3) \qquad\qquad \|(A - \lambda_{j-1})\{E(\lambda_j) - E(\lambda_{j-1})\}\| \leq (\lambda_j - \lambda_{j-1}).$$

Take $x \in H$. Then

$$\left\| Ax - \sum_{j=1}^{n} \lambda_{j-1}\{E(\lambda_j) - E(\lambda_{j-1}\}x \right\|^2$$

$$= \left\| \sum_{j=1}^{n} (A - \lambda_{j-1})\{E(\lambda_j) - E(\lambda_{j-1})\}x \right\|^2$$

$$= \sum_{j=1}^{n} \|(A - \lambda_{j-1})\{E(\lambda_j) - E(\lambda_{j-1})\}x\|^2$$

$$\leq \sum_{j=1}^{n} (\lambda_j - \lambda_{j-1})^2 \|\{E(\lambda_j) - E(\lambda_{j-1})\}x\|^2$$

$$\leq \max_{1 \leq j \leq n} (\lambda_j - \lambda_{j-1})^2 \cdot \sum_{j=1}^{n} \|\{E(\lambda_j) - E(\lambda_{j-1})\}x\|^2 = \nu(P)^2 \|x\|^2,$$

where $\nu(P)$ is the width of the partition P. The first inequality in the above calculation follows from (3). Furthermore, we have used that $\mathrm{Im}\{E(\lambda_j) - E(\lambda_{j-1})\}$ is invariant under A for $j = 1, \ldots, n$, and hence the vectors $(A - \lambda_{j-1})\{E(\lambda_j) - E(\lambda_{j-1})\}x$, $j = 1, \ldots, n$, are mutually orthogonal. We conclude that

$$\sum_{j=1}^{n} \lambda_{j-1}\{E(\lambda_j) - E(\lambda_{j-1})\} \to A, \qquad \nu(P) \to 0,$$

in the operator norm, and (1) is proved. $\square$

The next theorem shows that the resolution of the identity for a selfadjoint operator A is uniquely determined by formula (1).

THEOREM 4.2. *Let $\{E(t)\}_{t \in \mathbf{R}}$ be a resolution of the identity on H supported by the interval $[m, M]$. Then the operator*

$$(4) \qquad A := \int_{m-0}^{M+0} \lambda dE(\lambda)$$

is selfadjoint and $\{E(t)\}_{t \in \mathbf{R}}$ is the resolution of the identity for A.

PROOF. It is clear from Theorem 3.2 that A is selfadjoint. Let $\{F(t)\}_{t \in \mathbf{R}}$ be the resolution of the identity for A. Put $n = \min(m, m(A))$ and $N = \max(M, M(A))$. Then the two resolutions $\{E(t)\}_{t \in \mathbf{R}}$ and $\{F(t)\}_{t \in \mathbf{R}}$ are both supported by the interval $[n, N]$. Furthermore (cf. formula (4) in Section 3), we have

$$\int_{n-0}^{N+0} \lambda dE(\lambda) = A = \int_{n-0}^{N+0} \lambda dF(\lambda).$$

So we can apply Theorem 3.3 (i), (ii) and (iv) to show that

$$(5) \qquad \int_{n-0}^{N+0} f(\lambda)dE(\lambda) = \int_{n-0}^{N+0} f(\lambda)dF(\lambda)$$

for each complex polynomial $f(\lambda)$. But then Theorem 3.3(v) and the Weierstrass approximation theorem imply that (5) holds for each $f \in C([n, N])$. Take $\alpha < n$ and $\beta > N$. According to formula (4) in Section 3 and by what has been proved so far

$$(6) \qquad \int_{\alpha}^{\beta} f(\lambda)dE(\lambda) = \int_{\alpha}^{\beta} f(\lambda)dF(\lambda)$$

for each $f \in C([\alpha, \beta])$, the space of all complex valued continuous functions on $[\alpha, \beta]$. Take $x \in H$, and put $g(t) = \langle E(t)x, x \rangle$ and $h(t) = \langle F(t)x, x \rangle$ for $\alpha \leq t \leq \beta$. From (6) it follows that

$$(7) \qquad \int_{\alpha}^{\beta} f(\lambda)dg(\lambda) = \int_{\alpha}^{\beta} f(\lambda)dh(\lambda), \qquad f \in C([\alpha, \beta]).$$

Note that $g(\alpha) = h(\alpha) = 0$. Furthermore, both g and h are monotonely increasing and right continuous (by Proposition 3.1). But then (7) and the Stieltjes integration theory imply that $g = h$. In other words, for each $x \in H$

$$(8) \qquad \langle E(t)x, x \rangle = \langle F(t)x, x \rangle, \qquad \alpha \leq t \leq \beta.$$

Now use that for $T \in \mathcal{L}(H)$ and any x and y in H the following identity holds:

$$(9) \qquad \langle Tx, y \rangle = \frac{1}{4}\{\langle T(x+y), x+y \rangle - \langle T(x-y), x-y \rangle$$
$$+ i\langle T(x+iy), x+iy \rangle - i\langle T(x-iy), x-iy \rangle\}.$$

So (8) yields that for each x and $y \in H$

$$\langle E(t)x, y \rangle = \langle F(t)x, y \rangle, \qquad \alpha \leq t \leq \beta,$$

which implies that $E(t) = F(t)$ for $\alpha \leq t \leq \beta$. For $t \notin [\alpha, \beta]$ this equality is trivial. Thus $E(t) = F(t)$ for all $t \in \mathbf{R}$. $\square$

Let $A \in \mathcal{L}(H)$ be selfadjoint, and let $\{E(t)\}_{t \in \mathbf{R}}$ be the resolution of the identity for A. Theorems 4.1 and 3.3 imply that

$$p(A) = \int_{m(A)-0}^{M(A)+0} p(\lambda)dE(\lambda)$$

for any complex polynomial p. We extend this formula to any complex-valued continuous function f on $[m(A), M(A)]$ by setting

$$(10) \qquad f(A) := \int_{m(A)-0}^{M(A)+0} f(\lambda)dE(\lambda).$$

Note that the map $f \mapsto f(A)$ from $C([m(A), M(A)])$ into $\mathcal{L}(H)$ has the same properties as the map J in Theorem 3.3. The functional calculus defined by (10) involves a richer family of functions than the one defined in Section I.3.

The resolution of the identity for a selfadjoint operator A can be extended to a spectral measure defined on the Borel subsets of the spectrum of A. This fact allows one to extend the functional calculus defined by (10) to bounded Borel functions on $\sigma(A)$. We shall derive these results later (in Volume II) for the (larger) class of normal operators.

V.5 SPECTRUM AND RESOLVENT

In this section we describe the spectrum and resolvent of a bounded selfadjoint operator in terms of its resolution of the identity $\{E(t)\}_{t\in\mathbf{R}}$. In what follows $E(t-0)$ denotes the orthogonal projection onto the smallest closed linear manifold containing $\operatorname{Im} E(\lambda)$ for each $\lambda < t$.

THEOREM 5.1. *Let $A \in \mathcal{L}(H)$ be selfadjoint, and let $\{E(t)\}_{t\in\mathbf{R}}$ be the resolution of the identity for A. Then for $s \leq t$ the spectral subspace L of A associated with $[s,t]$ is equal to the image of $E(t) - E(s-0)$ and*

$$(1) \qquad \sigma(A|L) \cap (s,t) = \sigma(A) \cap (s,t).$$

PROOF. Put $N = \operatorname{Im}\{E(t)-E(s-0)\}$. The space $\operatorname{Im} E(t)$ is invariant under A. Also $\operatorname{Im} E(\lambda)$ is invariant under A for each $\lambda < s$. This implies that $\operatorname{Im} E(s - 0)$ is invariant under A. Since A is selfadjoint, we conclude that also $\operatorname{Ker} E(s - 0)$ is A-invariant, and thus the same holds true for N. Take $x \in N$. Choose $\alpha < m(A)$ and $\beta > M(A)$. Then

$$\langle Ax, x\rangle = \langle\left(\int_{\alpha}^{\beta} \lambda dE(\lambda)\right)x, x\rangle$$

$$= \int_{\alpha}^{\beta} \lambda d\langle E(\lambda)x, x\rangle.$$

Our choice of x yields

$$\langle E(\lambda)x, x\rangle = \begin{cases} \|x\|^2, & \lambda \geq t, \\ 0, & \lambda < s. \end{cases}$$

Hence, by the theory of Stieltjes integration,

$$\langle Ax, x\rangle = \int_s^t \lambda d\langle E(\lambda)x, x\rangle,$$

which implies that

$$\langle (A-s)x, x\rangle \geq 0, \qquad \langle (t-A)x, x\rangle \geq 0.$$

In other words, $m(A|N) \geq s$ and $M(A|N) \leq t$, and we can apply Theorem 2.1 to show that $\sigma(A|N) \subset [s,t]$. Since L is the largest A-invariant subspace with this property, we have proved that $N \subset L$.

We know that $\operatorname{Im} E(t)$ is the largest A-invariant subspace with the property that $\sigma(A|\operatorname{Im} E(t)) \subset (-\infty, t]$. This implies that $L \subset \operatorname{Im} E(t)$. Let us show that also $L \subset \operatorname{Ker} E(s-0)$. Take $\lambda < s$. Recall (see Corollary 2.5) that

$$\operatorname{Im} E(\lambda) = \operatorname{Ker}(\lambda - A) \oplus \overline{\operatorname{Im} \Omega_\lambda},$$

where Ω_λ is given by formula (33) in Section 2 (with λ instead of t). Since $\sigma(A|L) \subset [s,t]$, the functional calculus of Section I.3 yields $\Omega_\lambda x = 0$ for each $x \in L$, and thus

$$L \subset \operatorname{Ker} \Omega_\lambda = \left(\overline{\operatorname{Im} \Omega_\lambda}\right)^\perp,$$

because Ω_λ is selfadjoint. The inclusion $\sigma(A|L) \subset [s,t]$, also implies that $L = (\lambda - A)L$, and therefore

$$L \subset \operatorname{Im}(\lambda - A) \subset \operatorname{Ker}(\lambda - A)^\perp.$$

So $L^\perp \supset \operatorname{Im} E(\lambda)$ for each $\lambda < s$. But then $L^\perp \supset \operatorname{Im} E(s-0)$, and thus $L \subset \operatorname{Ker} E(s-0)$. We have now proved that $L = \operatorname{Im}\{E(t) - E(s-0)\}$.

It remains to prove (1). Since A is selfadjoint, $L^\perp$ is also invariant under A, and thus

$$\sigma(A) = \sigma(A|L) \cup \sigma(A|L^\perp),$$

because of Proposition I.2.4. So to prove (1) it suffices to show that $\sigma(A|L^\perp) \cap (s,t) = \emptyset$. Note that

$$L^\perp = \operatorname{Im} E(t)^\perp \oplus \operatorname{Im} E(s-0).$$

Both $\operatorname{Im} E(t)^\perp$ and $\operatorname{Im} E(s-0)$ are invariant under A. From the definition of the resolution $\{E(\lambda)\}_{\lambda \in \mathbf{R}}$ and Corollary 2.5 we know that $\langle (A-t)x, x\rangle \geq 0$ for each $x \in \operatorname{Im} E(t)^\perp$. Hence, by Theorem 2.1, we have $\sigma(A|\operatorname{Im} E(t)^\perp) \subset [t, \infty)$. Similarly, for each $\lambda < s$

$$\langle Ax, x\rangle \leq \lambda \|x\|^2 \leq s\|x\|^2, \qquad x \in \operatorname{Im} E(\lambda),$$

and thus $\langle (s-A)x, x\rangle \geq 0$ for each $x \in \operatorname{Im} E(s-0)$, which implies (by Theorem 2.1) that $\sigma(A|\operatorname{Im} E(s-0)) \subset (-\infty, s]$. Thus

$$\sigma(A|L^\perp) = \sigma(A|\operatorname{Im} E(t)^\perp) \cup \sigma(A|\operatorname{Im} E(s-0)) \subset [t, \infty) \cap (-\infty, s],$$

and hence $\sigma(A|L^{\perp}) \cap (s,t) = \emptyset$. $\square$

In (1) the open interval (s,t) may not be replaced by the closed interval $[s,t]$. For example, take

$$A : L_2([0,1]) \to L_2([0,1]), \qquad (Af)(t) = tf(t) \quad (0 \leq t \leq 1).$$

Consider the closed interval $[1,2]$. The spectral subspace L of A associated with $[1,2]$ consists of the zero element only, and hence $\sigma(A|L) = \emptyset$. But $\sigma(A) \cap [1,2]) = \{1\}$.

COROLLARY 5.2. *Let $A \in \mathcal{L}(H)$ be selfadjoint, and let $\{E(t)\}_{t\in\mathbf{R}}$ be the resolution of the identity for A. Then*

$$(2) \qquad\qquad \mathrm{Im}\{E(t) - E(t-0)\} = \mathrm{Ker}(t-A), \qquad t \in \mathbf{R}.$$

In particular, t is an eigenvalue of A if and only if $E(t) \neq E(t-0)$.

PROOF. Note that $\mathrm{Ker}(t-A)$ is the largest A-invariant subspace N such that $\sigma(A|N) \subset \{t\}$. Indeed, if $\sigma(A|N) \subset \{t\}$ and $N \neq \{0\}$, then $m(A|N) = M(A|N) = t$, and thus $\langle Ax,x \rangle = t\langle x,x \rangle$ for each $x \in N$, which implies (cf. [GG], Corollary III.4.2) that $A - t$ is zero on N. Now apply Theorem 5.1 with $s = t$. $\square$

COROLLARY 5.3. *Let $A \in \mathcal{L}(H)$ be selfadjoint, and let $\{E(t)\}_{t\in\mathbf{R}}$ be the resolution of the identity for A. Then $\mu \in \mathbf{R} \cap \rho(T)$ if and only if $E(\mu - \varepsilon) = E(\mu + \varepsilon)$ for some $\varepsilon > 0$.*

PROOF. We know (cf. Corollary 2.5) that

$$(3) \qquad\qquad \mathrm{Im}\, E(t) = \mathrm{Ker}(t-A) \oplus \overline{\mathrm{Im}\,\Omega_t},$$

where Ω_t is given by formula (33) in Section 2. Now, assume that $\mu \in \rho(A)$. Choose $\varepsilon > 0$ such that $[\mu - \varepsilon, \mu + \varepsilon] \subset \rho(A)$. Then the functional calculus of Section I.3 implies that $\Omega_t = \Omega_\mu$ for all $t \in [\mu - \varepsilon, \mu + \varepsilon]$. In particular, $\Omega_{\mu-\varepsilon} = \Omega_{\mu+\varepsilon}$. Furthermore, $\mathrm{Ker}(\mu + \varepsilon - A)$ and $\mathrm{Ker}(\mu - \varepsilon - A)$ both consist of the zero element only (because $\mu - \varepsilon$ and $\mu + \varepsilon$ are in $\rho(A)$). So (3) yields that $\mathrm{Im}\, E(\mu + \varepsilon) = \mathrm{Im}\, E(\mu - \varepsilon)$.

Conversely, assume that $E(\mu - \varepsilon) = E(\mu + \varepsilon)$ for some $\varepsilon > 0$. Then, by Theorem 5.1, the spectral subspace L of A associated with $[\mu - \frac{1}{2}\varepsilon, \mu + \frac{1}{2}\varepsilon]$ consists of the zero element only, and thus (1) implies that

$$\sigma(A) \cap \left(\mu - \frac{1}{2}\varepsilon, \mu + \frac{1}{2}\varepsilon\right) = \sigma(A|L) \cap \left(\mu - \frac{1}{2}\varepsilon, \mu + \frac{1}{2}\varepsilon\right) = \emptyset.$$

In particular, $\mu \in \rho(A)$. $\square$

V.6 SQUARE ROOT AND POLAR DECOMPOSITION

An operator $A \in \mathcal{L}(H)$ is called *non-negative* (notation: $A \geq 0$) if A is selfadjoint and $\langle Ax, x \rangle \geq 0$ for each $x \in H$. In that case $m(A) \geq 0$ and $M(A) = \|A\|$.

THEOREM 6.1. *If A is non-negative, then there exists a unique non-negative operator B such that $B^2 = A$. Furthermore, B commutes with each operator that commutes with A.*

The operator B in Theorem 6.1 is called the *square root* of A. To prove Theorem 6.1 we use the functional calculus introduced in the next to last paragraph of Section 4. Let $A \in \mathcal{L}(H)$ be selfadjoint. Recall that for a complex-valued continuous function f on $[m(A), M(A)]$

$$(1) \qquad f(A) := \int_{m(A)-0}^{M(A)+0} f(\lambda)dE(\lambda),$$

where $\{E(t)\}_{t \in \mathbb{R}}$ is the resolution of the identity for A. From Theorem 3.3 we know that

$$(2) \qquad (f + g)(A) = f(A) + g(A), \qquad (fg)(A) = f(A)g(A),$$

$$(3) \qquad f(A)^* = \overline{f}(A),$$

$$(4) \qquad \|f(A)\| \leq \max\{|f(t)| \mid m(A) \leq t \leq M(A)\},$$

whenever f and g are continuous on $[m(A), M(A)]$. Formula (2) and Theorems 4.1 and 3.3(iv) imply that formula (1) yields the usual expression for polynomials, i.e.,

$$(5) \qquad p(\lambda) = \sum_{j=0}^{n} \alpha_j \lambda^j \Rightarrow p(A) = \sum_{j=0}^{n} \alpha_j A^j.$$

LEMMA 6.2. *The operator $f(A)$ commutes with each operator that commutes with A.*

PROOF. By the Weierstrass approximation theorem there exists a sequence (p_n) of polynomials such that $p_n \to f$ uniformly on $[m(A), M(A)]$. The inequality (4) implies that $p_n(A) \to f(A)$ in the operator norm. Now, let S be an operator commuting with A. Then S commutes with $p_n(A)$ for each n, and thus

$$Sf(A) = \lim_{n \to \infty} Sp_n(A) = \lim_{n \to \infty} p_n(A)S = f(A)S. \qquad \square$$

PROOF OF THEOREM 6.1. Choose $\alpha < 0$ and $\beta > \|A\|$, put

$$g(t) = \begin{cases} 0 & \text{for} \quad \alpha \leq t \leq 0, \\ t^{1/2} & \text{for} \quad 0 \leq t \leq \beta, \end{cases}$$

and let g_0 be the restriction of g to $[m(A), M(A)]$. Then $B = g_0(A)$ is a well-defined operator, which is selfadjoint by formula (3). According to the second identity in (2) we have $B^2 = g_0^2(A) = A$. From

$$\langle Bx, x \rangle = \langle g_0(A)x, x \rangle = \int_{\alpha}^{\beta} g(t)d\langle E(t)x, x \rangle$$

we see that $B \geq 0$. Lemma 6.2 implies that B commutes with each operator that commutes with A.

To prove the uniqueness, let C be a second non-negative operator such that $C^2 = A$. Without loss of generality we may assume that $\beta > \|C\|$. Choose a sequence of polynomials $p_1, p_2, \ldots$ such that $p_n \to g$ $(n \to \infty)$ uniformly on $[\alpha, \beta]$. Then $p_n \to g_0$ uniformly on $[m(A), M(A)]$, and thus $p_n(A) \to B$ in the operator norm (by the inequality (4)). Put $q_n(t) = p_n(t^2)$ for $n = 1, 2, \ldots$. Then $q_1, q_2, \ldots$ is a sequence of polynomials which converges uniformly on $[0, \|C\|]$ to $q(t) = t$. It follows (apply the inequality (4) to C instead of A) that $q_n(C) \to C$ in the operator norm. From (5), applied to C, we see that $q_n(C) = p_n(A)$. Thus

$$B = \lim_{n \to \infty} p_n(A) = \lim_{n \to \infty} q_n(C) = C. \quad \square$$

Square roots can be used to extend the familiar representation $re^{i\varphi}$ of a complex number to operators on a Hilbert space. To do this, first a few preparations. An operator $U \in \mathcal{L}(H)$ is called a *partial isometry* if

$$(6) \qquad\qquad \|Ux\| = \|x\|, \qquad x \perp \operatorname{Ker} U.$$

In that case $\operatorname{Ker} U^{\perp}$ is called the *initial space* of U and $\operatorname{Im} U$ is called the *final space*. Note that the range of a partial isometry is always closed.

THEOREM 6.3. *Let $A \in \mathcal{L}(H)$. Then there exists a partial isometry U and a non-negative operator R such that*

$$(7) \qquad\qquad A = UR.$$

Furthermore, U and R may be chosen such that $\overline{\operatorname{Im} R}$ is the initial space of U, and in that case the decomposition (7) is unique.

The decomposition (7) is called the *polar decomposition* of A if the additional condition $\overline{\operatorname{Im} R} = \operatorname{Ker} U^{\perp}$ is fulfilled.

PROOF OF THEOREM 6.3. Let R be the square root of A^*A. Since A^*A is non-negative, R is well-defined by Theorem 6.1. Note that

$$(8) \qquad \begin{aligned} \|Ax\|^2 &= \langle Ax, Ax \rangle = \langle A^*Ax, x \rangle \\ &= \langle R^2 x, x \rangle = \langle Rx, Rx \rangle = \|Rx\|^2, \qquad x \in H, \end{aligned}$$

which implies that $\operatorname{Ker} R = \operatorname{Ker} A$. Now define an operator $U_0 : \operatorname{Im} R \to H$ by setting $U_0 y = Ax$, where x is any vector in H such that $y = Rx$. The particular choice of x is not important. Indeed, if $Rx_1 = Rx_2$, then $x_1 - x_2 \in \operatorname{Ker} R = \operatorname{Ker} A$, and thus $Ax_1 = Ax_2$. From (8) we see that for $y = Rx$

$$\|U_0 y\| = \|Ax\| = \|Rx\| = \|y\|.$$

So, by continuity, we may extend U_0 to a bounded linear operator from $\overline{\operatorname{Im} R}$ to H, which we shall also denote by U_0. This extension again preserves the norm, i.e., $\|U_0 y\| = \|y\|$ for $y \in \overline{\operatorname{Im} R}$. Next, put

$$(9) \qquad\qquad Uy = U_0 Py, \qquad y \in H,$$

where P is the orthogonal projection of H onto $\overline{\operatorname{Im} R}$. Then U is a partial isometry, the initial space of U is $\overline{\operatorname{Im} R}$, and (7) holds.

It remains to prove the uniqueness of the polar decomposition. Assume $A = U_1 R_1$, where R_1 is a non-negative operator and U_1 is a partial isometry with initial space $\overline{\operatorname{Im} R_1}$. The latter part of our assumptions implies that $\|U_1 y\| = \|y\|$ for all $y = R_1 x$. Thus

$$\langle A^* A x, x \rangle = \|Ax\|^2 = \|U_1 R_1 x\|^2 = \|R_1 x\|^2 = \langle R_1^2 x, x \rangle, \qquad x \in H.$$

But then $A^* A = R_1^2$, by [GG], Corollary III.4.2. Since $R_1 \geq 0$, the uniqueness statement in Theorem 6.1 implies that R_1 is the square root of $A^* A$. Let R and U be as in the first paragraph of the proof. Then $R_1 = R$ and $UR = U_1 R$. It follows that U_1 and U coincide on the subspace $M := \overline{\operatorname{Im} R_1} = \overline{\operatorname{Im} R}$. Since M is the initial space for both U_1 and U, we also know that U_1 and U coincide on $M^\perp$. Hence $U_1 = U$. $\square$

Let $A = UR$ be the polar decomposition of A. From the construction of U and R in the first paragraph of the proof of Theorem 6.3 it follows that

$$\overline{\operatorname{Im} A^*} = \operatorname{Ker} A^\perp = \operatorname{Ker} R^\perp = \overline{\operatorname{Im} R},$$

and hence $\overline{\operatorname{Im} A^*}$ is the initial space of the partial isometry U. One can also prove that $\overline{\operatorname{Im} A}$ is the final space of U. Indeed,

$$\operatorname{Im} U = \overline{\operatorname{Im} U} \supset \overline{\operatorname{Im} A} = \overline{U(\operatorname{Im} R)} \supset U(\overline{\operatorname{Im} R}) = \operatorname{Im} U,$$

and thus $\operatorname{Im} U = \overline{\operatorname{Im} A}$.

V.7 UNITARY OPERATORS

Let $\{E(t)\}_{t \in \mathbf{R}}$ be a resolution of the identity supported by $[-\pi, \pi]$ and let U be defined by

$$(1) \qquad U = \int_{-\pi-0}^{\pi+0} e^{i\lambda} dE(\lambda).$$

Then U is unitary, i.e., the operator U is invertible and $U^{-1} = U^*$ (see [GG], Section VIII.3 for the definition of a unitary operator). Indeed, the operator

$$(2) \qquad V = \int_{-\pi-0}^{\pi+0} e^{-i\lambda} dE(\lambda)$$

is well-defined, and by Theorem 3.3 we have $U^* = V$ and $VU = UV = I$, which shows that U is unitary.

Conversely, any unitary operator admits a representation as in (1). This is the spectral theorem for unitary operators.

THEOREM 7.1. *Let $U \in \mathcal{L}(H)$ be unitary. Then there exists a unique resolution of the identity $\{E(t)\}_{t \in \mathbf{R}}$ supported by $[-\pi, \pi]$ such that*

$$(3) \qquad U = \int_{-\pi-0}^{\pi+0} e^{i\lambda} dE(\lambda).$$

Furthermore, the resolution in (3) is obtained as follows. Put

$$(4) \qquad \Lambda_t = \frac{1}{2\pi i} \int_{\gamma_t} (e^{i\pi} - \zeta)(e^{it} - \zeta)(\zeta - U)^{-1} d\zeta,$$

where γ_t is the boundary of the set consisting of all $\lambda = re^{is}$ with $0 \leq r \leq 2$ and $-\pi \leq s \leq t$. Then $E(t)$ is the orthogonal projection onto the subspace

$$(5) \qquad \operatorname{Ker}(e^{i\pi} - U) \oplus \overline{\operatorname{Im} \Lambda_t} \oplus \operatorname{Ker}(e^{it} - U)$$

for $-\pi \leq t < \pi$, the projection $E(t) = 0$ for $t < -\pi$ and $E(t) = I$ for $t \geq \pi$.

The operator Λ_t in (4) is well-defined because of the following lemma.

LEMMA 7.2. *The spectrum of a unitary operator U lies on the unit circle and*

$$(6) \qquad \|(\lambda - U)^{-1}\| \leq |1 - |\lambda||^{-1}, \qquad |\lambda| \neq 1.$$

PROOF. Since U is an isometry (cf., Theorem VIII.3.1 in [GG]), $\|U\| \leq 1$ and hence $\lambda \in \rho(U)$ for $|\lambda| > 1$. Also $0 \in \rho(U)$, because U is invertible. Take $0 < |\lambda| < 1$. Then $\lambda - U = \lambda U(U^* - \lambda^{-1})$. Now use that U^* is an isometry. So $U^* - \lambda^{-1}$ is invertible and it follows that $\lambda \in \rho(U)$. This proves that $\sigma(U)$ lies in the unit circle.

Let $\lambda = re^{i\varphi}$ with $r > 1$. Then

$$\begin{aligned}
\|(\lambda - U)x\|^2 &= \langle (\lambda - U)x, (\lambda - U)x \rangle \\
&= \langle (\overline{\lambda} - U^*)(\lambda - U)x, x \rangle \\
&= (r^2 + 1)\|x\|^2 - \langle (\overline{\lambda}U + \lambda U^*)x, x \rangle \\
&\geq (r^2 + 1)\|x\|^2 - \|\overline{\lambda}U + \lambda U^*\| \|x\|^2 \\
&\geq (r^2 + 1)\|x\|^2 - 2r\|x\|^2 \\
&= (r - 1)^2 \|x\|^2
\end{aligned}$$

which proves (6) for $|\lambda| > 1$. For $\lambda = 0$ the inequality (6) is trivial. Take $0 < |\lambda| < 1$. Then $|\lambda^{-1}| > 1$, and thus

$$\begin{aligned}
\|(\lambda - U)^{-1}\| &= \|\lambda^{-1}(U^* - \lambda^{-1})^{-1} U^*\| \\
&\leq |\lambda^{-1}|(|\lambda^{-1}| - 1)^{-1} = (1 - |\lambda|)^{-1}. \quad \square
\end{aligned}$$

The proof of Theorem 7.1 can now be given using the same type of arguments as used in Sections 2–4. Again the subspace (5) may be identified with the largest U-invariant subspace N such that $\sigma(U|N)$ lies in the arc $\{e^{is} \mid -\pi \le s \le t\}$. We omit further details. We shall come back to the representation (3) in Volume II.

The approach to the spectral theorem developed in the present chapter applies to larger classes of operators. For example, it may also be applied to operators A that are normal (i.e., $AA^* = A^*A$) and have spectrum on a C^1-curve.

COMMENTS ON PART I

Chapter I contains standard material. One can find more about the topics of the last three sections of this chapter in the book Daleckii-Krein [1]. The operator differential equation $y' = Ay$, discussed in Section I.5, is of special interest when A is an unbounded operator. We shall return to this topic in Chapter XIX, where the connections between differential equations and semigroups will be explained. The material in Chapter II is taken from the corresponding material in Gohberg-Krein [1], [3], however the exposition is different and based on analogies with the finite dimensional case. The notion of equivalence, which is the main topic of Chapter III, has its roots in the analysis of matrix polynomials (see, e.g., Gelfand [1], Gohberg-Lancaster-Rodman [1]). In operator theory it was introduced in the seventies. The first three sections of Chapter III are based on the papers Gohberg-Kaashoek-Lay [1], [2] and Kaashoek-Van der Mee-Rodman [1]. Section III.4 is taken from Bart-Gohberg-Kaashoek [4]. The first section of Chapter IV, which extends the Riesz theory to operator pencils, contains results from Stummel [1]. The exposition of the spectral theorem for bounded selfadjoint operators given in Chapter V uses the main idea from the paper Lorch [1] (see also the book Lorch [2]). The approach used in this chapter may be viewed as the starting point of a spectral theory for more general classes of operators (see, e.g., the papers Lyubich-Macaev [1], Wolf [1], and the books Dunford-Schwartz [2], Colojara-Foias [1]).

EXERCISES TO PART I

In the exercises below, A is a bounded linear operator acting on a complex Banach space X. The symbol I stands for the identity operator on X.

1. Let N be a positive integer, and assume that $A^N = I$. Prove the following statements:

    (a) $\sigma(A)$ is a finite set;

    (b) each point of $\sigma(A)$ is an eigenvalue;

    (c) the resolvent of A has a *simple pole* at each $\mu \in \sigma(A)$, i.e., for λ near μ

$$(\lambda - A)^{-1} = \frac{1}{\lambda - \mu} B_{-1} + \sum_{j=0}^{\infty} (\lambda - \mu)^j B_j;$$

    (d) the operator B_{-1} in (c) is a projection with $\operatorname{Im} B_{-1} = \operatorname{Ker}(\mu - A)$ and $\operatorname{Ker} B_{-1}$ is the direct sum of $\operatorname{Ker}(\lambda - A)$ with $\lambda \neq \mu$.

2. Let N be a positive integer. Prove that $A^N = I$ if and only if

$$(*) \qquad\qquad A = \lambda_1 P_1 + \lambda_2 P_2 + \cdots + \lambda_N P_N,$$

where $\lambda_1, \lambda_2, \ldots, \lambda_N$ are the N-th roots of unity and $P_1, P_2, \ldots, P_N$ are *mutually disjoint* projections (i.e., $P_i P_j = \delta_{ij} P_j$ for all i and j) and $\sum_{j=1}^{N} P_j = I$.

3. Let A be an operator such that $p(A) = 0$, where p is a polynomial of degree N with N different roots. Generalize for this operator the statements in Exercises 1 and 2, which concern the polynomial $p(\lambda) = \lambda^N - 1$.

4. Let A be an operator such that $p(A) = 0$, where p is some non-zero polynomial. Prove the following statements:

    (a) $\sigma(A)$ is a finite set;

    (b) each point of $\sigma(A)$ is an eigenvalue;

    (c) the resolvent of A has a pole at each $\mu \in \sigma(A)$ of order less than or equal to the multiplicity of μ as a zero of $p(\lambda)$.

Is the representation $(*)$ in Exercise 2 again valid? If not, replace it by another representation.

5. Let $N \geq 2$ be an integer. Give an example of an operator A on ℓ_2 such that $A^N = 0$, $A^{N-1} \neq 0$ and $\dim X / \operatorname{Ker} A^{N-1} < \infty$.

6. Consider a direct sum decomposition $X = X_0 \oplus X_1 \oplus \cdots \oplus X_{N-1}$, where X_0 is a subspace of X and $X_1, \ldots, X_{N-1}$ are finite dimensional. Let $A_j : X_j \to X_{j-1}$ be an operator ($j = 1, \ldots, N - 1$), and assume that relative to the decomposition $X =$

$X_0 \oplus X_1 \oplus \cdots \oplus X_{N-1}$ the operator A admits the following representation:

$$(**) \qquad A = \begin{bmatrix} 0 & A_1 & 0 & \cdot & 0 & 0 \\ 0 & 0 & A_2 & \cdots & 0 & 0 \\ \cdot & \cdot & \cdot & \cdots & \cdot & \cdot \\ 0 & 0 & 0 & \cdots & 0 & A_{N-1} \\ 0 & 0 & 0 & \cdots & 0 & 0 \end{bmatrix}$$

Prove that $A^N = 0$ and $\dim \operatorname{Ker} A^{j+1}/\operatorname{Ker} A^j$ is finite for $j = 1, \dots, N-1$.

7. Let $N \geq 2$ be an integer. Assume that $A^N = 0$ and $\dim \operatorname{Ker} A^2/\operatorname{Ker} A$ is finite. Prove that A can be represented in the form $(**)$ of Exercise 6. Are the numbers $\dim X_j$ ($j = 1, \dots, N-1$) uniquely determined by A? If yes, prove this statement; if no, give additional conditions on $A_1, \dots, A_{N-1}$ that guarantee their uniqueness.

8. Assume that A has no spectrum on the unit circle $|\lambda| = 1$. Let P be the Riesz projection associated with the part of $\sigma(A)$ inside the unit circle. Prove that

    (a) $I - A^n$ is invertible;

    (b) $(I - A^n)^{-1} = \frac{1}{n} \sum_{k=0}^{n-1} (I - \varepsilon(n)^k A)^{-1}$, where $\varepsilon(n) = \exp(-2\pi i/n)$;

    (c) $P = \lim_{n \to \infty} \frac{\varepsilon(-n)-1}{2\pi i} \sum_{k=0}^{n-1} (I - \varepsilon(n)^k A)^{-1}$;

    (d) $P = \lim_{n \to \infty} (I - A^n)^{-1}$.

9. Let σ be an isolated part of $\sigma(A)$ and N an A-invariant subspace. Show that $\sigma(A|N) \subset \sigma$ implies that $N \subset \operatorname{Im} P_\sigma(A)$.

10. Let M be an A-invariant subspace of X. Prove that $\sigma(A|M) \subset \sigma(A)$ whenever A is compact. Is $\sigma(A|M)$ always a subset of $\sigma(A)$? If yes, prove this statement; if no, give an additional condition on $\sigma(A)$ that guarantees the inclusion.

11. Let $f \in \mathcal{F}(A)$ and $g \in \mathcal{F}(f(A))$. Show that the composition product $g \circ f$ is well-defined on a suitable open neighbourhood of $\sigma(A)$ and prove that $g(f(A)) = (g \circ f)(A)$.

12. Assume that the spectrum of A lies in the open left half plane. Prove that

$$(\lambda - A)^{-1} = \int_0^\infty e^{-\lambda t} e^{tA} dt, \qquad \Re\lambda \geq 0,$$

$$e^{tA} = \lim_{\substack{a \to \infty \\ a \in \mathbf{R}}} \int_{-ia}^{ia} e^{\lambda t} (\lambda - A)^{-1} d\lambda, \qquad t > 0.$$

13. Assume that $i\mathbf{R} \cap \sigma(A) = \emptyset$. Prove that

$$\lim_{\substack{a \to \infty \\ a \in \mathbf{R}}} \int_{-ia}^{ia} (\lambda - A)^{-1} d\lambda$$

exists in the operator norm. Let Q be the operator defined by this limit, and let P be the Riesz projection associated with the part of $\sigma(A)$ in the open left half plane. Which of the following statements is true?

(i) $Q = P$,

(ii) $Q = I - P$,

(iii) $Q = P - I$,

(iv) $Q = P - \frac{1}{2}I$.

14. Consider the non-homogeneous differential equation:

$$(\dagger) \qquad \begin{cases} x'(t) = Ax(t) + f(t), & t \geq 0, \\ x(0) = x_0, \end{cases}$$

where $f : [0, \infty) \to X$ is continuous. Prove that its solution is given by

$$x(t) = e^{tA}x_0 + \int_0^t e^{(t-s)A}f(s)ds, \qquad t \geq 0.$$

15. Assume that $i\mathbf{R} \cap \sigma(A) = \emptyset$. Show that for each bounded continuous X-valued function on $\mathbf{R}$ the equation

$$x'(t) = Ax(t) + f(t), \qquad -\infty < t < \infty,$$

has a unique bounded solution.

16. Consider the equation

$$\begin{cases} x''(t) = Ax(t) + f(t), & t \geq 0, \\ x(0) = x_0, & x'(0) = x_1, \end{cases}$$

where $f : [0, \infty) \to X$ is continuous. Extend the result of Exercise 14 for this equation. Hint: use linearization.

17. Compute the general solution of the equation

$$x_{n+1} = Ax_n + u_n, \qquad n = 0, 1, 2, \dots .$$

18. By definition the *spectral radius* is the number given by

$$r(A) = \max\{|\lambda| \mid \lambda \in \sigma(A)\}.$$

Prove that

(a) $r(A) \leq \|A^n\|^{1/n}$ for $n = 1, 2, \dots$;

(b) $\lim_{n \to \infty} \|A^n\|^{1/n}$ exists (use that for any sequence (α_n) of non-negative numbers $\lim_{n \to \infty} \sqrt[n]{\alpha_n}$ exists if $\alpha_{n+m} \leq \alpha_n \alpha_m$ for all n and m; see Problem 98 in Part I of Pólya-Szegö [1]);

(c) $r(A) = \lim_{n \to \infty} \|A^n\|^{1/n}$.

19. Show that

$$\lim_{t \to \infty} \frac{1}{t} \log \|e^{At}\| = \max\{\Re\lambda \mid \lambda \in \sigma(A)\}.$$

20. Assume that $\sigma(A)$ belongs to the open unit circle. Put

$$\|\|x\|\| = \sup\{\|A^n x\| \mid n \geq 0\}.$$

Show that $\|\| \cdot \|\|$ is an equivalent norm on X, and relative to this norm A is a contraction. Redefine the norm $\|\| \cdot \|\|$ so that A becomes a strict contraction, i.e.,

$$\|\|A\|\| = \sup_{x \neq 0} \frac{\|\|Ax\|\|}{\|\|x\|\|} < 1.$$

If the underlying space is a Hilbert space, then one can choose a norm $\|\| \cdot \|\|$, defined via an inner product, such that A is a strict contraction relative to this Hilbert space norm.

21. Assume that in the operator norm $A^n \to P$ if $n \to \infty$. Prove that

(a) $AP = PA = P$ and $P^2 = P$;

(b) 1 is an isolated point of $\sigma(A)$ if $P \neq 0$.

Determine the principal part of the series expansion of $(\lambda - A)^{-1}$ at the point 1.

22. Consider the equation

$$\sum_{j,k=0}^{n} c_{jk} A^j Z B^k = C,$$

where $A : X \to X$, $B : Y \to Y$ and $C : Y \to X$ are given operators acting between Banach spaces and c_{jk} $(j, k = 0, \ldots, n)$ are constants. Assume that

$$p(\lambda, \mu) = \sum_{j,k=0}^{n} c_{jk} \lambda^j \mu^k \neq 0, \qquad (\lambda, \mu) \in \sigma(A) \times \sigma(B).$$

Show that the equation has a unique solution which has the form:

$$Z = \frac{-1}{4\pi^2} \int_{\Gamma_A} \int_{\Gamma_B} \frac{1}{p(\lambda, \mu)} (\lambda - A)^{-1} C (\mu - B)^{-1} d\lambda d\mu.$$

Here Γ_A and Γ_B are Cauchy contours around $\sigma(A)$ and $\sigma(B)$, respectively, such that $p(\lambda, \mu) \neq 0$ for $(\lambda, \mu) \in \Delta_A \times \Delta_B$, where Δ_A (resp. Δ_B) is the closure of the inner domain of Γ_A (resp. Γ_B).

23. Let $A = \sum_{j=1}^{N} \lambda_j P_j$, where $\lambda_1, \ldots, \lambda_N$ are different complex numbers and $P_1, \ldots, P_N$ are mutually disjoint projections such that $\sum_{j=1}^{N} P_j = I$. Consider the operator $\mathcal{J} : \mathcal{L}(X) \to \mathcal{L}(X)$ defined by

$$\mathcal{J}(S) = AS - SA, \qquad S \in \mathcal{L}(X).$$

Prove that

$$\mathrm{Ker}(\mathcal{J}) = \{S \in \mathcal{L}(X) \mid P_j S P_k = 0 \ (j \neq k)\},$$

$$\mathrm{Im}(\mathcal{J}) = \{S \in \mathcal{L}(X) \mid P_j S P_j = 0, \ j = 1, \dots, N\}.$$

Determine the spectrum of $\mathcal{J}$ and show that $\mathcal{J}$ can be written in the form $\mathcal{J} = \sum_{j=1}^{r} \mu_j \mathcal{R}_j$, where $\mathcal{R}_1, \dots, \mathcal{R}_r$ are mutually disjoint projections and $\mu_1, \dots, \mu_r$ are complex numbers.

24. Let $T : X \to Y$ and $S : Y \to X$ be operators acting between Banach spaces, and assume that $\mu \neq 0$ is an eigenvalue of finite type of ST. Prove that μ is also an eigenvalue of finite type of TS. Show that the operators $ST|\operatorname{Im} P_{\{\mu\}}(ST)$ and $TS|\operatorname{Im} P_{\{\mu\}}(TS)$ are similar. Are the above results also true if one allows μ to be zero?

25. Assume that for some polynomial p the operator $p(A)$ is compact. Show that $\lambda \in \rho(A)$ or λ is an eigenvalue of finite type whenever $p(\lambda) \neq 0$. If $p(\lambda) = 0$, does it follow that λ is not an eigenvalue of finite type? If yes, prove this statement; if no, give an example.

26. Prove that in the infinite dimensional case A always has spectral points that are not eigenvalues of finite type.

27. Let A and B be operators acting on the Banach space X, and assume that $\sigma(A) \cap \sigma(B) = \emptyset$. Prove that the operator functions

$$\begin{bmatrix} (\lambda - A)(\lambda - B) & 0 \\ 0 & I \end{bmatrix}, \qquad \begin{bmatrix} \lambda - A & 0 \\ 0 & \lambda - B \end{bmatrix}$$

are globally equivalent on $\mathbf{C}$.

28. Let $A_0, A_1, A_2, \dots$ be a bounded sequence in $\mathcal{L}(X)$, and consider the entire operator function

$$A(\lambda) = \sum_{j=0}^{\infty} \frac{1}{j!} \lambda^j A_j.$$

The aim of this exercise is to construct a "linearization" of $A(\cdot)$. Let $\ell_1\{X\}$ be the Banach space of all sequences $x = (x_0, x_1, x_2, \dots)$ with elements x_j in X such that $\|x\| = \sum_{j=0}^{\infty} \|x_j\| < \infty$. Define T and S on $\ell_1\{X\}$ by setting

$$T(x_0, x_1, x_2, \dots) = \left(\sum_{j=0}^{\infty} A_j x_j, x_1, x_2, \dots \right),$$

$$S(x_0, x_1, x_2, \dots) = \left(0, x_0, \frac{1}{2}x_1, \frac{1}{3}x_2, \dots \right).$$

Prove that T and S are bounded linear operators, $\sigma(S) = \{0\}$ and for a suitable Banach space Z the Z-extension of $A(\cdot)$ is equivalent on $\mathbf{C}$ to the linear function $T - \lambda S$.

      In the remaining exercises the underlying space X is assumed to be a Hilbert space and is denoted by H.

29. Assume that A is selfadjoint, and let λ_0 be an isolated point in $\sigma(A)$. Prove that

    (a) λ_0 is an eigenvalue of A,

    (b) the resolvent of A has a simple pole at λ_0,

    (c) the Riesz projection $P_{\{\lambda_0\}}(A)$ is an orthogonal projection.

30. Assume that A is selfadjoint, and let σ be an isolated part of $\sigma(A)$. Show that the Riesz projection $P_\sigma(A)$ is selfadjoint.

31. Assume that A is selfadjoint, and let $\{E(t)\}_{t\in\mathbf{R}}$ be the resolution of the identity for A. If N is an A-invariant subspace, then N is also invariant under $E(t)$ for each $t \in \mathbf{R}$. Prove this statement.

32. Assume that A is selfadjoint, and let $\{E(t)\}_{t\in\mathbf{R}}$ be the resolution of the identity for A. For $s < t$ let $\Omega_{s,t}$ be the operator defined by formula (8) in Section V.2. Show that

$$\Omega_{s,t} = (t - A)(s - A)\{E(t) - E(s)\}.$$

33. Assume that A is selfadjoint, and let $\{E(t)\}_{t\in\mathbf{R}}$ be the resolution of the identity for A. Prove the following statements:

    (a) If the sequence $(A^n x)$ converges for each $x \in H$, then $[m(A), M(A)] \subset [-1, 1]$ and -1 is not an eigenvalue of A.

    (b) If $0 < \delta \le 1$, then

$$\lim_{n\to\infty} \|A^n\{E(1 - \delta) - E(-1 + \delta)\}\| = 0.$$

    (c) For each $x \in H$

$$A^n\{E(1 - 0) - E(-1)\}x \to 0 \qquad (n \to \infty).$$

    (d) The sequence $(A^n x)$ is convergent for $x \in \mathrm{Im}\{E(1) - E(1 - 0)\}$.

    (e) If $[m(A), M(A)] \subset [-1, 1]$ and -1 is not an eigenvalue of A, then the sequence $(A^n x)$ is convergent for each $x \in H$. In this case the map P, defined by

$$Px = \lim_{n\to\infty} A^n x \qquad (x \in H),$$

is the orthogonal projection of H onto $\mathrm{Ker}(I + A)$.

PART II

CLASSES OF COMPACT OPERATORS

This part contains the basic elements of the theory of non-selfadjoint compact operators. The topics are singular values (Chapter VI), trace and determinant (Chapter VII), Hilbert-Schmidt operators (Chapter VIII), evaluation of the resolvent and completeness theorems (Chapters VII, VIII and X). Some of the results are illustrated for integral operators with semi-separable kernel functions (Chapter IX). One of the main aims is to obtain the results (whenever possible) by a limit procedure from the corresponding results for matrices.

CHAPTER VI
SINGULAR VALUES OF COMPACT OPERATORS

The singular values of a compact operator A are by definition the eigenvalues of $(A^*A)^{1/2}$. These numbers are important characteristics for compact operators. In this chapter the main properties of the singular values are studied. In the last sectionthe trace class operators are introduced. For the latter operators the trace and determinant will be defined in the next chapter.

VI.1 THE SINGULAR VALUES

In the study of compact operators one of the difficulties is that there are non-trivial compact operators which do not have any eigenvalue (see the example in Section III.4 of [GG]). To analyze such operators one cannot use the methods of the previous chapters. The singular values, which form the main topic of the present section, provide alternative tools.

Let $A\colon H_1 \to H_2$ be a compact operator acting between Hilbert spaces. Note that A^*A is a compact (positive) selfadjoint operator on H_1. Let

$$(1) \qquad \lambda_1(A^*A) \geq \lambda_2(A^*A) \geq \lambda_3(A^*A) \geq \cdots$$

be the sequence of non-zero eigenvalues of A^*A where each eigenvalue is repeated as many times as the value of its multiplicity. The number of non-zero eigenvalues of A^*A is finite if and only if A has finite rank and in that case the sequence (1) is extended by zero elements so that in all cases (1) is an infinite sequence. By definition, for $j = 1, 2, \ldots$ the j-th *singular value* or j-th *s-number* is the number $s_j(A) := \big(\lambda_j(A^*A)\big)^{1/2}$. By using the min-max theorem ([GG], Theorem III.9.1), one sees that

$$(2) \qquad s_j(A) = \min_{\dim M = j-1} \ \max_{0 \neq x \perp M} \frac{\|Ax\|}{\|x\|}, \qquad j \geq 1.$$

In (2) the symbol M stands for a finite dimensional subspace of H_1.

In what follows we shall assume that A is a compact operator on a Hilbert space H. Thus we take $H_1 = H_2 = H$. However all results can easily be extended to the case when the operators act between different spaces.

THEOREM 1.1. *A compact operator A on a Hilbert space H admits a representation of the form*

$$(3) \qquad A = \sum_{j=1}^{\nu(A)} s_j(A)\langle \cdot, \varphi_j\rangle \psi_j,$$

where $\nu(A)$ is the number of non-zero s-numbers of A (counted according to multi-plicities), $(\varphi_j)_{j=1}^{\nu(A)}$ and $(\psi_j)_{j=1}^{\nu(A)}$ are orthonormal systems in H and the series in (3) *converges in the operator norm if $\nu(A) = \infty$. Furthermore, if $B = \sum_{j=1}^{\nu} \alpha_j \langle \cdot, \varphi_j \rangle \psi_j$, where $(\varphi_j)_{j=1}^{\nu}$ and $(\psi_j)_{j=1}^{\nu}$ are orthonormal systems and $(\alpha_j)_{j=1}^{\nu}$ is a non-increasing sequence of positive numbers which converges to zero if $\nu = \infty$, then B is a compact operator and $s_j(B) = \alpha_j$, $1 \le j < \nu + 1$, are the non-zero s-numbers of B.*

PROOF. Write $s_j = s_j(A)$ for $j \ge 1$. From the spectral theorem for self-adjoint compact operators we know that there exists an orthonormal system $\varphi_1, \varphi_2, \ldots$ such that $A^*A = \sum_{j=1}^{\nu(A)} s_j^2 \langle \cdot, \varphi_j \rangle \varphi_j$. Put $\psi_j = s_j^{-1} A \varphi_j$ for each j. Then

$$
\begin{aligned}
\langle \psi_j, \psi_k \rangle &= s_j^{-1} s_k^{-1} \langle A\varphi_j, A\varphi_k \rangle \\
&= s_j^{-1} s_k^{-1} \langle A^* A\varphi_j, \varphi_k \rangle \\
&= s_j s_k^{-1} \langle \varphi_j, \varphi_k \rangle = \delta_{jk},
\end{aligned}
$$

and thus $\psi_1, \psi_2, \ldots$ is also an orthonormal system. Any $x \in H$ can be written in the form

$$
(4) \qquad x = \sum_{j=1}^{\nu(A)} \langle x, \varphi_j \rangle \varphi_j + u,
$$

where $u \in \operatorname{Ker} A^*A$. From $\langle A^*Au, u \rangle = \|Au\|^2$ it follows that $\operatorname{Ker} A^*A = \operatorname{Ker} A$. Furthermore, $A\varphi_j = s_j \psi_j$. Applying A to (4) now yields $Ax = \sum_{j=1}^{\nu(A)} s_j \langle x, \varphi_j \rangle \psi_j$. To prove the first part of the theorem it remains to show that the series in (3) converges in the operator norm if $\nu(A) = \infty$. Assume $\nu(A) = \infty$, and let $A_n = \sum_{j=1}^{n} s_j \langle \cdot, \varphi_j \rangle \psi_j$. Then for each $x \in H$

$$
\|(A - A_n)x\|^2 \le \sum_{j=n+1}^{\infty} s_j^2 |\langle x, \varphi_j \rangle|^2 \le \left(\sup_{j>n} s_j^2 \right) \|x\|^2,
$$

and thus $A_n \to A$ in the operator norm.

To prove the converse statement, let B be as in the theorem. If ν is finite, then B is of finite rank, and hence compact. If $\nu = \infty$, put $B_n = \sum_{j=1}^{n} \alpha_j \langle \cdot, \varphi_j \rangle \psi_j$. As in the previous paragraph one shows that $B_n \to B$ in the operator norm. Since B_n is of finite rank, it follows that B is compact. Next, one computes that $B^* = \sum_{j=1}^{\nu} \alpha_j \langle \cdot, \psi_j \rangle \varphi_j$, and so $B^*B = \sum_{j=1}^{\nu} \alpha_j^2 \langle \cdot, \varphi_j \rangle \varphi_j$. Thus $s_j(B) = \left(\lambda_j(B^*B) \right)^{1/2} = \alpha_j$. $\square$

We call (3) a *Schmidt-representation* of A. Since $s_j(A) = 0$ for $j > \nu(A)$, we can represent A also in the form $A = \sum_{j=1}^{\nu} s_j(A) \langle \cdot, \varphi_j \rangle \psi_j$, where $\nu = \dim H$ (which is finite or equal to ∞) and $(\varphi_j)_{j=1}^{\nu}$ and $(\psi_j)_{j=1}^{\nu}$ are orthonormal systems (which are extensions of the orthonormal systems in (3)). For the finite dimensional case the latter representation of A means that an $n \times n$ matrix A can be factored as

$$
A = U \begin{bmatrix} s_1 & & & 0 \\ & s_2 & & \\ & & \ddots & \\ 0 & & & s_n \end{bmatrix} V,
$$

where U and V are $n \times n$ unitary matrices and $s_1 \geq s_2 \geq \cdots \geq s_n$ are non-negative numbers which are precisely the first n singular values of the operator induced by the canonical action of the matrix A on $\mathbf{C}^n$. The above matrix factorization is called a *singular value decomposition* of the matrix A.

The infinite dimensional version of the singular value decomposition is the representation $A = UDV$. Here A is a compact operator on an infinite dimensional Hilbert space H, the middle term D is a diagonal operator on ℓ_2 with diagonal elements $s_1(A), s_2(A), \ldots$ and $V \colon H \to \ell_2$ and $U \colon \ell_2 \to H$ are bounded linear operators such that U^*U and VV^* are the identity operators on ℓ_2.

By taking adjoints in (3), we get

$$
(5) \qquad A^* = \sum_{j=1}^{\nu(A)} s_j(A) \langle \cdot, \psi_j \rangle \varphi_j,
$$

and, by the second part of Theorem 1.1, formula (5) is a Schmidt-representation of A^*. This yields the following corollary.

COROLLARY 1.2. *The operator A and its adjoint A^* have the same singular values.*

PROPOSITION 1.3. *If $A \colon H \to H$ is a compact operator and $B, C \colon H \to H$ are bounded linear operators, then*

$$
(6) \qquad s_j(BAC) \leq \|B\| \cdot \|C\| \cdot s_j(A), \qquad j \geq 1.
$$

PROOF. By duality, using Corollary 1.2, it suffices to show that $s_j(BA) \leq \|B\| s_j(A)$. But the latter inequality is an immediate consequence of the min-max description of the s-numbers in (2). $\square$

COROLLARY 1.4. *For any orthogonal projection P on H*

$$
(7) \qquad s_j(PA| \operatorname{Im} P) \leq s_j(A), \qquad j \geq 1.
$$

PROOF. It is easy to check that the operators $PA|\operatorname{Im} P \colon \operatorname{Im} P \to \operatorname{Im} P$ and $PAP \colon H \to H$ have the same singular values. Since $\|P\| \leq 1$, formula (6) implies that $s_j(PAP) \leq s_j(A)$. Thus (7) holds. $\square$

The following theorem gives another interpretation of the singular values, namely as approximation numbers.

THEOREM 1.5. *Let A be a compact operator on the Hilbert space H. Then for $n = 1, 2, \ldots$*

$$
s_n(A) = \min\{\|A - K\| \,|\, K \in \mathcal{L}(H), \operatorname{rank} K \leq n - 1\}.
$$

PROOF. Assume $\operatorname{rank} K = m \leq n - 1$. Then $\dim(\operatorname{Ker} K)^{\perp} = m$, and so by the min-max formula for the s-numbers

$$
s_n(A) \leq s_{m+1}(A) \leq \max_{0 \neq x \in \operatorname{Ker} K} \frac{\|Ax\|}{\|x\|}
$$

$$
= \max_{0 \neq x \in \operatorname{Ker} K} \frac{\|(A - K)x\|}{\|x\|} \leq \|A - K\|.
$$

So $s_n(A) \leq \|A-K\|$ for any K with rank $K \leq n-1$. It remains to prove that the infimum is attained and is equal to $s_n(A)$. To do this, let $A = \sum_{j=1}^{\nu(A)} s_j(A)\langle \cdot, \varphi_j \rangle \psi_j$ be a Schmidt-representation of A. Take $n < \nu(A) + 1$, and consider $K_n = \sum_{j=1}^{n-1} s_j(A)\langle \cdot, \varphi_j \rangle \psi_j$ (the operator $K_1 = 0$). Then rank $K_n = n - 1$ and

$$\|A - K_n\| \leq \sup_{j \geq n} s_j(A) = s_n(A).$$

Thus the minimum is attained in $K = K_n$ and is equal to $s_n(A)$. If $n > \nu(A)$, then rank A must be $\leq n - 1$ and $s_n(A) = 0$, and thus in this case the minimum is attained at $K = A$. $\square$

COROLLARY 1.6. *For compact operators A, B on H*

(8) $$|s_n(A) - s_n(B)| \leq \|A - B\|, \qquad n \geq 1.$$

PROOF. Take rank $K \leq n-1$. Then $s_n(A) \leq \|A-K\| \leq \|A-B\|+\|B-K\|$. It follows that $s_n(A) - \|A - B\|$ is a lower bound for $\|B - K\|$ when K runs over all operators of rank $\leq n-1$. According to Theorem 1.5 this implies that $s_n(A)-\|A-B\| \leq s_n(B)$, and thus $s_n(A) - s_n(B) \leq \|A - B\|$. Interchanging in the latter inequality the roles of A and B yields (8). $\square$

We compute the s-numbers for the operator of integration. Consider

$$V: L_2([0,1]) \to L_2([0,1]), \qquad (Vf)(t) = 2i \int_t^1 f(s)ds.$$

To find the s-numbers of V we have to determine the non-zero eigenvalues of V^*V. Note that $(V^*f)(t) = -2i \int_0^t f(s)ds$, and thus

$$(V^*Vf)(t) = 4 \int_0^t \left(\int_s^1 f(u)du \right) ds.$$

Take $\lambda > 0$. We want to find a non-zero solution of the equation $V^*Vf = \lambda f$. By putting $g = V^*Vf$ one checks that the equation $V^*Vf = \lambda f$ is equivalent to the following boundary value problem:

(9) $$\lambda g'' + 4g = 0, \qquad g(0) = 0, \qquad g'(1) = 0.$$

The general solution of the differential equation in (9) is equal to

$$g(t) = \gamma_1 \exp(2it\lambda^{-1/2}) + \gamma_2 \exp(-2it\lambda^{-1/2}).$$

The boundary condition at 0 implies that $\gamma_1 = -\gamma_2 \ (= \gamma)$, and thus $g(t) = 2i\gamma \sin(2t\lambda^{-1/2})$. Now recall that we want $g \neq 0$ (which implies $\gamma \neq 0$) and $g'(1) = 0$. Such a solution exists if and only if $\cos 2\lambda^{-1/2} = 0$. It follows that the non-zero eigenvalues of V^*V are precisely the numbers $\left(\frac{1}{4}(2k + 1)\pi\right)^{-2}$, $k = 0, 1, 2, \ldots$, and hence

(10) $$s_j(V) = \frac{4}{(2j - 1)\pi}, \qquad j = 1, 2, \ldots .$$

VI.2 EIGENVALUES AND s-NUMBERS

This section concerns various connections between eigenvalues and s-numbers of compact operators. As in Section II.3 we let $\lambda_1(A_1), \lambda_2(A), \ldots$ denote the sequence of non-zero eigenvalues of the compact operator A. The eigenvalues are ordered according to decreasing absolute values and multiplicities are taken into account. We shall extend the sequence $\lambda_1(A), \lambda_2(A), \ldots$ to an infinite sequence by adding zero elements if necessary. In particular, if A is a Volterra operator, then $\lambda_j(A) = 0$ for $j = 1, 2, \ldots$. If A acts on $H = \mathbf{C}^m$, then

$$(1) \qquad \prod_{j=1}^{m} |\lambda_j(A)| = \prod_{j=1}^{m} s_j(A).$$

To see this recall that $\det A = \prod_{j=1}^{m} \lambda_j(A)$, and thus

$$\prod_{j=1}^{m} s_j(A)^2 = \det A^*A = (\det A^*)\det A = |\det A|^2 = \prod_{j=1}^{m} |\lambda_j(A)|^2.$$

Besides (1) what other connections exist between eigenvalues and s-numbers in the finite dimensional case? We shall prove (see Theorem 2.1 below) that

$$(2) \qquad \prod_{j=1}^{n} |\lambda_j(A)| \leq \prod_{j=1}^{n} s_j(A), \qquad n = 1, \ldots, m-1.$$

It turns out that (1) and (2) give a complete description of all possible connections between eigenvalues and s-numbers of operators on $\mathbf{C}^m$. That is, if $\alpha_1 \geq \cdots \geq \alpha_m \geq 0$ and $\beta_1 \geq \cdots \geq \beta_m \geq 0$ are two sets of non-negative numbers which are related in the following way:

$$\prod_{j=1}^{n} \alpha_j \leq \prod_{j=1}^{n} \beta_j \quad (n = 1, \ldots, m-1), \qquad \prod_{j=1}^{m} \alpha_j = \prod_{j=1}^{m} \beta_j,$$

then there exists A on $\mathbf{C}^m$ such that $|\lambda_j(A)| = \alpha_j$ and $s_j(A) = \beta_j$ (see A. Horn [1]).

THEOREM 2.1. *For a compact operator A on a Hilbert space*

$$(3) \qquad \prod_{j=1}^{n} |\lambda_j(A)| \leq \prod_{j=1}^{n} s_j(A), \qquad n \geq 1.$$

PROOF. We use Schur's lemma (Lemma II.3.3). Let E_A be the smallest closed linear subspace of H spanned by all eigenvectors and generalized eigenvectors of A corresponding to non-zero eigenvalues of A. We know that E_A has an orthonormal basis $(\psi_j)_{j=1}^{\nu}$ such that $A\psi_k = \sum_{j=1}^{k} a_{jk}\psi_j$ with $a_{kk} = \lambda_k(A)$. If $n > \nu$ (and hence $\dim E_A$ is finite), then $\lambda_n(A) = 0$ and in that case the inequality (3) holds true trivially. Next, take $n < \nu+1$, and let M be the subspace spanned by $\psi_1, \ldots, \psi_n$. Since $\dim M = n$ is finite and $AM \subset M$, formula (1) implies that

$$\prod_{j=1}^{n} |\lambda_j(A)| = \prod_{j=1}^{n} |\lambda_j(A|M)| = \prod_{j=1}^{n} s_j(A|M).$$

Next, use Corollary 1.4, to show that $s_j(A|M) \leq s_j(A)$ and the proof is finished. $\square$

By using the next lemma, Theorem 2.1 serves as a source for many other useful inequalities between eigenvalues and s-numbers.

LEMMA 2.2. *Let $\varphi = \varphi(t_1, \ldots, t_n)$ be a real-valued differentiable function on an open set $\mathcal{D}$ of $\mathbf{R}^n$, and let Ω be a convex subset of $\mathcal{D}$ such that*

$$(4) \qquad \frac{\partial \varphi}{\partial t_1}(t) \geq \frac{\partial \varphi}{\partial t_2}(t) \geq \cdots \geq \frac{\partial \varphi}{\partial t_n}(t) \geq 0, \qquad t \in \Omega.$$

If $a = (a_1, \ldots, a_n)$ and $b = (b_1, \ldots, b_n)$ are points in Ω such that $\sum_{j=1}^k a_j \leq \sum_{j=1}^k b_j$ for $k = 1, 2, \ldots, n$, then $\varphi(a) \leq \varphi(b)$.

PROOF. Define $E: \mathbf{R}^n \to \mathbf{R}^n$ by setting $Et = (t_1, t_1 + t_2, \ldots, t_1 + t_2 + \cdots t_n)$. Then E is invertible and relative to the standard basis of $\mathbf{R}^n$ the operators E and E^{-1} are given by the matrices

$$E = \begin{bmatrix} 1 & & & 0 \\ 1 & 1 & & \\ \vdots & \vdots & \ddots & \\ 1 & 1 & \cdots & 1 \end{bmatrix}, \qquad E^{-1} = \begin{bmatrix} 1 & & & 0 \\ -1 & 1 & & \\ & \ddots & \ddots & \\ 0 & & -1 & 1 \end{bmatrix}.$$

Since Ω is convex, $\lambda b + (1 - \lambda)a \in \Omega$ for $0 \leq \lambda \leq 1$. Define $\psi: [0, 1] \to \mathbf{R}$ by $\psi(\lambda) = \varphi(\lambda b + (1 - \lambda)a)$. Then ψ is differentiable and

$$\psi'(\lambda) = \sum_{j=1}^n \frac{\partial \varphi}{\partial t_j}(\lambda b + (1 - \lambda)a)(b_j - a_j) = \langle y_\lambda, b - a \rangle$$

where $y_\lambda = \left(\frac{\partial \varphi}{\partial t_1}(t_\lambda), \ldots, \frac{\partial \varphi}{\partial t_n}(t_\lambda) \right)$ with $t_\lambda = \lambda b + (1 - \lambda)a$ and $\langle \cdot, \cdot \rangle$ is the inner product on $\mathbf{R}^n$. Using the operators E and E^{-1} we may write $\psi'(\lambda) = \langle (E^{-1})^* y_\lambda, E(b - a) \rangle$. Because of our conditions on a and b the coordinates of the vector $E(b - a)$ are all non-negative. Formula (4) implies that the same is true for $(E^{-1})^* y_\lambda$. It follows that $\psi'(\lambda) \geq 0$ for $0 \leq \lambda \leq 1$, and thus $\varphi(a) = \psi(0) \leq \psi(1) = \varphi(b)$. $\square$

COROLLARY 2.3. *Let $f: \mathbf{R} \to \mathbf{R}$ be twice differentiable, and assume that $f'(t) \geq 0$ and $f''(t) \geq 0$ for each $t \in \mathbf{R}$. Let $a_1 \geq a_2 \geq \cdots \geq a_n$ and $b_1 \geq b_2 \geq \cdots \geq b_n$ be two systems of real numbers such that $\sum_{j=1}^k a_j \leq \sum_{j=1}^k b_j$ for $k = 1, \ldots, n$. Then*

$$\sum_{j=1}^k f(a_j) \leq \sum_{j=1}^k f(b_j), \qquad k = 1, \ldots, n.$$

PROOF. Apply Lemma 2.2 with $\varphi(t_1, \ldots, t_n) = \sum_{j=1}^n f(t_j)$, the set $\mathcal{D} = \mathbf{R}^n$ and $\Omega = \{t \in \mathbf{R}^n \mid t_1 \geq t_2 \geq \cdots \geq t_n\}$. $\square$

We return to the connections between eigenvalues and s-numbers.

COROLLARY 2.4. *For a compact operator A on H and $p > 0$*

$$(5) \qquad \sum_{j=1}^n |\lambda_j(A)|^p \leq \sum_{j=1}^n s_j(A)^p, \qquad n \geq 1.$$

PROOF. Choose ν such that $\lambda_j(A) \neq 0$ for $j < \nu + 1$ and $\lambda_j(A) = 0$ for $j \geq \nu + 1$. (Note that ν may be finite or infinite and $\nu \leq \nu(A)$.) It suffices to prove (5) for $1 \leq n < \nu + 1$. Put $a_j = \log|\lambda_j(A)|$ and $b_j = \log s_j(A)$. Then $a_1 \geq a_2 \geq \cdots \geq a_n$ and $b_1 \geq b_2 \geq \cdots \geq b_n$, and Theorem 2.1 yields that $\sum_{j=1}^{k} a_j \leq \sum_{j=1}^{k} b_j$ for $k = 1, 2, \ldots, n$. Now apply Corollary 2.3 with $f(t) = e^{pt}$, and (5) follows. $\square$

COROLLARY 2.5. *For a compact operator A on H and $r > 0$*

$$(6) \qquad \prod_{j=1}^{n}\bigl(1 + r|\lambda_j(A)|\bigr) \leq \prod_{j=1}^{n}\bigl(1 + rs_j(A)\bigr), \qquad n \geq 1.$$

PROOF. As in the proof of Corollary 2.4, put $a_j = \log|\lambda_j(A)|$, $b_j = \log s_j(A)$, and apply Corollary 2.3 with $f(t) = \log(1 + re^t)$. $\square$

COROLLARY 2.6. *Let A be a compact operator on H, and let n be a positive integer. Then*

$$(7) \qquad \sum_{1 \leq j_1 < \cdots < j_n \leq r} |\lambda_{j_1}(A) \cdots \lambda_{j_n}(A)| \leq \sum_{1 \leq j_1 < \cdots < j_n \leq r} s_{j_1}(A) \cdots s_{j_n}(A), \qquad r \geq n.$$

PROOF. Choose ν as in the proof of Corollary 2.4. It suffices to prove (7) for $r < \nu + 1$. Take $\mathcal{D} = \mathbf{R}^r$ and $\Omega = \{t \in \mathbf{R}^r \mid t_1 \geq \cdots \geq t_r\}$. Consider on $\mathcal{D}$ the function

$$\varphi(t_1, \ldots, t_r) = \sum_{1 \leq j_1 < \cdots < j_n \leq r} \exp(t_{j_1} + \cdots + t_{j_n}).$$

Then Ω is a convex subset of $\mathcal{D}$ and for $t \in \Omega$ the inequalities (4) hold true. Theorem 2.1 allows us to apply Lemma 2.2 with $a_j = \log|\lambda_j(A)|$ and $b_j = \log s_j(A)$, $j = 1, \ldots, n$. This yields the desired inequality. $\square$

VI.3 FURTHER PROPERTIES OF SINGULAR VALUES

THEOREM 3.1. *For a compact operator A on a Hilbert space H and $n = 1, 2, \ldots \; (\leq \dim H)$*

$$(1) \qquad \sum_{j=1}^{n} s_j(A) = \max_{U, \varphi_1, \ldots, \varphi_n} \left| \sum_{j=1}^{n} \langle U A \varphi_j, \varphi_j \rangle \right|,$$

where the maximum is taken over all unitary operators U on H and all orthonormal systems $\varphi_1, \ldots, \varphi_n$ in H.

PROOF. Let U be a unitary operator on H, and let $\varphi_1, \ldots, \varphi_n$ be an orthonormal system in H. Define P to be the orthogonal projection onto $\mathrm{span}\{\varphi_1, \ldots, \varphi_n\}$. Put $F = PUAP|\,\mathrm{Im}\,P \colon \mathrm{Im}\,P \to \mathrm{Im}\,P$. Since $\mathrm{Im}\,P$ is finite dimensional, we know that

$$\mathrm{tr}\, F = \sum_{j=1}^{n} \langle PUAP\varphi_j, \varphi_j \rangle = \sum_{j=1}^{n} \langle U A \varphi_j, \varphi_j \rangle.$$

But $\operatorname{tr} F = \sum_{j=1}^{n} \lambda_j(F)$, and so, using Corollaries 2.4, 1.4 and Proposition 1.3, we see that

$$\left| \sum_{j=1}^{n} \langle U A \varphi_j, \varphi_j \rangle \right| \leq \sum_{j=1}^{n} |\lambda_j(F)| \leq \sum_{j=1}^{n} s_j(F)$$

$$\leq \sum_{j=1}^{n} s_j(A).$$

So $\sum_{j=1}^{n} s_j(A)$ is an upper bound for the numbers $|\sum_{j=1}^{n} \langle U A \varphi_j, \varphi_j \rangle|$. It remains to show that this upper bound is attained.

To do this, write $A = \sum_{j=1}^{\nu} s_j(A) \langle \cdot, \widetilde{\varphi}_j \rangle \widetilde{\psi}_j$, where $(\widetilde{\varphi}_j)_{j=1}^{\nu}$ and $(\widetilde{\psi}_j)_{j=1}^{\nu}$ are orthonormal systems and $n \leq \nu$. Choose a unitary operator $\widetilde{U} : H \to H$ such that $\widetilde{U} \widetilde{\psi}_j = \widetilde{\varphi}_j$, $j = 1, \ldots, n$. Then

$$\langle \widetilde{U} A \widetilde{\varphi}_j, \widetilde{\varphi}_j \rangle = \langle A \widetilde{\varphi}_j, \widetilde{\psi}_j \rangle = s_j(A),$$

and it is clear that in (1) the maximum is attained for $U = \widetilde{U}$ and $\varphi_j = \widetilde{\varphi}_j$ $(j = 1, \ldots, n)$.
$\square$

The absolute value in the right hand side of (1) can also be put after the summation sign, that is,

$$(2) \qquad \sum_{j=1}^{n} s_j(A) = \max_{U, \varphi_1, \ldots, \varphi_n} \sum_{j=1}^{n} |\langle U A \varphi_j, \varphi_j \rangle|,$$

where as in (1) the maximum is taken over all unitary operators U on H and all orthonormal systems $\varphi_1, \ldots, \varphi_n$. To see this, fix U and $\varphi_1, \ldots, \varphi_n$. For $j = 1, \ldots n$ let θ_j be the argument of $\langle U A \varphi_j, \varphi_j \rangle$. Introduce a new unitary operator U_0 on H by setting $U_0^* \varphi_j = e^{i\theta_j} U^* \varphi_j$, $j = 1, \ldots, n$, and $U_0^* x = U^* x$ for $x \perp \operatorname{span}\{\varphi_1, \ldots, \varphi_n\}$. Then

$$\left| \sum_{j=1}^{n} \langle U A \varphi_j, \varphi_j \rangle \right| \leq \sum_{j=1}^{n} |\langle U A \varphi_j, \varphi_j \rangle|$$

$$= \sum_{j=1}^{n} e^{-i\theta_j} \langle U A \varphi_j, \varphi_j \rangle$$

$$= \sum_{j=1}^{n} \langle U_0 A \varphi_j, \varphi_j \rangle \leq \sum_{j=1}^{n} s_j(A),$$

and hence (2) follows from (1).

COROLLARY 3.2. *If A and B are compact operators on the Hilbert space H, then*

$$(3) \qquad \sum_{j=1}^{n} s_j(A + B) \leq \sum_{j=1}^{n} s_j(A) + \sum_{j=1}^{n} s_j(B), \qquad n \geq 1.$$

PROOF. Choose a unitary operator U on H and an orthonormal system $\varphi_1, \ldots, \varphi_n$ in H such that

$$\sum_{j=1}^{n} s_j(A+B) = \left| \sum_{j=1}^{n} \langle U(A+B)\varphi_j, \varphi_j \rangle \right|.$$

Such a choice can be made by Theorem 3.1. It follows (again using Theorem 3.1) that

$$\begin{aligned}
\sum_{j=1}^{n} s_j(A+B) &= \left| \sum_{j=1}^{n} \langle UA\varphi_j, \varphi_j \rangle + \sum_{j=1}^{n} \langle UB\varphi_j, \varphi_j \rangle \right| \\
&\leq \left| \sum_{j=1}^{n} \langle UA\varphi_j, \varphi_j \rangle \right| + \left| \sum_{j=1}^{n} \langle UB\varphi_j, \varphi_j \rangle \right| \\
&\leq \sum_{j=1}^{n} s_j(A) + \sum_{j=1}^{n} s_j(B). \quad \square
\end{aligned}$$

By taking U in (2) to be the identity operator on H one obtains the following corollary.

COROLLARY 3.3. *Let A be a compact operator on H, and let $\varphi_1, \ldots, \varphi_n$ be an orthonormal system in H. Then*

$$(4) \qquad \sum_{j=1}^{n} |\langle A\varphi_j, \varphi_j \rangle| \leq \sum_{j=1}^{n} s_j(A), \qquad n \geq 1.$$

Inequality (4) is an interesting inequality, even in the finite dimensional case. In the finite dimensional case (4) describes all the connections between the diagonal elements relative to an orthonormal basis and the s-numbers (see Mirsky [1]).

VI.4 TRACE CLASS OPERATORS

We want a reasonable class of operators for which a trace and a determinant can be defined. For an operator A acting on a finite dimensional Hilbert space H there is no problem:

$$\operatorname{tr} A := \sum_{j=1}^{n} \langle A\varphi_j, \varphi_j \rangle,$$

$$\det(I+A) := \det(\delta_{ij} + \langle A\varphi_j, \varphi_i \rangle)_{i,j=1}^{n},$$

where $\varphi_1, \ldots, \varphi_n$ is an arbitrary orthonormal basis in H. Choose now $\varphi_1, \ldots, \varphi_n$ such that the matrix of A with respect to $\varphi_1, \ldots, \varphi_n$ has an upper triangular form. Then the diagonal elements of the matrix are preciesly the eigenvalues (multiplicities taken into account) of A, and thus

$$(1) \qquad \operatorname{tr} A = \sum_{j} \lambda_j(A), \qquad \det(I+A) = \prod_{j} \big(1 + \lambda_j(A)\big).$$

In the infinite dimensional case the number of eigenvalues may be infinite, and hence in that case one has to worry about convergence in (1).

For a Hilbert space H we define

$$S_1 := \left\{ A \colon H \to H \mid A \text{ compact} , \sum_{j=1}^{\infty} s_j(A) < \infty \right\}.$$

The elements of S_1 are called the *trace class operators* on H. Note that because of Corollaries 2.4 and 2.5 for any trace class operator the series and infinite product in (1) converge. It turns out (as we shall see in the next chapter) that one can define $\operatorname{tr} A$ and $\det(I + A)$ for $A \in S_1$ and that for such an operator (1) holds true. In this section we give the first properties of the class S_1.

THEOREM 4.1. *The space S_1 endowed with the norm*

$$(2) \qquad \|A\|_1 = \sum_{j=1}^{\infty} s_j(A)$$

is a complex Banach space. The finite rank operators are dense in S_1 relative to the norm (2). Furthermore, $\|A\|_1 \geq \|A\|$ and equality holds if and only if A has rank ≤ 1.

PROOF. From $s_1(A) = \|A\|$ it follows that $\|A\|_1 \geq \|A\|$, and hence $\|A\|_1 = 0$ implies $A = 0$. Conversely, if $A = 0$, then all s-numbers of A are zero, and hence $\|A\|_1 = 0$. So $\|A\|_1 = 0$ if and only if $A = 0$.

From $s_j(\alpha A) = |\alpha| s_j(A)$ for each $j \geq 1$ and Corollary 3.2 it is clear that S_1 is a linear space over $\mathbb{C}$ and (using the result of the previous paragraph) we also see that $\|\cdot\|_1$ is a norm on S_1. Let us prove the completeness. Let $(A_n)_n$ be a Cauchy sequence in $(S_1, \|\cdot\|_1)$. From $\|A\|_1 \geq \|A\|$ we conclude that $(A_n)_n$ is a Cauchy sequence in the usual operator norm. But $\mathcal{L}(H)$ with the operator norm is complete, and so in the operator norm the sequence $(A_n)_n$ converges to some operator A. Since all A_n are compact, the same is true for A. Given $\varepsilon > 0$ there exists a positive integer $N = N(\varepsilon)$ such that

$$(3) \qquad \sum_{j=1}^{k} s_j(A_n - A_m) \leq \|A_n - A_m\|_1 < \varepsilon \qquad (n, m \geq N).$$

Here k is an arbitrary positive integer. Next use that the s-numbers are continuous in the operator norm (Corollary 1.6), and take the limit for $m \to \infty$ in the left hand side of (3). It follows that for each $k \geq 1$

$$\sum_{j=1}^{k} s_j(A_n - A) \leq \varepsilon \qquad (n \geq N),$$

and hence $\|A_n - A\|_1 \leq \varepsilon$ for $n \geq N$. This shows that the sequence $(A_n)_n$ converges to A in $\|\cdot\|_1$ and $(S_1, \|\cdot\|_1)$ is complete.

Obviously, the finite rank operators are in S_1. To see that they are dense in $(S_1, \|\cdot\|_1)$, let $A \in S_1$ have infinite rank. Consider its Schmidt-representation $A = \sum_{j=1}^{\infty} s_j(A)\langle\cdot, \varphi_j\rangle\psi_j$, and put $A_n = \sum_{j=1}^{n} s_j(A)\langle\cdot, \varphi_j\rangle\psi_j$. Then

$$A - A_n = \sum_{j=n+1}^{\infty} s_j(A)\langle\cdot, \varphi_j\rangle\psi_j,$$

and thus $\|A - A_n\|_1 = \sum_{j=n+1}^{\infty} s_j(A) \to 0$ if $n \to \infty$.

We already know that $\|A\|_1 \geq \|A\|$. If $\|A\|_1 = \|A\|$, then $s_j(A) = 0$ for $j \geq 2$, and from the Schmidt-representation of A it is clear that $\operatorname{rank} A \leq 1$. Conversely, if $\operatorname{rank} A \leq 1$, then we can use Theorem 1.5 to show that $s_j(A) = 0$ for $j \geq 2$, and hence $\|A\|_1 = \|A\|$. $\square$

The norm $\|\cdot\|_1$ is called the *trace class norm*. If the underlying Hilbert space H is separable, then the same is true for $(S_1, \|\cdot\|_1)$. To see this, we first note that for each x, y in H

$$(4) \qquad \|\langle\cdot, x\rangle y\|_1 = \|x\|\|y\|,$$

because the trace class norm and the operator norm coincide for rank one operators. Thus for two finite rank operators $F = \sum_{j=1}^{n}\langle\cdot, x_j\rangle y_j$ and $\widetilde{F} = \sum_{j=1}^{n}\langle\cdot, \widetilde{x}_j\rangle\widetilde{y}_j$ we have

$$\|F - \widetilde{F}\|_1 \leq \sum_{j=1}^{n} \|\langle\cdot, x_j\rangle y_j - \langle\cdot, \widetilde{x}_j\rangle\widetilde{y}_j\|_1$$

$$\leq \sum_{j=1}^{n} \|\langle\cdot, x_j\rangle(y_j - \widetilde{y}_j)\|_1 + \sum_{j=1}^{n} \|\langle\cdot, x_j - \widetilde{x}_j\rangle\widetilde{y}_j\|_1$$

$$= \sum_{j=1}^{n} \|x_j\| \cdot \|y_j - \widetilde{y}_j\| + \sum_{j=1}^{n} \|x_j - \widetilde{x}_j\|\|\widetilde{y}_j\|$$

$$\leq \sum_{j=1}^{n} (\|x_j\| \cdot \|y_j - \widetilde{y}_j\| + \|x_j - \widetilde{x}_j\|\|\widetilde{y}_j - y_j\| + \|x_j - \widetilde{x}_j\|\|y_j\|)$$

Now assume $H = \ell_2$, and consider the set $\mathcal{F}_\mathbf{Q}$ of all finite rank operators $F = \sum_{j=1}^{n}\langle\cdot, x_j\rangle y_j$, where n is arbitrary and the vectors $x_1, \ldots, x_n$ and $y_1 \ldots, y_n$ are ℓ_2-sequences of the form $(a_\nu + ib_\nu)_{\nu=1}^{\infty}$ with a_ν and b_ν rational for all ν. From what we proved so far it is clear that in the trace class norm the set $\mathcal{F}_\mathbf{Q}$ is dense in the set of all finite rank operators, and hence in the trace class norm $\mathcal{F}_\mathbf{Q}$ is dense in S_1. Since $\mathcal{F}_\mathbf{Q}$ is countable, this proves that S_1 is separable when $H = \ell_2$. But any separable Hilbert space is isometrically isomorphic to ℓ_2, and and thus S_1 is separable if H is separable.

The next proposition shows that S_1 is an ideal in the ring of all bounded linear operators on H.

PROPOSITION 4.2. *Let $A \in S_1$, and let B and C be bounded linear operators on H. Then $BAC \in S_1$ and*

$$(5) \qquad \|BAC\|_1 \leq \|B\|\|A\|_1\|C\|.$$

PROOF. Apply Proposition 1.3. $\square$

To give a non-trivial example of a trace class operator, consider, on $L_2([a,b])$, the integral operator

$$Kf = \int_a^b k(\cdot,s)f(s)ds.$$

If $k(\cdot,\cdot)$ is continuous and the operator K is non-negative, then K is a trace class operator. Indeed, since K is non-negative, $s_j(K) = \lambda_j(K)$ for $j \geq 1$. But then it follows from Theorem IV.4.1 in [GG] (which is a corollary of Mercer's theorem) that

$$\sum_{j=1}^{\infty} s_j(K) = \sum_{j=1}^{\infty} \lambda_j(K) = \int_a^b k(s,s)ds < \infty.$$

Proposition 4.2 allows us to produce many other examples of trace class operators. Consider on $L_2([a,b])$ the integral operator

$$Af = \int_a^b k(\cdot,s)\alpha(s)f(s)ds,$$

where $\alpha(\cdot)$ and $k(\cdot,\cdot)$ are continuous functions and $\int_a^b \int_a^b k(t,s)f(s)\overline{f(t)}dsdt \geq 0$ for each $f \in L_2([a,b])$. Then A is a trace class operator. Indeed, $A = KM$, where M is the operator of multiplication by $\alpha(\cdot)$ on $L_2([a,b])$ and K is the non-negative trace class operator considered in the previous paragraph. Hence $A = KM \in S_1$.

The operator of integration, $(Vf)(t) = 2i \int_t^1 f(s)ds$, on $L_2([0,1])$ is not a trace class operator, since formula (10) in Section 1 gives

$$\sum_{j=1}^{\infty} s_j(V) = \sum_{j=1}^{\infty} \frac{4}{(2j-1)\pi} = \infty.$$

The next theorem will turn out to be useful later.

THEOREM 4.3. *Let $A \in S_1$, and let (T_n) and (S_n) be sequences of bounded linear operators on H which converge pointwise to T and S, respectively (i.e., $T_n x \to Tx$ and $S_n x \to Sx$ for each $x \in H$). Then $T_n A S_n^* \to TAS^*$ in the trace class norm.*

PROOF. First, assume that $A = \langle\cdot,\varphi\rangle\psi$. Then $T_n A S_n^* = \langle\cdot, S_n\varphi\rangle T_n\psi$, and thus

$$\|T_n A S_n^* - TAS^*\|_1 \leq \|S_n\varphi - S\varphi\| \cdot \|T\psi\| + \|S\varphi\| \cdot \|T_n\psi - T\psi\| + \|S_n\varphi - S\varphi\| \cdot \|T_n\psi - T\psi\|.$$

Because of the pointwise convergence we see that $T_n A S_n^* \to TAS^*$. By taking finite linear combinations we may conclude that the theorem is proved for the case when A is a finite rank operator. The general case will be proved by approximation.

Take $A \in S_1$. Let $\varepsilon > 0$ be given. Since $T_n x \to Tx$ and $S_n x \to Sx$ for each $x \in H$, the uniform boundedness principle shows that there exists a positive constant $\gamma < \infty$ such that $\|T_n\| \leq \gamma$ and $\|S_n\| \leq \gamma$ for all $n \geq 1$. It follows that also $\|T\| \leq \gamma$ and $\|S\| \leq \gamma$. Choose a finite rank operator F such that $\|A - F\|_1 < \varepsilon/3\gamma^2$. Using the triangle inequality, Proposition 4.2 and $\|B^*\| = \|B\|$, one sees that

$$\|T_n A S_n^* - TAS^*\|_1 \leq \|T_n\|\|A - F\|_1\|S_n^*\| + \|T_n F S_n^* - TFS^*\|_1 + \|T\|\|A - F\|_1\|S^*\|$$

$$\leq \frac{2}{3}\varepsilon + \|T_n F S_n^* - TFS^*\|_1.$$

Next, apply the result of the first paragraph of the proof. So there exists a positive integer N such that $\|T_n F S_n^* - TFS^*\|_1 < \frac{1}{3}\varepsilon$ for $n \geq N$. But then $\|T_n A S_n^* - TAS^*\|_1 < \varepsilon$ for $n \geq N$. $\square$

CHAPTER VII
TRACE AND DETERMINANT

The first section of this chapter has an introductory character; it explains the principles we use to define the trace and determinant. The precise definitions are given in the next two sections where we also derive the first properties of the trace and determinant. In Section 4 the analyticity of $\det(I - \lambda A)$ as a function of λ is proved. The main theorem is given in the sixth section and expresses trace and determinant in terms of the eigenvalues. Some technical results from complex function theory, which are used in the proof of the main theorem, are derived in Section 5. The connections with the classical Fredholm determinant are described in Section 7. The last section contains as a first application two completeness theorems for eigenvectors and generalized eigenvectors.

VII.1 INTRODUCTION

Throughout this chapter we assume, for simplicity, that *the underlying Hilbert spaces are separable.* To introduce trace and determinant for trace class operators we follow two principles, namely that of permanency and of continuity. The principle of permanency means that trace and determinant will be introduced in such a way that for operators on finite dimensional spaces the new definitions agree with what is already known from matrix theory. This leads in a natural way to definitions of trace and determinant for operators of finite rank. Next we use continuity to extend the definitions of trace and determinant to all trace class operators.

Let us start with a finite rank operator $F\colon H \to H$ acting on the Hilbert space H. Given F there exists a finite dimensional subspace H_1 of H such that

$$(1) \qquad\qquad FH_1 \subset H_1, \qquad H_1^{\perp} \subset \operatorname{Ker} F.$$

Indeed, if $F = \sum_{j=1}^{n} \langle \cdot, \varphi_j \rangle \psi_j$, then $H_1 = \operatorname{span}\{\varphi_1, \cdots, \varphi_n, \psi_1, \ldots, \psi_n\}$ has the desired properties. It follows that with respect to the decomposition $H = H_1 \oplus H_1^{\perp}$ the operators F and $I + F$ can be written as 2×2 operator matrices of the following form:

$$(2) \qquad\qquad F = \begin{bmatrix} F_1 & 0 \\ 0 & 0 \end{bmatrix}, \qquad I + F = \begin{bmatrix} I_1 + F_1 & 0 \\ 0 & I_2 \end{bmatrix}.$$

Here F_1 is the restriction of F to the finite dimensional subspace H_1 and for $i = 1, 2$ the symbol I_i stands for the identity operator H_i, where $H_2 = H_1^{\perp}$. Since H_1 is a finite dimensional space, $\operatorname{tr} F_1$ and $\det(I_1 + F_1)$ are well-defined. We use this and define

$$(3) \qquad\qquad \operatorname{tr} F := \operatorname{tr} F_1, \qquad \det(I + F) := \det(I_1 + F_1).$$

Note that the definitions do not depend on the particular choice of the space H_1. To see this first of all recall that

$$(4) \qquad\qquad \operatorname{tr} F_1 = \sum_{j=1}^{n} \lambda_j, \qquad \det(I_1 + F_1) = \prod_{j=1}^{n} (1 + \lambda_j),$$

where $n = \dim H_1$ and $\lambda_1, \ldots, \lambda_n$ are the eigenvalues of F_1 counted according to their algebraic multiplicities. In the two identities in (4) the zero eigenvalues do not give a contribution. Furthermore, it is clear from (2) that the non-zero eigenvalues of F_1 coincide with the non-zero eigenvalues of F (multiplicities taken into account). Thus

$$(5) \qquad \operatorname{tr} F = \sum_j \lambda_j(F), \qquad \det(I + F) = \prod_j (1 + \lambda_j(F)),$$

VII.2 DEFINITION OF THE TRACE

On the operators of finite rank the trace acts as a linear functional which is continuous in the trace class norm. Indeed, let F and G be operators of finite rank acting on the Hilbert space H. Then

$$(1a) \qquad \operatorname{tr}(F + G) = \operatorname{tr} F + \operatorname{tr} G,$$

$$(1b) \qquad \operatorname{tr}(\alpha F) = \alpha \operatorname{tr} F \qquad (\alpha \in \mathbb{C}),$$

$$(1c) \qquad |\operatorname{tr} F - \operatorname{tr} G| \le \|F - G\|_1.$$

To prove (1a) and (1b) we use the corresponding properties of the trace on matrices. Choose finite dimensional subspaces H_1 and H_2 such that $FH_1 \subset H_1$, $H_1^\perp \subset \operatorname{Ker} F$, $GH_2 \subset H_2$ and $H_2^\perp \subset \operatorname{Ker} G$. Put $H_0 = H_1 + H_2$. Then H_0 is finite dimensional and

$$(2) \qquad (F + G)H_0 \subset H_0, \qquad H_0^\perp \subset (\operatorname{Ker} F) \cap (\operatorname{Ker} G) \subset \operatorname{Ker}(F + G).$$

Thus

$$\operatorname{tr}(F + G) = \operatorname{tr}((F + G)|H_0) = \operatorname{tr} F|H_0 + \operatorname{tr} G|H_0 = \operatorname{tr} F + \operatorname{tr} G.$$

This proves (1a). Further

$$\operatorname{tr}(\alpha F) = \operatorname{tr}(\alpha F)|H_1 = \operatorname{tr} \alpha(F|H_1) = \alpha \operatorname{tr} F|H_1 = \alpha \operatorname{tr} F.$$

To prove (1c) we use the linearity of the trace and Corollary VI.2.4 to obtain

$$(3) \qquad |\operatorname{tr} F| \le \sum_j |\lambda_j(F)| \le \sum_j s_j(F) = \|F\|_1.$$

The continuity of the trace on operators of finite rank allows us to extend the definition of the trace to all trace class operators. In fact, given a trace class operator A we define

$$(4) \qquad \operatorname{tr} A := \lim_{n \to \infty} \operatorname{tr} F_n,$$

where $F_1, F_2, F_3, \ldots$ is an arbitrary sequence of finite rank operators converging in trace class norm to A. Such a sequence always exists because the finite rank operators are dense in S_1. Note that formula (1c) guarantees that the limit exists and does not depend on the particular choice of the sequence $F_1, F_2, \ldots$. From definition (4) it is clear that formulas (1a), (1b), (1c) and (3) also hold for arbitrary trace class operators. So we have the following theorem.

THEOREM 2.1. *On the trace operators the trace is a linear functional which is continuous in the trace class norm and*

$$|\operatorname{tr} A| \leq \|A\|_1 \qquad (A \in S_1).$$

The trace of a square matrix is equal to the sum of its diagonal elements. The next theorem is the infinite dimensional analogue of this result.

THEOREM 2.2. *Let A be a trace class operator on H. Then*

$$(5) \qquad \operatorname{tr} A = \sum_{j=1}^{\infty} \langle A\varphi_j, \varphi_j \rangle$$

for any orthonormal basis $\varphi_1, \varphi_2, \ldots$ of H.

PROOF. Let P_n be the orthogonal projection on the space spanned by $\varphi_1, \ldots, \varphi_n$. Then $P_n x \to x$ for each x. It follows (cf. Theorem VI.4.3) that $P_n A P_n \to A$ in trace class norm, and thus, by the definition of the trace,

$$(6) \qquad \operatorname{tr} A = \lim_{n \to \infty} \operatorname{tr} P_n A P_n.$$

From matrix theorem we know that $\operatorname{tr} P_n A P_n = \sum_{j=1}^{n} \langle A\varphi_j, \varphi_j \rangle$. Inserting this in (6) yields the desired formula (5). $\square$

For a finite rank operator F we know that $\operatorname{tr} F = \sum_j \lambda_j(F)$. This identity also holds for an arbitrary trace class operator, but at this stage it cannot be obtained by a simple continuity argument. We shall prove the equality in Section 6 (see Theorem 6.1).

According to Theorem 2.2, the trace of A is precisely the sum of the diagonal elements in the matrix A relative to an orthonormal basis. The next theorem may be viewed as a continuous analogue of this result.

THEOREM 2.3. *Let A be the integral operator on $L_2([a, b])$ defined by*

$$(7) \qquad (Af)(t) = \int_a^b k(t, s) f(s) ds, \qquad a \leq t \leq b,$$

and assume that kernel function k is continuous on $[a, b] \times [a, b]$. If, in addition, A is a trace class operator, then

$$(8) \qquad \operatorname{tr} A = \int_a^b k(t, t) dt.$$

PROOF. If the operator A is non-negative, then the theorem follows immediately from Mercer's theorem, see [GG], Theorem IV.4.1. The proof of the general case also uses Mercer's theorem; in fact, it is reduced to Mercer's theorem by using the averaging operator M_h $(h > 0)$, which on $L_2([a, b])$ is defined by

$$(9) \qquad (M_h f)(t) = \int_a^b \Delta_h(t - s) f(s), \qquad a \leq t \leq b,$$

where

$$\Delta_h(t) = \begin{cases} (2h)^{-1} & \text{for } |t| \leq h, \\ 0 & \text{for } |t| > h. \end{cases}$$

Obviously, M_h is a selfadjoint integral operator on $L_2([a, b])$ and its range consists of continuous functions only. Furthermore

$$(10) \qquad \lim_{h \downarrow 0} \|M_h f - f\|_2 = 0, \qquad f \in L_2([a, b]).$$

Here $\| \cdot \|_2$ denotes the usual norm on $L_2([a, b])$.

To prove (10), take $f \in L_2([a, b])$. It will be convenient to consider f on $\mathbf{R}$ by setting $f(t) = 0$ for $t \notin [a, b]$. Put $f_\alpha(t) = f(t - \alpha)$. Now let $\varepsilon > 0$ be given. Then (see [R], Theorem 9.5) there exists $\delta > 0$ such that $\|f_\alpha - f\|_2 < \varepsilon$ for $|\alpha| < \delta$. Choose $|h| < \delta$. Then, for each $g \in L_2([a, b])$, we have

$$\int_a^b |(M_h f)(t) - f(t)||g(t)|dt$$

$$= \int_a^b \left| \int_a^b \Delta_h(\alpha)\big(f(t - \alpha) - f(t)\big)\, d\alpha \right| |g(t)|dt$$

$$\leq \int_a^b \int_a^b \Delta_h(\alpha)|f_\alpha(t) - f(t)||g(t)|d\alpha dt$$

$$= \int_a^b \Delta_h(\alpha)\left(\int_a^b |f_\alpha(t) - f(t)||g(t)|dt \right) d\alpha$$

$$\leq \frac{1}{2h} \int_{-h}^h \|f_\alpha - f\|_2 \|g\|_2 d\alpha < \varepsilon \|g\|_2.$$

It follows that $\|M_h f - f\|_2 < \varepsilon$ for $|h| < \delta$, which proves (10).

For any $g \in L_2([a, b] \times [a, b])$ we define

$$(11) \qquad g_h(t, s) = \frac{1}{4h^2} \int_{t-h}^{t+h} \int_{s-h}^{s+h} g(\alpha, \beta)d\alpha d\beta.$$

In (11) the function g is assumed to be zero outside $[a, b] \times [a, b]$. Note that g_h is continuous on $[a, b] \times [a, b]$. Let G and G_h be the integral operators on $L_2([a, b])$ with kernel functions g and g_h, respectively. Since

$$g_h(t, s) = \int\limits_a^b \int\limits_a^b \Delta(t - \alpha) g(\alpha, \beta) \Delta(\beta - s) d\alpha d\rho,$$

we see that $G_h = M_h G M_h$. We shall prove that for any $h > 0$ the operator G_h is a trace class operator and

$$(12) \qquad\qquad \operatorname{tr} G_h = \int\limits_a^b g_h(t, t) dt.$$

First we show that G can be written as a sum $G_{[1]} - G_{[2]} + i G_{[3]} - i G_{[4]}$, where $G_{[1]}$, $G_{[2]}$, $G_{[3]}$ and $G_{[4]}$ are non-negative integral operators with kernel functions in $L_2([a, b] \times [a, b])$. Indeed, $G = G_{\Re} + i G_{\Im}$, where $G_{\Re}$ (resp., $G_{\Im}$) is the integral operator whose kernel function is equal to $2^{-1}(g + g_*)$ (resp., $(2i)^{-1}(g - g_*)$), where $g_*(t, s) = \overline{g(s, t)}$. Since $G_{\Re}$ and $G_{\Im}$ are selfadjoint, we may write

$$G_{\Re} f = \sum_j a_j \langle f, \varphi_j \rangle \varphi_j, \qquad f \in L_2([a, b]),$$

$$G_{\Im} f = \sum_j b_j \langle f, \psi_j \rangle \psi_j, \qquad f \in L_2([a, b]),$$

where $\{(\varphi_j), (a_j)\}$ and $\{(\psi_j), (b_j)\}$ are basic systems of eigenvectors and eigenvalues for $G_{\Re}$ and $G_{\Im}$, respectively. Now put

$$G_{[1]} f = \sum_j \max(a_j, 0) \langle f, \varphi_j \rangle \varphi_j,$$

$$G_{[2]} f = \sum_j \min(-a_j, 0) \langle f, \varphi_j \rangle \varphi_j,$$

$$G_{[3]} f = \sum_j \max(b_j, 0) \langle f, \psi_j \rangle \psi_j,$$

$$G_{[4]} f = \sum_j \max(-b_j, 0) \langle f, \psi_j \rangle \psi_j.$$

Since (a_j) and (b_j) are sequences in ℓ_2, the operators $G_{[1]}$, $G_{[2]}$, $G_{[3]}$ and $G_{[4]}$ are again integral operators with kernel functions from $L_2([a, b] \times [a, b])$ and they are non-negative. Obviously, $G = G_{[1]} - G_{[2]} + i G_{[3]} - i G_{[4]}$, which is the desired decomposition. Since $G_h = M_h G M_h$ and M_h is selfadjoint, the map $G \mapsto G_h$ preserves the non-negativity of the operator G. Thus for $\nu = 1, 2, 3, 4$ the operator $(G_{[\nu]})_h$ is a non-negative integral operator with a continuous kernel function. Hence we can apply [GG], Theorem IV.4.1 to show that $G_{[\nu]}$ is a trace class operator and that (12) holds for $G_{[\nu]}$ instead of G ($\nu = 1, 2, 3, 4$). But then, by linearity, G_h is a trace class operator and (12) is proved.

Next, we apply the result of the previous paragraph to k. Since k is continuous,

$$k_h(t, s) \to k(t, s) \qquad (h \to 0)$$

uniformly on any square $[a', b'] \times [a', b']$ with $a < a' < b' < b$ and

$$\max_{a \leq t, s \leq b} |k_h(t, s)| \leq \max_{a \leq t, s \leq b} |k(t, s)|$$

for any $h > 0$. Now use that A is a trace class operator. So according to Theorem VI.4.3, formula (10) implies that $M_h A M_h \to A$ in the trace class norm if $h \to 0$. It follows (cf. Theorem 2.1) that

$$\operatorname{tr} A = \lim_{h \to 0} \operatorname{tr} M_h A M_h = \lim_{h \to 0} \operatorname{tr} A_h$$

$$= \lim_{h \to 0} \int_a^b k_h(t, t)dt = \int_a^b k(t, t)dt,$$

which proves (8).   □

VII.3 DEFINITION OF THE DETERMINANT

We return to the determinant. To define the determinant for $I + A$, where A is an arbitrary trace class operator on a Hilbert space H, we follow the same procedure as for the trace. However, we shall see that for the determinant the continuity argument is more involved.

First note that for finite rank operators F and G on H the following equality holds:

$$(1) \qquad \det(I + F)(I + G) = \det(I + F) \cdot \det(I + G).$$

Observe that $(I + F)(I + G) = I + C$, where $C = F + G + FG$ is an operator of finite rank. Thus the left-hand side of (1) is well-defined. To prove (1) we use that this product formula holds for matrices and for operators on a finite dimensional space. Take H_0 as in formula (2.2). Then

$$(F + G + FG)H_0 \subset H_0, \qquad H_0^{\perp} \subset \operatorname{Ker}(F + G + FG).$$

Let I_0 be the identity on H_0. According to the definition of the determinant for $I + F$, where F is of finite rank, we have

$$\det(I + F)(I + G) = \det[(I + F)(I + G)|H_0]$$
$$= \det[(I_0 + F|H_0)(I_0 + G|H_0)]$$
$$= \det(I_0 + F|H_0) \cdot \det(I_0 + G|H_0)$$
$$= \det(I + F) \cdot \det(I + G).$$

This proves (1).

We shall also need that for a finite rank operator F the following inequalities hold true:

$$(2) \qquad |\det(I + F)| \leq \exp \|F\|_1.$$

$$(3) \qquad |\det(I + F) - 1| \leq (\exp \|F\|_1) - 1.$$

To prove these inequalities we use that $1 + t \leq \exp t$ for $t \geq 0$ and we apply the second equality in (1.5) and Corollaries VI.2.5 and VI.2.6. Indeed,

$$\begin{aligned}
|\det(I + F)| &\leq \prod_j (1 + |\lambda_j(F)|) \\
&\leq \prod_j (1 + s_j(F)) \leq \exp\left(\sum_j s_j(F)\right) \\
&= \exp \|F\|_1,
\end{aligned}$$

which proves (2), and

$$\begin{aligned}
|\det(I + F) - 1| &= \left|\prod_j (1 + \lambda_j(F)) - 1\right| \\
&\leq \sum_n \sum_{j_1 < j_2 < \cdots < j_n} |\lambda_{j_1}(F)\lambda_{j_2}(F) \cdots \lambda_{j_j}(F)| \\
&\leq \sum_n \sum_{j_1 < j_2 < \cdots < j_n} s_{j_1}(F)s_{j_2}(F) \cdots s_{j_n}(F) \\
&= \prod_j (1 + s_j(F)) - 1 \\
&\leq -1 + \exp\left(\sum_j s_j(F)\right) = -1 + \exp \|F\|_1,
\end{aligned}$$

which proves (3). $\square$

LEMMA 3.1. *Let A be a trace class operator, and let $F_1 F_2, \ldots$ be a sequence of finite rank operators converging to A in trace class norm. Then*

$$(4) \qquad \lim_{n \to \infty} \det(I + F_n)$$

exists and does not depend on the particular choice of the sequence $F_1, F_2, \ldots$.

PROOF. First assume that $I + A$ is invertible. Then $I + F_n$ is invertible for n sufficiently large. So without loss of generality we may assume that $I + F_n$ is invertible for all n. We have

$$I + F_m = (I + F_n)[I + (I + F_n)^{-1}(F_m - F_n)].$$

From the multiplicativity property (1) and the inequalities (2) and (3) we conclude that

$$|\det(I + F_m) - \det(I + F_n)| = |\det(I + F_n)| \cdot |\det[I + (I + F_n)^{-1}(F_m - F_n)] - 1|$$
$$\leq e^{\|F_n\|_1}\left(e^{\|(I+F_n)^{-1}(F_m-F_n)\|_1} - 1\right) \to 0 \qquad (m, n \to \infty).$$

Here we used that $(\|F_n\|_1)_n$ is a bounded sequence and

$$\|(I + F_n)^{-1}(F_m - F_n)\|_1 \to 0 \qquad (m, n \to \infty).$$

So we see that $\big(\det(I + F_n)\big)_n$ is a Cauchy sequence. Furthermore, if $F_1', F_2', \ldots$ is a second sequence of finite rank operators converging to A in trace class norm, then by the same arguments

$$\det(I + F_n) - \det(I + F_n') \to 0 \qquad (n \to \infty).$$

So the limit (4) exists and does not depend on the particular choice of the sequence $F_1, F_2, \ldots$.

　　　　Next we consider the case when $I + A$ is not invertible. We shall prove that in this case

$$(5) \qquad\qquad \det(I + F_n) \to 0 \qquad (n \to \infty).$$

Since $I + A$ is not invertible, the point -1 is an isolated eigenvalue of finite type for A. Let P be the corresponding Riesz projection, and put $H_1 = \operatorname{Im} P$, $H_2 = \operatorname{Ker} P$. With respect to the decomposition $H = H_1 \oplus H_2$ we write $F_1, F_2, \ldots$ and the operator A as 2×2 operator matrices:

$$F_n = \begin{bmatrix} K_{11}^{(n)} & K_{12}^{(n)} \\ K_{21}^{(n)} & K_{22}^{(n)} \end{bmatrix}, \qquad A = \begin{bmatrix} A_{11} & 0 \\ 0 & A_{22} \end{bmatrix}.$$

Since $F_n \to A$ in trace class norm, we have

$$\|K_{11}^{(n)} - A_{11}\|_1 \to 0, \qquad \|K_{12}^{(n)}\|_1 \to 0,$$
$$\|K_{21}^{(n)}\|_1 \to 0, \qquad \|K_{22}^{(n)} - A_{22}\|_1 \to 0$$

for $n \to \infty$. For $i = 1, 2$ let I_i be the identity operator on H_i. Note that $I_2 + A_{22}$ is invertible. So for n sufficiently large the operator $I_2 + K_{22}^{(n)}$ will be invertible too. Without loss of generality we may assume that $I_2 + K_{22}^{(n)}$ is invertible for all n. This allows us to factor $I + F_n$ in the following way:

$$(6) \qquad\qquad I + F_n = (I + C_n)(I + D_n)(I + E_n),$$

where

$$C_n = \begin{bmatrix} 0 & K_{12}^{(n)}(I_2 + K_{22}^{(n)})^{-1} \\ 0 & 0 \end{bmatrix},$$

$$D_n = \begin{bmatrix} K_{11}^{(n)} - K_{12}^{(n)}(I_2 + K_{22}^{(n)})^{-1}K_{21}^{(n)} & 0 \\ 0 & K_{22}^{(n)} \end{bmatrix},$$

$$E_n = \begin{bmatrix} 0 & 0 \\ (I_2 + K_{22}^{(n)})^{-1}K_{21}^{(n)} & 0 \end{bmatrix}.$$

For each n the operators C_n, D_n and E_n are operators of finite rank. From the structure of C_n and E_n it is clear that C_n and E_n have no non-zero eigenvalues. It follows that $\det(I + C_n) = \det(I + E_n) = 1$. So, using formula (1), we see from (6) that

$$(7) \qquad \det(I + F_n) = \det(I + D_n).$$

Now

$$(8) \qquad \det(I + D_n) = \det[I_1 + K_{11}^{(n)} - K_{12}^{(n)}(I_2 + K_{22}^{(n)})^{-1}K_{21}^{(n)}] \cdot \det(I_2 + K_{22}^{(n)}).$$

By the first part of the proof, $\det(I_2 + K_{22}^{(n)}) \to \det(I + A_{22})$. Since H_1 is finite dimensional and

$$K_{11}^{(n)} - K_{12}^{(n)}(I_2 + K_{22}^{(n)})^{-1}K_{21}^{(n)} \to A_{11}$$

in H_1, we know from the continuity of the determinant on matrices that

$$\det[I_1 + K_{11}^{(n)} - K_{12}^{(n)}(I_2 + K_{22}^{(n)})^{-1}K_{21}^{(n)}] \to \det(I_1 + A_{11}).$$

But $\det(I_1 + A_{11}) = 0$. So (7) and (8) together show that $\det(I + F_n) \to 0$ if $n \to \infty$. $\square$

Lemma 3.1 and the fact that the finite rank operators are dense in S_1 allow us to extend the definition of the determinant to operators $I + A$, where A is an arbitrary trace class operator. Indeed, given a trace class operator A we define

$$(9) \qquad \det(I + A) := \lim_{n \to \infty} \det(I + F_n),$$

where $F_1, F_2, \ldots$ is a sequence of finite rank operators which converges to A in the trace class norm. Lemma 3.1 tells us that $\det(I + A)$ is well-defined.

THEOREM 3.2. *Let A be a trace class operator on H, and let $\varphi_1, \varphi_2, \ldots$ be an orthonormal basis of H. Then*

$$\det(I + A) = \lim_{n \to \infty} \det\big(\delta_{jk} + (A\varphi_k, \varphi_j)\big)_{j,k=1}^n.$$

PROOF. Let P_n be the orthogonal projection onto the space spanned by $\varphi_1, \ldots, \varphi_n$. Then $(P_n A P_n)$ is a sequence of finite rank operators which converges in the trace class norm to A. Thus

$$\det(I + A) = \lim_{n \to \infty} \det(I + P_n A P_n)$$
$$= \lim_{n \to \infty} \det\big((I + P_n A P_n)|\operatorname{Im} P_n\big).$$

With respect to the basis $\varphi_1, \ldots, \varphi_n$ the operator $(I + P_n A P_n)|\operatorname{Im} P_n$ has the following matrix representation:

$$\operatorname{mat}\big((I + P_n A P_n)|\operatorname{Im} P_n\big) = \big(\delta_{jk} + (A\varphi_k, \varphi_j)\big)_{j,k=1}^{n},$$

which proves the theorem. $\square$

  THEOREM 3.3. *For trace class operators A and B the following formulas hold true:*

(10) $$\det(I + A)(I + B) = \det(I + A)\det(I + B),$$

(11) $$|\det(I + A)| \le \prod_j (1 + s_j(A)) \le \exp \|A\|_1,$$

(12) $$|\det(I + A) - 1| \le (\exp \|A\|_1) - 1.$$

  PROOF. To prove (10) and (12) we use the definition of the determinant and the fact that these formulas hold true for operators of finite rank (cf. (1) and (3)). To prove the first inequality in (11), let $\varphi_1, \varphi_2, \ldots$ be an orthonormal basis in H, and let P_n be the orthogonal projection onto the space spanned by $\varphi_1, \ldots, \varphi_n$. According to the second identity in formula (1.5) we have

$$\begin{aligned}
|\det(I + P_n A P_n)| &= \left|\prod_j (1 + \lambda_j(P_n A P_n))\right| \\
&\le \prod_j (1 + s_j(P_n A P_n)) \\
&\le \prod_j (1 + s_j(A)),
\end{aligned}$$

where Proposition VI.1.3 has been used to obtain the last inequality. Note that the last term in the preceding formula does not depend on n. Since

$$\det(I + A) = \lim_{n \to \infty} \det(I + P_n A P_n),$$

the first inequality in (11) is proved. The second inequality in (11) follows from $1 + x \le \exp x$ for $x \ge 0$ and the definition of the trace class norm. $\square$

  From the second part of the proof of Lemma 3.1 (cf. formula (5)) it follows that $\det(I + A) = 0$ whenever $I + A$ is not invertible. The converse statement is also true. To see this assume that $I + A$ is invertible. Then $(I + A)^{-1} = I + C$, where $C = -A(I + A)^{-1}$ is a trace class operator. So, according to formula (10), we have

$$\det(I + A) \cdot \det(I + C) = \det(I + A)(I + C) = 1.$$

It follows that $\det(I = A) \ne 0$. So we proved that

(13) $$\det(I + A) = 0 \Leftrightarrow I + A \text{ not invertible.}$$

We shall come back to this statement with more details in Section 6.

The map $A \mapsto \det(I + A)$ is continuous in the trace class norm, i.e.,

$$(14) \qquad \lim_{n \to \infty} \det(I + A_n) = \det(I + A)$$

for any sequence $A_1, A_2, \ldots$ of trace class operators which converges in the trace class norm to A. To see this, we choose for each n a finite rank operator F_n such that

$$\|A_n - F_n\|_1 < \frac{1}{n}, \qquad |\det(I + A_n) - \det(I + F_n)| < \frac{1}{n}.$$

Then $\|F_n - A\|_1 \to 0$ if $n \to \infty$ and thus

$$\det(I + A) = \lim_{n \to \infty} \det(I + F_n) = \lim_{n \to \infty} \det(I + A_n),$$

which proves (14).

VII.4 ANALYTICITY OF THE DETERMINANT

For a trace class operator A the determinant $\det(I - \lambda A)$ is well-defined for each $\lambda \in \mathbf{C}$. In this section we describe the properties of $\det(I - \lambda A)$ as a function of λ. If F is an operator of finite rank, then

$$(1) \qquad \det(I - \lambda F) = \prod_{j=1}^{r} \big(1 - \lambda \lambda_j(F)\big),$$

where $\lambda_1(F), \ldots, \lambda_r(F)$ are the non-zero eigenvalues of F counted according to their algebraic multiplicity. In particular, $\det(I - \lambda F)$ is a polynomial in λ when F is of finite rank.

LEMMA 4.1. *Let A be a trace class operator, and let $F_1, F_2, \ldots$ be a sequence of operators of finite rank which converges to A in trace class norm. Then*

$$(2) \qquad \det(I - \lambda F_2) \to \det(I - \lambda A) \qquad (n \to \infty)$$

and the convergence in (2) is uniform on compact subsets of $\mathbf{C}$.

PROOF. Formula (2) is clear from the definition of the determinant. The main point is to prove that in (2) the convergence is uniform on compact subsets of $\mathbf{C}$. Let $\Omega \subset \mathbf{C}$ be compact. Take $R > 0$ such that Ω is contained in the open disc $|\lambda| < R$. Further we choose R in such a way that A has no spectrum on the circle $|\zeta| = R^{-1}$ (i.e., $I - \lambda A$ is invertible for $|\lambda| = R$). From our hypotheses it follows that $F_n \to A$ in the operator norm. Hence for n sufficiently large the operator F_n has no spectrum on $|\zeta| = R^{-1}$ and for $|\lambda| = R$

$$(3) \qquad (I - \lambda F_n)^{-1} \to (I - \lambda A)^{-1} \qquad (n \to \infty).$$

Furthermore, in (3) the convergence is uniform on $|\lambda| = R$. For n sufficiently large we have (see the first paragraph of the proof of Lemma 3.1)

$$\max_{|\lambda|=R} |\det(I - \lambda F_m) - \det(I - \lambda F_n)| \le e^{R\|F_n\|_1} \{ e^{R\|(I-\lambda F_n)^{-1}(F_m - F_n)\|_1} - 1 \}.$$

Now observe that $(\|F_n\|_1)_n$ is a bounded sequence. Furthermore, since, in (3), the convergence is uniform on $|\lambda| = R$, there exists a constant C_R such that

$$\sup_{|\lambda|=R} \|(I - \lambda F_n)^{-1}(F_m - F_n)\|_1 \leq C_r \|F_m - F_n\|_1.$$

But then it is clear that the sequence $\big(\det(I - \lambda F_n)\big)$ converges uniformly on $|\lambda| = R$. As F_n has finite rank, $\det(I - \lambda F_n)$ is a polynomial in λ. So, by the maximum modulus principle, the sequence $\big(\det(I - \lambda F_n)\big)_n$ converges uniformly on $|\lambda| \leq R$ and hence also on Ω. $\square$

THEOREM 4.2. *Let A be a trace class operator, and let $\lambda_1(A), \lambda_2(A), \ldots$ be the sequence of non-zero eigenvalues of A. Then the function $\Delta(\lambda) = \det(I - \lambda A)$ is an entire function which satisfies the identity*

$$(4) \qquad \Delta'(\lambda) = -\Delta(\lambda)\operatorname{tr} A(I - \lambda A)^{-1}, \qquad \lambda \neq \lambda_j(A)^{-1},$$

and $\Delta(\lambda) = \sum_{k=0}^{\infty} \Delta_k \lambda^k$ with

$$(5) \qquad \Delta_k = \frac{(-1)^k}{k!} \det \begin{bmatrix} \sigma_1 & 1 & & & 0 \\ \sigma_2 & \sigma_1 & 2 & & \\ \vdots & \vdots & \ddots & \ddots & \\ & & & & k-1 \\ \sigma_k & \sigma_{k-1} & \cdots & & \sigma_1 \end{bmatrix}, \qquad k \geq 1,$$

where $\sigma_j = \operatorname{tr} A^j$ $(j = 1, 2, \ldots)$.

PROOF. Choose a sequence $F_1, F_2, \ldots$ of operators of finite rank which converges to A in trace class norm. Put $d_n(\lambda) = \det(I - \lambda F_n)$. We know that $d_n(\lambda)$ is a polynomial in λ and $d_n(\lambda) \to \Delta(\lambda)$ uniformly on $|\lambda| < R$ for each $R > 0$. It follows that $\Delta(\lambda)$ is analytic on $|\lambda| < R$. Since R is arbitrary it follows that Δ is an entire function.

Choose $0 < R < |\lambda_1(A)|^{-1}$. Then $I - \lambda A$ is invertible for $|\lambda| \leq R$. Hence for n sufficiently large the operator $I - \lambda F_n$ is invertible for each $|\lambda| \leq R$ and

$$(6) \qquad (I - \lambda F_n)^{-1} \to (I - \lambda A)^{-1} \qquad (n \to \infty)$$

on $|\lambda| \leq R$. Without loss of generality we may assume that for each n and $|\lambda| \leq R$ the operator $I - \lambda F_n$ is invertible. This implies that $d_n(\lambda) \neq 0$ for $|\lambda| \leq R$. From (1) we see that

$$\frac{d_n'(\lambda)}{d_n(\lambda)} = \sum_{j=1}^{r_n} \frac{-\lambda_j(F_n)}{1 - \lambda \lambda_j(F_n)}, \qquad |\lambda| \leq R.$$

Now, $\lambda_j(F_n)\big(1 - \lambda \lambda_j(F_n)\big)^{-1}$, $j = 1, \ldots, r_n$, are precisely the non-zero eigenvalues (counted according to their algebraic multiplicity) of the finite rank operator $F_n(I - \lambda F_n)^{-1}$. It follows that

$$(7) \qquad d_n'(\lambda) = -d_n(\lambda)\operatorname{tr} F_n(I - \lambda F_n)^{-1}, \qquad |\lambda| \leq R.$$

Note that $F_n(I - \lambda F_n)^{-1} \to A(I - \lambda A)^{-1}$ in trace class norm. Furthermore, $d_n(\lambda) \to \Delta(\lambda)$, and the fact that the latter convergence is uniform on compact subsets of $\mathbb{C}$ implies that also $d'_n(\lambda) \to \Delta'(\lambda)$. So in (7) we may take the limit for $n \to \infty$, and hence we see that the identity in (4) holds for $|\lambda| \leq R$.

Next we observe that $\operatorname{tr} A(I - \lambda A)^{-1}$ is analytic on the open set $\Omega = \{\lambda \in \mathbb{C} \mid \lambda \neq \lambda_j(A)^{-1}, j = 1, 2, \ldots\}$. Indeed, for $\lambda_0 \in \Omega$ and $|\lambda - \lambda_0|$ sufficiently small we have

$$A(I - \lambda A)^{-1} = \sum_{n=0}^{\infty} (\lambda - \lambda_0)^n A^{n+1} (I - \lambda_0 A)^{-n-1},$$

and the series converges in the trace class norm. But the trace is a continuous linear functional on S_1. So

$$\operatorname{tr} A(I - \lambda A)^{-1} = \sum_{n=0}^{\infty} (\lambda - \lambda_0)^n \alpha_n,$$

where $\alpha_n = \operatorname{tr}[A^{n+1}(I - \lambda_0 A)^{-n-1}]$, $n = 0, 1, 2, \ldots$. This proves the analyticity of $\operatorname{tr} A(I - \lambda A)^{-1}$ on Ω. We already know that Δ and its derivative Δ' are analytic on Ω. So, by analytic continuation, the identity in (4) holds for all $\lambda \neq \lambda_j(A)^{-1}$, $j = 1, 2, \ldots$.

To prove (5) we compare the coefficients of the Taylor expansions at zero of the functions appearing in (4). By taking $\lambda_0 = 0$ in the previous paragraph one sees that

$$\operatorname{tr} A(I - \lambda A)^{-1} = \sum_{j=1}^{\infty} \lambda^j \operatorname{tr} A^{j+1}$$

for $|\lambda|$ sufficiently small. Furthermore, $\Delta(\lambda) = \sum_{j=0}^{\infty} \Delta_j \lambda^j$ with $\Delta_0 = 1$ and $\Delta'(\lambda) = \sum_{j=0}^{\infty} (j+1)\Delta_{j+1} \lambda^j$. So, according to (4),

$$(8) \qquad k\Delta_k = -\sum_{j=0}^{k-1} \Delta_j \operatorname{tr} A^{k-j}, \qquad k = 1, 2, \ldots .$$

Put $\sigma_j = \operatorname{tr} A^j$, $j = 1, 2, \ldots$, and recall that $\Delta_0 = 1$. This allows us to rewrite (8) in the form:

$$(9) \qquad \left\{ \begin{array}{l} -\sigma_1 = 1\Delta_1, \\ -\sigma_2 = \sigma_1 \Delta_1 + 2\Delta_2, \\ \quad \vdots \\ -\sigma_k = \sigma_{k-1}\Delta_1 + \sigma_{k-2}\Delta_2 + \cdots + \sigma_1 \Delta_{k-1} + k\Delta_k. \end{array} \right.$$

Now consider (9) as a system of k linear equations with k unknowns $\Delta_1, \ldots, \Delta_k$. Note that the determinant of the coefficient matrix is equal to $k!$. So, by applying Cramer's rule, we see that Δ_K is given by (5). $\square$

The next theorem describes the behaviour of $\det(I - \lambda A)$ for λ near infinity.

THEOREM 4.3. *Let A be a trace class operator. Given $\varepsilon > 0$ there exists a positive constant C_ε such that*

$$|\det(I - \lambda A)| \leq C_\varepsilon e^{\varepsilon|\lambda|}, \qquad \lambda \in \mathbf{C}.$$

PROOF. Since A is a trace class operator, there exists N such that $\sum_{j=N+1}^\infty s_j(A) > \frac{\varepsilon}{2}$. Moreover, there exists a positive constant C_ε such that

$$\prod_{j=1}^N \big(1 + |\lambda| s_j(A)\big) \leq C_\varepsilon \exp\left(\frac{\varepsilon}{2}|\lambda|\right),$$

because $t\exp(-\eta t)$ is bounded on $0 \leq t < \infty$ for each $\eta > 0$. Next we use the first inequality in (3.11), which yields

$$|\det(I - \lambda A)| \leq \prod_{j=1}^\infty \big(1 + |\lambda| s_j(A)\big)$$

$$\leq \left\{\prod_{j=1}^N \big(1 + |\lambda| s_j(A)\big)\right\} \exp\left(|\lambda| \sum_{j=N+1}^\infty s_j(A)\right)$$

$$\leq C_\varepsilon e^{\varepsilon|\lambda|}. \quad \square$$

In the terminology of the next section the property described in Theorem 4.3 means that as a function of λ the determinant $\det(I - \lambda A)$ is an entire function of exponential type zero.

VII.5 INTERMEZZO ABOUT ENTIRE FUNCTIONS OF EXPONENTIAL TYPE

This section may be omitted in the first reading; it plays a role in the second proof of the fundamental theorem for the trace given in the next section.

Let f be an entire complex-valued function. We say that f is of *exponential type zero* if given $\varepsilon > 0$ there exists a positive constant C_ε such that

$$(1) \qquad\qquad |f(\lambda)| \leq C_\varepsilon e^{\varepsilon|\lambda|}, \qquad \lambda \in \mathbf{C}.$$

The function $f(\lambda) = \det(I - \lambda A)$, where A is a trace class operator, is an example of such a function. The condition that f is of exponential type zero is equivalent to the requirement that

$$(2) \qquad\qquad \lim_{R \to \infty} \frac{\log M(R)}{R} = 0,$$

where $M(R) = \max\{|f(\lambda)| \, | \, |\lambda| = R\}$. We shall prove the following theorem.

THEOREM 5.1. *Let $a_1, a_2, \ldots$ be the zeros of the entire function f ordered according to increasing absolute values and multiplicities taken into account. Assume that*

$$(3) \qquad\qquad \sum_{j=1}^\infty \frac{1}{|a_j|} < \infty.$$

If f is of exponential type zero and $f(0) = 1$, then f admits the representation

$$(4) \qquad f(\lambda) = \prod_{j=1}^{\infty} \left(1 - \frac{\lambda}{a_j}\right).$$

We begin with a lemma.

LEMMA 5.2. *Let $g(\lambda) = \sum_{k=0}^{\infty} c_k \lambda^k$ for $|\lambda| < R$, and take $0 < r < R$. Then*

$$(5) \qquad |c_k| \leq \frac{2}{r^k}(\gamma(r) - \Re g(0)), \qquad k \geq 1,$$

where $\gamma(r) = \max\{\Re g(\lambda) \mid |\lambda| = r\}$.

PROOF. From the definition of g it follows that

$$\Re g(re^{it}) = \Re c_0 + \sum_{k=1}^{\infty} \frac{1}{2} c_k r^k e^{ikt} + \sum_{k=-\infty}^{-1} \frac{1}{2} \overline{c}_{|k|} r^{|k|} e^{-ikt}, \qquad 0 \leq t \leq 2\pi.$$

Therefore

$$(6a) \qquad \Re c_0 = \frac{1}{2\pi} \int_0^{2\pi} \Re g(re^{it}) \, dt,$$

$$(6b) \qquad c_k = \frac{1}{\pi r^k} \int_0^{2\pi} \Re g(re^{it}) e^{-itk} \, dt, \qquad k \geq 1.$$

Take $k \geq 1$. Recall that $\int_0^{2\pi} e^{-itk} \, dt = 0$. So we have

$$-c_k = \frac{1}{\pi r^k} \int_0^{2\pi} [\gamma(r) - \Re g(re^{it})] e^{-itk} \, dt.$$

Next we use that $\gamma(r) \geq \Re g(re^{it})$. It follows that

$$|c_k| \leq \frac{1}{\pi r^k} \int_0^{2\pi} [\gamma(r) - \Re g(re^{it})] \, dt$$

$$= \frac{1}{\pi r^k} \left(2\pi \gamma(r) - 2\pi \Re c_0\right) = \frac{2}{r^k}(\gamma(r) - \Re g(0)). \qquad \square$$

The above lemma already allows us to prove Theorem 5.1 for the case when the entire function f has no zeros. We have the following corollary.

COROLLARY 5.3. *An entire function of exponential type zero which has no zeros is a constant function.*

PROOF. Let f be an entire function of exponential type zero. We may assume that $f(0) = 1$. Our hypotheses imply that $f(\lambda) = \exp(g(\lambda))$ for an entire function g. Indeed, put

$$g(\lambda) = \int_0^\lambda \frac{f'(t)}{f(t)} \, dt,$$

where the integration is along any rectifiable curve joining the points 0 and λ. Then g is well-defined, and one checks easily that the derivative of $f(\lambda) \exp(-g(\lambda))$ vanishes identically. Since $g(0) = 0$ and $f(0) = 1$, this implies that $f(\lambda) = \exp(g(\lambda))$.

From $|f(\lambda)| \leq M(R)$ for $|\lambda| \leq R$ we conclude that

$$\Re g(\lambda) \leq \log M(R), \qquad |\lambda| \leq R.$$

So we apply Lemma 5.2 to show that for $k \geq 1$ the k-th coefficient in the Taylor expansion of g at 0 satisfies the following inequality:

$$(7) \qquad \frac{1}{k!} |g^{(k)}(0)| \leq 2 \frac{\log M(R) - \Re g(0)}{R^k}.$$

Next we use that f is of exponential type 0. So, according to formula (2), the right hand side of (7) converges to 0 if $R \to \infty$. Hence g is a constant function and thus so is f. $\square$

PROOF OF THEOREM 5.1. Because of the condition (3) the infinite product in (4) is well-defined and converges uniformly on compact subsets of $\mathbf{C}$. Put

$$f(\lambda) = h(\lambda) \prod_{j=1}^\infty \left(1 - \frac{\lambda}{a_j}\right).$$

Then h is an entire function which has no zeros, and thus (cf. the first part of the proof of Corollary 5.3) $h(\lambda) = \exp(g(\lambda))$ for some entire function g. Note that $g(0) = 0$. Take $R > 1$ and put

$$(8) \qquad f_R(\lambda) = e^{g(\lambda)} \prod_{|a_j| > R} \left(1 - \frac{\lambda}{a_j}\right).$$

Obviously $f_R(\lambda)$ has no zeros in $|\lambda| < R$, and we may write

$$(9) \qquad f_R(\lambda) = e^{g_R(\lambda)}, \qquad |\lambda| < R,$$

where

$$(10) \qquad g_R(\lambda) = g(\lambda) + \sum_{|a_j| > R} \log\left(1 - \frac{\lambda}{a_j}\right).$$

Note that the series in (10) converges uniformly on $|\lambda| < R$. This allows us to compare in a simple way the coefficients of the Taylor expansions at zero of the various functions appearing in (10). For $k \geq 1$ we have

$$\frac{1}{k!} g_R^{(k)}(0) = \frac{1}{k!} g^{(k)}(0) - \sum_{|a_j| > R} \frac{1}{k a_j^k}.$$

It follows that

$$(11) \qquad \frac{1}{k!}|g^{(k)}(0)| \leq \frac{1}{k!}|g_R^{(k)}(0)| + \sum_{|a_j|>R} \frac{1}{|a_j|} \qquad (k \geq 1).$$

Because of (3) the second term in the right hand side of (11) converges to zero if $R \to \infty$. We shall prove that the same is true for the first term.

According to (8),

$$f(\lambda) = f_R(\lambda) \prod_{|a_j| \leq R} \left(1 - \frac{\lambda}{a_j}\right).$$

Note that $|1 - \frac{\lambda}{a_j}| \geq 1$ whenever $|\lambda| = 2R$ and $|a_j| \leq R$. It follows that $|f(\lambda)| \geq |f_R(\lambda)|$ for $|\lambda| = 2R$. In particular, $|f_R(\lambda)| \leq M(2R)$ if $|\lambda| = 2R$, which by the maximum modulus principle implies that

$$|f_R(\lambda)| \leq M(2R), \qquad |\lambda| \leq R.$$

It follows that

$$(12) \qquad \Re g_R(\lambda) \leq \log M(2R), \qquad |\lambda| \leq R.$$

By Lemma 5.2, this implies that

$$\left| \frac{1}{k!} g_R^{(k)}(0) \right| \leq \frac{2}{R^k} \{\log M(2R) - \Re g_R(0)\}, \qquad k \geq 1.$$

But f is of exponential type zero. So formula (2) holds true, and we may conclude that for each $k \geq 1$ the number $\frac{1}{k!} g_R^{(k)}(0) \to 0$ if $R \to \infty$. Hence, returning to (11), we see that $g^{(k)}(0) = 0$ for each $k \geq 0$, and thus $g \equiv 0$. The theorem is proved. $\square$

Theorem 5.1 is a special case of the Hadamard factorization theorem (see, e.g., [C], Section XI.3.1). The converse of Theorem 5.1 also holds true, i.e., if (3) holds and f is given by (4), then f is of exponential type zero, $f(0) = 1$ and $a_1, a_2, \ldots$ are the zeros of f.

VII.6 THE FUNDAMENTAL THEOREM FOR TRACE AND DETERMINANT

The main theorem of this section identifies trace and determinant in terms of eigenvalues and their multiplicities. The result for the trace is due to V.B. Lidskii [2]. Two proofs are given. The first follows the main idea of the proof of Gohberg-Kreĭn [3]. It is based on Theorems 4.2 and 4.3 and does not require the result of the preceding section. The second proof focusses on the connection with complex analysis and is based on the representation theorem for entire functions of exponential type zero which has been proved in the previous section.

THEOREM 6.1. *Let A be a trace class operator, and let $\lambda_1(A), \lambda_2(A),\dots$ be the sequence of non-zero eigenvalues of A (multiplicities taken into account). Then*

(i) $\operatorname{tr} A = \sum_j \lambda_j(A)$,

(ii) $\det(I - A) = \prod_j (1 - \lambda_j(A))$.

FIRST PROOF OF THEOREM 6.1. Let E_A be the closed linear hull of the eigenvectors and generalized eigenvectors corresponding to the non-zero eigenvalues of A. We consider three cases.

(I) The case $E_A = \{0\}$. Thus A has no non-zero eigenvalues. We have to prove that $\operatorname{tr} A = 0$ and $\det(I - A) = 1$. First we shall prove that $\operatorname{tr} A^k = 0$ for $k \geq 2$. To do this, choose an orthonormal basis $\varphi_1, \varphi_2, \dots$ in H. Let P_n be the orthogonal projection onto $\operatorname{span}\{\varphi_1, \dots, \varphi_n\}$. Put $A_n = P_n A P_n$, and fix $k \geq 2$. Since $P_n A P_n$ has finite rank,

$$\operatorname{tr} A_n^k = \sum_j \lambda_j(A_n)^k, \qquad n = 1, 2, \dots .$$

By Corollary VI.2.6 and Proposition VI.1.3,

$$\sum_j |\lambda_j(A_n)| \leq \sum_j s_j(A_n) \leq \sum_j s_j(A) = \|A\|_1,$$

and hence

$$|\operatorname{tr} A_n^k| \leq \sum_j |\lambda_j(A_n)|^k$$

$$\leq |\lambda_1(A_n)|^{k-1} \sum_j |\lambda_j(A_n)|$$

$$\leq |\lambda_1(A_n)|^{k-1} \|A\|_1.$$

From Theorem VI.4.3 we know that $A_n \to A$ in the trace class norm, and hence $A_n \to A$ in the operator norm (because $\|T\| \leq \|T\|_1$). Since $\sigma(A) = \{0\}$, we can apply Theorem II.4.1 to show that $\lambda_1(A_n) \to 0$ if $n \to \infty$. But then we may conclude that $\operatorname{tr} A_n^k \to 0$ if $n \to \infty$. On the other hand (by Theorem VI.4.3) $A_n^k \to A^k$ in the trace class norm if $n \to \infty$, and thus the continuity of the trace on S_1 yields

$$(1) \qquad \operatorname{tr} A^k = \lim_{n \to \infty} \operatorname{tr} A_n^k = 0, \qquad k \geq 2.$$

From (1) and Theorem 4.2 we see that

$$(2) \qquad \det(I - \lambda A) = \sum_{k=0}^{\infty} \frac{(-1)^k}{k!} (\operatorname{tr} A)^k \lambda^k = e^{-\lambda \operatorname{tr} A}.$$

This, together with Theorem 4.3, implies that $\operatorname{tr} A = 0$. Indeed, assume $\operatorname{tr} A \neq 0$. Take $\varepsilon = \frac{1}{2}|\operatorname{tr} A|$. Then, by Theorem 4.3, there exists a constant $C \geq 0$ such that

$$e^{-\varepsilon|\lambda|}|e^{-\lambda \operatorname{tr} A}| = e^{-\varepsilon|\lambda|}|\det(I - \lambda A)| \leq C$$

for all $\lambda \in \mathbb{C}$, which is impossible. Thus $\operatorname{tr} A = 0$, and, by formula (2), also $\det(I-A) = 1$.

(II) The case $E_A = H$. By Schur's lemma in Section II.3, the space H has an orthonormal basis $\varphi_1, \varphi_2, \ldots$ such that

$$A\varphi_j = a_{1j}\varphi_1 + \cdots + a_{jj}\varphi_j, \qquad a_{jj} = \lambda_j(A).$$

Let Q_n be the orthogonal projection on the space spanned by $\varphi_1, \ldots, \varphi_n$. Obviously, $\operatorname{Im} Q_n$ is invariant under A. The matrix of $A|\operatorname{Im} Q_n$ is upper triangular with respect to the basis $\varphi_1, \ldots, \varphi_n$ and has $\lambda_1(A), \ldots, \lambda_n(A)$ on the main diagonal. Thus

$$\operatorname{tr}(A|\operatorname{Im} Q_n) = \sum_{j=1}^{n} \lambda_j(A), \qquad \det(I_{\operatorname{Im} Q_n} - A|\operatorname{Im} Q_n) = \prod_{j=1}^{n}\big(1 - \lambda_j(A)\big).$$

Since $Q_n A Q_n \to A$ in the trace class norm, the definitions of trace and determinant yield

$$\operatorname{tr} A = \lim_{n \to \infty} \operatorname{tr}(Q_n A Q_n)$$
$$= \lim_{n \to \infty} \operatorname{tr}(A|\operatorname{Im} Q_n) = \sum_j \lambda_j(A),$$

and

$$\det(I - A) = \lim_{n \to \infty} \det(I - Q_n A Q_n)$$
$$= \lim_{n \to \infty} \det(I_{\operatorname{Im} Q_n} - A|\operatorname{Im} Q_n) = \prod_j \big(1 - \lambda_j(A)\big),$$

which proves the theorem for the case when $E_A = H$.

(III) The general case. Put $H_1 = E_A$ and $H_2 = E_A^\perp$, and let I_1 and I_2 be the identity operators on H_1 and H_2, respectively. Since E_A is invariant under A, the operator A has a 2×2 operator matrix representation of the following type:

$$(3) \qquad A = \begin{bmatrix} A_{11} & A_{12} \\ 0 & A_{22} \end{bmatrix} : H_1 \oplus H_2 \to H_1 \oplus H_2.$$

The entries A_{11}, A_{12} and A_{22} are trace class operators, by Proposition VI.4.2. For the operator A_{11} the system of eigenvectors and generalized eigenvectors corresponding to the non-zero eigenvalues is dense in H_1. Furthermore, $\lambda_j(A_{11}) = \lambda_j(A)$, because of the definition of E_A. So, by what has been proved under (II),

$$(4) \qquad \operatorname{tr} A_{11} = \sum_j \lambda_j(A), \qquad \det(I_1 - A_{11}) = \prod_j \big(1 - \lambda_j(A)\big).$$

According to Lemma II.3.4, the operator A_{22} has no non-zero eigenvalues. So we may use the result proved under (I) to conclude that

$$(5) \qquad \operatorname{tr} A_{22} = 0, \qquad \det(I_2 - A_{22}) = 1.$$

Finally, by Theorems 2.2 and 3.2,

$$\operatorname{tr} A = \operatorname{tr} A_{11} + \operatorname{tr} A_{22},$$

$$\det(I - A) = \det(I_1 - A_{11}) \det(I_2 - A_{22}),$$

which together with (4) and (5), proves the theorem. $\square$

SECOND PROOF OF THEOREM 6.1. We know that the determinant $\Delta(\lambda) = \det(I - \lambda A)$ is an entire function. First we shall show that

$$(6) \qquad \lambda_1(A)^{-1}, \ \lambda_2(A)^{-1}, \ \ldots$$

is the sequence of zeros of Δ ordered according to increasing absolute values and multiplicities taken into account.

Take $z \neq \lambda_j(A)^{-1}, j = 1, 2, \ldots$. Then $I - zA$ is invertible. Hence, by (3.13), $\det(I - zA) \neq 0$. So z is not a zero of Δ.

Next, assume that $z = \lambda_0^{-1}$, where λ_0 is a non-zero eigenvalue of A. Let P be the Riesz projection of A corresponding to λ_0. Since λ_0 is an eigenvalue of finite type, we know that P has finite rank. Moreover, $\operatorname{Im} P$ has a basis such that, with respect to this basis, the matrix of $A|\operatorname{Im} P$ is a Jordan matrix which has λ_0 on the main diagonal. It follows that

$$\det(I - \lambda PAP) = \det(I_{\operatorname{Im} P} - \lambda A|\operatorname{Im} P)$$
$$= (1 - \lambda\lambda_0)^m,$$

where $m = \dim \operatorname{Im} P$. Note that by definition m is the algebraic multiplicity of λ_0 as an eigenvalue of A. Since λ_0 is not an eigenvalue of $(I - P)A(I - P)$, the result proved in the previous paragraph implies that $z = \lambda_0^{-1}$ is not a zero of $\det\big(I - \lambda(I - P)A(I - P)\big)$. Now observe that

$$I - \lambda A = (I - \lambda PAP)\big(I - \lambda(I - P)A(I - P)\big).$$

So, using the multiplicativity of the determinant, we see that

$$\det(I - \lambda A) = (1 - \lambda\lambda_0)^m \det[I - \lambda(I - P)A(I - P)],$$

which shows that $z = \lambda_0^{-1}$ is a zero of Δ whose multiplicity is equal to the algebraic multiplicity of λ_0 as an eigenvalue of A. Hence we know the zeros of $\Delta(\lambda) = \det(I - \lambda A)$ and their multiplicities.

Next we prove (ii). From Theorems 4.2 and 4.3 we know that Δ is an entire function of exponential type zero. Obviously $\Delta(0) = 1$. Since

$$\sum_j |\lambda_j(A)| \leq \sum_j s_j(A) < \infty,$$

the sequence of zeros of Δ satisfies the inequality (5.3). So we can apply Theorem 5.1 to show that

$$(7) \qquad \det(I - \lambda A) = \prod_j \big(1 - \lambda\lambda_j(A)\big).$$

By taking $\lambda = 1$ in (7) we obtain (ii).

From (7) it also follows that the coefficient Δ_1 of λ in the Taylor expansion of $\Delta(\lambda)$ at zero is equal to $-\sum_j \lambda_j(A)$. On the other hand, according to Theorem 4.2, we have $\Delta_1 = -\operatorname{tr} A$. This proves (i). $\square$

COROLLARY 6.2. *Let A be a trace class operator and B be an arbitrary bounded linear operator, both acting on the Hilbert space H. Then*

(i) $\operatorname{tr} AB = \operatorname{tr} BA$

(ii) $\det(I - AB) = \det(I - BA)$.

PROOF. From Proposition VI.4.2 we know that AB and BA are both trace class operators. Now use (see Section III.2) that the operator functions

$$\begin{bmatrix} \lambda - AB & 0 \\ 0 & I \end{bmatrix}, \qquad \begin{bmatrix} \lambda - BA & 0 \\ 0 & I \end{bmatrix}$$

are globally equivalent on $\mathbb{C}\backslash\{0\}$. It follows (see Theorem III.3.1) that AB and BA have the same non-zero eigenvalues multiplicities taken into account. But then (i) and (ii) are clear from Theorem 6.1. $\square$

VII.7 CRAMER'S RULE AND FREDHOLM FORMULAS FOR THE RESOLVENT

From matrix theory we know that for an $n \times n$ matrix A the matrix $I - \lambda A$ is invertible if and only if $\Delta(\lambda) = \det(I - \lambda A) \neq 0$, and in that case Cramer's rule gives a formula for the inverse matrix, namely:

$$(1) \qquad (I - \lambda A)^{-1} = \frac{1}{\Delta(\lambda)} M(\lambda),$$

where $M(\lambda)$ is an $n \times n$ matrix whose (i,j)-th entry is the cofactor of the (j,i)-th entry in the matrix $I - \lambda A$. In particular, the entries of $M(\lambda)$ are polynomials in λ. Since $(I - \lambda A)^{-1} = I + \lambda A(I - \lambda A)^{-1}$, we can rewrite (1) in the form

$$(2) \qquad (I - \lambda A)^{-1} = I + \frac{\lambda}{\Delta(\lambda)} D(\lambda),$$

where $D(\lambda)$ is an $n \times n$ matrix whose entries are polynomials in λ.

THEOREM 7.1. *Let $A \in S_1$. Then $I - \lambda A$ is invertible if and only if $\Delta(\lambda) = \det(I - \lambda A) \neq 0$, and in that case*

$$(3) \qquad (I - \lambda A)^{-1} = I + \frac{\lambda}{\Delta(\lambda)} D(\lambda),$$

where $D(\lambda) = \sum_{j=0}^{\infty} \lambda^j D_j$ is an entire operator function whose coefficients are trace class operators which are uniquely determined by the recurrence relation:

$$(4) \qquad D_0 = A, \qquad D_m = D_{m-1}A - \frac{1}{m}(\operatorname{tr} D_{m-1})A \qquad (m \geq 1).$$

The sequence $\sum_{j=0}^{\infty} \lambda^j D_j$ converges in the trace class norm and

$$(5) \qquad D_m = \frac{(-1)^m}{m!} \det \begin{bmatrix} \sigma_1 & 1 & 0 & \cdots & 0 & 0 \\ \sigma_2 & \sigma_1 & 2 & \cdots & 0 & 0 \\ \vdots & \vdots & \vdots & \vdots & \vdots & \vdots \\ \sigma_{m-1} & \sigma_{m-2} & \sigma_{m-3} & \cdots & m-1 & 0 \\ \sigma_m & \sigma_{m-1} & \sigma_{m-2} & \cdots & \sigma_1 & m \\ A^{m+1} & A^m & A^{m-1} & \cdots & A^2 & A \end{bmatrix},$$

where $\sigma_j = \operatorname{tr} A^j$ $(j \geq 1)$ and in (5) the determinant has to be understood as the operator which is obtained by formal expansion according to the last row.

PROOF. From Theorem 6.1 we know that $\Delta(\lambda) = \det(I - \lambda\Delta) = 0$ if and only if $\lambda = \lambda_j(A)^{-1}$ for some j. It follows that $I - \lambda A$ is invertible if and only if $\Delta(\lambda) \neq 0$.

To prove (3) we first assume that A is an operator of finite rank. Then there exists a finite dimensional subspace H_1 of H such that $AH_1 \subset H_1$ and $H_1^{\perp} \subset \operatorname{Ker} A$. With respect to the decomposition $H = H_1 \oplus H_1^{\perp}$ the operator $(I - \lambda A)^{-1}$ can be written in the form:

$$(6) \qquad (I - \lambda A)^{-1} = \begin{bmatrix} (I_1 - \lambda A_1)^{-1} & 0 \\ 0 & I_2 \end{bmatrix}.$$

Since H_1 is finite dimensional, we know from matrix theory (see the first paragraph of this section) that

$$(7) \qquad (I_1 - \lambda A_1)^{-1} = I_1 + \frac{\lambda}{\Delta_1(\lambda)} D_1(\lambda),$$

where $\Delta_1(\lambda) = \det(I_1 - \lambda A_1)$ and $D_1(\lambda)$ is a polynomial in λ. By definition, $\Delta(\lambda) = \Delta_1(\lambda)$. Put

$$D(\lambda) = \begin{bmatrix} D_1(\lambda) & 0 \\ 0 & 0 \end{bmatrix}.$$

It is clear from (6) and (7) that formula (3) holds true. Note that in this case $D(\lambda)$ is a polynomial in λ whose coefficients are operators of finite rank.

To prove (3) for an arbitrary $A \in S_1$ we take a sequence $F_1, F_2, \ldots$ of operators of finite rank which converges in the trace class norm to A. Put $\Delta_n(\lambda) = \det(I - \lambda F_n)$, and write

$$(I - \lambda F_n)^{-1} = I + \frac{\lambda}{\Delta_n(\lambda)} S_n(\lambda).$$

We know that $S_n(\lambda)$ is a polynomial in λ. Furthermore, by Lemma 4, $\Delta_n(\lambda) \to \Delta(\lambda)$ $(n \to \infty)$ uniformly on compact subsets of $\mathbf{C}$.

Choose a large circle Γ around 0 such that $I - \lambda A$ is invertible for each $\lambda \in \Gamma$. Since $F_n \to A$ in S_1-norm and hence also in the ordinary operator norm, it follows that for n sufficiently large the operator $I - \lambda F_n$ is invertible for each $\lambda \in \Gamma$,

$$(8) \qquad (I - \lambda F_n)^{-1} \to (I - \lambda A)^{-1} \qquad (n \to \infty),$$

and in (8) the convergence is uniform on Γ. Next consider

$$S_n(\lambda) = \Delta_n(\lambda)F_n(I - \lambda F_n)^{-1}.$$

From what has been proved so far, it is clear that for each $\lambda \in \Gamma$,

$$(9) \qquad S_n(\lambda) \to \Delta(\lambda)A(I - \lambda A)^{-1} \qquad (n \to \infty)$$

and in (9) the convergence is uniform on Γ.

Since $S_n(\lambda)$ is a polynomial in λ, we may conclude that $\big(S_n(\lambda)\big)_n$ converges uniformly on $|\lambda| \leq R$, where R is the radius of the circle Γ. Note that R can be taken arbitrary large. Thus

$$(10) \qquad D(\lambda) = \lim_{n \to \infty} S_n(\lambda)$$

exists for each λ and $D(\lambda)$ is an entire operator function. By combining (9) and (10) we see that (3) holds for $\lambda \in \Gamma$. But then, using analytic continuation, it is clear that formula (3) holds for each λ with $\Delta(\lambda) \neq 0$.

Since $D(\lambda) = \Delta(\lambda)A(I - \lambda A)^{-1}$, we have

$$(11) \qquad D(\lambda) = \sum_{m=0}^{\infty} \lambda^m D_m = \sum_{m=0}^{\infty} \lambda^m \left(\sum_{j=0}^{m} \Delta_j A^{m+1-j}\right),$$

where $\Delta_0, \Delta_1, \ldots$ are the Taylor coefficients of $\Delta(\lambda) = \det(I - \lambda A)$ at zero. It follows that

$$D_m = \sum_{j=0}^{m} \Delta_j A^{m+1-j} \qquad (m \geq 0).$$

Hence $D_0 = A$ and

$$(12) \qquad D_m = D_{m-1}A + \Delta_m A \qquad (m \geq 1).$$

From formula (4) in Section 4 we know that $\Delta'(\lambda) = -\Delta(\lambda)\operatorname{tr} A(I - \lambda A)^{-1}$. Thus

$$(13) \qquad \operatorname{tr} D(\lambda) = -\Delta'(\lambda).$$

Since $\sum_{n=0}^{\infty} \lambda^n D_n$ converges in the operator norm, we may conclude (cf. Theorem VI.4.3) that $\sum_{n=0}^{\infty} \lambda^n D_n A$ converges in the trace class norm, and hence, by (12), the series in (11) converges in the trace class norm. This implies that $\operatorname{tr} D_{m-1} = -m\Delta_m$ for $m \geq 1$. Inserting this in (12) yields the desired recurrence relation (4).

It remains to prove (5). Put $E_0 = A$, and for $m \geq 1$ let E_m be the operator defined by the right hand side of (5). Note that

$$\operatorname{tr} E_n = \frac{(-1)^n}{n!} \det M_{n+1} \qquad (n \geq 0),$$

where M_k is the $k \times k$ matrix defined by

$$M_k = \begin{bmatrix} \sigma_1 & 1 & 0 & \cdots & 0 \\ \sigma_2 & \sigma_1 & 2 & \cdots & 0 \\ \vdots & \vdots & \vdots & \cdots & \vdots \\ \sigma_{k-1} & \sigma_{k-2} & \sigma_{k-3} & \cdots & k-1 \\ \sigma_k & \sigma_{k-1} & \sigma_{k-2} & \cdots & \sigma_1 \end{bmatrix} \qquad (k \geq 1).$$

From the definition of E_n it is clear that

$$E_n = \left(\frac{(-1)^n}{n!} \det M_n \right) A + E_{n-1} A \qquad (n \geq 1).$$

It follows that the sequence $E_0, E_1, E_2, \ldots$ satisfies the same recurrence relation as the sequence $D_0, D_1, D_2, \ldots$. As $E_0 = D_0$, we may conclude that $D_n = E_n$ for $n \geq 1$, and the theorem is proved. $\square$

If a trace class operator A admits a particular representation (e.g., as an integral operator), then the recurrence relation (4) can often be used to give explicit formulas for the resolvent $(I - \lambda A)^{-1}$ in terms of the particular representation of A. We shall illustrate this with two examples.

First consider the case when $A = (\alpha_{ij})_{i,j=1}^k$. Consider the subdeterminants:

$$(14) \qquad \alpha \begin{pmatrix} r_1 \cdots r_p \\ s_1 \cdots s_p \end{pmatrix} = \det(\alpha_{r_i s_j})_{i,j=1}^p.$$

Here $1 \leq r_i \leq k$, $1 \leq s_j \leq k$ for $i, j = 1, \ldots, p$. Note that the numbers defined by (14) are equal to zero for $p \geq k + 1$. We claim that the (i,j)-th entry of $(I - \lambda A)^{-1}$ is given by

$$(15) \qquad \delta_{ij} + \frac{\lambda}{\Delta(\lambda)} \left\{ \alpha_{ij} + \sum_{n=1}^{k-1} \frac{(-1)^n}{n!} \lambda^n \left(\sum_{r_1,\ldots,r_n=1}^{k} \alpha \begin{pmatrix} i & r_1 \cdots r_n \\ j & r_1 \cdots r_n \end{pmatrix} \right) \right\},$$

where $\Delta(\lambda) = \det(I - \lambda A)$. To see this, put $E_0 = A$ and let E_n be the $k \times k$ matrix whose (i,j)-th entry is equal to

$$E_n^{ij} = \frac{(-1)^n}{n!} \sum_{r_1,\ldots,r_n=1}^{k} \alpha \begin{pmatrix} i & r_1 \cdots r_n \\ j & r_1 \cdots r_n \end{pmatrix}.$$

Note that

$$(16) \qquad \operatorname{tr} E_n = \frac{(-1)^n}{n!} \sum_{r_1,\ldots,r_{n+1}=1}^{k} \alpha \begin{pmatrix} r_1 \cdots r_{n+1} \\ r_1 \cdots r_{n+1} \end{pmatrix}, \qquad n \geq 0.$$

By expanding

$$\alpha \begin{pmatrix} i & r_1 \cdots r_n \\ j & r_1 \cdots r_n \end{pmatrix}$$

in terms of its first column, one sees that

$$E_n^{ij} = \left(-\frac{1}{n}\operatorname{tr}E_{n-1}\right)\alpha_{ij} + \sum_{r=1}^{k}E_{n-1}^{ir}\alpha_{rj} \qquad (n \geq 1).$$

It follows that

$$E_n = \left(-\frac{1}{n}\operatorname{tr}E_{n-1}\right)A + E_{n-1}A \qquad (n \geq 1),$$

and hence the operators $E_0, E_1, E_2, \ldots$ satisfy the recurrence relation (4). Thus

$$(I - \lambda A)^{-1} = I + \frac{\lambda}{\Delta(\lambda)}\sum_{n=0}^{\infty}\lambda^n E_n.$$

Since $E_n = 0$ for $n \geq k$, this proves (15). Next, recall that the m-th Taylor coefficients of $\Delta(\lambda)$ at zero is given by

$$\Delta_m = -\frac{1}{m}\operatorname{tr}E_{m-1}, \qquad m \geq 1.$$

Together with (16), this yields the following formula for the determinant:

$$\Delta(\lambda) = 1 + \sum_{n=1}^{k}\frac{(-1)^n}{n!}\lambda^n \left(\sum_{r_1,\ldots,r_n=1}^{k}\alpha\left(\begin{array}{c} r_1\cdots r_n \\ r_1\cdots r_n \end{array}\right)\right).$$

Our second example concerns an integral operator on $L_2([a,b])$ of the form

$$(Af)(t) = \int_a^b k(t,s)f(s)ds, \qquad a \leq t \leq b.$$

We assume that the kernel k is a continuous function on $[a,b] \times [a,b]$. Such an operator A does not have to be a trace class operator. Therefore we require, in addition, that A is a trace class operator.

Consider the following continuous function of $2n$ variables:

$$K\left(\begin{array}{c} t_1\cdots t_n \\ s_1\cdots s_n \end{array}\right) = \det\big(k(t_i,s_j)\big)_{i,j=1}^{n}.$$

We claim that

$$(17) \qquad (I - \lambda A)^{-1} = I + \frac{\lambda}{\Delta(\lambda)}\sum_{n=0}^{\infty}\lambda^n H_n,$$

where $\Delta(\lambda) = \det(I - \lambda A)$ and H_n is the integral operator on $L_2([a,b])$ with kernel function

$$(18) \qquad h_n(t,s) = \frac{(-1)^n}{n!}\int_a^b\cdots\int_a^b K\left(\begin{array}{c} s \ \ t_1\cdots t_n \\ s \ \ t_1\cdots t_n \end{array}\right)dt_1\cdots dt_n.$$

To see this we use Theorem 2.3.

First we expand the integrand in (18) in terms of its first column. This gives

$$(19) \qquad h_n(t,s) = \varepsilon_n k(t,s) + \int_a^b h_{n-1}(t,u)k(u,s)\,du,$$

where

$$(20) \qquad \varepsilon_n = \frac{(-1)^n}{n!} \int_a^b \cdots \int_a^b K\left(\begin{array}{c} t_1 \cdots t_n \\ t_1 \cdots t_n \end{array}\right) dt_1 \cdots dt_n.$$

From (19) we see that the operators $H_0, H_1, H_2, \ldots$ satisfy the following recurrence relation:

$$(21) \qquad H_0 = A, \qquad H_n = \varepsilon_n A + H_{n-1}A \qquad (n \geq 1).$$

It follows that all operators H_n are trace class operators. Since h_n is a continuous kernel we can apply Theorem 2.3 to show that

$$\operatorname{tr} H_n = -(n+1)\varepsilon_{n+1} \qquad (n \geq 0),$$

and thus $\varepsilon_n = -\frac{1}{n}\operatorname{tr} H_{n-1}$. Inserting this in (21) shows that the operators $H_0, H_1, \ldots$ satisfy the recurrence relation (4). Hence $H_n = D_n$ and (17) is proved. Since the m-th Taylor coefficient of $\Delta(\lambda)$ at zero is given by $\Delta_m = -m^{-1}\operatorname{tr} H_{m-1} = \varepsilon_m$, we see that

$$(22) \qquad \det(I - \lambda A) = 1 + \sum_{n=1}^{\infty} \frac{(-\lambda)^n}{n!} \int_a^b \cdots \int_a^b K\left(\begin{array}{c} t_1 \cdots t_n \\ t_1 \cdots t_n \end{array}\right) dt_1 \cdots dt_n.$$

Formula (17) suggests to write the resolvent $(I - \lambda A)^{-1}$ in the form of an integral operator:

$$((I - \lambda A)^{-1}f)(t) = f(t) + \frac{\lambda}{\Delta(\lambda)} \int_a^b R(t,s;\lambda)f(s)\,ds,$$

where

$$(23) \qquad R(t,s;\lambda) = \sum_{n=0}^{\infty} \frac{(-\lambda)^n}{n!} \int_a^b \cdots \int_a^b K\left(\begin{array}{cc} t & t_1 \cdots t_n \\ s & s_1 \cdots s_n \end{array}\right) dt_1 \cdots dt_n.$$

It turns out that the series in (23) is convergent for each λ. The representation of the resolvent kernel $R(t,s;\lambda)$ given above holds for an arbitrary continuous kernel. This goes back to the work of I. Fredholm [1]. With some modifications the representation is also true for non-continuous L_2-kernels (see D. Hilbert [1], T. Carleman [1]).

VII.8 COMPLETENESS OF EIGENVECTORS AND GENERALIZED EIGENVECTORS

Recall (see Section II.2) that a compact operator A on a Hilbert space H is said to have a *complete system of eigenvectors and generalized eigenvectors* if the smallest linear manifold in H containing all eigenvectors and generalized eigenvectors of A is dense in H.

THEOREM 8.1. *Let A be a trace class operator, and assume that $A_\Im = \frac{1}{2i}(A-A^*)$ is non-negative. Then A has a complete system of eigenvectors and generalized eigenvectors.*

Theorem 8.1 is sharp in the following sense. If, in Theorem 8.1, the condition $A \in S_1$ is replaced by the weaker condition that

$$(1) \qquad \sum_{j=1}^{\infty} s_j(A)^{1+\varepsilon} < \infty$$

for each $\varepsilon > 0$, then the theorem is no longer true. To see this, consider the operator of integration:

$$V\colon L_2([0,1]) \to L_2([0,1]), \qquad (Vf)(t) = 2i \int_t^1 f(s)ds.$$

We know (see formula (10) in Section VI.1) that $s_j(V) = 4/(2j-1)\pi$ for $j = 1, 2, \dots$. It follows that (1) is satisfied by $A = V$, but $V \notin S_1$. Note that the imaginary part of V is non-negative; in fact,

$$(V_\Im f)(t) = \int_0^1 f(s)ds,$$

and thus

$$\langle V_\Im f, f \rangle = \left| \int_0^1 f(s)ds \right|^2 \geq 0.$$

Obviously, since V has no non-zero eigenvalues, the operator V does not have a complete system of eigenvectors and generalized eigenvectors.

We shall prove Theorem 8.1 as a corollary of the following more general theorem.

THEOREM 8.2. *Let A be a compact operator, and assume that $A_\Im = \frac{1}{2i}(A - A^*)$ is a non-negative trace class operator. Then*

$$(2) \qquad \sum_j \Im \lambda_j(A) \leq \operatorname{tr} A_\Im,$$

and we have equality in (2) if and only if A has a complete system of eigenvectors and generalized eigenvectors.

PROOF. Let H be the space on which A acts, and let E_A be the smallest subspace of H containing all eigenvectors and generalized eigenvectors corresponding to non-zero eigenvalues of A. We first show (under the condition $A_{\Im} \geq 0$) the operator A has a complete system of eigenvectors and generalized eigenvectors if and only if $E_A^{\perp} \subset \operatorname{Ker} A$.

Take $x \in \operatorname{Ker} A$. Then $\langle Ax, x \rangle = 0$ and $\langle A^*x, x \rangle = \langle x, Ax \rangle = 0$. It follows that $\langle A_{\Im} x, x \rangle = 0$. The fact that $A_{\Im} \geq 0$ implies that exists a unique non-negative operator S such that $S^2 = A_{\Im}$ ([GG], Corollary VI.1.2). Note that

$$\|Sx\|^2 = \langle Sx, Sx \rangle = \langle S^2 x, x \rangle = \langle A_{\Im} x, x \rangle = 0,$$

and thus $Sx = 0$. But then $A_{\Im} x = S(Sx) = 0$, and hence $A^*x = 0$. We may conclude that $\operatorname{Ker} A \subset \operatorname{Ker} A^*$. Repeating the above argument for A^* instead of A shows that $\operatorname{Ker} A = \operatorname{Ker} A^*$, and thus

$$(3) \qquad\qquad H = \operatorname{Ker} A \oplus \overline{\operatorname{Im} A}.$$

The latter identity implies that A has no generalized eigenvectors corresponding to the eigenvalue $\lambda = 0$. Indeed, assume $x_0, x_1, \ldots, x_{r-1}$ is a Jordan chain of A corresponding to $\lambda = 0$. Then $Ax_1 = x_0$ and $Ax_0 = 0$. Thus $0 \neq x_0 \in \operatorname{Ker} A \cap \operatorname{Im} A$, which contradicts (3). Note that $E_A \subset \overline{\operatorname{Im} A}$. It follows that A has a complete system of eigenvectors and generalized eigenvectors if and only if $E_A = \overline{\operatorname{Im} A}$, which happens if and only if $E_A^{\perp} \subset \operatorname{Ker} A$.

Next, we take a Schur basis in E_A, i.e., an orthonormal basis $\varphi_1, \varphi_2, \ldots$ of E_A such that

$$(4) \qquad\qquad A\varphi_j = a_{1j}\varphi_1 + \cdots + a_{jj}\varphi_j, \qquad a_{jj} = \lambda_j(A).$$

Let $\psi_1, \psi_2, \ldots$ be an orthonormal basis of $E_A^{\perp}$. By Theorem 2.2,

$$(5) \qquad\qquad \operatorname{tr} A_{\Im} = \sum_{j=1}^{\infty} \langle A_{\Im}\varphi_j, \varphi_j \rangle + \sum_{j=1}^{\infty} \langle A_{\Im}\psi_j, \psi_j \rangle.$$

Note that $\langle A_{\Im}\varphi, \varphi \rangle = \Im\langle A\varphi, \varphi \rangle$ for each $\varphi \in H$. In particular, $\langle A_{\Im}\varphi_j, \varphi_j \rangle = \Im\langle A\varphi_j, \varphi_j \rangle$ for each j. According to (4), we have $\langle A\varphi_j, \varphi_j \rangle = \lambda_j(A)$. We conclude that

$$\sum_{j=1}^{\infty} \langle A_{\Im}\varphi_j, \varphi_j \rangle = \sum_j \Im\lambda_j(A).$$

Since $A_{\Im} \geq 0$, all terms in the right hand side of (5) are non-negative. It follows that (2) holds and we have equality in (2) if and only if

$$(6) \qquad\qquad \langle A_{\Im}\psi_j, \psi_j \rangle = 0 \qquad (j = 1, 2, \ldots).$$

Let S be the unique non-negative operator such that $S^2 = A_{\Im}$. Then $\langle A_{\Im}\psi_j, \psi_j \rangle = 0$ implies that $S\psi_j = 0$, and hence also $A_{\Im}\psi_j = S^2\psi_j = 0$. So (6) is equivalent to the requirement that $A_{\Im}\varphi = 0$ for each $\varphi \in E_A^{\perp}$.

Assume $A_\Im$ is zero on $E_A^\perp$. Since $A^* E_A^\perp \subset E_A^\perp$, we conclude that A leaves $E_A^\perp$ invariant and the operators A and A^* coincide on $E_A^\perp$. Then $B := A|E_A^\perp$ is a compact selfadjoint operator which has no non-zero eigenvalues (cf. Lemma II.3.4). Hence $B = 0$, and we see that $E_A^\perp \subset \operatorname{Ker} A$.

Conversely, if $E_A^\perp \subset \operatorname{Ker} A$, then $A_\Im$ is zero on $E_A^\perp$. Thus we have equality in (2) if and only if $E_A^\perp \subset \operatorname{Ker} A$, which, as we pointed out, occurs if and only if $E_A = \overline{\operatorname{Im} A}$. $\square$

PROOF OF THEOREM 8.1. According to the first part of Theorem 6.1,

$$(7) \qquad\qquad \operatorname{tr} A = \sum_j \lambda_j(A).$$

Since A is a trace class operator, the same is true for A^*. Hence $A_\Im = \frac{1}{2i}(A - A^*)$ is a trace class operator and

$$\operatorname{tr} A_\Im = \frac{1}{2i} \operatorname{tr} A - \frac{1}{2i} \operatorname{tr} A^* = \sum_j \Im \lambda_j(A).$$

But then we can apply Theorem 8.2 to show that A has a complete system of eigenvectors and generalized eigenvectors. $\square$

Note that the proof of Theorem 8.1 requires the fundamental theorem for the trace (i.e., formula (7)), but to apply Theorem 8.1 no information about the eigenvalues of A is needed.

CHAPTER VIII

HILBERT-SCHMIDT OPERATORS

This chapter concerns another important set of compact operators, namely the Hilbert-Schmidt operators. The class of Hilbert-Schmidt operators has a natural Hilbert space structure and it contains the trace class operators. In the first section some additional properties of s-numbers are derived. In the second section the Hilbert-Schmidt operators are introduced and their main properties are established. The last section contains a completeness theorem for eigenvectors and generalized eigenvectors of Hilbert-Schmidt operators.

VIII.1 FURTHER INEQUALITIES ABOUT s-NUMBERS

Throughout the chapter we assume that *the underlying spaces are separable Hilbert spaces*. In this section we shall prove the following basic inequality for s-numbers of a compact operator A:

$$(1) \qquad \sum_{j=1}^{k} s_j(A^q)^{r/q} \leq \sum_{j=1}^{k} s_j(A)^r, \qquad k = 1, 2, \ldots .$$

Here $r > 0$ is an arbitrary positive number and q is an arbitrary positive integer. The proof of this inequality is based on the following theorem.

THEOREM 1.1. *Let A be a compact operator on the Hilbert space H, and let $\varphi_1, \ldots, \varphi_k$ be vectors in H. Then*

$$(2) \qquad \det((A\varphi_i, A\varphi_j))_{i,j=1}^{k} \leq \left(\prod_{j=1}^{k} s_j(A)^2 \right) \det((\varphi_i, \varphi_j))_{i,j=1}^{k}.$$

Furthermore, if $\varphi_1, \ldots, \varphi_k$ are the first k eigenvectors of A^*A in the Schmidt form of A, then equality holds in (2).

PROOF. If $\varphi_1, \ldots, \varphi_k$ are linearly dependent, then the same is true of $A\varphi_1, \ldots, A\varphi_k$, and hence in that case the left and right hand sides of (2) are both equal to zero. So, we may assume that $\varphi_1, \ldots, \varphi_k$ are linearly independent.

Let P be the orthogonal projection onto the space M spanned by $\varphi_1, \ldots, \varphi_k$. Choose an orthonormal basis $\psi_1, \ldots, \psi_k$ in $M = \operatorname{Im} P$. There exist an invertible linear operator $E \colon M \to M$ such that $E\psi_i = \varphi_i$ for $i = 1, \ldots, k$. Note that

$$\langle A\varphi_i, A\varphi_j \rangle = \langle E^* P A^* A P E \psi_i, \psi_j \rangle.$$

Define $D \colon M \to M$ by setting $D\varphi = PA^*AP\varphi$ for $\varphi \in M$. As $\psi_1, \ldots, \psi_k$ is an orthonormal basis, we have

$$(3) \qquad \begin{aligned} \det((A\varphi_i, A\varphi_j)) &= \det(E^* P A^* A P E) \\ &= \det(E^* D E) = (\det D)(\det E^* E). \end{aligned}$$

Again, using the fact that $\psi_1, \ldots, \psi_k$ is an orthonormal basis for M, we have

$$(4) \qquad \det E^* E = \det(\langle E^* E \psi_i, \psi_j \rangle) = \det(\langle \varphi_i, \varphi_j \rangle).$$

Furthermore,

$$(5) \qquad \det D = \prod_{j=1}^{k} \lambda_j(D) = \prod_{j=1}^{k} s_j(PA^*AP) \le \prod_{j=1}^{k} s_j(A)^2.$$

By inserting (4) and (5) in (3) one obtains the desired inequality (2).

If $\varphi_1, \ldots, \varphi_k$ are the first k eigenvectors of A^*A in a Schmidt representation of A, then $s_j(PA^*AP) = s_j(A^*A) = s_j(A)^2$ for $j = 1, \ldots, k$. So in that case we have equality in formula (5), and thus in (2) also. $\square$

Formula (2) has an interesting geometric interpretation. Recall that $\det(\langle \varphi_i, \varphi_j \rangle)$ is just equal to the Gram determinant of the vectors $\varphi_1, \ldots, \varphi_k$. Hence, if $\varphi_1, \ldots, \varphi_k$ are linearly independent, then $\det(\langle \varphi_i, \varphi_j \rangle)$ is equal to the square of the volume of the parallelopiped spanned by $\varphi_1, \ldots, \varphi_k$. Now assume that A is an invertible linear transformation. Then formula (2) says that after applying the linear transformation A the volume of the parallelopiped is at most equal to the volume of the original parallelopiped multiplied by the product of the first k singular values of A.

As a consequence of Theorem 1.1 we derive the following inequality.

COROLLARY 1.2. *Let A and B be compact operators on a Hilbert space H. Then*

$$(6) \qquad \prod_{j=1}^{k} s_j(AB) \le \prod_{j=1}^{k} s_j(A)s_j(B), \qquad k = 1, 2, \ldots .$$

PROOF. Let $\varphi_1, \ldots, \varphi_k$ be the first k eigenvectors of B^*A^*AB in the Schmidt form of AB. According to Theorem 1.1, we have

$$(7) \qquad \det(\langle AB\varphi_i, AB\varphi_j \rangle)_{i,j=1}^{k} = \left(\prod_{j=1}^{k} s_j(AB)^2 \right) \det(\langle \varphi_i, \varphi_j \rangle)_{i,j=1}^{k}.$$

Now apply the inequality (2), first for A with respect to the vectors $B\varphi_1, \ldots, B\varphi_k$ and next for B with respect to $\varphi_1, \ldots, \varphi_k$. One obtains

$$(8) \qquad \begin{aligned} \det(\langle AB\varphi_i, AB\varphi_j \rangle)_{i,j=1}^{k} &\le \left(\prod_{j=1}^{k} s_j(A)^2 \right) \det(\langle B\varphi_i, B\varphi_j \rangle)_{i,j=1}^{k} \\ &\le \left(\prod_{j=1}^{k} s_j(A)^2 s_j(B)^2 \right) \det(\langle \varphi_i, \varphi_j \rangle)_{i,j=1}^{k}. \end{aligned}$$

As $\varphi_1, \ldots, \varphi_k$ is an orthonormal system, the determinant $\det(\langle \varphi_i, \varphi_j \rangle) = 1$. So (7) and (8) together give the desired inequality (6). $\square$

We are now able to prove the inequality (1). Let q be an arbitrary positive integer. By repeatedly applying (6) it is clear that

$$(9) \qquad \prod_{j=1}^{k} s_j(A^q) \leq \prod_{j=1}^{k} s_j(A)^q, \qquad k = 1, 2, \ldots .$$

Put $a_j = \log s_j(A^q)$ and $b_j = \log s_j(A)^q$. Obviously $a_1 \geq a_2 \geq a_3 \geq \cdots$ and $b_1 \geq b_2 \geq b_3 \geq \cdots$. Furthermore, from (9) we have

$$\sum_{j=1}^{k} a_j \leq \sum_{j=1}^{k} b_j, \qquad k = 1, 2, \ldots .$$

Take $r > 0$, and put $f(t) = e^{\frac{r}{q}t}$. The first and second derivative of f are positive functions on the real line. So by Corollary VI.2.3,

$$\sum_{j=1}^{k} f(a_j) \leq \sum_{j=1}^{k} f(b_j).$$

But $f(a_j) = s_j(A^q)^{r/q}$ and $f(b_j) = s_j(A)^r$, and thus inequality (1) is satisfied.

COROLLARY 1.3. *Let A and B be compact operators on a Hilbert space H. Then*

$$\sum_{j=1}^{k} s_j(AB) \leq \sum_{j=1}^{k} s_j(A)s_j(B), \qquad k = 1, 2, \ldots .$$

PROOF. One uses the same arguments as in the paragraph preceding the present corollary. Put $a_j = \log s_j(AB)$ and $b_j = \log\big(s_j(A)s_j(B)\big)$, and apply Corollary VI.2.3 for the case when $f(t) = e^t$. $\square$

VIII.2 HILBERT-SCHMIDT OPERATORS

A compact linear operator A on a Hilbert space is said to be a *Hilbert-Schmidt operator* if A^*A is a trace class operator.

THEOREM 2.1. *For a compact linear operator A the following statements are equivalent:*

  (i)   *A is a Hilbert-Schmidt operator;*

  (ii) *$\sum_{j=1}^{\infty} \|A\varphi_j\|^2 < \infty$ for some (for any) orthonormal basis $\varphi_1, \varphi_2, \varphi_3, \ldots$;*

  (iii) *$\sum_{j,k=1}^{\infty} |\langle A\varphi_j, \varphi_k\rangle|^2 < \infty$ for some (for any) orthonormal basis*

      $\varphi_1, \varphi_2, \varphi_3, \ldots$;

  (iv) *$\sum_{j=1}^{\infty} s_j(A)^2 < \infty$.*

PROOF. Let $\lambda_1(A^*A), \lambda_2(A^*A), \ldots$ be the sequence of nonzero eigenvalues of A^*A. We extend this sequence to an infinite sequence by adding zero elements if necessary. For any orthonormal basis $\varphi_1, \varphi_2, \ldots$ one has

$$\begin{aligned}
\sum_{j=1}^{\infty} \lambda_j(A^*A) &= \sum_{j=1}^{\infty} \langle A^*A\varphi_j, \varphi_j \rangle \\
&= \sum_{j=1}^{\infty} \|A\varphi_j\|^2 = \sum_{j,k=1}^{\infty} |\langle A\varphi_j, \varphi_k \rangle|^2.
\end{aligned} \tag{1}$$

In formula (1) we allow the sums to be infinite. Note that the last equality in (1) is a direct consequence of the identity of Parseval. The second equality in (1) is obvious from $\langle A^*A\varphi, \varphi \rangle = \|A\varphi\|^2$. To prove the first equality in (1) we use the spectral theorem for A^*A. Write

$$A^*A = \sum_{\nu=1}^{\infty} \lambda_\nu(A^*A) \langle \cdot, \psi_\nu \rangle \psi_\nu,$$

where $\psi_1, \psi_2, \ldots$ is an orthonormal system of eigenvectors of A^*A. Note that

$$\langle A^*A\varphi_j, \varphi_j \rangle = \sum_{\nu=1}^{\infty} \lambda_\nu(A^*A) |\langle \varphi_j, \psi_\nu \rangle|^2.$$

As all terms are nonnegative we have

$$\begin{aligned}
\sum_{j=1}^{\infty} \langle A^*A\varphi_j, \varphi_j \rangle &= \sum_{\nu=1}^{\infty} \lambda_\nu(A^*A) \left(\sum_{j=1}^{\infty} |\langle \varphi_j, \psi_\nu \rangle|^2 \right) \\
&= \sum_{\nu=1}^{\infty} \lambda_\nu(A^*A).
\end{aligned}$$

This proves formula (1).

From formula (1) the equivalence of the statements (i), (ii) and (iii) is evident. Moreover, as $\lambda_j(A^*A) = s_j(A)^2$, the implications (i) $\Leftrightarrow$ (iv) are trivial. $\square$

Examples of Hilbert-Schmidt operators are easy to find. Let $(a_{ij})_{i,j=1}^{\infty}$ be an infinite matrix with complex entries, and assume that $A: \ell_2 \to \ell_2$ is defined by

$$(Ax)_i = \sum_{j=1}^{\infty} a_{ij}x_j, \qquad i = 1, 2, \ldots . \tag{2}$$

Then A is a Hilbert-Schmidt operator if and only if

$$\sum_{i,j=1}^{\infty} |a_{ij}|^2 < \infty. \tag{3}$$

To see this, let $e_1, e_2, \ldots$ be the standard orthonormal basis of ℓ_2. Note that

$$\langle Ae_j, e_i \rangle = (Ae_j)_i = a_{ij}. \tag{4}$$

Thus, if A is Hilbert-Schmidt, then (3) holds true because of Theorem 2.1(iii). Conversely, if (3) is true, then A is a compact operator, and applying Theorem 2.1, again, one sees that A is Hilbert-Schmidt.

The second example concerns the integral operator

$$(5) \qquad (K\varphi)(t) = \int_0^1 k(t,s)\varphi(s)ds, \quad 0 \le t \le 1, \quad \text{a.e.,}$$

on $L_2([0,1])$. If the kernel function k is square integrable, then the operator K is Hilbert-Schmidt. To prove this, we choose an orthonormal basis $(\varphi_j)_{j=1}^\infty$ of $L_2([0,1])$. Define $\psi_{ij}(t,s) = \varphi_i(t)\overline{\varphi_j(s)}$. We know that $(\psi_{ij})_{i,j=1}^\infty$ is an orthonormal basis of $L_2([0,1] \times [0,1])$. We have

$$\langle K\varphi_\alpha, \varphi_\beta \rangle = \int_0^1 \int_0^1 k(t,s)\varphi_\alpha(s)\overline{\varphi_\beta(t)}dsdt$$

$$= \int_0^1 \int_0^1 k(t,s)\overline{\psi_{\beta\alpha}(t,s)}dsdt$$

$$= \langle k, \psi_{\beta\alpha} \rangle,$$

where the latter inner product denotes the inner product on $L_2([0,1] \times [0,1])$. As $k \in L_2([0,1] \times [0,1])$, it follows that

$$(6) \qquad \sum_{\alpha,\beta} |\langle K\varphi_\alpha, \varphi_\beta \rangle|^2 = \|k\|^2 < \infty.$$

Since K is compact, Theorem 2.1 implies that K is Hilbert-Schmidt.

The preceding example is, in a certain sense, a universal model for a Hilbert-Schmidt operator. More precisely, given a Hilbert-Schmidt operator $A\colon H \to H$, there exists a unitary operator $U\colon H \to L_2([0,1])$ such that the operator UAU^{-1} is an integral operator on $L_2([0,1])$ with a square integrable kernel function. Take an orthonormal basis (ω_j) in H and an orthonormal basis (φ_j) in $L_2([0,1])$. Define $U\colon H \to L_2([0,1])$ by setting $U\omega_j = \varphi_j$, $j = 1, 2, \dots$. Consider $\psi_{ij}(t,s) = \varphi_i(t)\overline{\varphi_j(s)}$, and put

$$k = \sum_{\alpha,\beta} \langle A\omega_\alpha, \omega_\beta \rangle \psi_{\beta\alpha}.$$

The fact that $\sum_{\alpha,\beta} |\langle A\omega_\alpha, \omega_\beta \rangle|^2 < \infty$ implies that k is square integrable. Let $K\colon L_2([0,1]) \to L_2([0,1])$ be the corresponding integral operator. From what we proved above for the integral operator K we know that

$$\langle K\varphi_\alpha, \varphi_\beta \rangle = \langle k, \psi_{\beta\alpha} \rangle = \langle A\omega_\alpha, \omega_\beta \rangle.$$

But then

$$UAU^{-1}\varphi_\alpha = UA\omega_\alpha = U\left(\sum_\beta \langle A\omega_\alpha, \omega_\beta\rangle\omega_\beta\right)$$

$$= \sum_\beta \langle A\omega_\alpha, \omega_\beta\rangle\varphi_\beta$$

$$= \sum_\beta \langle K\varphi_\alpha, \varphi_\beta\rangle\varphi_\beta = K\varphi_\alpha,$$

which proves that $UAU^{-1} = K$.

The set of all Hilbert-Schmidt operators will be denoted by S_2. On S_2 we define a norm:

$$(7) \qquad \|A\|_2 = \left(\sum_{j=1}^\infty s_j(A)^2\right)^{1/2}.$$

We shall refer to (7) as the *Hilbert-Schmidt norm* of A. From formula (1) it is clear that

$$(8) \qquad \|A\|_2 = \left(\sum_{j,k=1}^\infty |\langle A\varphi_j, \varphi_k\rangle|^2\right)^{1/2}$$

for any orthonormal basis $\varphi_1, \varphi_2, \ldots$ in the underlying Hilbert space. This identity can be used to compute the Hilbert-Schmidt norm in concrete cases. For example, if $A: \ell_2 \to \ell_2$ is given by (2), then

$$\|A\|_2 = \left(\sum_{i,j=1}^\infty |a_{ij}|^2\right)^{1/2}.$$

This follows immediately from (4) and the identity (8). Also, the Hilbert-Schmidt norm of the integral operator K in (5) is given by

$$\|K\|_2 = \left(\int_0^1 \int_0^1 |k(t,s)|^2 dt ds\right)^{1/2}.$$

This is clear from (6) and (8).

The set S_2 is a normed operator ideal, whose norm comes from a Hilbert space structure. This is the contents of the next theorem. First we prove the following lemma.

LEMMA 2.2. *If A and B are in S_2, then $AB \in S_1$ and*

$$(9) \qquad \|AB\|_1 \leq \|A\|_2\|B\|_2.$$

If, in addition, $B = A^*$, then equality holds in (9).

PROOF. From Corollary 1.3 it is clear that

$$\sum_{j=1}^{\infty} s_j(AB) \le \sum_{j=1}^{\infty} s_j(A)s_j(B) \le \|A\|_2\|B\|_2.$$

This proves (9). If, in addition, $B = A^*$, then (by Corollary VI.1.2)

$$\|AA^*\|_1 = \sum_{j=1}^{\infty} s_j(AA^*) = \sum_{j=1}^{\infty} \lambda_j(AA^*) = \sum_{j=1}^{\infty} s_j(A^*)^2$$
$$= \sum_{j=1}^{\infty} s_j(A)^2 = \|A\|_2^2. \quad \square$$

THEOREM 2.3. *The set of all Hilbert-Schmidt operators on a Hilbert space H is a Hilbert space with respect to the inner product*

$$\langle A, B \rangle = \operatorname{tr} AB^*, \tag{10}$$

and the corresponding norm is equal to the Hilbert-Schmidt norm $\| \cdot \|_2$. Furthermore, if $A\colon H \to H$ is Hilbert-Schmidt and $B\colon H_0 \to H$ and $C\colon H \to H_0$ are bounded linear operators, then CAB is Hilbert-Schmidt and

$$\|CAB\|_2 \le \|C\|\|A\|_2\|B\|. \tag{11}$$

For operators of rank one the Hilbert-Schmidt norm $\|\cdot\|_2$ coincides with the usual operator norm, while in general

$$\|A\| \le \|A\|_2 = \|A^*\|_2, \qquad A \in S_2. \tag{12}$$

PROOF. We first prove the second part of the theorem. From the inequality $s_j(CAB) \le \|C\|s_j(A)\|B\|$ formula (11) is clear. If A has rank one, then $s_j(A) = 0$ for $j \ge 2$, and hence in that case

$$\|A\|_2 = \left(\sum_j s_j(A)^2\right)^{1/2} = s_1(A) = \|A\|.$$

In general, we have $\|A\|_2 \ge s_1(A) = \|A\|$. From $s_j(A) = s_j(A^*)$ it follows that $\|A\|_2 = \|A^*\|_2$. So (12) is proved.

Take $A, B\colon H \to H$ in S_2, and let $\varphi_1, \varphi_2, \ldots$ be an orthonormal basis in H. Note that $(\|A\varphi_j\|)_j$ and $(\|B\varphi_j\|)_j$ belong to ℓ_2. So $(\|A\varphi_j\| + \|B\varphi_j\|)_j$ is a square summable sequence, and hence

$$\sum_j \|(A + B)\varphi_j\|^2 \le \sum_j (\|A\varphi_j\| + \|B\varphi_j\|)^2 < \infty.$$

This shows that $A + B \in S_2$. As $\|(\alpha A)\varphi_j\| = |\alpha|\|A\varphi_j\|$, one also sees that $\alpha A \in S_2$. So the set of all Hilbert-Schmidt operators on H is a vector space.

If $A, B\colon H \to H$ are in S_2, then, by Lemma 2.2, the operator $AB^* \in S_1$. So the sesquilinear form (10) is well-defined. From the properties of the trace it is clear that $\langle A, B \rangle$ is an inner product. Moreover,

$$\langle A, A \rangle = \operatorname{tr} AA^* = \operatorname{tr} A^*A = \sum_{j=1}^{\infty} \lambda_j(A^*A)$$

$$= \sum_{j=1}^{\infty} s_j(A)^2 = \|A\|_2^2.$$

To prove that the inner product of S_2 is complete, let $A_n\colon H \to H$, $n = 1, 2, \ldots$, be a Cauchy sequence in the Hilbert-Schmidt norm $\|\cdot\|_2$. As $\|A\| \leq \|A\|_2$, the sequence (A_n) is a Cauchy sequence in the usual operator norm, and hence

(13) $$A = \lim_{n \to \infty} A_n$$

exists in $\mathcal{L}(H)$. From (9) and (12) we see that

$$\|A_n A_n^* - A_k A_k^*\|_1 \leq \|A_n - A_k\|_2 \|A_k^*\|_2 + \|A_n\|_2 \|A_n^* - A_k^*\|_2,$$

and hence $(A_n A_n^*)$ is a Cauchy sequence in S_1. But S_1 is a Banach space. So $\lim_{n \to \infty} A_n A_n^*$ exists in S_1. Since (13) holds in the usual operator norm, the operator A is compact and

$$AA^* = \lim_{n \to \infty} A_n A_n^* \in S_1.$$

Hence $A \in S_2$. Now $(A_n A^*)$ is also a Cauchy sequence in S_1. So, using the same arguments as before, we conclude that $A_n A^* \to AA^*$ in S_1. Taking adjoints we see that also $AA_n^* \to AA^*$ in S_1. Hence, by Lemma 2.2,

$$\|A_n - A\|_2^2 = \|(A_n - A)(A_n^* - A^*)\|_1$$
$$\leq \|A_n A_n^* - AA^*\|_1 + \|A_n A^* - AA^*\|_1 + \|AA_n^* - AA^*\|_1$$

tends to zero if $n \to \infty$. Hence (A_n) converges in S_2. $\square$

The fact that the integral operator (5) is a universal model of a Hilbert-Schmidt operator implies that the Hilbert space S_2 is linearly isometric to $L_2([0,1] \times [0,1])$. Indeed, let $\omega_1, \omega_2, \ldots$ and $\varphi_1, \varphi_2, \ldots$ be orthonormal bases in H and $L_2([0,1])$, respectively, and define

(14) $$J\colon L_2([0,1] \times [0,1]) \to S_2, \qquad Jk = U^{-1}KU,$$

where K is the integral operator given by (5) and $U\colon H \to L_2([0,1])$ is the unitary operator defined by $U\omega_j = \varphi_j$. Obviously, J is linear and

$$\|Jk\|_2 = \|K\|_2 = \|k\|_{L_2([0,1] \times [0,1])}.$$

Thus J is a linear isometry. To see that J is onto, note that $J^{-1}(A) = k$ with

$$k = \sum_{\alpha,\beta} \langle A\omega_\alpha, \omega_\beta \rangle \psi_{\beta\alpha},$$

where $\psi_{\beta\alpha}(t,s) = \varphi_\beta(t)\overline{\varphi_\alpha(s)}$.

Since $L_2([0,1] \times [0,1])$ is a Hilbert space, one could use the isometry J in (14) to give a quick proof of the fact that any Cauchy sequence in S_2 is convergent.

COROLLARY 2.4. *The operators of finite rank are dense in* S_2.

PROOF. We use the notation introduced in the two paragraphs preceding this corollary. Let M be the linear space of all functions of the form

$$h = \sum_{\alpha,\beta=1}^{n} c_{\alpha\beta}\psi_{\beta\alpha},$$

where n is an arbitrary positive integer and $c_{\alpha\beta}$ are arbitrary complex numbers. Obviously, M is dense in $L_2([0,1] \times [0,1])$. Now apply the isometry J defined by (14). It follows that $J(M)$ is dense in S_2. Since $J(h)$ is a finite rank operator for each h in M, we conclude that the finite rank operators are dense in S_2. $\square$

VIII.3 COMPLETENESS FOR HILBERT-SCHMIDT OPERATORS

THEOREM 3.1. *Let A be a Hilbert-Schmidt operator, and assume that* $A_\Re = \frac{1}{2}(A + A^*)$ *and* $A_\Im = \frac{1}{2i}(A - A^*)$ *are nonnegative operators. Then the system of eigenvectors and generalized eigenvectors of A is complete.*

PROOF. Consider the Schur decomposition of H corresponding to A, that is, $H = E_A \oplus E_A^\perp$, where E_A is the smallest subspace containing all the eigenvectors and generalized eigenvectors of A corresponding to non-zero eigenvalues. Since $A_\Im$ is nonnegative, we have to show (see the first paragraph of the proof of Theorem VII.8.2) that $E_A^\perp \subset \operatorname{Ker} A$.

Let P be the orthogonal projection of H onto $E_A^\perp$, and consider the operator

$$B := PA|E_A^\perp : E_A^\perp \to E_A^\perp.$$

We know that B is a Volterra operator (Lemma II.3.4). As $s_j(B) \leq s_j(A)$, the operator B is also Hilbert-Schmidt. Furthermore,

$$\langle B\varphi, \varphi \rangle = \langle A\varphi, \varphi \rangle, \qquad \langle B^*\varphi, \varphi \rangle = \langle A^*\varphi, \varphi \rangle$$

for each $\varphi \in E_A^\perp$, and thus $B_\Re = \frac{1}{2}(B + B^*)$ and $B_\Im = \frac{1}{2i}(B - B^*)$ are nonnegative operators as well. So B is an operator of the same type as A. It suffices to show that $B = 0$. Indeed, if B is the zero operator, then

$$0 = \langle B\varphi, \varphi \rangle = \langle A\varphi, \varphi \rangle = \langle A_\Re\varphi, \varphi \rangle + i\langle A_\Im\varphi, \varphi \rangle$$

for each $\varphi \in E_A^\perp$. In particular, $\langle A_\Re\varphi, \varphi \rangle = 0$ and $\langle A_\Im\varphi, \varphi \rangle = 0$ for $\varphi \in E_A^\perp$. As $A_\Re$ is nonnegative, there exists a unique nonnegative operator C such that $C^2 = A_\Re$. Obviously, $\|C\varphi\|^2 = \langle A_\Re\varphi, \varphi \rangle$. It follows that $C\varphi = 0$ for each $\varphi \in E_A^\perp$. But then

$A_{\mathfrak{R}}\varphi = C^2\varphi = 0$ for $\varphi \in E_A^\perp$. Similarly, $A_{\mathfrak{S}}\varphi = 0$ for each $\varphi \in E_A^\perp$. Thus $A\varphi = A_{\mathfrak{R}}\varphi + iA_{\mathfrak{S}}\varphi = 0$ for $\varphi \in E_A^\perp$, which proves that $E_A^\perp \subset \operatorname{Ker} A$.

Let us prove that $B = 0$. As B is Hilbert-Schmidt and Volterra, the operator B^2 is trace class and Volterra. So $\operatorname{tr} B^2 = 0$ by Theorem VII.6.1(i). Note that

$$(1) \qquad B^2 = (B_{\mathfrak{R}}^2 - B_{\mathfrak{S}}^2) + i(B_{\mathfrak{R}}B_{\mathfrak{S}} + B_{\mathfrak{S}}B_{\mathfrak{R}}).$$

The fact that B is Hilbert-Schmidt, implies that $B_{\mathfrak{R}}$ and $B_{\mathfrak{S}}$ are Hilbert-Schmidt, and hence all the terms in (1) are trace class operators. Using the linearity of the trace we have

$$(2) \qquad 0 = \operatorname{tr} B^2 = (\operatorname{tr} B_{\mathfrak{R}}^2 - \operatorname{tr} B_{\mathfrak{S}}^2) + i\big(\operatorname{tr}(B_{\mathfrak{R}}B_{\mathfrak{S}}) + \operatorname{tr}(B_{\mathfrak{S}}B_{\mathfrak{R}})\big).$$

Note that $B_{\mathfrak{R}}^2$ and $B_{\mathfrak{S}}^2$ are selfadjoint. So $\operatorname{tr} B_{\mathfrak{R}}^2 - \operatorname{tr} B_{\mathfrak{S}}^2$ is a real number. Furthermore,

$$\operatorname{tr}(B_{\mathfrak{R}}B_{\mathfrak{S}}) = \operatorname{tr}(B_{\mathfrak{S}}B_{\mathfrak{R}}) = \operatorname{tr}\big((B_{\mathfrak{R}}B_{\mathfrak{S}})^*\big) = \overline{\operatorname{tr}(B_{\mathfrak{R}}B_{\mathfrak{S}})}.$$

So $\operatorname{tr}(B_{\mathfrak{R}}B_{\mathfrak{S}}) + \operatorname{tr}(B_{\mathfrak{S}}B_{\mathfrak{R}}) = 2\operatorname{tr}(B_{\mathfrak{R}}B_{\mathfrak{S}})$ is a real number too. But then (2) yields $\operatorname{tr}(B_{\mathfrak{R}}B_{\mathfrak{S}}) = 0$. We shall prove that this implies that $B_{\mathfrak{R}}[\operatorname{Im} B_{\mathfrak{S}}] = (0)$.

Using the spectral theorem we can write

$$B_{\mathfrak{S}} = \sum_j \lambda_j(B_{\mathfrak{S}})(\cdot, \psi_j)\psi_j,$$

where $\psi_1, \psi_2, \ldots$ is an orthonormal system in $E_A^\perp$. Put $M = \operatorname{Im} B_{\mathfrak{S}}$. Note that M is the closed linear span of the vectors $\psi_1, \psi_2, \ldots$. As $B_{\mathfrak{S}}\varphi = 0$ for each φ in the orthogonal complement of M in $E_A^\perp$, we have

$$\operatorname{tr}(B_{\mathfrak{R}}B_{\mathfrak{S}}) = \sum_\nu \langle B_{\mathfrak{R}}B_{\mathfrak{S}}\psi_\nu, \psi_\nu \rangle.$$

But $B_{\mathfrak{S}}\psi_\nu = \lambda_\nu(B_{\mathfrak{S}})\psi_\nu$. So

$$(3) \qquad 0 = \operatorname{tr}(B_{\mathfrak{R}}B_{\mathfrak{S}}) = \sum_\nu \lambda_\nu(B_{\mathfrak{S}})\langle B_{\mathfrak{R}}\psi_\nu, \psi_\nu \rangle.$$

Note that all terms in the right hand side of (3) are nonnegative and $\lambda_\nu(B_{\mathfrak{S}}) > 0$. It follows that $\langle B_{\mathfrak{R}}\psi_\nu, \psi_\nu \rangle = 0$ for all ν. Since $B_{\mathfrak{R}}$ is nonnegative, we can use the square root argument as before to show that $B_{\mathfrak{R}}\psi_\nu = 0$ for all ν. Hence $B_{\mathfrak{R}}\varphi = 0$ for each $\varphi \in M = \operatorname{Im} B_{\mathfrak{S}}$.

Note that $B_{\mathfrak{S}}M \subset M$. The fact that $B_{\mathfrak{R}}[M] = (0)$, implies that $BM \subset M$ and

$$B_{\mathfrak{S}}|M = (-iB)|M.$$

Hence $B_{\mathfrak{S}}|M$ is a compact selfadjoint operator with no nonzero eigenvalues. It follows that $B_{\mathfrak{S}}$ is zero on M, and hence $0 = B_{\mathfrak{S}}\psi_\nu = \lambda_\nu(B_{\mathfrak{S}})\psi_\nu$ for all ν. But then $B_{\mathfrak{S}}$ must be the zero operator. So $B = B_{\mathfrak{R}}$ is a selfadjoint Volterra operator, and thus $B = 0$. This completes the proof of the theorem. $\square$

Theorem 3.1 is due to V.B. Lidskii [1]. To get completeness theorems for other classes of compact operators one has to use more powerful results than the trace formula in Theorem VII.6.1(i). In fact, one has to study the growth of the resolvent operator $(I - \lambda A)^{-1}$ at infinity. This will be done in Chapter X.

CHAPTER IX

INTEGRAL OPERATORS WITH SEMI-SEPARABLE KERNELS

In this chapter integral operators with semi-separable kernels are introduced and analyzed. Such operators turn out to be Hilbert-Schmidt, and their inversion properties may be read off from certain finite dimensional operators. In the case when the operators are also trace class, their trace and determinant may be computed explicitly in terms of the associated finite dimensional operators.

IX.1 DEFINITION AND EXAMPLES

Let $-\infty < a < b < \infty$. By $L_2^m([a,b])$ we denote the space of all $\mathbf{C}^m$-valued functions that are square integrable (relative to the Lebesgue measure) on $[a,b]$. Thus $\varphi\colon [a,b] \to \mathbf{C}^m$ belongs to $L_2^m([a,b])$ if and only if for each j the j-th component of φ is square integrable on $[a,b]$. As usual, two functions that are equal almost everywhere are identified. The space $L_2^m([a,b])$ is a Hilbert space with inner product

$$(1) \qquad \langle \varphi, \psi \rangle = \sum_{j=1}^{n} \int_a^b \varphi_j(t)\overline{\psi_j(t)}\,dt.$$

Here φ_j and ψ_j are the j-th components of φ and ψ, respectively.

An operator $K\colon L_2^m([a,b]) \to L_2^m([a,b])$ will be called an *integral operator* if its action is given by

$$(2) \qquad (K\varphi)(t) = \int_a^b k(t,s)\varphi(s)\,ds, \qquad a \le t \le b.$$

Here k is an $m \times m$ matrix function which is called the *kernel function* of K. Note that K may be written as

$$(3) \qquad K = \sum_{i=1}^{m} \sum_{j=1}^{m} \tau_i K_{ij} \pi_j,$$

where

$$\pi_j \colon L_2^m([a,b]) \to L_2([a,b]), \qquad \pi_j \varphi := \varphi_j\,;$$

$$\tau_i \colon L_2([a,b]) \to L_2^m([a,b]), \qquad \tau_i f := \begin{bmatrix} \delta_{1i} f \\ \vdots \\ \delta_{mi} f \end{bmatrix};$$

$$K_{ij} \colon L_2([a,b]) \to L_2([a,b]),$$

$$(K_{ij}f)(t) := \int_a^b k_{ij}(t,s)f(s)ds, \qquad a \le t \le b.$$

Here φ_j is the j-th component of φ, the symbol δ_{ki} stands for the Kronecker delta, and $k_{ij}(t,s)$ is the (i,j)-th entry of $k(t,s)$.

We say that K has a *semi-separable* kernel function if k admits the following representation:

$$(4) \qquad k(t,s) = \begin{cases} F_1(t)G_1(s), & a \le s \le t \le b, \\ F_2(t)G_2(s), & a \le t < s \le b. \end{cases}$$

Here $F_\nu(t)$ and $G_\nu(t)$ $(\nu = 1,2)$ are matrices of sizes $m \times n_\nu$ and $n_\nu \times m$, respectively, and as functions of t the entries of $F_\nu(t)$ and $G_\nu(t)$ are square integrable on $[a,b]$. In that case the right hand side of (4) is called a *semi-separable representation* of k, the integer $n := n_1 + n_2$ is called the *order* of the representation (4), and n_1 and n_2 will be referred to as the *lower order* and *upper order*, respectively. If (4) holds, then (2) can be rewritten as

$$(5) \qquad (K\varphi)(t) = F_1(t)\int_a^t G_1(s)\varphi(s)ds + F_2(t)\int_t^b G_2(s)\varphi(s)ds, \qquad a \le t \le b.$$

The kernel function

$$(6) \qquad \ell(t,s) = \sum_{\nu=1}^n c_\nu e^{d_\nu|t-s|}$$

is an example of a (scalar) semi-separable kernel function. Indeed, ℓ can be brought into the form (4) by taking

$$F_1(t) = [c_1 e^{d_1 t} \cdots c_n e^{d_n t}]$$

$$F_2(t) = [c_1 e^{-d_1 t} \cdots c_n e^{-d_n t}]$$

$$G_1(t) = \begin{bmatrix} e^{-d_1 t} \\ \vdots \\ e^{-d_n t} \end{bmatrix}, \qquad G_2(t) = \begin{bmatrix} e^{d_1 t} \\ \vdots \\ e^{d_n t} \end{bmatrix}.$$

As we shall see in Section XIV.3, semi-separable kernel functions also arise from Green's functions corresponding to certain differential operators.

PROPOSITION 1.1. *An integral operator with a semi-separable kernel function is a Hilbert-Schmidt operator.*

PROOF. We first consider the scalar case, i.e., $m = 1$. Then the semi-separability of the kernel function k implies that k is square integrable on each of the triangles $a \le s \le t \le b$ and $a \le t < s \le b$. It follows that k is a square integrable over $[a, b] \times [a, b]$, and hence the corresponding integral operator is a Hilbert-Schmidt operator (cf., Section VIII.2). The general case is reduced to the scalar case by using formula (3) and Theorem VIII.2.3. $\square$

In general, an integral operator with a semi-separable kernel function does not belong to the trace class operators. For example, consider the operator of integration:

$$(7) \qquad V\colon L_2([0, 1]) \to L_2([0, 1]), \quad (Vf)(t)\colon = 2i \int_t^1 f(s)ds, \qquad 0 \le t \le 1.$$

In Section VI.1 it was shown that the j-th singular value of V is equal to $4/(2j-1)\pi$, and thus V is not a trace class operator. However, the kernel function of V is semi-separable. Indeed, for $V = K$ formula (5) holds true with

$$F_1(t) = G_1(t) = 0, \quad F_2(t) = 2i, \quad G_2(t) = 1, \qquad 0 \le t \le 1.$$

All operators $K\colon L_2^m([a, b]) \to L_2^m([a, b])$ of finite rank are integral operators with a semi-separable kernel function. To prove this, assume $\operatorname{rank} K = n < \infty$. Then (see [GG], Theorem II.6.1) there exist vectors $g_1, \ldots, g_n$ in $L_2^m([a, b])$ and $f_1, \ldots, f_n$ in $L_2^m([a, b])$ such that for every $\varphi \in L_2^m([a, b])$:

$$(8) \qquad (K\varphi)(t) = \sum_{\nu=1}^n \langle \varphi, g_\nu \rangle f_\nu(t).$$

Now, let $F(t)$ be the $m \times n$ matrix of which the j-th column is equal to $f_j(t)$, and let $G(t)$ be the $n \times m$ matrix of which the i-th row is equal to $g_i(t)^*$. Here, for $x \in \mathbf{C}^n$ the symbol x^* denotes the row vector of which the j-th entry is equal to the complex conjugate of the j-th entry of x. Then (8) implies that

$$(K\varphi)(t) = F(t) \int_a^b G(s)\varphi(s)ds, \qquad a \le t \le b.$$

In particular, the kernel function k of K can be written as in (4) with $F_1 = F_2 = F$ and $G_1 = G_2 = G$.

The converse statement is also true: if in (4) we have $F_1 = F_2$ and $G_1 = G_2$, then the corresponding integral operator has finite rank. In general, as examples (6) and (7) show, integral operators with semi-separable kernel do not have finite rank.

IX.2 INVERSION

This section concerns the inversion of an operator $I - K$, where $K\colon L_2^m([a, b]) \to L_2^m([a, b])$ is an integrable operator with a semi-separable kernel function k. Let

$$(1) \qquad k(t, s) = \begin{cases} F_1(t)G_1(s), & a \le s \le t \le b, \\ F_2(t)G_2(s), & a \le t < s \le b, \end{cases}$$

be a semi-separable representation of k. With (1) we associate the following matrix

$$(2) \qquad A(t) = \begin{bmatrix} G_1(t)F_1(t) & G_1(t)F_2(t) \\ -G_2(t)F_1(t) & -G_2(t)F_2(t) \end{bmatrix}, \qquad a \leq t \leq b.$$

Since the entries of F_1, F_2 and G_1, G_2 are square integrable on $a \leq t \leq b$, it follows that entries of A are (Lebesgue) integrable on $a \leq t \leq b$. Note that $A(t)$ is a square matrix of size $n \times n$, where n is the order of the representation (1).

THEOREM 2.1. *Let* $K \colon L_2^m([a,b]) \to L_2^m([a,b])$ *be an integral operator with a semi-separable kernel function* k, *and let* (1) *be a semi-separable representation of* k *of order* n, *say.- Let* $U(t)$ *be the unique continuous* $n \times n$ *matrix function such that*

$$(3) \qquad U(t) = I_n + \int_a^t A(s)U(s)ds, \qquad a \leq t \leq b,$$

where $A(t)$ *is given by* (2) *and* I_n *is the* $n \times n$ *identity matrix. Partition* $U(t)$ *as a* 2×2 *block matrix according to the partitioning of the right hand side of* (2):

$$(4) \qquad U(t) = \begin{bmatrix} U_{11}(t) & U_{12}(t) \\ U_{21}(t) & U_{22}(t) \end{bmatrix}, \qquad a \leq t \leq b.$$

Then $I - K$ *is invertible if and only if* $\det U_{22}(b) \neq 0$. *Furthermore, in that case*

$$(5) \qquad ((I - K)^{-1}\varphi)(t) = \varphi(t) + \int_a^b \gamma(t,s)\varphi(s)ds, \qquad a \leq t \leq b,$$

with resolvent kernel

$$(6) \qquad \gamma(t,s) = \begin{cases} C(t)U(t)(I - P)U(s)^{-1}B(s), & s \leq t, \\ -C(t)U(t)PU(s)^{-1}B(s), & s > t, \end{cases}$$

where

$$(7) \qquad C(t) = [F_1(t) \quad F_2(t)], \qquad B(t) = \begin{bmatrix} G_1(t) \\ -G_2(t) \end{bmatrix},$$

$$(8) \qquad P = \begin{bmatrix} 0 & 0 \\ U_{22}(b)^{-1}U_{21}(b) & I_{n_2} \end{bmatrix}.$$

In (8) the integer n_2 is the upper order of the representation (1). The operator $V \colon \mathbb{C}^{n_2} \to \mathbb{C}^{n_2}$ defined by $Vx = U_{22}(b)x$ will be called the *indicator of* $I - K$ *associated with the representation* (1).

The proof of Theorem 2.1 goes in a number of steps. First we prove the existence of the matrix $U(t)$ and derive its properties (Lemma 2.2 below). Next we show that the operator $I - K$ and its indicator V are matricially coupled (Theorem 2.3 below), and finally we apply Corollary III.4.3 to get the desired results about the inversion of $I - K$.

LEMMA 2.2. *Let A be an $n \times n$ matrix of which the entries are Lebesgue integrable on $a \leq t \leq b$. Then there exists a unique continuous $n \times n$ matrix function $U(t)$ such that*

$$(9) \qquad U(t) = I_n + \int_a^t A(s)U(s)ds, \qquad a \leq t \leq b.$$

Furthermore, $\det U(t) \neq 0$ and

$$(10) \qquad U(t)^{-1} = I_n - \int_a^t U(s)^{-1}A(s)ds, \qquad a \leq t \leq b.$$

PROOF. Given an $n \times n$ matrix A the symbol $\|A\|$ denotes the norm of the operator on $\mathbf{C}^n$ induced by A. For $a \leq t \leq b$, put

$$U_1(t) = \int_a^t A(s)ds, \quad U_{k+1}(t) = \int_a^t A(s)U_k(s)ds, \qquad k \geq 1.$$

Then for $k \geq 1$,

$$(11) \qquad \|U_k(t)\| \leq \frac{1}{k!}\left(\int_a^t \|A(s)\|ds\right)^k, \qquad a \leq t \leq b.$$

This inequality can be proved by induction. Indeed, (11) holds for $k = 1$. Let

$$\mathcal{J}(t) = \int_a^t \|A(s)\|ds, \qquad a \leq t \leq b.$$

Suppose that (11) holds for $k = p - 1$. Then $\mathcal{J}'(t) = \|A(t)\|$ a.e. and

$$\|U_p(t)\| \leq \int_a^t \|A(s)\|\|U_{p-1}(s)\|ds$$

$$\leq \frac{1}{(p-1)!}\int_a^t \mathcal{J}'(s)\mathcal{J}(s)^{p-1}ds$$

$$= \frac{1}{p!}\mathcal{J}(t)^p = \frac{1}{p!}\left(\int_a^t \|A(s)\|ds\right)^p.$$

Thus (11) is established. Put

$$(12) \qquad U(t) = I_n + \sum_{k=1}^{\infty} U_k(t), \qquad a \le t \le b.$$

It follows from (11) that for each $a \le t \le b$

$$\|U_k(t)\| \le \frac{1}{k!} \mathcal{J}(b)^k, \qquad k \ge 1.$$

Thus the series in (12) converges uniformly on $a \le t \le b$. Since all U_k are continuous on $[a, b]$, we conclude that the same holds true for U. Furthermore, by Lebesgue's dominated convergence theorem,

$$\int_a^t A(t)U(s)ds = \int_a^t A(s)ds + \sum_{k=1}^{\infty} \int_a^t A(s)U_k(s)ds$$

$$= U_1(t) + \sum_{k=1}^{\infty} U_{k+1}(t)$$

$$= U(t) - I, \qquad a \le t \le b.$$

We have now proved that U is a continuous $n \times n$ matrix function satisfying (9). Let $\widetilde{U}$ be a second function with the same properties. Put $W = U - \widetilde{U}$. Then W is a continuous $n \times n$ matrix function and

$$W(t) = \int_a^t A(s)W(s)ds, \qquad a \le t \le b.$$

Let $m = \max\{\|W(s)\| \mid a \le s \le b\}$. An argument similar to the one used to establish inequality (11) gives

$$\|W(t)\| \le \frac{m}{k!} \mathcal{J}(b)^k, \qquad a \le t \le b, \quad k = 1, 2, \dots .$$

By taking the limit for $k \to \infty$ one sees that $\|W(t)\| = 0$ for $a \le t \le b$. Hence $U = \widetilde{U}$, and we have proved that U is uniquely determined by (9).

Using similar arguments (or duality), one proves that there exists a continuous $n \times n$ matrix function V such that

$$(13) \qquad V(t) = I_n - \int_a^t V(s)A(s)ds, \qquad a \le t \le b.$$

From the theory of integration (see, e.g., [R], Section 8.15) we know that formulas (9) and (13) imply that U and V are absolutely continuous on $[a, b]$ and, except for a set of measure zero, $U'(t) = A(t)U(t)$ and $V'(t) = -V(t)A(t)$. Put $Z(t) = V(t)U(t)$ for $a \le t \le b$. Then Z is also absolutely continuous and

$$Z'(t) = V'(t)U(t) + V(t)U'(t)$$
$$= -V(t)A(t)U(t) + V(t)A(t)U(t) = 0$$

almost everywhere on $[a, b]$. But then

$$V(t)U(t) = V(a)U(a) = I_n, \qquad a \le t \le b.$$

Since $U(t)$ and $V(t)$ are square matrices, we may conclude that $\det U(t) \ne 0$ and $U(t)^{-1} = V(t)$ for $a \le t \le b$. $\square$

The matrix $U(t)$ defined by (9) is called the *fundamental matrix* (normalized to I_n at $t = a$) of the differential equation

$$(14) \qquad\qquad x'(t) = A(t)x(t), \qquad a \le t \le b.$$

The general solution of (14) is given by $U(t)x$, where x is an arbitrary vector in $\mathbf{C}^n$.

THEOREM 2.3. *Let $K: L_2^m([a, b]) \to L_2^m([a, b])$ be an integral operator with a semi-separable kernel function k, and let (1) be a semi-separable representation of k. Let $V: \mathbf{C}^{n_2} \to \mathbf{C}^{n_2}$ be the indicator of $I - K$ associated with the representation (1). Then $I - K$ and V are matricially coupled. More precisely,*

$$(15) \qquad \begin{bmatrix} I - K & -R \\ Q & I_{n_2} \end{bmatrix}^{-1} = \begin{bmatrix} (I - H)^{-1} & (I - H)^{-1}R \\ -Q(I - H)^{-1} & V \end{bmatrix},$$

where H, Q and R are defined as follows:

$$(16) \qquad\qquad H: L_2^m([a, b]) \to L_2^m([a, b]),$$

$$(H\varphi)(t) = F_1(t) \int_a^t G_1(s)\varphi(s)ds - F_2(t) \int_a^t G_2(s)\varphi(s)ds, \qquad a \le t \le b,$$

$$(17) \qquad\qquad Q: L_2^m([a, b]) \to \mathbf{C}^{n_2}, \qquad Q\varphi = \int_a^b G_2(s)\varphi(s)ds,$$

$$(18) \qquad\qquad R: \mathbf{C}^{n_2} \to L_2^m([a, b]), \quad (Rx)(t) = F_2(t)x, \qquad a \le t \le b.$$

PROOF. Let $C(t)$ and $B(t)$ be defined by (7), and let $U(t)$ be as in (3), where $A(t)$ is given by (2). First we shall prove that $I - H$ is invertible and

$$(19) \qquad ((I - H)^{-1}\psi)(t) = \psi(t) + C(t)U(t) \int_a^t U(s)^{-1}B(s)\psi(s)ds, \qquad a \le t \le b.$$

To do this, let $T: L_2^m([a,b]) \to L_2^m([a,b])$ be the operator defined by the right hand side of (19). Take $\varphi \in L_2^m([a,b])$. Then

$$(H\varphi)(t) = C(t) \int_a^t B(s)\varphi(s)ds, \qquad a \le t \le b,$$

and hence for $a \le t \le b$ we have

$$(T(I-H)\varphi)(t) = \varphi(t) - C(t)\int_a^t B(s)\varphi(s)ds + C(t)U(t)\int_a^t U(s)^{-1}B(s)\varphi(s)ds$$

$$- C(t)U(t)\int_a^t U(s)^{-1}B(s)C(s)\left(\int_a^s B(\alpha)\varphi(\alpha)d\alpha\right)ds.$$

Note that $B(s)C(s) = A(s)$, and hence we can use formula (10) and partial integration to show that

$$\int_a^t U(s)^{-1}B(s)C(s)\left(\int_a^s B(\alpha)\varphi(\alpha)d\alpha\right)ds =$$

$$= -U(t)^{-1}\int_a^t B(\alpha)\varphi(\alpha)d\alpha + \int_a^t U(s)^{-1}B(s)\varphi(s)ds.$$

It follows that $T(I-H)\varphi = \varphi$. In a similar way one shows that $(I-H)T\varphi = \varphi$. Thus (19) holds.

Next, we use formula (10) to compute that for $a \le t \le b$,

$$((I-H)^{-1}Rx)(t) = (Rx)(t) + C(t)U(t)\int_a^t U(s)^{-1}B(s)(Rx)(s)ds$$

$$= C(t)\begin{bmatrix} 0 \\ x \end{bmatrix} + C(t)U(t)\int_a^t U(s)^{-1}B(s)C(s)\begin{bmatrix} 0 \\ x \end{bmatrix}ds$$

$$= C(t)\begin{bmatrix} 0 \\ x \end{bmatrix} + C(t)U(t)\int_a^t U(s)^{-1}A(s)\begin{bmatrix} 0 \\ x \end{bmatrix}ds$$

$$= C(t)\begin{bmatrix} 0 \\ x \end{bmatrix} + C(t)U(t)(I - U(t)^{-1}]\begin{bmatrix} 0 \\ x \end{bmatrix}ds.$$

It follows that

$$(20) \qquad\qquad ((I-H)^{-1}Rx)(t) = C(t)U(t)\begin{bmatrix} 0 \\ x \end{bmatrix}, \qquad a \le t \le b,$$

and hence

$$Q(I-H)^{-1}Rx = \int_a^b G_2(s)C(s)U(s)\begin{bmatrix}0\\x\end{bmatrix}ds$$

$$= -[0 \quad I]\int_a^b B(s)C(s)U(s)\begin{bmatrix}0\\x\end{bmatrix}ds$$

$$= -[0 \quad I]\int_a^b A(s)U(s)\begin{bmatrix}0\\x\end{bmatrix}ds$$

$$= x - U_{22}(b)x = x - Vx.$$

This proves that $V = I_{n_2} - Q(I-H)^{-1}R$.

Note that $K = H + RQ$. Thus

$$I - K = I - H - RQ = (I-H)\{I - (I-H)^{-1}RQ\}.$$

Now apply formula (4) in Section III.4 with $\lambda = 1$, $A = -(I-H)^{-1}R$ and $B = -Q$. It follows that

$$(21) \qquad \begin{bmatrix} I & (I-H)^{-1}R \\ -Q & V \end{bmatrix}^{-1} = \begin{bmatrix} I - (I-H)^{-1}RQ & -(I-H)^{-1}R \\ Q & I_{n_2} \end{bmatrix}.$$

Multiplying both sides of (21) on the left by $(I-H)\oplus I_{n_2}$ and then taking inverses yields (15). $\square$

PROOF OF THEOREM 2.1. Let V be the indicator of $I-K$ associated with the representation (1). Then (Theorem 2.3) the operators $I-K$ and V are matricially coupled via (15). So we can apply Corollary III.4.3 to show that $I-K$ is invertible if and only if V is invertible. Furthermore, in that case (see formula (16) in Section III.4)

$$(22) \qquad (I-K)^{-1} = (I-H)^{-1} + (I-H)^{-1}RV^{-1}Q(I-H)^{-1},$$

where H, Q and R are defined by the formulas (16)–(18). Since $U_{22}(b)$ is the matrix of V relative to the standard basis in $\mathbf{C}^n$, the operator $I-K$ is invertible if and only if $\det U_{22}(b) \neq 0$.

To derive an explicit formula for $(I-K)^{-1}$ we use (22). We already have explicit formulas for $(I-H)^{-1}$ and $(I-H)^{-1}R$ (see (19) and (20)). Let us compute $Q(I-H)^{-1}$. Take $\varphi \in L_2^m([a,b])$. Then

$$Q(I-H)^{-1}\varphi = -[0 \quad I]\int_a^b B(s)((I-H)^{-1}\varphi)(s)ds$$

$$= -[0 \quad I]\int_a^b B(s)\varphi(s)ds - [0 \quad I]\int_a^b B(s)C(s)U(s)$$

$$\times \left(\int_a^s U(\alpha)^{-1}B(\alpha)\varphi(\alpha)d\alpha\right)ds.$$

Since $B(s)C(s) = A(s)$, we can apply (3) and partial integration to show that

$$\int_a^b B(s)C(s)U(s)\left(\int_a^s U(\alpha)^{-1}B(\alpha)\varphi(\alpha)d\alpha\right)ds =$$

$$= U(b)\int_a^b U(\alpha)^{-1}B(\alpha)\varphi(\alpha)d\alpha - \int_a^b B(s)\varphi(s)ds.$$

It follows that

$$(23) \qquad Q(I - H)^{-1}\varphi = -[0 \quad I]U(b)\int_a^b U(s)^{-1}B(s)\varphi(s)ds,$$

and hence (use (20))

$$(24) \qquad \begin{aligned} ((I - H)^{-1}RV^{-1}Q(I - H)^{-1}\varphi)(t) &= \\ -C(t)U(t)P\int_a^b U(s)^{-1}B(s)\varphi(s)ds, &\qquad a \le t \le b, \end{aligned}$$

where P is given by (8). Finally use (19), (22) and (24) to derive the formulas (5) and (6). $\square$

COROLLARY 2.4. *Let* $K\colon L_2^m([a,b]) \to L_2^m([a,b])$ *be an integral operator with a semi-separable kernel function* k, *and let* (1) *be a semi-separable representation of* k. *Let* $V\colon \mathbb{C}^{n_2} \to \mathbb{C}^{n_2}$ *be the indicator of* $I - K$ *associated with the representation* (1). *Then*

$$(25) \qquad \mathrm{Ker}(I - K) = \left\{\varphi \in L_2^m([a,b]) \mid \varphi(t) = C(t)U(t)\begin{bmatrix} 0 \\ x \end{bmatrix}, x \in \mathrm{Ker}\,V\right\},$$

$$(26) \qquad \mathrm{Im}(I - K) = \left\{\psi \in L_2^m([a,b]) \mid [0 \quad I]U(b)\int_a^b U(s)^{-1}B(s)\psi(s)ds \in \mathrm{Im}\,V\right\}.$$

Here $U(t)$, $B(t)$ *and* $C(t)$ *are as in* (3) *and* (7).

PROOF. From Theorem 2.3 and Corollary III.4.3 we conclude that

$$\mathrm{Ker}(I - K) = \{\varphi \mid \varphi = (I - H)^{-1}Rx, x \in \mathrm{Ker}\,V\},$$

$$\mathrm{Im}(I - K) = \{\psi \mid Q(I - H)^{-1}\psi \in \mathrm{Im}\,V\},$$

where H, Q and R are defined by (16)–(18). But then we can use formulas (20) and (23) to get (25) and (26). $\square$

Let us illustrate Theorem 2.1 and Corollary 2.4 with an example. Consider the integral equation

$$(27) \qquad f(t) - \int_0^\tau e^{-|t-s|} f(s)\,ds = g(t), \qquad 0 \le t \le \tau.$$

The right hand side $g \in L_2([0,\tau])$, and we want to solve (27) in $L_2([0,\tau])$. The corresponding kernel is semi-separable (cf., formula (1.6)); in fact

$$(28) \qquad e^{-|t-s|} = \begin{cases} F_1(t)G_1(s), & 0 \le s \le t \le \tau, \\ F_2(t)G_2(s), & 0 \le t < s \le \tau, \end{cases}$$

where $F_1(t) = G_2(t) = e^{-t}$ and $F_2(t) = G_1(t) = e^t$. To solve (27) we first determine the indicator associated with the representation (28). Note that the upper order $n_2 = 1$, and hence in this case the indicator is just a number. One computes that for the representation (28) the matrices $A(t)$ and $U(t)$ in (2) and (3) are given by

$$A(t) = \begin{bmatrix} 1 & e^{2t} \\ -e^{-2t} & -1 \end{bmatrix}, \qquad U(t) = \begin{bmatrix} e^t \cos t & e^t \sin t \\ -e^{-t} \sin t & e^{-t} \cos t \end{bmatrix}.$$

Hence $V = e^{-\tau} \cos \tau$.

Take $0 < \tau \neq \frac{\pi}{2} + k\pi$ for $k = 0, 1, 2, \ldots$. Then $V \neq 0$, and hence equation (27) is uniquely solvable in $L_2([0,\tau])$. To get the solution we apply formulas (5) and (6). One computes that in this case

$$U(t)^{-1} = \begin{bmatrix} e^{-t} \cos t & -e^t \sin t \\ e^{-t} \sin t & e^t \cos t \end{bmatrix},$$

$$C(t)U(t) = [\, \cos t - \sin t \quad \sin t + \cos t \,],$$

$$U(s)^{-1}B(s) = \begin{bmatrix} \sin s + \cos s \\ \sin s - \cos s \end{bmatrix},$$

$$P = \begin{bmatrix} 0 & 0 \\ -\tan \tau & 1 \end{bmatrix},$$

and hence the unique solution of (27) is given by

$$f(t) = g(t) + \{(\cos t - \sin t) + \tan \tau (\sin t + \cos t)\} \int_0^t (\cos s + \sin s)g(s)\,ds$$

$$- (\sin t + \cos t) \int_t^\tau \{(1 - \tan \tau)\sin s - (1 + \tan \tau)\cos s\}g(s)\,ds, \qquad 0 \le t \le \tau.$$

Next, assume that $\tau = \frac{\pi}{2} + k\pi$ for some non-negative integer k. Then the indicator V is equal to zero, and hence equation (27) is not uniquely solvable. By applying

Corollary 2.4 one sees (use formula (26)) that (27) is solvable if and only if

$$\int_0^\tau (\sin s + \cos s)g(s)ds = 0,$$

and in that case (use formulas (25), (19) and the coupling relation (15)) the general solution of equation (27) is given by

$$f(t) = g(t) + (\sin t + \cos t)\int_0^t (\sin s - \cos s)g(s)ds$$

$$+ (\cos t - \sin t)\int_0^t (\sin s + \cos s)g(s)ds + c(\sin t + \cos t), \qquad c \in \mathbf{C}.$$

IX.3 EIGENVALUES AND DETERMINANT

This section concerns the eigenvalues of an integral operator with a semi-separable kernel. As before,

$$(1) \qquad k(t,s) = \begin{cases} F_1(t)G_1(s), & a \le s \le t \le b, \\ F_2(t)G_2(s), & a \le t < s \le b, \end{cases}$$

is a semi-separable representation and

$$(2) \qquad A(t) = \begin{bmatrix} G_1(t)F_1(t) & G_1(t)F_2(t) \\ -G_2(t)F_1(t) & -G_2(t)F_2(t) \end{bmatrix}, \qquad a \le t \le b.$$

THEOREM 3.1. *Let* $K \colon L_2^m([a,b]) \to L_2^m([a,b])$ *be an integral operator with a semi-separable kernel function* k, *and let* (1) *be a semi-separable representation of* k *of order* n, *say. For each* $\mu \in \mathbf{C}$ *let* $U(t;\mu)$ *be the unique* $n \times n$ *matrix function determined by*

$$(3) \qquad U(t;\mu) = I_n + \mu \int_a^t A(s)U(s;\mu)ds, \qquad a \le t \le b,$$

where $A(t)$ *is given by* (2) *and* I_n *is the* $n \times n$ *identity matrix. Partition* $U(t;\mu)$ *as a* 2×2 *block matrix according to the partitioning of the right hand side of* (2):

$$(4) \qquad U(t;\mu) = \begin{bmatrix} U_{11}(t;\mu) & U_{12}(t;\mu) \\ U_{21}(t;\mu) & U_{22}(t;\mu) \end{bmatrix}, \qquad a \le t \le b.$$

Put

$$(5) \qquad \delta_K(\mu) := \det U_{22}(b;\mu), \qquad \mu \in \mathbf{C}.$$

Then δ_K is an entire function, and $\lambda \neq 0$ is an eigenvalue of K if and only if $\delta_K(\lambda^{-1}) = 0$. Furthermore, the geometric multiplicities of the non-zero eigenvalues of K are bounded above by the minimum of the lower and upper order of the semi-separable representation (1).

PROOF. For $a \leq t \leq b$ put

$$U_1(t) = \int_a^t A(s)ds, \quad U_{k+1}(t) = \int_a^t A(s)U_k(s)ds, \qquad k \geq 2.$$

From the proof of Lemma 2.2 it follows that for every $\mu \in \mathbb{C}$

$$U(t;\mu) = I_n + \sum_{k=1}^{\infty} \mu^k U_k(t), \qquad a \leq t \leq b.$$

But then the estimate (2.11) implies that for each $t \in [a,b]$ the entries of $U(t;\mu)$ are analytic in μ on the entire complex plane. In particular, the entries of $U_{22}(b;\mu)$ are entire functions in μ, and hence the same is true for $\delta_K(\mu)$.

Note that μK is again an integral operator with a semi-separable kernel function. In fact,

$$(6) \qquad \mu k(t,s) = \begin{cases} \mu F_1(t)G_1(s), & a \leq s \leq t \leq b, \\ \mu F_2(t)G_2(s), & a \leq t < s \leq b, \end{cases}$$

and the right hand side of (6) is a semi-separable representation for the kernel function of μK. Let $V(\mu)$ be the indicator of $I - \mu K$ associated with the representation (6). Then

$$(7) \qquad \det V(\mu) = \det U_{22}(b;\mu) = \delta_K(\mu).$$

Take $\lambda \neq 0$. Then λ is an eigenvalue of K if and only if $I - \lambda^{-1}K$ is not invertible. But then Theorem 2.1 and formula (7) imply that λ is an eigenvalue of K if and only if $\delta_K(\lambda^{-1}) = 0$. Furthermore, we can apply Corollary 2.4 to show that

$$\dim \operatorname{Ker}(\lambda - K) = \dim \operatorname{Ker}(I - \lambda^{-1}K) \leq \dim \operatorname{Ker} V(\lambda^{-1}) \leq n_2,$$

where n_2 is the upper order of the representation (1). This shows that the latter number is an upper bound for the geometric multiplicities of the non-zero eigenvalues of K. By interchanging the roles of the triangles $a \leq s \leq t \leq b$ and $a \leq t \leq s \leq b$ one sees that the lower order is also an upper bound. $\square$

Theorem 3.1 states that the function $\delta_K(\lambda)$ has the properties of a determinant function. The next theorem makes this statement more precise for trace class operators.

THEOREM 3.2. *Let $K: L_2^m([a,b]) \to L_2^m([a,b])$ be an integral operator with a semi-separable kernel function k, and let (1) be a semi-separable representation of k. Assume K is a trace class operator. Then*

$$(8) \qquad \operatorname{tr} K = \int_a^b \operatorname{tr} G_2(s)F_2(s)ds,$$

$$(9) \qquad \det(I - \mu K) = \delta_K(\mu), \qquad \mu \in \mathbf{C},$$

where $\delta_K(\cdot)$ is the entire function defined by (5).

PROOF. Let H, Q, and R be the operators defined by the formulas (16)–(18) in the previous section. We know that $K = H + RQ$. Furthermore, RQ is an operator of finite rank. It follows that $H = K - RQ$ is a trace class operator. In the proof of Theorem 2.3 it was shown that the operator $I - H$ is invertible. Now, replace K by μK, and use the semi-separable representation (6) for the kernel function of μK. Then H has to be replaced by μH, and we may conclude that $I - \mu H$ is invertible. This holds for each $\mu \in \mathbf{C}$. Hence H has no non-zero eigenvalues. But then Theorem VII.6.1 implies that

$$(10) \qquad \operatorname{tr} H = 0, \quad \det(I - \mu H) = 1 \qquad (\mu \in \mathbf{C}).$$

The first equality in (10) implies that

$$\operatorname{tr} K = \operatorname{tr}(H + RQ) = \operatorname{tr} RQ = \operatorname{tr} QR,$$

by Corollary VII.6.2(i). Note that QR is the operator on $\mathbf{C}^n$ given by $\int_a^b G_2(s)F_2(s)ds$. Since the trace class norm and the usual operator norm are equivalent norms on operators on $\mathbf{C}^n$, the continuity of the trace in the trace class norm implies that the integral and trace may be interchanged. This proves (8).

To prove (9) we use the second identity in (10) and Corollary VII.6.2(ii). Take $\mu \in \mathbf{C}$. Since μK, μH and $\mu(I - \mu H)^{-1}RQ$ are trace class operators, the identity

$$I - \mu K = (I - \mu H)\{I - \mu(I - \mu H)^{-1}RQ\}$$

implies (see Theorem VII.3.3) that

$$\begin{aligned}
\det(I - \mu K) &= \det(I - \mu H)\det\{I - \mu(I - \mu H)^{-1}RQ\} \\
&= \det\{I - \mu(I - \mu H)^{-1}RQ\} \\
&= \det\{I_{n_2} - \mu Q(I - \mu H)^{-1}R\}.
\end{aligned}$$

Here n_2 is the upper order of the semi-separable representation (1). Let $V(\mu)$ be the indicator of $I - \mu K$ associated with the representation (6). From the proof of Theorem 2.3 (with K replaced by μK) we know that

$$V(\mu) = I_{n_2} - \mu Q(I - \mu H)^{-1}R.$$

Thus $\det(I - \mu K) = \det V(\mu) = \delta_K(\mu)$, by formula (7). $\square$

Theorem 3.2 may be used to compute trace and determinant of an integral operator. For example, let $K\colon L_2([0,\tau]) \to L_2([0,\tau])$ be the integral operator with kernel

$$(11) \qquad k(t,s) = e^{-|t-s|}, \qquad 0 \le t \le \tau, \quad 0 \le s \le \tau.$$

First, let us prove that K is a trace class operator. To do this, let

$$T\colon L_2([0,\tau]) \to L_2([0,\tau]), \qquad (Tf)(t) = \int_0^t f(s)ds.$$

The operator T is an integral operator with a semi-separable kernel function, and hence T is Hilbert-Schmidt. Also, for each $\mu \in \mathbf{C}$ the operator $I - \mu T$ is invertible and

$$((I - \mu T)^{-1} T g)(t) = \int_0^t e^{\mu(t-s)} g(s) ds, \qquad 0 \le t \le \tau.$$

Let $H \colon L_2([0,\tau]) \to L_2([0,\tau])$ be defined by

$$(Hf)(t) = \int_0^t e^{-(t-s)} f(s) ds - \int_0^t e^{(t-s)} f(s) ds.$$

Then

$$H = (I + T)^{-1} T - (I - T)^{-1} T = -2(I + T)^{-1}(I - T)^{-1} T^2,$$

and hence H is a trace class operator. Since $K - H$ is an operator of rank 1, it follows that K is also a trace class operator.

The kernel function (11) is semi-separable. In fact,

$$(12) \qquad e^{-|t-s|} = \begin{cases} F_1(t) G_1(s), & 0 \le s \le t \le \tau, \\ F_2(t) G_2(s), & 0 \le t < s \le \tau, \end{cases}$$

where $F_1(t) = G_2(t) = e^{-t}$ and $F_2(t) = G_1(t) = e^t$. Thus, according to formula (8),

$$\operatorname{tr} K = \int_0^\tau \operatorname{tr} G_2(s) F_2(s) ds = \int_0^\tau ds = \tau.$$

For the representation (12) the entries of the matrix

$$U(t; \mu) = \begin{bmatrix} U_{11}(t; \mu) & U_{12}(t; \mu) \\ U_{21}(t; \mu) & U_{22}(t; \mu) \end{bmatrix},$$

which is defined by (3), are as follows:

$$U_{11}(t; \mu) = e^t \left\{ \frac{1}{2}(e^{-\alpha t} + e^{\alpha t}) + \frac{1-\mu}{2\alpha}(e^{-\alpha t} - e^{\alpha t}) \right\},$$

$$U_{12}(t; \mu) = e^t \left\{ \frac{\mu}{2\alpha}(e^{\alpha t} - e^{-\alpha t}) \right\},$$

$$U_{21}(t; \mu) = e^{-t} \left\{ \frac{\mu}{2\alpha}(e^{-\alpha t} - e^{\alpha t}) \right\},$$

$$U_{22}(t; \mu) = e^{-t} \left\{ \frac{1}{2}(e^{-\alpha t} + e^{\alpha t}) + \frac{1-\mu}{2\alpha}(e^{\alpha t} - e^{-\alpha t}) \right\},$$

where $\alpha^2 = 1 - 2\mu$. Note that the choice of the square root is not essential here because the formulas do not change if α is replaced by $-\alpha$. Formula (9) in Theorem 3.2 now implies that

$$\det(I - \mu K) = U_{22}(\tau; \mu)$$

$$= e^{-\tau} \left\{ \cosh(\tau \sqrt{2\mu - 1}) + (1 - \mu)\tau \frac{\sinh(\tau \sqrt{2\mu - 1})}{\tau \sqrt{2\mu - 1}} \right\}.$$

CHAPTER X

THE GROWTH OF THE RESOLVENT OPERATOR
AND APPLICATIONS TO COMPLETENESS

The behaviour of the resolvent in a neighbourhood of the spectrum is one of the important characteristics of an operator. The first section presents a basic theorem in which a sharp evaluation is given of the norm of an inverse operator in terms of the determinant and the singular values. The theorem is used in the second section to evaluate the growth of the resolvent of a Volterra operator. The applications concern two completeness theorems, which are presented in the last two sections.

X.1 MAIN THEOREM

Throughout this chapter *the underlying spaces are separable Hilbert spaces.* We begin with a simple observation which will be used in the proof of the main theorem. If A is an $n \times n$ matrix, then

$$(1) \qquad s_j(I + A) \leq 1 + s_j(A), \qquad j = 1, \ldots, n.$$

For $j = 1$ the inequality (1) is an immediate consequence of triangle inequality for the norm and the fact that $\|I\| = 1$. For arbitrary j we use Theorem VI.1.5. Indeed, choose an $n \times n$ matrix F of rank $j - 1$ such that $s_j(A) = \|A - F\|$. Then

$$s_j(I + A) \leq \|I + A - F\| \leq 1 + \|A - F\| = 1 + s_j(A),$$

and formula (1) is proved.

THEOREM 1.1. *Let A be a trace class operator, and assume $\det(I + A) \neq 0$. Then $I + A$ is invertible and*

$$(2) \qquad \|(I + A)^{-1}\| \leq \frac{1}{|\det(I + A)|} \prod_{j=1}^{\infty} (1 + s_j(A)).$$

PROOF. We have already seen that $\det(I + A) \neq 0$ implies that $I + A$ is invertible. So we have to prove (2). This will be done in three steps. First we show that (2) may be proved by reduction to operators of finite rank. So let us assume that (2) holds for operators of finite rank. Choose a sequence (P_n) of orthogonal projections such that $P_n \to I$ $(n \to \infty)$ pointwise. Then $P_n A P_n \to A$ in the trace class norm by Theorem VI.4.3.

As $\det(I + A) = \lim_{n \to \infty} \det(I + P_n A P_n)$ (see the remark at the end of Section VII.3), we have $\det(I + P_n A P_n) \neq 0$ for n sufficiently large. So $I + P_n A P_n$ is invertible and

$$(3) \qquad \|(I + P_n A P_n)^{-1}\| \leq \frac{1}{|\det(I + P_n A P_n)|} \prod_j (1 + s_j(P_n A P_n))$$

for $n \geq n_0$, say. The fact that $P_n A P_n \to A$ in the trace class norm implies that $P_n A P_n \to A$ in the usual operator norm, and hence

$$\|(I + P_n A P_n)^{-1}\| \to \|(I + A)^{-1}\| \qquad (n \to \infty).$$

As $s_j(P_n A P_n) \leq s_j(A)$ for all j, we see from (3) that

$$\|(I + P_n A P_n)^{-1}\| \leq \frac{1}{|\det(I + P_n A P_n)|} \prod_j (1 + s_j(A)).$$

Taking limits, as $n \to \infty$, on the left and right yields (2).

Next, assume that A is of finite rank. Then there exists a finite dimensional subspace M of H such that

$$AM \subset M, \qquad AM^\perp = (0).$$

Let $A_0 \colon M \to M$ be the restriction of A to M. Note that the right hand side of (2) does not change if A is replaced by A_0. Furthermore

$$\|(I + A)^{-1}\| \leq \max\{1, \|(I + A_0)^{-1}\|\}.$$

As the right hand side of (2) is larger than or equal to 1 (see Theorem VII.3.3), it suffices to prove (2) for A_0 instead of A. In other words, without loss of generality we may assume that $\dim H < \infty$.

Assume $\dim H = n$, and consider $R = [\det(I + A)](I + A)^{-1}$. To compute the norm of R, we compute $s_1(R)$:

$$s_1^2(R) = \lambda_1(R^* R) = |\det(I + A)|^2 \lambda_1\big((I + A^*)^{-1}(I + A)^{-1}\big).$$

Observe that

$$|\det(I + A)|^2 = \det[(I + A)(I + A^*)] = \prod_{j=1}^{n} \lambda_j\big((I + A)(I + A^*)\big),$$

$$\lambda_1\big((I + A^*)^{-1}(I + A)^{-1}\big) = \frac{1}{\lambda_n\big((I + A)(I + A^*)\big)}.$$

So

$$s_1^2(R) = \prod_{j=1}^{n-1} \lambda_j\big((I + A)(I + A^*)\big) = \prod_{j=1}^{n-1} s_j^2(I + A).$$

Next apply the inequalities (1). One obtains that

$$s_1(R) = \prod_{j=1}^{n-1} s_j(I + A) \leq \prod_{j=1}^{n-1} (1 + s_j(A)).$$

So, if $\dim H = n$, then

$$(4) \qquad \|(I + A)^{-1}\| \leq \frac{1}{|\det(I + A)|} \prod_{j=1}^{n-1} (1 + s_j(A)),$$

and the theorem is proved. $\square$

The inequalities (1), (2) and (4) cannot be sharpened. In fact, if A is a non-negative $n \times n$ matrix, then we have equality in (1) as well as in (4). This follows immediately from the fact that for a non-negative operator the j-th singular number is equal to the j-th eigenvalue. In (2) we have equality if A is a non-negative trace class operator with an infinite number of eigenvalues.

X.2 COROLLARIES TO THE MAIN THEOREM

In this section we use the main theorem to describe the behaviour of $\|(I - zA)^{-1}\|$ for $|z|$ large. First of all, if A is trace class and $\det(I - zA) \neq 0$, then $I - zA$ is invertible and

$$\|(I - zA)^{-1}\| \leq \frac{1}{\det(I - zA)} e^{|z|\|A\|_1},$$

because of (1.2) and $1 + t \leq e^t$ for $t \geq 0$. Recall that $\|A\|_1$ stands for the trace class norm of the operator A. It follows that if A is trace class and a Volterra operator, then $\det(I - zA) = 1$ for each z, and hence

$$\|(I - zA)^{-1}\| \leq e^{|z|\|A\|_1}, \qquad z \in \mathbf{C}.$$

This inequality can be sharpened. In fact, the next theorem shows that for a trace class Volterra operator the entire function $\|(I - zA)^{-1}\|$ is a function of exponential type zero (cf. Section VII.5).

THEOREM 2.1. *Let A be a trace class Volterra operator. Then, given $\delta > 0$, there exists a positive constant C such that*

$$(1) \qquad \|(I - zA)^{-1}\| \leq Ce^{\delta|z|}, \qquad z \in \mathbf{C}.$$

PROOF. Choose m such that $\sum_{j=m+1}^{\infty} s_j(A) < \frac{1}{2}\delta$. By applying the main theorem of the previous section, one obtains

$$\|(I - zA)^{-1}\| \leq \prod_{j=1}^{\infty} (1 + |z|s_j(A)) \leq \left[\prod_{j=1}^{m} (1 + |z|s_j(A)) \right] e^{\frac{1}{2}\delta|z|}.$$

Now use the fact that a polynomial is of exponential type zero. So there exists a constant C such that

$$\prod_{j=1}^{m} (1 + |z|s_j(A)) \leq Ce^{\frac{1}{2}\delta|z|},$$

and (1) is proved. $\square$

      If not A but a certain power of A is trace class, then still one can obtain an estimation of the growth of the resolvent at ∞. This is the content of the next theorem.

      THEOREM 2.2. *Let A be a Volterra operator, and assume that*

$$(2) \qquad \sum_{j=1}^{\infty} s_j(A)^r < \infty$$

for some $r > 0$. Then, given $\delta > 0$, there exists a positive constant C such that

$$(3) \qquad \|(I - zA)^{-1}\| \leq C e^{\delta |z|^r}, \qquad z \in \mathbf{C}.$$

      PROOF. First, take $0 < r \leq 1$. We begin with an elementary inequality:

$$(4) \qquad 1 + |z| \leq e^{\beta |z|^r}, \qquad z \in \mathbf{C},$$

where β is some constant depending on r only. To prove (4), note that for $|z| < 1$ one has $|z| \leq |z|^r$, and thus

$$1 + |z| \leq e^{|z|} \leq e^{|z|^r}, \qquad |z| < 1.$$

As $|z|^{-r} \log(1 + |z|) \to 0$ if $|z| \to \infty$, the function $|z|^{-r} \log(1 + |z|)$ is bounded on $|z| \geq 1$. So there exists $\beta \geq 1$ such that $\log(1 + |z|) \leq \beta |z|^r$ for $|z| \geq 1$. This proves (4).

      Let $\delta > 0$ be given. Choose m such that $\sum_{j=m+1}^{\infty} s_j(A)^r < \frac{\delta}{2}\beta^{-1}$, where $\beta > 0$ is as in (4). Using (4) and Theorem 1.1, we get

$$\|(I - zA)^{-1}\| \leq \prod_{j=1}^{\infty} \left(1 + |z|s_j(A)\right) \leq \left[\prod_{j=1}^{m} \left(1 + |z|s_j(A)\right)\right] e^{\frac{1}{2}\delta |z|^r}.$$

Proceeding as in the proof of Theorem 2.1, one obtains the desired inequality (3).

      Next, assume $r \geq 1$. Choose a positive integer q such that $r \leq q < r + 1$, and put $p = r/q$. Note that $0 < p \leq 1$. From the inequality,

$$\sum_{j=1}^{\infty} s_j(A^q)^{r/q} \leq \sum_{j=1}^{\infty} s_j(A)^r$$

(see Section VIII.1), it follows that $\sum_{j=1}^{\infty} s_j(A^q)^p < \infty$. So we can apply the first part of the proof to A^q. Hence there exists a positive constant C_1 such that

$$\|(I - \lambda A^q)^{-1}\| \leq C_1 e^{\frac{1}{2}\delta |\lambda|^p}, \qquad \lambda \in \mathbf{C}.$$

In particular,

$$(5) \qquad \|(I - z^q A^q)^{-1}\| \leq C_1 e^{\frac{1}{2}\delta |z|^r}, \qquad \lambda \in \mathbf{C}.$$

Now, we use that

$$(6) \qquad (I - zA)^{-1} = (I + zA + \cdots + z^{q-1}A^{q-1})(I - z^q A^q)^{-1}.$$

Since any polynomial is of exponential type zero, there exists a constant C_2 such that

$$\|I + zA + \cdots + z^{q-1}A^{q-1}\| \leq C_2 e^{\frac{1}{2}\delta|z|}.$$

Using $|z| \leq |z|^r$ for $|z| \geq 1$, one gets

$$(7) \qquad \|I + zA + \cdots + z^{q-1}A^{q-1}\| \leq C_3 e^{\frac{1}{2}\delta|z|^r}$$

for some constant C_3 and each $z \in \mathbf{C}$. Now take the norm in (6) and insert the inequalities (5) and (7). One obtains the desired formula (3). $\square$

X.3 APPLICATIONS TO COMPLETENESS

Theorem VII.8.1 tells us that the system of eigenvectors and generalized eigenvectors of a trace class operator A is complete whenever for each φ the number $\langle A\varphi, \varphi \rangle$ is in the closed upper halfplane. One has also completeness if A is Hilbert-Schmidt and the numbers $\langle A\varphi, \varphi \rangle$ are in the first closed quadrant of the complex plane (Theorem VIII.3.1). In the next completeness theorem the condition that A is trace class or Hilbert-Schmidt is weakened, while on the other hand the numbers $\langle A\varphi, \varphi \rangle$ are required to be in a certain angular sector with a sharper opening.

THEOREM 3.1. *Let A be a compact operator such that $\sum_{j=1}^{\infty} s_j(A)^r < \infty$ for some $r \geq 1$. Assume that the set*

$$(1) \qquad W_A = \{\langle A\varphi, \varphi \rangle \mid \varphi \in H, \|\varphi\| = 1\}$$

lies in a closed angle with vertex at zero and opening $\frac{\pi}{r}$. Then the system of eigenvectors and generalized eigenvectors of A is complete.

For $r = 1$ the above theorem is just Theorem VII.8.1. For $r = 2$ it contains the completeness theorem for Hilbert-Schmidt operators. The set W_A, defined by (1), is called the *numerical range* of A. We begin with a lemma that relates the location of the numerical range to that of the spectrum.

LEMMA 3.2. *Let W_A be the numerical range of an arbitrary bounded linear operator A acting on a Hilbert space H. If $\lambda \notin \overline{W}_A$, then $\lambda - A$ is invertible and*

$$(2) \qquad \|(\lambda - A)^{-1}\| \leq [\mathrm{dist}(\lambda, \overline{W}_A)]^{-1}.$$

PROOF. Here $\overline{W}_A$ denotes the closure of the set W_A. Take $\lambda \notin \overline{W}_A$, and put $\delta = \mathrm{dist}(\lambda, \overline{W}_A)$. Obviously, $\delta > 0$. For $x \in H$ with $\|x\| = 1$ we have

$$(3) \qquad \|(\lambda - A)x\| \geq |\langle (\lambda - A)x, x \rangle| = |\lambda - \langle Ax, x \rangle| \geq \delta > 0.$$

So $\lambda - A$ is injective and has closed range. Similarly

$$\|(\lambda - A)^* x\| \geq |\overline{\lambda} - \langle A^* x, x \rangle| = |\lambda - \langle Ax, x \rangle| \geq \delta > 0,$$

and thus $\mathrm{Ker}(\lambda - A)^* = (0)$. It follows that

$$\mathrm{Im}(\lambda - A) = \overline{\mathrm{Im}(\lambda - A)} = [\mathrm{Ker}(\lambda - A)^*]^{\perp} = H.$$

Thus $\lambda - A$ is invertible and (3) implies (2). $\square$

LEMMA 3.3. *If the numerical range W_A lies in a closed angle with vertex at zero and opening $\leq \pi$, then $\mathrm{Ker}\, A = \mathrm{Ker}\, A^*$.*

PROOF. We may choose ω such that $B := e^{i\omega} A$ has its numerical range in the closed upper half plane. Note that this implies that the imaginary part $B_{\Im} = \frac{1}{2i}(B - B^*)$ is non-negative. But then (see the second paragraph of the proof of Theorem VII.8.2) we know that $\mathrm{Ker}\, B = \mathrm{Ker}\, B^*$. Obviously, $\mathrm{Ker}\, A = \mathrm{Ker}\, B$ and $\mathrm{Ker}\, A^* = \mathrm{Ker}\, B^*$, and hence the lemma is proved. $\square$

PROOF OF THEOREM 3.1. Let Ω be an open angle with vertex at zero and opening $\frac{\pi}{r}$ such that

$$(4) \qquad\qquad\qquad\qquad W_A \subset \overline{\Omega}.$$

The sides of Ω we denote by ℓ and m. First we show that (4) implies the existence of an open angle Ω' with the same opening as Ω such that $I - zA$ is invertible for $z \notin \Omega'$ and

$$(5) \qquad\qquad\qquad \|(I - zA)^{-1}\| \leq 2|z|, \qquad z \notin \Omega'.$$

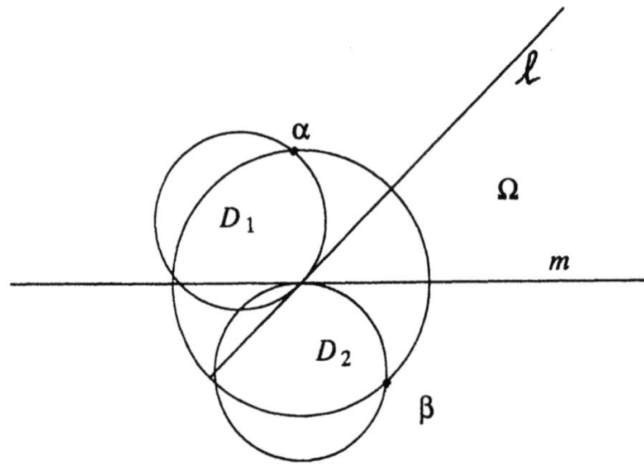

Figure 1

To prove (5), let D_1 and D_2 be closed discs of radius 1 that are outside Ω and are tangent to the lines ℓ and m at the point 0, respectively. Put $D = D_1 \cup D_2$. Take $0 \neq \lambda \in D$. From Lemma 3.2 and condition (4) we know that $\lambda - A$ is invertible and

$$\|(\lambda - A)^{-1}\| \leq \{\mathrm{dist}(\lambda, \overline{\Omega})\}^{-1}.$$

Put $\rho = |\lambda|$, and let $\Gamma_\rho = \{z \in D \mid |z| = \rho\}$. Let us compute $d_0 = \min\{\mathrm{dist}(z, \overline{\Omega}) \mid z \in \Gamma_\rho\}$. Of course this minimum d_0 is attained in the points α and β (see Figure 1). One easily computes (see Figure 2) that the distance of α to the line ℓ is equal to $\frac{1}{2}\rho^2$. So $d_0 = \frac{1}{2}\rho^2$, and hence $\mathrm{dist}(\lambda, \overline{\Omega}) \geq \frac{1}{2}\rho^2$. It follows that

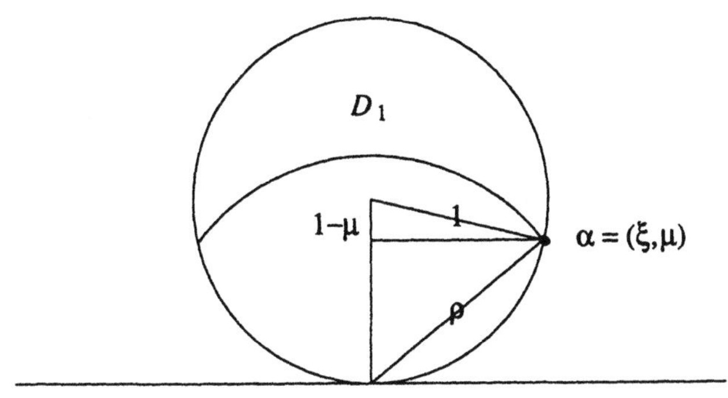

Figure 2

$$(6) \qquad \|(\lambda - A)^{-1}\| \leq 2|\lambda|^{-2}, \qquad 0 \neq \lambda \in D.$$

Next one makes the transformation $z \mapsto z^{-1}$. This transformation maps the closed discs D_1 and D_2 onto closed halfplanes D_1' and D_2', respectively, and thus the complement of D is mapped by $z \mapsto z^{-1}$ onto an open angle Ω'. As D_1 and D_2 are tangent at 0 to the lines ℓ and m, respectively, the opening of Ω' is equal to the opening of Ω. Now take $z \notin \Omega'$. Then $\lambda = z^{-1} \neq 0$ and belongs to D. So (6) holds. But then $I - zA$ is invertible and inequality (5) holds true.

Next, assume that A is a Volterra operator which satisfies the conditions of Theorem 3.1. We shall prove that $A = 0$. Introduce the operator function $A(z) = A(I - zA)^{-1}$. Obviously, $A(z)$ is an entire function, and

$$A(z) = z^{-1}\{(I - zA)^{-1} - I\}, \qquad z \neq 0.$$

So we apply the inequality (5) to show that

$$(7) \qquad \sup_{z \notin \Omega'} \|A(z)\| < \infty.$$

Furthermore, according to the evaluation of the resolvent given in Theorem 2.2, we know that for each $\delta > 0$ there exists a constant C such that

$$(8) \qquad \|A(z)\| \leq C e^{\delta|z|^r}, \qquad z \in \mathbb{C}.$$

Now, consider the function $A(z)$ on Ω'. We know that $A(z)$ is analytic in Ω' and continuous up to its boundary. On the sides of the angle Ω' the function $A(z)$ is uniformly bounded by (7). Furthermore, inside Ω' we have the inequality (8). As the opening of Ω' is equal to $\frac{\pi}{r}$, we can apply the Phragmén-Lindelöf theorem (see [C], Corollary VI.4.4) to show $A(z)$ is bounded on Ω'. Using (7), we conclude that $A(z)$ is bounded on the entire complex plane. So by Liouville's theorem $A(z)$ is a constant function. It follows that

$$A(z) = A + zA^2 + \cdots = A,$$

and hence $A^2 = 0$.

The identity $A^2 = 0$ implies $A = 0$. Suppose not. Then there exists a vector x such that $A^2x = 0$ and $y := Ax \neq 0$. According to Lemma 3.3 the vector $y \in \operatorname{Ker} A = \operatorname{Ker} A^* = \operatorname{Im} A^{\perp}$. On the other hand, $y \in \operatorname{Im} A$. So we conclude that $y \perp y$, and hence $y = 0$, which is a contradiction. Thus $A = 0$.

Finally, let A be an arbitrary compact operator that satisfies the conditions of Theorem 3.1. Let $H = E_A \oplus E_A^{\perp}$, where E_A is the smallest subspace containing all the eigenvectors and generalized eigenvectors of A corresponding to the non-zero eigenvalues. Since, for some ω, the operator $e^{i\omega}A$ has a non-negative imaginary part, it suffices to show (cf. the first paragraph of the proof of Theorem VII.8.2) that $E_A^{\perp} \subset \operatorname{Ker} A$. Let P be the orthogonal projection of H onto $E_A^{\perp}$. We know (Lemma II.3.4) that the operator

$$B := PA|E_A^{\perp} : E_A^{\perp} \to E_A^{\perp}$$

is a compact Volterra operator. For $\varphi \in E_A^{\perp}$ one has

$$\langle B\varphi, \varphi \rangle = \langle PA\varphi, \varphi \rangle = \langle A\varphi, \varphi \rangle.$$

So $W_B \subset W_A$. Furthermore, $s_j(B) \leq s_j(A)$ for $j = 1, 2, \ldots$. So B satisfies the condition of Theorem 3.1. But then $B = 0$. Now, note that

$$B^* = A^*|E_A^{\perp}.$$

So $E_A^{\perp} \subset \operatorname{Ker} A^*$. But then we can apply Lemma 3.3 to show that $E_A^{\perp} \subset \operatorname{Ker} A$, and the proof is complete. $\square$

X.4 THE KELDYSH THEOREM FOR COMPLETENESS

In this section we shall prove the following completeness theorem, which is due to V.M. Keldysh [1].

THEOREM 4.1. *Let K be a compact selfadjoint operator with $\operatorname{Ker} K = (0)$, and assume that*

$$(1) \qquad \sum_{j=1}^{\infty} |\lambda_j(K)|^r < \infty$$

for some $r \geq 1$. Furthermore, let S be a compact operator such that $I + S$ is invertible. Then the system of eigenvectors and generalized eigenvectors of the operator

$$(2) \qquad A = K(I + S)$$

is complete. Moreover, given $\varepsilon > 0$, all the eigenvalues of A, with a possible exception of a finite number, are in the sector

$$(3) \qquad \Delta = \{\rho e^{i\varphi} \mid \rho \geq 0, |\pi - \varphi| < \varepsilon \text{ or } |\varphi| < \varepsilon\}.$$

The conditions of the Keldysh theorem appear naturally in certain differential equations. For example, consider the differential operator

$$\frac{dx}{dt} + b(t)x.$$

Write $B = \frac{d}{dt}$, and let C be the operator of multiplication by $b(t)$. Choose boundary conditions such that B is selfadjoint, invertible and B^{-1} is Hilbert-Schmidt. Taking inverses, one obtains the compact operator

$$A = B^{-1}(I + CB^{-1})^{-1},$$

which is an operator of the type considered in the Keldysh theorem. In the chapter on unbounded Fredholm operators (see Section XVII.5) we shall return to examples of this type and make the statements more precise.

Theorem 4.1 does not hold true if the compactness condition on K is dropped. To see this, take $K = I$, and let S be the operator of integration. Also, note that the invertibility of $I + S$ is equivalent to the statement that $\operatorname{Ker} A = (0)$. Indeed, since S is compact, the operator $I + S$ is invertible if and only if $\operatorname{Ker}(I + S) = (0)$. But $\operatorname{Ker}(I + S) = \operatorname{Ker} A$ because K is injective by assumption.

To prove Theorem 4.1 we need the following lemma.

LEMMA 4.2. *Let K be a compact selfadjoint operator with $\operatorname{Ker} K = (0)$, and let T be an arbitrary compact operator. Furthermore, let Ω be a closed angle with vertex at zero that does not contain non-zero real points. Then*

$$\lim_{z \in \Omega, z \to \infty} \|T(I - zK)^{-1}\| = 0,$$

and the convergence is uniform on Ω.

PROOF. First we consider the case that

$$T = \langle \cdot, f \rangle g,$$

where f and g are fixed vectors of norm one. According to the spectral theorem for compact selfadjoint operators,

$$K = \sum_{j=1}^{\infty} \lambda_j \langle \cdot, \varphi_j \rangle \varphi_j,$$

where $\varphi_1, \varphi_2, \ldots$ is an orthonormal set in H and $\lambda_j = \lambda_j(K)$ for $j = 1, 2, \ldots$. For $\varphi \in H$ and $z \in \Omega$ we have

$$T(I - zK)^{-1}\varphi = \sum_{j=1}^{\infty} (1 - z\lambda_j)^{-1} \langle \varphi, \varphi_j \rangle \langle \varphi_j, f \rangle g.$$

It follows that

$$\|T(I - zK)^{-1}\varphi\| \le \sum_{j=1}^{\infty} |(1 - z\lambda_j)^{-1}\langle f, \varphi_j\rangle| \cdot |\langle \varphi, \varphi_j\rangle|$$

$$\le \left(\sum_{j=1}^{\infty} |(1 - z\lambda_j)^{-1}\langle f, \varphi_j\rangle|^2\right)^{1/2} \cdot \|\varphi\|,$$

and so for $z \in \Omega$

$$(4) \qquad \|T(I - zK)^{-1}\| \le \left(\sum_{j=1}^{\infty} |(1 - z\lambda_j)^{-1}\langle f, \varphi_j\rangle|^2\right)^{1/2}.$$

Now let Ω' be the image of the angle Ω under $z \to z^{-1}$. Again, Ω' is a closed angle with origin at zero. Without loss of generality we may assume that Ω' is as in Figure 1. From Figure 1 it is clear that for $0 \ne z \in \Omega$

$$(5) \qquad \left|\frac{1}{1 - \lambda_j z}\right| = \frac{|z^{-1}|}{|z^{-1} - \lambda_j|} \le \frac{1}{\sin \omega}, \qquad j = 1, 2, \dots .$$

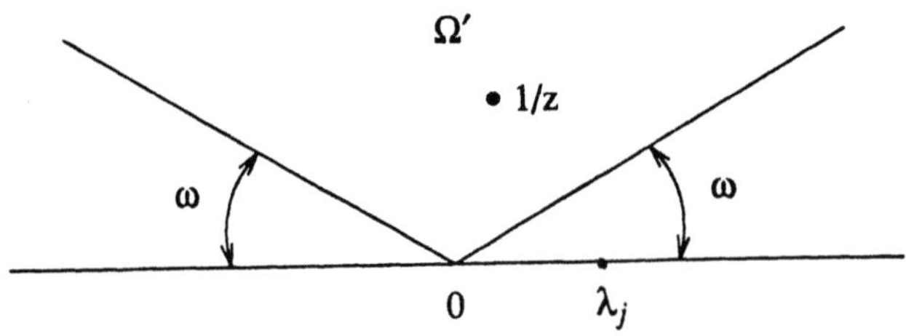

Figure 1

Now, let $\varepsilon > 0$ be given. Choose N such that

$$\sum_{j=N+1}^{\infty} |\langle f, \varphi_j\rangle|^2 < \frac{1}{2}\varepsilon^2 \sin \omega.$$

Next, choose $R > 0$ such that for $|z| \ge R$ and $z \in \Omega$

$$\sum_{j=1}^{N} |(1 - z\lambda_j)^{-1}\|\langle f, \varphi_j\rangle|^2 < \frac{1}{2}\varepsilon^2.$$

Employing formulas (4) and (5), one obtains that

$$\|T(I - zK)^{-1}\| < \varepsilon \qquad (z \in \Omega, |z| \ge R).$$

This proves the lemma for the case that T has rank one.

Using finite linear combinations of operators of rank one, one sees that the lemma holds true for any operator T of finite rank. The case when T is an arbitrary compact operator is proved via approximation by operators of finite rank. In order to do this, note that formula (5) also implies that

$$(6) \qquad \|(I - zK)^{-1}\| \le \frac{1}{\sin\omega} \qquad (z \in \Omega),$$

and thus

$$\|T(I - zK)^{-1}\| \le \frac{1}{\sin\omega}\|T - F\| + \|F(I - zK)^{-1}\|$$

for any $z \in \Omega$. From this inequality, and the first part of the proof, it is clear that the lemma holds for an arbitrary compact operator T. $\square$

PROOF OF THEOREM 4.1. Let Ω and Ω' be the closed angles with origin at zero described by the following figure:

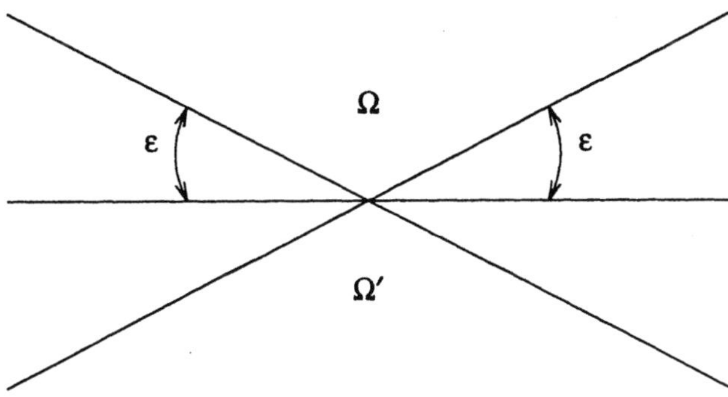

Figure 2

So the complement of $\Omega \cup \Omega'$ in $\mathbf{C}$ is the sector described by formula (3). Note that $\Omega \cup \Omega' \cup \{\infty\}$ is invariant under the map $z \mapsto 1/z$.

Define $T = S(I + S)^{-1}$. The operator T is compact and $I - T = (I + S)^{-1}$. So

$$I - zA = I - zK(I + S) = (I - T - zK)(I + S).$$

Now take $z \in \Omega \cup \Omega'$. Then $I - zK$ is invertible, and thus

$$(7) \qquad I - zA = [I - T(I - zK)^{-1}](I - zK)(I + S).$$

According to Lemma 4.2, there exists $R > 0$ such that

$$(8) \qquad \|T(I - zK)^{-1}\| \le \frac{1}{2} \qquad (z \in \Omega \cup \Omega', |z| \ge R).$$

It follows that for $z \in \Omega \cup \Omega'$ and $|z| \geq R$ the operator $I - zA$ is invertible. So the non-zero eigenvalues of A in $\Omega \cup \Omega'$ lie outside the disc $|z| \leq 1/R$, and thus A has only a finite number of eigenvalues in $\Omega \cup \Omega'$. This shows that all the eigenvalues of A, except for a finite number, are in the set Δ defined in (3).

Let E_A be the smallest subspace containing all eigenvectors and generalized eigenvectors corresponding to non-zero eigenvalues of A. Since $I + S$ and K are injective, the operator A is injective, and hence the system of eigenvectors and generalized eigenvectors of A is complete if and only if $E_A = H$. Let P be the orthogonal projection onto $E_A^\perp$, and put

$$B = PA|E_A^\perp \colon E_A^\perp \to E_A^\perp.$$

To prove completeness it suffices to show that the Volterra operator B is the zero operator. Indeed, assume $B = 0$. Then $B^* = 0$. But $B^* = A^*|E_A^\perp$. So $E_A^\perp \subset \operatorname{Ker} A^*$. However $A^* = (I + S)^* K$ has a trivial kernel. It follows that $E_A^\perp = (0)$ and hence $H = E_A$.

To prove that $B = 0$, we first show that, for some $\gamma \geq 0$,

$$(9) \qquad \|(I - zB)^{-1}\| \leq \gamma < \infty, \qquad z \in \Omega \cup \Omega'.$$

From the inequality (8) it is clear that

$$\|[I - T(I - zK)^{-1}]^{-1}\| \leq 2 \qquad (z \in \Omega \cup \Omega', |z| \geq R).$$

Furthermore, we know from the proof of Lemma 4.2 (see formula (6)) that

$$\|(I - zK)^{-1}\| \leq \frac{1}{\sin \varepsilon} \qquad (z \in \Omega \cup \Omega').$$

It follows that

$$(10) \qquad \|(I - zA)^{-1}\| \leq \frac{2}{\sin \varepsilon} \|(I + S)^{-1}\| \qquad (z \in \Omega \cup \Omega', |z| \geq R).$$

Now

$$(I - zB)^{-1} = P(I - zA)^{-1}|E_A^\perp,$$

whenever $I - zA$ is invertible. It follows that formula (10) remains true if in the left hand side of (10) the operator A is replaced by B. Next, observe that $\|(I - zB)^{-1}\|$ is a continuous function on the compact set

$$(11) \qquad \{z \in \Omega \cap \Omega' \mid |z| \leq R\},$$

and thus $\|(I - zB)^{-1}\|$ is bounded on the set (11). So $\|(I - zB)^{-1}\|$ is bounded on $\Omega \cup \Omega'$, and (9) is proved.

According to formula (1), we have $\sum_{j=1}^\infty s_j(K)^r < \infty$. As

$$B = PK(I + S)|E_A^\perp,$$

we also have $\sum_{j=1}^{\infty} s_j(B)^r < \infty$ (see Proposition VI.1.3 and Corollary VI.1.4). So we can apply Theorem 2.2 to show that given $\delta > 0$ there exists a constant C such that

$$(12) \qquad \|(I - zB)^{-1}\| \leq Ce^{\delta|z|^r}, \qquad z \in \mathbb{C}.$$

Now consider $(I - zB)^{-1}$ on the closed angle

$$\Delta_1 = \{\rho e^{i\varphi} \mid \rho \geq 0, |\varphi| \leq \varepsilon\}.$$

We may assume that $0 < \varepsilon \leq \frac{\pi}{2r}$. On the sides of Δ_1 the function $(I - zB)^{-1}$ is bounded in norm (see formula (9)) and for the behaviour at infinity we have formula (12). So we may apply the Phragmén-Lindelöf theorem (see [C], Corollary VI.4.4) to show that $(I - zB)^{-1}$ is bounded on Δ_1. In exactly the same way one can prove that $(I - zB)^{-1}$ is bounded on

$$\Delta_2 = \{\rho e^{i\varphi} \mid \rho \geq 0, |\pi - \varphi| \leq \varepsilon\}.$$

This, together with (9), implies that $(I - zB)^{-1}$ is a bounded entire function. Hence, by the Liouville theorem, the function $(I - zB)^{-1}$ must be constant. As its value at zero is equal to I, we see that $I = I - zB$, and hence B is the zero operator. $\square$

We conclude with a few remarks. If the system of eigenvectors and generalized eigenvectors of an operator A is complete, then it does not follow that one has also completeness for the adjoint operator A^*. But, if A is as in Theorem 4.1, then we have also completeness for A^*. Indeed, first of all note that completeness is invariant under similarity. So, if K and S are as in Theorem 4.1, then we also have completeness for the operator

$$(I + S)K = (I + S)[K(I + S)](I + S)^{-1}.$$

As S^* has the same properties as S, Theorem 4.1 and the previous remark imply completeness for $(I + S^*)K$, which is just the adjoint of $A = K(I + S)$.

In Gohberg-Krein [3] it is shown that in Theorem 4.1 one does not need the full strength of condition (1). Also, condition (1) may be replaced by

$$\sum_{j=1}^{\infty} s_j(S)^r < \infty$$

for some $r \geq 1$. Furthermore, Macaev [1] (see also Macaev [2]) has proved that the conclusion of Theorem 4.1 remains valid if instead of (1) one has

$$\sum_{j=1}^{\infty} j s_j(S) < \infty,$$

but one cannot go beyond this class of operators.

COMMENTS ON PART II

Except for Chapter IX the present part consists of a selection from the main results in the book Gohberg-Krein [3]. The exposition, however, is considerably different. Trace and determinant are introduced from their finite dimensional analogues by a continuity argument. Also the evaluation of the growth of the resolvent in Chapter X is based on the finite dimensional case. The material in Chapter IX is taken from the paper Gohberg-Kaashoek [1]. The problem of completeness of eigenvectors and generalized eigenvectors, considered in Chapters VII, VIII and X, is an important topic which is mainly discussed in the Soviet literature. For recent developments in this area (which are many) we refer to the excellent book by A.S. Markus (Markus [1]). The latter contains also an extended list of references.

EXERCISES TO PART II

In the exercises below, H is a separable Hilbert space and, unless mentioned explicitly otherwise, all operators are bounded linear operators acting on H. The sequence of s-numbers $s_1(A), s_2(A), \ldots$ of an operator A is considered to be an infinite sequence by adding zeros if necessary.

1. Let A and B be compact operators and r a positive number. Show that

$$s_n(A) = o(n^{-r}), \quad s_n(B) = o(n^{-r}), \qquad n \to \infty,$$

implies that $s_n(A + B) = o(n^{-r})$ for $n \to \infty$.

2. Let A, B and r be as in Exercise 1, and let γ be a nonnegative integer. Show that

$$\lim_{n \to \infty} n^r s_n(A) = 0, \qquad \lim_{n \to \infty} n^r s_n(B) = \gamma$$

implies that $n^r s_n(A + B) \to \gamma$ if $n \to \infty$.

3. Let A be a compact operator, and let $\varphi_1, \varphi_2, \ldots$ be an orthonormal basis of H. Assume that

$$|\langle A\varphi_j, \varphi_j \rangle| = s_j(A), \qquad j = 1, 2, \ldots .$$

Prove that

(i) $A^* A\varphi_j = s_j(A)^2 \varphi_j$, $j = 1, 2, \ldots$,

(ii) $A\varphi_j = \langle A\varphi_j, \varphi_j \rangle \varphi_j$, $j = 1, 2, \ldots$.

4. Assume $s_j(A) = |\lambda_j(A)|$ for $j = 1, 2, \ldots$. Show that there exists an orthonormal basis consisting of eigenvectors of A. What happens additionally if $s_j(A) = \lambda_j(A)$ for $j = 1, 2, \ldots$?

5. Let A be a compact operator and P an orthogonal projection. Show that

$$s_j\big(PAP + (I - P)A(I - P)\big) = s_j(A), \qquad j = 1, 2, \ldots,$$

implies that $PAP + (I - P)A(I - P) = A$.

6. Let the operator A be given by

$$Ax = \sum_{j=1}^{\infty} \langle x, \varphi_j \rangle \psi_j, \qquad x \in H,$$

where $\sum_{j=1}^{\infty} \|\varphi_j\|^2 < \infty$, $\sum_{j=1}^{\infty} \|\psi_j\|^2 < \infty$. Show that A is a trace class operator.

7. Let T be the operator on ℓ_2 given by the infinite matrix

$$\big(\langle \psi_j, \varphi_i \rangle\big)_{i,j=1}^{\infty} ,$$

where (φ_i) and (ψ_i) are as in the previous exercise. Show that T is a trace class operator, and prove that T and the operator A of the previous exercise have the same non-zero eigenvalues.

8. Let A be an operator of finite rank. Show that all the non-zero s-numbers of A are equal to 1 if $A^* = A^*AA^*$ and $A = AA^*A$. Prove also the reverse implication. What can you say about these implications if A is compact but not of finite rank?

9. Let $A_1, A_2, \ldots$ be a sequence in S_1 which converges in the trace class norm to A. Prove that

$$(*) \qquad\qquad \lim_{n \to \infty} \det(I + \lambda A_n) = \det(I + \lambda A)$$

and the convergence in $(*)$ is uniform on compact subsets of C.

10. Let A be compact. Determine linearly independent vectors $y_1, \ldots, y_n$ such that

$$g(Ay_1, \ldots, Ay_n) = \left(\prod_{j=1}^{n} s_j(A)^2 \right) g(y_1, \ldots, y_n).$$

Here $g(z_1, \ldots, z_n)$ denotes the Gram determinant of the vectors $z_1, \ldots, z_n$.

11. Let T be the operator on ℓ_2 given by the infinite matrix

$$\left(2^{-\frac{1}{2}(i+j)} \right)_{i,j=1}^{\infty}.$$

Compute the s-numbers of T.

12. Let T be the operator on ℓ_2 given by the infinite matrix

$$(\alpha_i \beta_j)_{i,j=1}^{\infty},$$

where $(\alpha_1, \alpha_2, \ldots)$ and $(\beta_1, \beta_2, \ldots)$ are in ℓ_2. Compute the s-numbers of T.

13. Let K be the operator on $L_2([0,1])$ defined by

$$(K\varphi)(t) = \int_0^1 \operatorname{sgn}(t - s)\varphi(s)ds, \qquad 0 \le t \le 1.$$

Compute the s-numbers of K.

14. Assume that A has N $(< +\infty)$ non-zero s-numbers. Show that the number of non-zero eigenvalues of A, multiplicities taken into account, is at most equal to N.

15. Let T on ℓ_2 be given by the infinite matrix

$$\begin{bmatrix} 0 & 0 & 0 & 0 & \cdots \\ 1 & 0 & 0 & 0 & \\ 0 & \frac{1}{2} & 0 & 0 & \\ 0 & 0 & \frac{1}{3} & 0 & \\ \vdots & & & & \ddots \end{bmatrix}.$$

Prove that T is a Hilbert-Schmidt operator, but not a trace class operator.

16. Let r be an integer, and let T on ℓ_2 be given by the infinite matrix

$$\left(\delta_{i-r,j}\frac{1}{j}\right)^{\infty}_{i,j=1},$$

where $\delta_{k,j}$ is the Kronecker delta. Prove that T is not a trace class operator.

17. Let r be an integer, and let T on ℓ_2 be given by the infinite matrix

$$\left(\delta_{i-r,j}j^{-p}\right)^{\infty}_{i,j=1},$$

where $\delta_{k,j}$ is the Kronecker delta. For which p is T a trace class operator?

18. Let r be an integer, and let A on ℓ_2 be given by the infinite matrix

$$\left(\delta_{i-r,j}\alpha_j\right)^{\infty}_{i,j=1},$$

where $\delta_{k,j}$ is the Kronecker delta. Show that $A \in S_1$ if and only if $\sum_j |\alpha_j| < \infty$, and prove that in that case

$$\|A\|_1 = \sum_{j=\min(1,r+1)}^{\infty} |\alpha_j|.$$

19. Let A on ℓ_2 be given by the infinite matrix $(a_{ij})^{\infty}_{i,j=1}$, and assume that

$$\gamma_1 := \sum_{i=1}^{\infty}\sum_{j=1}^{\infty} |a_{ij}| < \infty.$$

Prove that $A \in S_1$ and $\|A\|_1 \leq \gamma_1$.

20. Let q be a positive integer, and let T on ℓ_2 be given by the infinite matrix $(a_{ij})^{\infty}_{i,j=1}$, where

$$a_{ij} = \begin{cases} \alpha_i & \text{for} \quad j = q, \\ 0 & \text{for} \quad j \neq q. \end{cases}$$

Prove that $T \in S_1$ and $\|T\|_1 = \left(\sum_{i=1}^{\infty} |\alpha_i|^2\right)^{1/2}$.

21. Let T on ℓ_2 be given by the infinite matrix $(a_{ij})^{\infty}_{i,j=1}$, and assume that

$$\gamma_2 := \sum_{j=1}^{\infty}\left(\sum_{i=1}^{\infty} |a_{ij}|^2\right)^{1/2} < \infty.$$

Prove that $T \in S_1$ and $\|T\|_1 \leq \gamma_2$. How is the number γ_1 in Exercise 19 related to γ_2?

22. Assume that

$$A = \begin{bmatrix} A_{11} & A_{12} \\ A_{21} & A_{22} \end{bmatrix} : H \oplus H \to H \oplus H$$

is a trace class operator. Prove:

(a) each A_{ij} is a trace class operator;

(b) $\operatorname{tr} A = \operatorname{tr} A_{11} + \operatorname{tr} A_{22}$;

(c) $\det(I - \lambda A) = \det(I - \lambda A_{22}) \det\big(I - \lambda A_{11} - \lambda^2 A_{12}(I - \lambda A_{22})^{-1} A_{21}\big)$.

23. Let T_0, T_1 and T_2 be trace class operators. Show that the sequence of zeros of the function

$$\lambda \mapsto \det(I + \lambda T_0 + \lambda^2 T_1 T_2)$$

is absolutely summable. (Hint: use linearization and global equivalence.)

24. Show that a trace class operator can be written as the product of two Hilbert-Schmidt operators.

25. For $i, j = 1, \ldots, n$ let A_{ij} be a Hilbert-Schmidt operator on H. Prove that the operator

$$(*) \qquad\qquad T = \begin{bmatrix} A_{11} & \cdots & A_{1n} \\ \vdots & & \vdots \\ A_{n1} & \cdots & A_{nn} \end{bmatrix}$$

is a Hilbert-Schmidt operator on the direct sum H^n of n copies of H and

$$\|T\|_2 = \left(\sum_{i,j=1}^{n} \|A_{ij}\|_2^2 \right)^{1/2}.$$

Let T_{pq} be the operator on H^n which one obtains if in $(*)$ for $(i, j) \neq (p, q)$ the operator A_{ij} is replaced by the zero operator. Prove that $\{T_{pq} \mid p, q = 1, \ldots, n\}$ is an orthonormal system relative to the inner product on S_2.

26. Let the operator A on ℓ_2 be given by an infinite upper triangular matrix

$$(*) \qquad\qquad \begin{bmatrix} a_{11} & a_{12} & a_{13} & \cdots \\ 0 & a_{22} & a_{23} & \cdots \\ 0 & 0 & a_{33} & \cdots \\ \vdots & \vdots & & \end{bmatrix}.$$

Prove that the linear span of eigenvectors and generalized eigenvectors of A is dense in ℓ_2.

27. Let $H = H_1 \oplus H_2$, and assume that the operator

$$A = \begin{bmatrix} A_{11} & A_{12} \\ 0 & A_{22} \end{bmatrix} : H_1 \oplus H_2 \to H_1 \oplus H_2$$

is compact. Show that the linear span of eigenvectors and generalized eigenvectors of A is dense in H whenever this property holds for A_{11} and A_{22} in the corresponding spaces.

28. Does the statement in Exercise 26 remain true if "upper" is replaced by "lower"? What happens if we add compactness to the conditions?

29. Let A be a compact operator. If the linear span of eigenvectors and generalized eigenvectors of A is dense in H, does it follow that the same is true for A^*?

A family $\mathcal{F}$ of operators on H is called *projectionally invariant* if given $A \in \mathcal{F}$ the operator $PAP \in \mathcal{F}$ for each orthogonal projection P.

30. Let $\mathcal{F}$ be a projectionally invariant set of compact operators on H, and assume that each non-zero A in $\mathcal{F}$ has a non-zero eigenvalue. Then the linear span of eigenvectors and generalized eigenvectors of A is dense in H for each A in $\mathcal{F}$. Prove this statement.

31. Let $\mathcal{F}$ be the family of the trace class operators A on H for which $A_\Im \geq 0$. Show that $\mathcal{F}$ is projectionally invariant and each non-zero A in $\mathcal{F}$ has a non-zero eigenvalue.

32. Derive the completeness theorems of Lidskii (Theorems VII.8.1 and VIII.3.1) by using the preceding exercises.

33. Let K_1, K_2 be integral operators on $L_2^m([a, b])$ with semi-separable kernel functions. Prove that $K_1 + K_2$ and $K_1 K_2$ are again integral operators with semi-separable kernel functions.

34. Let K be the integral operator on $L_2([0, 1])$ with kernel function $k(t, s) = \mathrm{sgn}(t - s)$. Determine a semi-separable representation of k, and use this representation to compute the eigenvalues, the eigenvectors and the resolvent of K. Is K a trace class operator?

35. Let K be the integral operator on $L_2([0, 1])$ with kernel function

$$k(t, s) = \begin{cases} \alpha & \text{for} \quad 0 \leq s \leq t \leq 1, \\ \beta & \text{for} \quad 0 \leq t < s \leq 1, \end{cases}$$

where α and β are two different complex numbers. Compute the eigenvalues, the eigenvectors and the resolvent of K by using a semi-separable representation of k. Is K a trace class operator?

36. Consider the following system of equations:

$$(*) \qquad \begin{cases} \dot{\rho}(t) = A(t)\rho(t) + B(t)u(t), \quad a \leq t \leq b, \\ y(t) = C(t)\rho(t), \\ N_1 x(a) + N_2 x(b) = 0. \end{cases}$$

Here $A(\cdot)$, $B(\cdot)$ and $C(\cdot)$ are matrix functions of sizes $n \times n$, $n \times m$ and $m \times n$, respectively, the entries of $A(\cdot)$ are integrable on $[a, b]$ and those of $B(\cdot)$ and $C(\cdot)$ are square integrable on the same interval. The matrices N_1, N_2 are $n \times n$ matrices such that

$$\det\bigl(N_1 + N_2 U(b)\bigr) \neq 0,$$

where $U(\cdot)$ is the fundamental matrix (normalized to I_n at $t = a$) of the differential equation

$$x'(t) = A(t)x(t), \qquad a \leq t \leq b.$$

Show that for each $u \in L_2^m([a, b])$, there exists a unique $y \in L_2^m([a, b])$ such that $(*)$ is satisfied for some $\rho \in L_2^n([a, b])$. Prove that the map $u \mapsto y$ determined by $(*)$ is an integral operator on $L_2^m([a, b])$ with a semi-separable kernel function and determine a semi-separable representation for this kernel function. Show that any integral operator on $L_2^m([a, b])$ with a semi-separable kernel function may be represented in this way.

PART III

FREDHOLM OPERATORS: GENERAL THEORY AND
WIENER-HOPF INTEGRAL OPERATORS

This part presents an introduction to the general theory of Fredholm operators. It also contains elements of the theory of Wiener-Hopf integral operators. The latter are treated here as examples of Fredholm operators. A main part of the Wiener-Hopf theory developed in this part concerns systems of equations with a rational matrix symbol. This restriction allows one to include recent results which provide explicit formulas for the inverse and Fredholm characteristics.

CHAPTER XI

FREDHOLM OPERATORS

This chapter presents a concise introduction to the abstract theory of Fredholm operators. It also contains some examples; however the main applications will concern Wiener-Hopf integral operators and Toeplitz operators which we shall deal with in the next chapters and in Volume II. The first section contains the definition of a Fredholm operator and the first examples. In Section 2 we pay special attention to operators with closed range. The basic perturbation theorems and the properties of the index are given in Sections 3 and 4. In Section 5 Fredholm operators are studied in the framework of the Calkin algebra. Connections with generalized invertibility appear in Section 6. Index formulas in terms of trace and determinant are given in Section 7. Sections 8 and 9 are devoted to Fredholm operator valued functions that are analytic and to equivalence of such functions. An operator theory generalization of Rouché's theorem appears here. The last section concerns singular values of bounded operators and their connections with the essential spectrum.

XI.1 DEFINITION AND FIRST EXAMPLES

A bounded linear operator $A\colon X \to Y$, acting between complex Banach spaces X and Y, is called a *Fredholm operator* if its range $\operatorname{Im} A$ is closed and the numbers

$$(1) \qquad n(A) = \dim \operatorname{Ker} A, \qquad d(A) = \dim(Y/\operatorname{Im} A)$$

are finite. In that case $\operatorname{ind} A = n(A) - d(A)$ is said to be the *index* of A. In the next section (Corollary 2.3) we shall see that the condition "$\operatorname{Im} A$ is closed" is automatically fulfilled if the quotient space $Y/\operatorname{Im} A$ is finite dimensional. Let us consider a few examples.

(i) If X and Y are both finite dimensional spaces, then any operator $A\colon X \to Y$ is Fredholm and

$$\operatorname{ind} A = \dim X - \dim Y.$$

(ii) If $K\colon X \to X$ is a compact operator on the Banach space X, then $A = I - K$ is a Fredholm operator and $\operatorname{ind} A = 0$. This follows from the Fredholm theory for compact operators (see [GG], Theorem XI.4.1). We shall come back to this example in Sections 2 and 4.

(iii) Consider the following two point boundary value problem:

$$(2) \qquad \begin{cases} f'(t) = Mf(t) + g(t), & a \leq t \leq b, \\ N_1 f(a) + N_2 f(b) = x. \end{cases}$$

Here M is an $n \times n$ matrix, the function g is a given function in $L_2^n([a,b])$ (thus g is an $\mathbb{C}^n$-valued function of which the components are in $L_2([a,b])$) and x is a given vector in

$\mathbf{C}^n$. The boundary conditions are defined in terms of two $n \times n$ matrices N_1 and N_2. The problem is to find a solution in the space

$$(3) \qquad X = \{f \in L_2^n([a,b]) \mid f \text{ absolutely continuous}, f' \in L_2^n([a,b])\}.$$

Note that X is a Hilbert space with the norm $\|\|f\|\| = (\|f\|^2 + \|f'\|^2)^{1/2}$, where $\|\cdot\|$ denotes the usual norm on $L_2^n([a,b])$. In operator form, (2) may be summarized by

$$(4) \qquad Af := \begin{bmatrix} f' - Mf \\ N_1 f(a) + N_2 f(b) \end{bmatrix} = \begin{bmatrix} g \\ x \end{bmatrix}.$$

We claim that $A : X \to L_2^n([a,b]) \oplus \mathbf{C}^n$ is a Fredholm operator of index zero. To see this let T be the $n \times n$ matrix defined by

$$(5) \qquad T = N_1 + N_2 e^{(b-a)M},$$

and introduce the following auxiliary operators

$$(6) \qquad E = \begin{bmatrix} 0 & I_{L_2} \\ I_n & V \end{bmatrix} : \mathbf{C}^n \oplus L_2^n([a,b]) \to L_2^n([a,b]) \oplus \mathbf{C}^n,$$

$$(7) \qquad F = \begin{bmatrix} \pi \\ D \end{bmatrix} : X \to \mathbf{C}^n \oplus L_2^n([a,b]),$$

where

$$Vg = N_2 \int_a^b e^{(b-s)M} g(s) ds, \qquad g \in L_2^n([a,b]),$$

$$\pi f = f(a), \qquad Df = f' - Mf, \quad f \in X,$$

and I_{L_2} and I_n are the identity operators on $L_2^n([a,b])$ and $\mathbf{C}^n$, respectively. Then E and F are invertible,

$$E^{-1} = \begin{bmatrix} -V & I_n \\ I_{L_2} & 0 \end{bmatrix},$$

$$F^{-1} \begin{bmatrix} x \\ g \end{bmatrix} = e^{(t-a)M} x + \int_a^t e^{(t-s)M} g(s) ds, \qquad a \le t \le b,$$

and the following equivalence relation holds true:

$$(8) \qquad A = E \begin{bmatrix} T & 0 \\ 0 & I_{L_2} \end{bmatrix} F.$$

Since E and F are invertible, (8) implies that A is a Fredholm operator and $\operatorname{ind} A = 0$. We shall return to this example in Section 8.

(iv) Consider on ℓ_2 the shift operators S_r and S_ℓ defined by

$$S_r(x_1, x_2, x_3, \ldots) = (0, x_1, x_2, \ldots),$$
$$S_\ell(x_1, x_2, x_3, \ldots) = (x_2, x_3, x_4, \ldots).$$

Both operators are Fredholm operators, $\operatorname{ind} S_r = -1$ and $\operatorname{ind} S_\ell = +1$. With S_r and S_ℓ it is easy to build a Fredholm operator whose index is equal to an arbitrary prescribed integer. Indeed, put $X = \ell_2 \oplus \ell_2$, and define $A\colon X \to X$ by the following 2×2 operator matrix

$$A = \begin{bmatrix} S_r^p & 0 \\ 0 & S_\ell^q \end{bmatrix}.$$

Then A is a Fredholm operator, $n(A) = q$, $d(A) = p$ and $\operatorname{ind} A = q - p$.

XI.2 OPERATORS WITH CLOSED RANGE

As before, $\mathcal{L}(X, Y)$ denotes the space of all bounded linear operators acting from X into Y, where X and Y are complex Banach spaces. Given $A \in \mathcal{L}(X, Y)$, we write $d(x, \operatorname{Ker} A)$ for the distance of a vector $x \in X$ to the kernel of A, i.e.,

$$d(x, \operatorname{Ker} A) = \inf\{\|x - y\| \,|\, Ay = 0\}.$$

THEOREM 2.1. *The operator $A \in \mathcal{L}(X, Y)$ has closed range if and only if there exists $c > 0$ such that*

$$(1) \qquad\qquad \|Ax\| \geq cd(x, \operatorname{Ker} A), \qquad x \in X.$$

PROOF. An element of the quotient space $\hat{X} = X/\operatorname{Ker} A$ will be denoted by $\hat{x}$. The space $\hat{X}$ is a Banach space with norm $\|\hat{x}\| = d(x, \operatorname{Ker} A)$. Define $\hat{A}\colon \hat{X} \to Y$ by $\hat{A}\hat{x} = Ax$. The operator $\hat{A}$ is $1 - 1$, linear, bounded and $\operatorname{Im} \hat{A} = \operatorname{Im} A$. Suppose $\operatorname{Im} A$ is closed. Then $\hat{A}^{-1}$, considered as a map from the Banach space $\operatorname{Im} A$ into the Banach space $\hat{X}$, is a closed operator. Hence, by the closed graph theorem, $\hat{A}^{-1}$ is bounded and

$$\|Ax\| = \|\hat{A}\hat{x}\| \geq \|\hat{A}^{-1}\|^{-1}\|\hat{x}\| = \|\hat{A}^{-1}\|^{-1}d(x, \operatorname{Ker} A).$$

Thus (1) holds with $c = \|\hat{A}^{-1}\|^{-1}$.

Conversely, assume (1) holds. Suppose $Ax_n \to y$. It follows from (1) that $(\hat{x}_n)$ is a Cauchy sequence in the Banach space $\hat{X}$ which therefore converges to some $\hat{x} \in \hat{X}$. Hence

$$Ax_n = \hat{A}\hat{x}_n \to \hat{A}\hat{x} = Ax,$$

and $y = Ax$, which proves that $\operatorname{Im} A$ is closed. $\square$

THEOREM 2.2. *The operator $A \in \mathcal{L}(X, Y)$ has a closed range whenever there exists a subspace (i.e. a closed linear manifold) Y_0 such that $\operatorname{Im} A \oplus Y_0$ is closed.*

PROOF. Define $A_0\colon X \times Y_0 \to Y$ by $A_0(x, y_0) = Ax + y_0$. The space $X \times Y_0$ is a Banach space with the norm defined by $\|(x, y)\| = \|x\| + \|y\|$. Clearly,

A_0 is a bounded linear operator and $\operatorname{Im} A_0 = \operatorname{Im} A \oplus Y_0$ which is closed by hypothesis. Moreover, $\operatorname{Ker} A_0 = \operatorname{Ker} A \times \{0\}$. Theorem 2.1 asserts that there exists $c > 0$ such that for all $x \in X$

$$\|Ax\| = \|A_0(x,0)\| \geq cd((x,0), \operatorname{Ker} A_0) = cd(x, \operatorname{Ker} A).$$

Hence $\operatorname{Im} A$ is closed by the same theorem. $\square$

COROLLARY 2.3. *If the range of $A \in \mathcal{L}(X,Y)$ is complemented (in particular, if $\dim(Y/\operatorname{Im} A)$ is finite), then $\operatorname{Im} A$ is closed.*

PROOF. The range of A is complemented means that there exists a closed linear manifold Y_0 of Y such that $\operatorname{Im} A \oplus Y_0 = Y$. According to Theorem 2.2, this implies that $\operatorname{Im} A$ is closed. $\square$

To see the importance of the previous corollary, note that for a linear manifold M of a Banach space Y the statement

$$(2) \qquad\qquad\qquad Y = M \oplus Y_0$$

for some subspace Y_0 does not imply that M is closed. To see this, take a non-continuous linear functional f on Y and put $M = \operatorname{Ker} f$. Then there exists a one-dimensional subspace Y_0 such that (2) holds true. But $M = \operatorname{Ker} f$ cannot be closed because f is not continuous.

THEOREM 2.4. *If the operator $A \in \mathcal{L}(X,Y)$ maps bounded closed sets in X onto closed sets in Y, then $\operatorname{Im} A$ is closed.*

PROOF. Suppose $\operatorname{Im} A$ is not closed. Then, by Theorem 2.1, there exists a sequence (x_n) such that $Ax_n \to 0$ $(n \to \infty)$ and $d(x_n, \operatorname{Ker} A) = 1$ for $n = 1, 2, \ldots$. For each n choose $z_n \in \operatorname{Ker} A$ such that $\|x_n - z_n\| < 2$, and let V be the closure of the set $\{x_n - z_n \mid n = 1, 2, \ldots\}$. Since V is closed and bounded in X, its image AV is closed in Y. Note that $Ax_n = A(x_n - z_n) \in AV$. So $0 \in AV$, and thus there exists $u \in V \cap \operatorname{Ker} A$. From the definition of V it follows that

$$(3) \qquad\qquad\qquad \|u - (x_{n_0} - z_{n_0})\| < \frac{1}{2}$$

for some n_0. But (3) implies that $d(x_{n_0}, \operatorname{Ker} A) < \frac{1}{2}$, which contradicts the fact that $d(x_n, \operatorname{Ker} A) = 1$ for all n. So $\operatorname{Im} A$ is closed. $\square$

If $K \colon X \to X$ is compact, then the operator $A = I - K$ satisfies the condition of Theorem 2.1. To see this, let V be a closed bounded set in X, and let

$$(4) \qquad\qquad\qquad y = \lim_{n \to \infty} (I - K)x_n,$$

where $x_1, x_2, \ldots$ is a sequence in V. We have to prove that $y = (I - K)x$ for some $x \in V$. Since V is bounded and K is compact, the sequence $(Kx_n)_n$ has a convergent subsequence $(Kx_{n_i})_i$, say. Using (4), we see that

$$x_0 = \lim_{i \to \infty} x_{n_i} = \lim_{i \to \infty} \left((I - K)x_{n_i} + Kx_{n_i}\right)$$

exists. But then $y = (I - K)x_0 \in (I - K)V$, and $(I - K)V$ is closed. We have the following corollary.

COROLLARY 2.5. *If $K: X \to X$ is compact, then $I - K$ is a Fredholm operator.*

PROOF. From the previous paragraph we know that $A = I - K$ satisfies the condition of Theorem 2.4. So $\operatorname{Im}(I - K)$ is closed. Since $x = Kx$ for $x \in \operatorname{Ker}(I - K)$, it follows that the identity operator acts as a compact operator on $\operatorname{Ker}(I - K)$. Thus $n(I - K) < \infty$.

To prove that $d(I - K) < \infty$, we use that the conjugate operator $K': X' \to X'$ is also compact. Since $\operatorname{Im}(I - K)$ is closed, we have $\operatorname{Im}(I - K) = \operatorname{Ker}(I - K')^{\perp}$, and thus $d(I - K) = n(I - K') < \infty$. $\square$

In Section 4 we shall see that $\operatorname{ind}(I - K) = 0$ if the operator K is compact.

XI.3 PRODUCT OF FREDHOLM OPERATORS

To study the properties of a Fredholm operator we shall employ bijective operators which are closely related to Fredholm operators. Let $A: X \to Y$ be a bounded linear operator acting between the Banach spaces X and Y. Suppose A has the property that $\operatorname{Ker} A$ and $\operatorname{Im} A$ are complemented by subspaces (i.e., closed linear manifolds) X_0 and Y_0, respectively. Define $\tilde{A}: X_0 \times Y_0 \to Y$ by

$$\tilde{A}(x_0, y_0) = Ax_0 + y_0.$$

The space $X_0 \times Y_0$ is a Banach space with the norm defined by $\|(x, y)\| = \|x\| + \|y\|$ and the operator $\tilde{A}$ is a bijective bounded linear operator. We call $\tilde{A}$ the *bijection associated with* A (and the subspaces X_0 and Y_0). If A is Fredholm, then such a bijection always exists and Y_0 is finite dimensional. If we identify the space X_0 with $X_0 \times \{0\}$, then the operator

$$A_0: X_0 \to Y, \qquad A_0 x = Ax,$$

is a common restriction of A and $\tilde{A}$.

LEMMA 3.1. *Suppose $A_0: M \to Y$ is a restriction of $A \in \mathcal{L}(X, Y)$ to a subspace M of X with $\operatorname{codim} M = n < \infty$. Then A is Fredholm if and only if A_0 is Fredholm, in which case $\operatorname{ind} A = \operatorname{ind} A_0 + n$.*

PROOF. It suffices to prove the lemma for $n = 1$. Put $X = M \oplus \operatorname{sp}\{x_1\}$. We consider two cases.

Case 1. Assume $Ax_1 \notin \operatorname{Im} A_0$. Then $AX = A_0 M \oplus \operatorname{sp}\{Ax_1\}$ and $\operatorname{Ker} A = \operatorname{Ker} A_0$. Hence

$$(1) \qquad\qquad d(A_0) = d(A) + 1, \qquad n(A_0) = n(A).$$

Case 2. Assume $Ax_1 \in \operatorname{Im} A_0$. Then $\operatorname{Im} A = \operatorname{Im} A_0$ and there exists $u \in M$ such that $Ax_1 = A_0 u$. Consequently, $\operatorname{Ker} A = \operatorname{Ker} A_0 \oplus \operatorname{sp}\{x_1 - u\}$. Thus

$$(2) \qquad\qquad d(A_0) = d(A), \qquad n(A_0) = n(A) - 1.$$

The lemma follows immediately from (1) or (2). $\square$

THEOREM 3.2. *If $A: X \to Y$ and $B: Y \to Z$ are Fredholm operators, then BA is a Fredholm operator and*

$$\operatorname{ind}(BA) = \operatorname{ind} B + \operatorname{ind} A.$$

PROOF. Let $\tilde{A}$ be a bijection associated with A and the subspaces X_0 and Y_0, and let A_0 be the restriction of A to X_0. Since $\tilde{A}$ is invertible, the operator $B\tilde{A}$ is Fredholm and $\operatorname{ind} B\tilde{A} = \operatorname{ind} B$. By identifying the spaces X_0 and $X_0 \times \{0\}$, we see that BA_0 is a common restriction of $B\tilde{A}$ and BA. So, according to the previous lemma, BA is Fredholm and

$$\begin{aligned}
\operatorname{ind} BA &= \operatorname{ind} BA_0 + \dim(X/X_0) \\
&= \operatorname{ind} B\tilde{A} - \dim(X_0 \times Y_0/X_0 \times \{0\}) + n(A) \\
&= \operatorname{ind} B + \operatorname{ind} A. \quad \square
\end{aligned}$$

XI.4 PERTURBATION THEOREMS

THEOREM 4.1. *Suppose $A: X \to Y$ is a Fredholm operator, and let $\tilde{A}$ be a bijection associated with A. If $B: X \to Y$ is a bounded linear operator with $\|B\| < \|\tilde{A}^{-1}\|^{-1}$, then $A + B$ is Fredholm and*

(i) $n(A + B) \le n(A)$,

(ii) $d(A + B) \le d(A)$,

(iii) $\operatorname{ind}(A + B) = \operatorname{ind} A$.

PROOF. Let X_0 and Y_0 be the subspaces corresponding to the bijection $\tilde{A}$. Put $C = A + B$, and define $\tilde{C}: X_0 \times Y_0 \to Y$ by $\tilde{C}(x_0, y_0) = Cx_0 + y_0$. By definition, $\tilde{A}(x_0, y_0) = Ax_0 + y_0$. Since $\tilde{A}$ is invertible and

$$\|\tilde{A} - \tilde{C}\| \le \|A - C\| = \|B\| < \|\tilde{A}^{-1}\|^{-1},$$

the operator $\tilde{C}$ is also invertible. Note that the operator $C_0: X_0 \to Y$, defined by $C_0 x = Cx$, is a common restriction of C and $\tilde{C}$. So, by Lemma 3.1, the operator C is Fredholm and

$$\begin{aligned}
\operatorname{ind} C &= \operatorname{ind} C_0 + n(A) \\
&= \operatorname{ind} \tilde{C} - d(A) + n(A) = \operatorname{ind} A
\end{aligned}$$

The invertibility of $\tilde{C}$ implies that $X_0 \cap \operatorname{Ker} C = \{0\}$. Thus

$$n(C) \le \dim X/X_0 = n(A),$$

which proves (i). Finally, note that (ii) is a simple consequence of (i) and (iii). $\square$

THEOREM 4.2. *If $A: X \to Y$ is a Fredholm operator and $K: X \to Y$ is compact, then $A + K$ is a Fredholm operator and $\operatorname{ind} A = \operatorname{ind}(A + K)$.*

PROOF. Let $X = X_0 \oplus \operatorname{Ker} A$ and $Y = Y_0 \oplus \operatorname{Im} A$. On $X_0 \times Y_0$ define $\tilde{A}$ and $\tilde{K}$ by

$$\tilde{A}(x_0, y_0) = Ax_0 + y_0, \qquad \tilde{K}(x_0, y_0) = Kx_0 + y_0.$$

The operator $\tilde{K}$ is compact, since K is compact and Y_0 is finite dimensional. From $(\tilde{A} + \tilde{K})(x_0, 0) = (A + K)x_0$ and Lemma 3.1 it follows that $A + K$ is Fredholm if and only if $\tilde{A} + \tilde{K}$ is Fredholm. But $\tilde{A}$ is invertible. So

$$\tilde{A} + \tilde{K} = \tilde{A}[I + \tilde{A}^{-1}\tilde{K}].$$

Observe that $\tilde{A}^{-1}\tilde{K}$ is compact. So, by Corollary 2.5, the operator $I + \tilde{A}^{-1}\tilde{K}$ is Fredholm. Hence $A + K$ is Fredholm.

To prove the statement about the index, consider the integer-valued function $f(\lambda) = \operatorname{ind}(A + \lambda K)$. Applying Theorem 4.1 to $A + \lambda K$ in place of A shows that f is continuous on $[0, 1]$. Consequently, f is constant. In particular,

$$\operatorname{ind} A = f(0) = f(1) = \operatorname{ind}(A + K). \quad \square$$

COROLLARY 4.3. *If* $K: X \to X$ *is compact, then* $I - K$ *is Fredholm and* $\operatorname{ind}(I - K) = 0$.

PROOF. Apply the preceding theorem with $A = I$ and note that $\operatorname{ind} I = 0$. $\square$

XI.5 INVERTIBILITY MODULO COMPACT OPERATORS (CALKIN ALGEBRA)

THEOREM 5.1. *An operator* $A \in \mathcal{L}(X, Y)$ *is Fredholm if and only if there exists an operator* $T \in \mathcal{L}(Y, X)$ *such that* $I - TA$ *and* $I - AT$ *are operators of finite rank.*

PROOF. Suppose A is Fredholm, and let $X = X_0 \oplus \operatorname{Ker} A$, $Y = Y_0 \oplus \operatorname{Im} A$ where X_0 and Y_0 are subspaces of X and Y, respectively. Define A_0 to be the restriction of A to X_0. Since A_0 is $1 - 1$ and $\operatorname{Im} A_0$ is closed, the operator A_0^{-1}, considered as a map from $\operatorname{Im} A$ onto X_0, is bounded (cf. Theorem 2.1). Put $T = A_0^{-1}Q$, where Q is the projection from Y onto $\operatorname{Im} A$ along Y_0. Obviously, $\operatorname{Im} T = X_0$ and $\operatorname{Ker} T = Y_0$. It is easy to verify that $I - TA$ is the projection of X onto $\operatorname{Ker} A$ along X_0 and $I - AT$ is the projection of Y onto Y_0 along $\operatorname{Im} A$. In particular, $I - TA$ and $I - AT$ are operators of finite rank.

Conversely, assume that $TA = I - K_1$ and $AT = I - K_2$, where K_1 and K_2 are operators of finite rank. Since $\operatorname{Ker} A \subset \operatorname{Ker} TA$ and $\operatorname{Im} A \supset \operatorname{Im} AT$, we see that

$$(1) \qquad n(A) \leq n(TA) = n(I - K_1) < \infty, \qquad d(A) \leq d(AT) = d(I - K_2) < \infty.$$

Thus A is Fredholm. $\square$

Theorem 5.1 remains true if the statement "$I - TA$ and $I - AT$ are operators of finite rank" is replaced by $I - TA$ and $I - AT$ are compact operators (cf. formula

(1)). In other words, an operator A is a Fredholm operator if and only if A is invertible modulo compact operators. We make this more precise for the case when $X = Y$.

Let $\mathcal{K}(X)$ denote the space of all compact operators on X. Note that $\mathcal{K}(X)$ is a subspace of $\mathcal{L}(X) = \mathcal{L}(X, X)$. On the quotient space $\mathcal{L}(X)/\mathcal{K}(X)$ define the product $[C][D] = [CD]$, where $[C]$ is the coset $C + \mathcal{K}(X)$. The space $\mathcal{L}(X)/\mathcal{K}(X)$ with this additional operation is an algebra, called the *Calkin algebra*, with unit $[I]$.

THEOREM 5.2. *An operator $A \in \mathcal{L}(X)$ is a Fredholm operator if and only if $[A]$ has an inverse in the Calkin algebra $\mathcal{L}(X)/\mathcal{K}(X)$.*

PROOF. If A is Fredholm, then there exists, by Theorem 5.1, an operator $T \in \mathcal{L}(X)$ such that $AT - I$ and $TA - I$ are compact operators. Hence $[A][T] = [T][A] = [I]$, and thus $[T]$ is the inverse of $[A]$ in the Calkin algebra.

On the other hand, if $[A][T] = [T][A] = [I]$, then $AT = I - K_1$ and $TA = I - K_2$, where K_1 and K_2 are compact operators. Then it follows from (1) that A is Fredholm. $\square$

Let $A \in \mathcal{L}(X)$. The *essential spectrum* (notation: $\sigma_{\mathrm{ess}}(A)$) of A is, by definition, the set of all $\lambda \in \mathbf{C}$ such that $\lambda - A$ is not a Fredholm operator. The essential spectrum $\sigma_{\mathrm{ess}}(A)$ may also be understood as the spectrum of the coset $A + \mathcal{K}(X)$ in the Calkin algebra. (The latter interpretation will be made more precise in the Banach algebra part in Volume II.) Obviously, $\sigma_{\mathrm{ess}}(A) \subset \sigma(A)$, and thus $\sigma_{\mathrm{ess}}(A)$ is a bounded set. From Theorem 4.1 it follows that $\mathbf{C}\backslash\sigma_{\mathrm{ess}}(A)$ is open, and hence $\sigma_{\mathrm{ess}}(A)$ is compact. If $\dim X < \infty$, then all operators on X are Fredholm operators, and hence, in that case, $\sigma_{\mathrm{ess}}(A) = \emptyset$. At the end of Section 8 it will be shown that $\sigma_{\mathrm{ess}}(A)$ is always nonempty if X is infinite dimensional. Theorem 4.2 implies that

$$(1) \qquad \sigma_{\mathrm{ess}}(A) = \sigma_{\mathrm{ess}}(A + K), \qquad K \in \mathcal{K}(X).$$

For later purposes we include the following result

THEOREM 5.3. *An operator $A \in \mathcal{L}(X, Y)$ is a Fredholm operator with $\mathrm{ind}\, A = 0$ if and only if there exists an operator $F \in \mathcal{L}(X, Y)$ of finite rank such that $A + F$ is invertible.*

PROOF. Suppose A is Fredholm with $\mathrm{ind}\, A = 0$. Put $X = X_0 \oplus \mathrm{Ker}\, A$ and $Y = Y_0 \oplus \mathrm{Im}\, A$. Since $\mathrm{ind}\, A = 0$, we have $\dim \mathrm{Ker}\, A = \dim Y_0$. So there exists an invertible operator $F_0 \colon \mathrm{Ker}\, A \to Y_0$. Define $F = F_0(I - P)$, where P is the projection of X onto X_0 along $\mathrm{Ker}\, A$. Obviously, $A + F$ is invertible.

Conversely, assume $S = A + F$ is invertible, where F has finite rank. Then, by Theorem 4.2, the operator A is Fredholm and $\mathrm{ind}\, A = \mathrm{ind}\, S = 0$. $\square$

XI.6 GENERALIZED INVERSES

The operator $T = A_0^{-1}Q$ which was defined in the first part of the proof of Theorem 5.1, has the following properties:

$$(1) \qquad ATA = A, \qquad TAT = T.$$

Given $A \in \mathcal{L}(X,Y)$, any operator $T \in \mathcal{L}(Y,X)$ satisfying the two identities in (1) is called a *generalized inverse* of A. If A is invertible, then A^{-1} is the only generalized inverse of A.

Generalized inverses are useful in solving linear equations. Suppose A has a generalized inverse T. If the equation $Ax = y$ is solvable for a given $y \in Y$, then Ty is a solution (not necessarily the only one). Indeed, since $Ax = y$ is solvable, there exists x_0 such that $Ax_0 = y$, and hence $ATy = ATAx_0 = Ax_0 = y$.

THEOREM 6.1. *An operator $A \in \mathcal{L}(X,Y)$ has a generalized inverse if and only if $\operatorname{Ker} A$ and $\operatorname{Im} A$ are complemented in X and Y, respectively.*

PROOF. If $X = X_0 \oplus \operatorname{Ker} A$ and $Y = Y_0 \oplus \operatorname{Im} A$, then the operator $T\colon Y \to X$ defined by $T(Ax_0 + y_0) = x_0$, where $x_0 \in X_0$, $y_0 \in Y_0$, is a generalized inverse of A.

Conversely, if A has a generalized inverse $T \in \mathcal{L}(Y,X)$, then it is easy to check that AT and TA are projections. Indeed

$$(AT)^2 = (ATA)T = AT, \qquad (TA)^2 = T(ATA) = TA.$$

Obviously, $\operatorname{Im}(AT) \subset \operatorname{Im} A$. From $A = (AT)A$ it follows that $\operatorname{Im} A \subset \operatorname{Im}(AT)$, and thus AT is a projection onto $\operatorname{Im} A$. Similarly, $\operatorname{Ker} A \subset \operatorname{Ker} TA$ and $A(TA) = A$ implies that $\operatorname{Ker} TA \subset \operatorname{Ker} A$. Thus TA is a projection whose kernel coincides with $\operatorname{Ker} A$. Thus $\operatorname{Im} A$ and $\operatorname{Ker} A$ are complemented. $\square$

COROLLARY 6.2. *Every Fredholm operator has a generalized inverse.*

Recall (see Corollary 2.3) that $\operatorname{Im} A$ is complemented in Y implies that $\operatorname{Im} A$ is closed. Thus, if X and Y are Hilbert spaces, then $A \in \mathcal{L}(X,Y)$ has a generalized inverse if and only if $\operatorname{Im} A$ is closed.

Every generalized inverse T of A is of the form described in the first part of the proof of Theorem 6.1. To see this, put $X_0 = \operatorname{Im} TA$ and $Y_0 = \operatorname{Ker} AT$. Since AT is a projection onto $\operatorname{Im} A$ and TA is a projection whose kernel coincides with $\operatorname{Ker} A$, it follows that

$$X = \operatorname{Ker} TA \oplus \operatorname{Im} TA = \operatorname{Ker} A \oplus X_0,$$
$$Y = \operatorname{Im} AT \oplus \operatorname{Ker} AT = \operatorname{Im} A \oplus Y_0.$$

Any $y \in Y$ is of the form $y = Ax_0 + y_0$ for some $x_0 \in X_0$ and $y_0 \in Y_0$. Now $Ty = TAx_0 + Ty_0$ and $TAx_0 = v_0 \in X_0$. Since $Ax_0 = ATAx_0 = Av_0$ and A is $1-1$ on X_0, we get $x_0 = v_0 = TAx_0$. Also, $y_0 \in Y_0 = \operatorname{Ker} AT$ gives $Ty_0 = TATy_0 = T0 = 0$. Thus $Ty = x_0$. Note that the foregoing also shows that T is uniquely determined by the spaces $X_0 = \operatorname{Im} TA$ and $Y_0 = \operatorname{Ker} AT$. Indeed, if S is a generalized inverse of A such that $\operatorname{Im} TA = \operatorname{Im} SA$ and $\operatorname{Ker} AT = \operatorname{Ker} AS$, then

$$S(Ax_0 + y_0) = T(Ax_0 + y_0),$$

and thus S and T coincide.

Another way to describe all generalized inverses of A is as follows.

THEOREM 6.3. *Suppose T_0 is a generalized inverse of A. The set of generalized inverses of A consists of all operators of the form*

$$(2) \qquad\qquad\qquad\qquad T = PT_0Q,$$

where Q is a projection onto $\operatorname{Im} A$ *and P is a projection whose kernel coincides with* $\operatorname{Ker} A$.

PROOF. Suppose T_0 and T are generalized inverses of A. Then $ATA = A = AT_0A$ and therefore

$$T = TAT = (TA)T_0(AT).$$

We observed previously that AT is a projection onto $\operatorname{Im} A$ and TA is a projection along $\operatorname{Ker} A$.

Conversely, suppose T is of the form (2). Since $\operatorname{Im}(I - P) = \operatorname{Ker} A$ we have $A = AP$. Also $QA = A$. Thus

$$A(PT_0Q)A = AT_0QA = AT_0A = A,$$
$$(PT_0Q)A(PT_0Q) = PT_0AT_0Q = PT_0Q. \quad \square$$

Let $A \in \mathcal{L}(X,Y)$. An operator $T \in \mathcal{L}(Y,X)$ is called a *generalized inverse of A in the weak sense* if the first identity in (1) holds (but not necessarily the second). The remark preceding Theorem 6.1 is also true for generalized inverses in the weak sense. Furthermore, if T is a generalized inverse of A in the weak sense, then the operator $S := TAT$ is a generalized inverse of A in the ordinary sense. Indeed

$$ASA = (ATA)TA = ATA = A,$$
$$SAS = T(ATA)TAT = T(ATA)T = TAT = S.$$

It follows that Theorem 6.1 remains true if the term generalized inverse is understood in the weak sense.

XI.7 INDEX, TRACE AND DETERMINANT

In this section all operators are bounded linear operators acting on (separable) Hilbert spaces. If $A: H \to H$ is a Fredholm operator, then there exists an operator $T: H \to H$ such that $I - AT$ and $I - TA$ are trace class operators or even operators of finite rank. The next theorem expresses the index of A in terms of the traces of $I - AT$ and $I - TA$.

THEOREM 7.1. *Let $A: H \to H$ be a Fredholm operator, and let $T: H \to H$ be such that $I - AT$ and $I - TA$ are trace class operators. Then $AT - TA$ is a trace class operator and*

$$\operatorname{ind} A = \operatorname{tr}[AT - TA].$$

PROOF. Let A^+ be a generalized inverse of A. Then $T = A^+ + G$, where

$$G = (I - A^+A)T - A^+(I - AT).$$

Since $I - A^+A$ has finite rank and $I - AT$ is a trace class operator, it follows that G is a trace class operator. Now

$$TA - AT = A^+A - AA^+ + GA - AG.$$

Note that $\operatorname{tr} GA = \operatorname{tr} AG$ (see Corollary VII.6.2). Thus it suffices to prove the theorem for $T = A^+$.

Recall that $I - AA^+$ and $I - A^+A$ are projections of finite rank. Thus $I - AA^+$ and $I - A^+A$ are trace class operators. But then $AA^+ - A^+A$ is a trace class operator and

$$\begin{aligned} \operatorname{tr}[AA^+ - A^+A] &= \operatorname{tr}(I - A^+A) - \operatorname{tr}(I - AA^+) \\ &= \operatorname{rank}(I - A^+A) - \operatorname{rank}(I - AA^+) \\ &= n(A) - d(A) = \operatorname{ind} A. \end{aligned}$$

Here we used that the trace of a finite rank projection is equal to the rank of the projection. $\square$

Consider the $n \times n$ operator matrix

$$(1) \qquad A = \begin{bmatrix} A_{11} & \cdots & A_{1n} \\ \vdots & & \vdots \\ A_{n1} & \cdots & A_{nn} \end{bmatrix} : H \oplus \cdots \oplus H \to H \oplus \cdots \oplus H.$$

Here A_{ij}, $1 \le i, j \le n$, are operators acting on H, and we view A as an operator acting on the direct sum of n copies of H. Given A in (1), we set

$$(2) \qquad \det A = \sum_\sigma (\operatorname{sgn} \sigma) A_{1\sigma_1} A_{2\sigma_2} \cdots A_{n\sigma_n}.$$

In (2) the summation is over all permutations σ of the numbers $1, 2, \ldots, n$ and $\operatorname{sgn} \sigma$ denotes the sign of the permutation σ. Note that $\det A$ is an operator on H; its definition depends not only on A but also on the given partitioning of A.

PROPOSITION 7.2. *Assume that all the entries of the operator matrix* $A = [A_{ij}]_{i,j=1}^n$ *commute with one another. Then A is invertible if and only if $\det A$ is invertible, and in that case A^{-1} is given by the analogue of Cramer's rule.*

PROOF. Note that $\det A$ commutes with all the entries A_{ij}. Assume that $\det A$ is invertible. Let M_{ij} be the $(n-1) \times (n-1)$ operator matrix which one obtains from $A = [A_{ij}]_{i,j=1}^n$ by deleting the i-th row and j-th column. For each i and j put

$$B_{ij} = (-1)^{i+j}(\det M_{ji})(\det A)^{-1},$$

and let $B = [B_{ij}]_{i,j=1}^n$. The usual argument for scalar determinants shows that $AB = BA$ is equal to the identity operator on H^n. So A is invertible and $A^{-1} = B$ is obtained by applying the operator matrix analogue of Cramer's rule.

Next, assume that A is invertible, and let $S = A^{-1}$. Write $S = [S_{ij}]_{i,j=1}^n$ where S_{ij} acts on H. Consider

$$D = \begin{bmatrix} A_{rs} & & & 0 \\ & A_{rs} & & \\ & & \ddots & \\ 0 & & & A_{rs} \end{bmatrix}$$

for a fixed pair r, s. From our hypothesis on A it follows that $DA = AD$. But then $A^{-1}D = DA^{-1}$, and thus A_{rs} commutes with each entry of $S = A^{-1}$. Next consider

$$
E = \begin{bmatrix} S_{ij} & & & 0 \\ & S_{ij} & & \\ & & \ddots & \\ 0 & & & S_{ij} \end{bmatrix},
$$

where S_{ij} is a fixed entry of S. According to what has been proved so far, $EA = AE$. Hence $A^{-1}E = EA^{-1}$, and thus S_{ij} commutes with each entry of S. It follows that the entries of S commute with one another and with the entries of A. But then we may conclude that

$$
det\, A\, det\, S = det\, AS = I, \qquad det\, S\, det\, A = det\, SA = I.
$$

Thus $det\, A$ is invertible. $\square$

THEOREM 7.3. *Assume that the entries of the operator matrix* $A = [A_{ij}]_{i,j=1}^{n}$ *commute modulo the compact operators. Then A is a Fredholm operator if and only if $det\, A$ is a Fredholm operator.*

PROOF. Go to the Calkin algebra and apply the same reasoning as in the proof of the previous proposition. $\square$

In general, under the conditions of Theorem 7.3, one may not conclude that $\mathrm{ind}\, A = \mathrm{ind}(det\, A)$. In fact, given integers p and q, there exists (see Kozak [1]) a 2×2 operator matrix whose entries commute modulo the compact operators such that $\mathrm{ind}\, A = p$ and $\mathrm{ind}(det\, A) = q$. The next theorems present positive results in this direction.

THEOREM 7.4. *Assume that the entries of the operator matrix* $A = [A_{ij}]_{i,j=1}^{n}$ *commute modulo the compact operators, and let the operators*

$$
D_k = det\big([A_{ij}]_{i,j=1}^{k}\big), \qquad k = 1, \ldots, n,
$$

be Fredholm operators. Then A and $det\, A$ are Fredholm operators with

$$
(3) \qquad\qquad \mathrm{ind}\, A = \mathrm{ind}(det\, A).
$$

PROOF. Since $D_n = det\, A$, the operator $det\, A$ is Fredholm. So we know from Theorem 7.3 that A is Fredholm. To prove (3) we go to the Calkin algebra $\mathcal{B} = \mathcal{L}(H)/\mathcal{K}(H)$, where $\mathcal{K}(H)$ denotes the set of all compact operators on H. Given $T \in \mathcal{L}(H)$, we denote by $\hat{T}$ the coset $T + \mathcal{K}(H)$. From our hypotheses it follows that

$$
\hat{D}_k = det\big([\hat{A}_{ij}]_{i,j=1}^{k}\big), \qquad k = 1, \ldots, n,
$$

is invertible in the Calkin algebra $\mathcal{B}$. As in the scalar case (see also Chapter XXI in Volume II) this allows us to make an LU-factorization for the matrix $[\hat{A}_{ij}]_{i,j=1}^{n}$, that is,

$$
(4) \qquad \begin{bmatrix} \hat{A}_{11} & \cdots & \hat{A}_{1n} \\ \vdots & & \\ \hat{A}_{n1} & \cdots & \hat{A}_{nn} \end{bmatrix} = \begin{bmatrix} \hat{L}_{11} & \cdots & \hat{L}_{1n} \\ \vdots & & \\ \hat{L}_{n1} & \cdots & \hat{L}_{nn} \end{bmatrix} \begin{bmatrix} \hat{U}_{11} & \cdots & \hat{U}_{1n} \\ \vdots & & \\ \hat{U}_{n1} & \cdots & \hat{U}_{nn} \end{bmatrix},
$$

where L_{ij} and U_{ij} are operators on H for each i and j,

$$(5) \qquad L_{ij} \in \mathcal{K}(H) \quad (j > i), \qquad U_{ij} \in \mathcal{K}(H) \quad (j < i),$$

and the two matrices in the right hand side of (4) are invertible in the algebra $\mathcal{B}^{n \times n}$ of all $n \times n$ matrices with entries in the Calkin algebra $\mathcal{B}$. Moreover, since the entries A_{ij} commute modulo the compact operators, the construction of the LU-factorization implies that all the operators L_{ij} and U_{ij} $(i, j = 1, \ldots, n)$ commute modulo the compact operators. Put

$$L = \begin{bmatrix} L_{11} & & 0 \\ \vdots & \ddots & \\ L_{n1} & \cdots & L_{nn} \end{bmatrix}, U = \begin{bmatrix} U_{11} & \cdots & U_{1n} \\ & \ddots & \vdots \\ 0 & & U_{nn} \end{bmatrix}.$$

Then L and U are Fredholm operators on H^n and $A - LU$ is compact, because of (4) and (5). Thus

$$\operatorname{ind} A = \operatorname{ind}(LU) = \operatorname{ind} L + \operatorname{ind} U.$$

Furthermore, $det\,A$, $det\,L$ and $det\,U$ are Fredholm operators and, again by (4) and (5), the operator $det\,A - (det\,L)(det\,U)$ is compact. Therefore

$$\operatorname{ind}(det\,A) = \operatorname{ind}(det\,L) + \operatorname{ind}(det\,U).$$

It remains to show that $\operatorname{ind} L = \operatorname{ind}(det\,L)$ and $\operatorname{ind} U = \operatorname{ind}(det\,U)$.

Let us prove that $\operatorname{ind} U = \operatorname{ind}(det\,U)$ (the equality for L instead of U is proved in a similar way). First observe that from the construction of the LU-factorization it follows that the product $U_{11}U_{22} \cdots U_{kk}$ is Fredholm for each k. Since the elements $U_{11}, \ldots, U_{nn}$ commute modulo the compact operators, each diagonal entry U_{ii} is Fredholm. Take $0 \leq t \leq 1$, and let $U(t)$ be the $n \times n$ operator matrix which one obtains if for each $i > j$ the entry U_{ij} in U is replaced by tU_{ij}. Note that $U(t)$ is upper triangular and has the same diagonal entries as U. So $U(t)$ is also Fredholm. Obviously, the map $t \mapsto U(t)$ is continuous in the operator norm. Theorem 4.1 implies that $\operatorname{ind} U(t)$ is independent of t. In particular,

$$\operatorname{ind} U = \operatorname{ind} U(1) = \operatorname{ind} U(0).$$

Note that $U(0)$ is a block diagonal matrix. Hence

$$\operatorname{ind} U(0) = \sum_{j=1}^{n} \operatorname{ind} U_{jj} = \operatorname{ind}(U_{11}U_{22} \cdots U_{nn}).$$

To complete the proof it remains to observe that, because of the triangular form of U, the product $U_{11}U_{22} \cdots U_{nn}$ is equal to $det\,U$. $\square$

COROLLARY 7.5. *Assume that the entries of the operator matrix* $B = [B_{ij}]_{i,j=1}^{n}$ *commute modulo the compact operators, and let* B *be a Fredholm operator. If*

B can be approximated sufficiently close in the operator norm on $\mathcal{L}(H^n)$ by an operator $A = [A_{ij}]_{i,j=1}^n$ with the properties described in the previous theorem, then

$$(6) \qquad\qquad \text{ind}\, B = \text{ind}(det\, B).$$

PROOF. We apply Theorem 4.1. Since B is Fredholm, there exists a constant $\gamma > 0$ such that $A \in \mathcal{L}(H^n)$ and $\|A - B\| < \gamma$ implies that A is Fredholm operator and $\text{ind}\, A = \text{ind}\, B$. Also $det\, B$ is Fredholm. Therefore there exists a constant $\gamma_1 > 0$ such that $T \in \mathcal{L}(H)$ and $\|T - det\, B\| < \gamma_1$ implies that T is a Fredholm operator and $\text{ind}\, T = \text{ind}(det\, B)$. Now use that the map $A \mapsto det\, A$ is continuous in the operator norm. So there exists a constant ρ, $0 < \rho \le \gamma$ such that $\|A-B\| < \rho$ yields $\|det\, A - det\, B\| < \gamma_1$. Thus, if the ball

$$(7) \qquad\qquad \{A \in \mathcal{L}(H^n) \mid \|A - B\| < \rho\}$$

contains an operator $A = [A_{ij}]_{i,j=1}^n$ with the properties described in the previous theorem, then

$$\text{ind}\, B = \text{ind}\, A = \text{ind}(det\, A) = \text{ind}(det\, B),$$

and hence (6) holds true. $\square$

The above corollary and the next theorem will be used later (in Section XII.3 and in Volume II) to derive index theorems for Wiener-Hopf and Toeplitz operators.

THEOREM 7.6. *Assume that the entries of the operator matrix $A = [A_{ij}]_{i,j=1}^n$ commute modulo the trace class operators, and let A be a Fredholm operator. Then $det\, A$ is a Fredholm operator and $\text{ind}(det\, A) = \text{ind}\, A$.*

PROOF. We already know that $det\, A$ is a Fredholm operator. Choose an operator $T\colon H \to H$ such that $T(det\, A) - I$ and $(det\, A)T - I$ are trace class operators. Theorem 5.2 shows that T is a Fredholm operator. Note that, modulo the trace class operators, $det\, A$ commutes with each entry A_{rs} of A. It follows that

$$TA_{rs} - A_{rs}T = TA_{rs}\{I - (det\, A)T\} + \{T(det\, A) - I\}A_{rs}T,$$

and hence, modulo the trace class operators, T commutes with each entry of A. Put

$$(8) \qquad\qquad B = \begin{bmatrix} T & & & 0 \\ & I & & \\ & & \ddots & \\ 0 & & & I \end{bmatrix} A.$$

Note that the entries of B commute modulo the trace class operators. Furthermore $det\, B = T\, det\, A$, and hence $det\, B - I$ is a trace class operator. From (8) we see that

$$\text{ind}\, B = \text{ind}\, T + \text{ind}\, A = -\,\text{ind}(det\, A) + \text{ind}\, A.$$

So it suffices to prove that $\text{ind}\, B = 0$.

Let M_{jk} be the $(n-1) \times (n-1)$ operator matrix which one obtains from B by deleting in B the j-th row and the k-th column. Put $C_{jk} = (-1)^{j+k}\, det\, M_{kj}$, and let $C = [C_{jk}]_{j,k=1}^n$. Note that modulo the trace class operators the entries C_{jk} commute with one another and with the entries of B. By the operator matrix analogue of Cramer's rule it follows that

$$BC - [(det\, B)\delta_{ij}]_{i,j=1}^n, \qquad CB - [(det\, B)\delta_{ij}]_{i,j=1}^n$$

are trace class operators. But $det\, B - I$ is a trace class operator. So $BC - I_{H^n}$ and $CB - I_{H^n}$ are operators of trace class. But then we can apply Theorem 7.1 to show that $\mathrm{ind}\, B = \mathrm{tr}(BC - CB)$.

Note that $BC - CB$ is an $n \times n$ operator matrix whose entries are trace class operators. A simple application of Theorem VII.2.2 shows that

$$\mathrm{tr}(BC - CB) = \mathrm{tr}\left(\sum_{j,k=1}^n B_{jk} C_{kj} - C_{jk} B_{kj} \right).$$

Next one checks that

$$(9a) \qquad B_{jk} C_{kj} = \sum_{\sigma_j = k} (\mathrm{sgn}\, \sigma) B_{j\sigma_j} B_{1\sigma_1} \cdots B_{j-1,\sigma_{j-1}} B_{j+1,\sigma_{j+1}} \cdots B_{n\sigma_n};$$

$$(9b) \qquad C_{jk} B_{kj} = \sum_{\sigma_k = j} (\mathrm{sgn}\, \sigma) B_{1\sigma_1} \cdots B_{k-1,\sigma_{k-1}} B_{k+1,\sigma_{k+1}} \cdots B_{n\sigma_n} B_{k\sigma_k}.$$

In (9a) the summation is over all permutations $\sigma = (\sigma_1, \ldots, \sigma_n)$ of the numbers $1, \ldots, n$ with $\sigma_j = k$, and in (9b) the summation is over all $\sigma = (\sigma_1, \ldots, \sigma_n)$ such that $\sigma_k = j$. Given a permutation $\sigma = (\sigma_1, \ldots, \sigma_n)$ write

$$E_j(\sigma) = B_{1\sigma_1} \cdots B_{j-1,\sigma_{j-1}} B_{j+1,\sigma_{j+1}} \cdots B_{n\sigma_n}.$$

Then

$$(10) \qquad \mathrm{ind}\, B = \sum_{\sigma} (\mathrm{sgn}\, \sigma)\, \mathrm{tr}\left(\sum_{j=1}^n B_{j\sigma_j} E_j(\sigma) - E_j(\sigma) B_{j\sigma_j} \right).$$

For $j = 2, \ldots, n-1$ the operator

$$E_j(\sigma) - B_{j+1,\sigma_{j+1}} \cdots B_{n\sigma_n} B_{1\sigma_1} \cdots B_{j-1\sigma_{j-1}}$$

is a trace class operator. So we can apply Corollary VII.6.2(i) to show that

$$\mathrm{tr}\{ B_{j\sigma_j} E_j(\sigma) - B_{j\sigma_j} \cdots B_{n\sigma_n} B_{1\sigma_1} \cdots B_{j-1\sigma_{j-1}} \}$$
$$= \mathrm{tr}\{ E_j(\sigma) B_{j\sigma_j} - B_{j+1\sigma_{j+1}} \cdots B_{n\sigma_n} B_{1\sigma_1} \cdots B_{j\sigma_j} \}.$$

So

$$\mathrm{tr}\{ B_{j\sigma_j} E_j(\sigma) - E_j(\sigma) B_{j\sigma_j} \}$$
$$= \mathrm{tr}\{ B_{j\sigma_j} \cdots B_{n\sigma_n} B_{1\sigma_1} \cdots B_{j-1\sigma_{j-1}} - B_{j+1\sigma_{j+1}} \cdots B_{n\sigma_n} B_{1\sigma_1} \cdots B_{j\sigma_j} \}$$

for $j = 2, \ldots, n-1$. Note that

$$B_{1\sigma_1} E_1(\sigma) - E_1(\sigma) B_{1\sigma_1} + B_{n\sigma_n} E_n(\sigma) - E_n(\sigma) B_{n\sigma_n}$$
$$= B_{n\sigma_n} B_{1\sigma_1} \cdots B_{n-1\sigma_{n-1}} - B_{2\sigma_2} \cdots B_{n-1\sigma_{n-1}} B_{1\sigma_1}.$$

By using this in (10), one obtains that $\mathrm{ind}\, B = 0$.　□

XI.8 ANALYTIC FREDHOLM OPERATOR VALUED FUNCTIONS

In this section, Ω is an open, connected subset of C, and $W \colon \Omega \to \mathcal{L}(X, Y)$ is an operator function, which is analytic on Ω. We assume that the values of W are Fredholm operators acting between the complex Banach spaces X and Y. Since Ω is connected and the index is an integer-valued continuous function, it follows that $\operatorname{ind} W(\lambda)$ is constant and does not depend on λ. We begin with the case when the index is zero.

To give an example of such an operator function, assume that the matrices M, N_1 and N_2 appearing in the two point boundary value problem (2) in Section 1 depend analytically on a parameter λ. Thus

$$\begin{cases} f'(t) = M(\lambda)f(t) + g(t), & a \le t \le b, \\ N_1(\lambda)f(a) + N_2(\lambda)f(b) = x, \end{cases}$$

where $M(\cdot)$, $N_1(\cdot)$ and $N_2(\cdot)$ are $n \times n$ matrix functions whose entries are analytic on Ω, say. Let $A(\lambda)$ be the associated operator,

$$A(\lambda) \colon X \to L_2^n([a, b]) \oplus \mathbf{C}^n, \qquad A(\lambda)f = \begin{bmatrix} f' - M(\lambda)f \\ N_1(\lambda)f(a) + N_2(\lambda)f(b) \end{bmatrix},$$

where X is the space defined in formula (3) of Section 1. Then $A(\cdot)$ is an operator function which is analytic on Ω and the values of $A(\cdot)$ are Fredholm operators of index 0. Put

$$T(\lambda) = N_1(\lambda) + N_2(\lambda)e^{(b-a)M(\lambda)}.$$

By adding the parameter λ to the equivalence relation in formula (8) of Section 1 one sees that the $L_2^n([a, b])$-extension of the matrix function $T(\cdot)$ is globally equivalent on Ω to the Fredholm operator valued function $A(\cdot)$ (cf. Section III.2).

Let $W \colon \Omega \to \mathcal{L}(X, Y)$ be an operator function which is analytic on Ω. Take $\lambda_0 \in \Omega$, and assume that $W(\lambda_0)$ is a Fredholm operator of index zero. Then (see Theorem 5.3) there exists an operator $F \colon X \to Y$ of finite rank such that $W(\lambda_0) + F$ is invertible. Since $W(\lambda)$ is continuous in λ, this implies that $E(\lambda) = W(\lambda) + F$ is invertible for λ in some open disc $|\lambda - \lambda_0| < \delta_0$. So

$$(1) \qquad W(\lambda) = E(\lambda) - F = E(\lambda)[I - E(\lambda)^{-1}F], \qquad |\lambda - \lambda_0| < \delta_0.$$

The fact that F is an operator of finite rank implies that $\operatorname{Ker} F$ has a finite dimensional complement X_0 in X. Let P be the projection of X along $\operatorname{Ker} F$ onto X_0. Note that P has finite rank and $FP = F$. It follows that

$$(2) \qquad I - E(\lambda)^{-1}F = [I - PE(\lambda)^{-1}FP][I - (I - P)E(\lambda)^{-1}FP].$$

Put $G(\lambda) = I - (I - P)E(\lambda)^{-1}FP$. Note that G is well-defined and analytic on the disc $|\lambda - \lambda_0| < \delta_0$. Furthermore, the values of G are invertible operators on X; in fact

$$G(\lambda)^{-1} = I + (I - P)E(\lambda)^{-1}FP, \qquad |\lambda - \lambda_0| < \delta_0.$$

By combining (1) and (2) we see that

$$(3) \qquad W(\lambda) = E(\lambda)\{I - PE(\lambda)^{-1}FP\}G(\lambda), \qquad |\lambda - \lambda_0| < \delta_0,$$

where E and F are analytic operator functions on $|\lambda - \lambda_0| < \delta_0$ and their values are invertible operators. In other words, using the terminology of Section III.3, the operator functions $W(\cdot)$ and $I - PE(\cdot)^{-1}FP$ are equivalent at λ_0. This leads to the following theorem.

THEOREM 8.1. *Let* $W: \Omega \to \mathcal{L}(X,Y)$ *be an analytic operator function, and assume that for some* $\lambda_0 \in \Omega$ *the operator* $W(\lambda_0)$ *is a Fredholm operator of index zero. Then* W *is equivalent at* λ_0 *to an analytic operator function* D *of the form*

$$(4) \qquad D(\lambda) = P_0 + (\lambda - \lambda_0)^{\kappa_1} P_1 + \cdots + (\lambda - \lambda_0)^{\kappa_r} P_r,$$

where $P_0, P_1, \ldots, P_r$ *are mutually disjoint projections of the Banach space* X, *the projections* $P_1, \ldots, P_r$ *have rank one, the projection* $I - P_0$ *has finite rank and* $\kappa_1 \leq \kappa_2 \leq \cdots \leq \kappa_r$ *are positive integers.*

PROOF. According to formula (3) the operator function W is equivalent at λ_0 to an operator function of the form

$$(5) \qquad \begin{bmatrix} W_0(\cdot) & 0 \\ 0 & I_{\operatorname{Ker} P} \end{bmatrix} : \operatorname{Im} P \oplus \operatorname{Ker} P \to \operatorname{Im} P \oplus \operatorname{Ker} P.$$

Here $W_0(\cdot)$ is holomorphic on $|\lambda - \lambda_0| < \delta$ and $W_0(\lambda)$ acts on the finite dimensional space $\operatorname{Im} P$. Now assume the theorem has been proved for W_0. Thus the operator function W_0 is equivalent at λ_0 to an operator function D_0 of the form

$$D_0(\lambda) = \pi_0 + (\lambda - \lambda_0)^{\kappa_1} \pi_1 + \cdots + (\lambda - \lambda_0)^{\kappa_r} \pi_r,$$

where $\pi_0, \pi_1, \ldots, \pi_r$ are mutually disjoint projections of $\operatorname{Im} P$ and $\operatorname{rank} \pi_j = 1$ for $j = 1, \ldots, r$. Put $P_j = \pi_j P$ for $j = 1, \ldots, r$, and let

$$P_0 = \begin{bmatrix} \pi_0 & 0 \\ 0 & I_{\operatorname{Ker} P} \end{bmatrix} : \operatorname{Im} P \oplus \operatorname{Ker} P \to \operatorname{Im} P \oplus \operatorname{Ker} P.$$

Then the operator function (5) (and hence W) is equivalent at λ_0 to the function

$$D(\lambda) = P_0 + (\lambda - \lambda_0)^{\kappa_1} P_1 + \cdots + (\lambda - \lambda_0)^{\kappa_r} P_r,$$

and the projections $P_0, P_1, \ldots, P_r$ have the desired properties. It follows that it suffices to prove the theorem for the case when $X = Y$ is finite dimensional.

Assume $X = Y = \mathbf{C}^n$. As usual, we identify an operator on $\mathbf{C}^n$ with its matrix corresponding to the standard basis of $\mathbf{C}^n$. So we assume that

$$(6) \qquad W(\lambda) = [a_{ij}(\lambda)]_{i,j=1}^n,$$

where a_{ij} are scalar functions that are analytic at λ_0. If all entries a_{ij} are identically zero in a neighbourhood of λ_0, then the theorem is true trivially. Therefore assume that for at least one pair (i, j) the function a_{ij} does not vanish identically in a neighbourhood of λ_0. In that case we may write

$$a_{ij}(\lambda) = (\lambda - \lambda_0)^{\ell(i,j)} b_{ij}(\lambda),$$

where $b_{ij}(\lambda_0) \neq 0$. Choose (i_0, j_0) in such a way that the number $\ell(i_0, j_0)$ is minimal. By interchanging rows and columns in (6) (that is, by applying a number of equivalence operations) we may assume that $i_0 = 1$, $j_0 = 1$. Furthermore, by multiplying $W(\lambda)$ on the left by the diagonal matrix

$$E(\lambda) = \begin{bmatrix} b_{11}(\lambda)^{-1} & & & 0 \\ & 1 & & \\ & & \ddots & \\ 0 & & & 1 \end{bmatrix},$$

we may suppose that $a_{11}(\lambda) = (\lambda - \lambda_0)^{\kappa_1}$ and $a_{ij}(\lambda) = (\lambda - \lambda_0)^{\kappa_1} c_{ij}(\lambda)$, where c_{ij} is analytic at λ_0. Note that the diagonal matrix $E(\lambda)$ is invertible and $E(\lambda)$ depends analytically on λ in a neighbourhood of λ_0. Thus multiplication by $E(\lambda)$ is an equivalence operation.

Next subtract $c_{i1}(\lambda)$ times the first row from the i-th row. That is, multiply $W(\lambda)$ on the left by

$$E_i(\lambda) = \begin{bmatrix} 1 & & & & \\ & \ddots & & & \\ -c_{i1}(\lambda) & & 1 & & \\ & & & \ddots & \\ & & & & 1 \end{bmatrix}.$$

Here blanks denote zero entries. Again $E_i(\lambda)$ is invertible and $E_i(\lambda)$ depends analytically on λ in a neighbourhood of λ_0. Thus multiplication by $E_i(\lambda)$ is an allowed operation. Also, subtract c_{1j} times the first column from the j-th column, which is also an equivalence operation. Do this for $1 \leq i, j \leq n$. It follows that W is equivalent at λ_0 to an operator function of the form

$$\begin{bmatrix} (\lambda - \lambda_0)^{\kappa} & 0 & \cdots & 0 \\ 0 & \alpha_{22}(\lambda) & \cdots & \alpha_{2n}(\lambda) \\ \vdots & \vdots & & \vdots \\ 0 & \alpha_{n2}(\lambda) & \cdots & \alpha_{nn}(\lambda) \end{bmatrix},$$

where $\alpha_{ij}(\lambda) = (\lambda - \lambda_0)^{\kappa_1} \beta_{ij}(\lambda)$ with β_{ij} analytic at λ_0. Apply induction and the theorem is proved. $\square$

One can show (see Gohberg-Sigal [1], Gohberg-Kaashoek-Lay [2], and the references therein) that the integers $k_1, \ldots, k_r$ appearing in (4) are uniquely determined by W. Furthermore, since W is equivalent at λ_0 to the function (4), we have

$$\dim \operatorname{Ker} W(\lambda_0) = \operatorname{rank}(I - P_0),$$

and hence $\operatorname{rank}(I - P_0)$ is also uniquely determined by W.

In Theorem 8.1 the condition that $\operatorname{ind} W(\lambda_0) = 0$ cannot be omitted. This stems from the fact that in (4) the operator P_0 is a Fredholm operator of index zero. Thus for any operator function W which is equivalent to D at λ_0, the operator $W(\lambda_0)$ has to be Fredholm with index zero, because the latter property is invariant under equivalence. However, if one allows the operator function D to be slightly more complicated, then the analogue of Theorem 8.2 holds for arbitrary Fredholm operator valued functions (see Gohberg-Kaashoek-Lay [2]).

THEOREM 8.2. *Let $W: \Omega \to \mathcal{L}(X, Y)$ be analytic, and assume that $W(\lambda)$ is Fredholm for each $\lambda \in \Omega$. Then there exists a finite or countable subset Σ of Ω, which has no accumulation points inside Ω, and there exist constants n_0 and d_0 such that*

$$(7a) \qquad \dim \operatorname{Ker} W(\lambda) = n_0, \qquad \operatorname{codim} \operatorname{Im} W(\lambda) = d_0 \quad (\lambda \in \Omega \backslash \Sigma),$$

$$(7b) \qquad \dim \operatorname{Ker} W(\lambda) > n_0, \qquad \operatorname{codim} \operatorname{Im} W(\lambda) > d_0 \quad (\lambda \in \Sigma).$$

PROOF. We already observed (see the beginning of this section) that $\operatorname{ind} W(\lambda)$ is constant on Ω. Put

$$n_0 = \min_{\lambda \in \Omega} \dim \operatorname{Ker} W(\lambda), \qquad d_0 = \min_{\lambda \in \Omega} \operatorname{codim} \operatorname{Im} W(\lambda),$$

and let $\Sigma = \{\lambda \in \Omega \mid \dim \operatorname{Ker} W(\lambda) > n_0\}$. Since $\operatorname{ind} W(\lambda)$ is constant on Ω, we also have $\Sigma = \{\lambda \in \Omega \mid \operatorname{codim} \operatorname{Im} W(\lambda) > d_0\}$. We have to prove that Σ is a finite or countable subset of Ω without accumulation points in Ω. In other words, it suffices to show that Σ consists of isolated points only.

First assume that $\operatorname{ind} W(\lambda) = 0$ for each $\lambda \in \Omega$. Take $\lambda_0 \in \Omega$. According to Theorem 8.1, the function W is equivalent at λ_0 to an operator function D of the form (4). Note that

$$(8a) \qquad \operatorname{Ker} D(\lambda) = \begin{cases} \operatorname{Ker} P_0 & \text{for} \quad \lambda = \lambda_0, \\ \operatorname{Ker}(P_0 + P_1 + \cdots + P_r) & \text{for} \quad \lambda \neq \lambda_0; \end{cases}$$

$$(8b) \qquad \operatorname{Im} D(\lambda) = \begin{cases} \operatorname{Im} P_0 & \text{for} \quad \lambda = \lambda_0, \\ \operatorname{Im}(P_0 + P_1 + \cdots + P_r) & \text{for} \quad \lambda \neq \lambda_0; \end{cases}$$

It follows that $\dim \operatorname{Ker} D(\lambda)$ and $\operatorname{codim} \operatorname{Im} D(\lambda)$ are constant on the punctured disc $0 < |\lambda - \lambda_0| < \delta_0$. Furthermore, $\dim \operatorname{Ker} D(\lambda_0) \geq \dim \operatorname{Ker} D(\lambda)$ and $\operatorname{codim} \operatorname{Im} D(\lambda_0) \geq \operatorname{codim} \operatorname{Im} D(\lambda)$ for $|\lambda - \lambda_0| < \delta_0$. Since W is equivalent at λ_0 to D, we have

$$\dim \operatorname{Ker} W(\lambda) = \dim \operatorname{Ker} D(\lambda), \qquad \operatorname{codim} \operatorname{Im} W(\lambda) = \operatorname{codim} \operatorname{Im} D(\lambda)$$

for λ in a neighbourhood of λ_0. So $\dim \operatorname{Ker} W(\lambda)$ and $\operatorname{codim} \operatorname{Im} W(\lambda)$ are constant on punctured discs and at the centers the values of these functions can only increase. It follows that the set Σ, which has been introduced in the first paragraph of the proof, consists of isolated points only. Indeed, let Λ be the set of discontinuity points of the function

$\dim \operatorname{Ker} W(\cdot)$. We have proved that Λ consists of isolated points only, and hence $\Omega \backslash \Lambda$ is open and connected. Since $\dim \operatorname{Ker} W(\cdot)$ is integer-valued, the function $\dim \operatorname{Ker} W(\cdot)$ is constant on $\Omega \backslash \Lambda$. We know that in the points of Λ the values of $\dim \ker W(\cdot)$ increase. Thus $\Sigma \subset \Lambda$ and Σ consists of isolated points only.

Next, we assume that $p = \operatorname{ind} W(\lambda) \neq 0$. First consider the case when $p > 0$. Put $\tilde{Y} = Y \oplus \mathbf{C}^p$, and define $\tilde{W} : \Omega \to \mathcal{L}(X, \tilde{Y})$ by setting

$$\tilde{W}(\lambda)x = \left[\begin{matrix} W(\lambda)x \\ 0 \end{matrix} \right], \qquad x \in X.$$

Then $\operatorname{Ker} \tilde{W}(\lambda) = \operatorname{Ker} W(\lambda)$ and $\operatorname{Im} \tilde{W}(\lambda) = \operatorname{Im} W(\lambda) \oplus \{0\}$. It follows that $\tilde{W}$ is analytic on Ω, its values are Fredholm operators and $\operatorname{ind} \tilde{W}(\lambda) = \operatorname{ind} W(\lambda) - p = 0$ for each $\lambda \in \Omega$. Now apply to $\tilde{W}$ the result proved in the previous paragraph and the theorem is proved for $p = \operatorname{ind} W(\lambda) > 0$. The case when $p = \operatorname{ind} W(\lambda) < 0$ is treated in a similar way by using the function

$$\hat{W} : \Omega \to \mathcal{L}(X \oplus \mathbf{C}^{-p}, Y), \qquad \hat{W}(\lambda) \left[\begin{matrix} x \\ z \end{matrix} \right] = W(\lambda)x. \quad \square$$

COROLLARY 8.3. *Let $A : X \to Y$ be a Fredholm operator, and let $B : X \to Y$ be a bounded linear operator. Then there exist $\varepsilon > 0$ and integers n_0 and d_0 such that*

$$\begin{aligned} n(A) \geq n_0 = n(A + \lambda B), &\qquad 0 < |\lambda| < \varepsilon, \\ d(A) \geq d_0 = d(A + \lambda B), &\qquad 0 < |\lambda| < \varepsilon. \end{aligned}$$

PROOF. Apply the previous theorem to $W(\lambda) = A + \lambda B$. $\quad \square$

COROLLARY 8.4. *Let $W : \Omega \to \mathcal{L}(X, Y)$ be an analytic Fredholm operator valued function, and assume that $W(z)$ is invertible for some $z \in \Omega$. Then the set*

$$\tag{9} \Sigma = \{ \lambda \in \Omega \mid W(\lambda) \text{ is not invertible} \}$$

is at most countable and has no accumulation point inside Ω. Furthermore, for $\lambda_0 \in \Sigma$ and $\lambda \in \Omega \backslash \Sigma$ sufficiently close to λ_0, we have

$$\tag{10} W(\lambda)^{-1} = \sum_{n=-q}^{\infty} (\lambda - \lambda_0)^n A_n,$$

where A_0 is a Fredholm operator of index zero and $A_{-1}, \ldots, A_{-q}$ are operators of finite rank.

PROOF. The fact that the set (9) is at most countable and has no accumulation point inside Ω is clear from Theorem 8.2. Take $\lambda_0 \in \Sigma$. Since $W(\lambda)$ is invertible for λ close to λ_0, it is clear that $\operatorname{ind} W(\lambda_0) = 0$. So according to Theorem 8.1 the function W is equivalent at λ_0 to an operator function D of the form (4). Thus

$$\tag{11} W(\lambda) = E(\lambda)D(\lambda)G(\lambda), \qquad |\lambda - \lambda_0| < \delta,$$

where $E(\lambda)$ and $G(\lambda)$ are invertible and depend analytically on λ. Write

$$E(\lambda)^{-1} = \sum_{n=0}^{\infty} (\lambda - \lambda_0)^n E_n, \qquad G(\lambda)^{-1} = \sum_{n=0}^{\infty} (\lambda - \lambda_0)^n G_n.$$

Note that $D(\lambda)^{-1} = P_0 + (\lambda - \lambda_0)^{-\kappa_1} P_1 + \cdots + (\lambda - \lambda_0)^{-\kappa_r} P_r$. Thus $D(\lambda)^{-1} = \sum_{n=-q}^{\infty} (\lambda - \lambda_0)^n D_n$, where $D_0 = P_0$ is a Fredholm operator of index zero and $D_{-1}, \ldots, D_{-q}$ are operators of finite rank. From (11) we see that

$$W(\lambda)^{-1} = G(\lambda)^{-1} D(\lambda)^{-1} E(\lambda)^{-1}, \qquad 0 < |\lambda - \lambda_0| < \delta_0.$$

Thus the operator A_n in (10) is given by

$$A_n = \sum_{\substack{k+\ell+m=n \\ k \geq 0, m \geq 0}} G_k D_\ell E_m.$$

For $n = 0$ this shows that $A_0 = G_0 D_0 E_0 + K_0$, where K_0 is a finite sum of operators of the form $G_k D_\ell E_m$ with $\ell < 0$. Thus K_0 is an operator of finite rank. Note that G_0 and E_0 are invertible. Thus $G_0 D_0 E_0$ is a Fredholm operator of index zero, and hence the same is true for A_0. For $n < 0$ the operator A_n is of finite rank, because in that case A_n is a finite sum of operators $G_k D_\ell E_m$ with $\ell < 0$. $\square$

Corollary 8.4 has interesting consequences for the essential spectrum of an operator.

COROLLARY 8.5. *Let $A \in \mathcal{L}(X)$, and assume that the complement in $\mathbf{C}$ of the essential spectrum $\sigma_{\mathrm{ess}}(A)$ is connected. Then $\sigma(A)\backslash\sigma_{\mathrm{ess}}(A)$ consists of eigenvalues of finite type only.*

PROOF. We apply Corollary 8.4 to the function $W(\lambda) = \lambda I - A$ with $\Omega = \mathbf{C}\backslash\sigma_{\mathrm{ess}}(A)$. Obviously, $W(\lambda)$ is a Fredholm operator which depends analytically on $\lambda \in \Omega$. For $|\lambda| > \|A\|$ the operator $W(\lambda)$ is invertible. Since Ω is an open, connected subset of $\mathbf{C}$, Corollary 8.4 shows that the set

$$\{\lambda \in \Omega \mid \lambda \in \sigma(A)\}$$

has no accumulation point in Ω. It follows that the points of $\sigma(A)\backslash\sigma_{\mathrm{ess}}(A)$ are isolated points of $\sigma(A)$. Take $\lambda_0 \in \sigma(A)\backslash\sigma_{\mathrm{ess}}(A)$. Then for $\lambda \neq \lambda_0$, λ sufficiently close to λ_0, we have

$$(\lambda I - A)^{-1} = \sum_{n=-q}^{\infty} (\lambda - \lambda_0)^n B_n,$$

where B_0 is a Fredholm operator of index zero and $B_{-1}, \ldots, B_{-q}$ are operators of finite rank (see the second part of Corollary 8.4). Recall that B_{-1} is equal to the Riesz projection $P_{\{\lambda_0\}}(A)$. Thus the latter operator has finite rank, and λ_0 is an eigenvalue of A of finite type because of Theorem II.1.1. $\square$

If X is a Hilbert space and $A \in \mathcal{L}(X)$ selfadjoint, then $\sigma_{\mathrm{ess}}(A)$ is a compact subset of the real line, and hence $\mathbf{C}\backslash\sigma_{\mathrm{ess}}(A)$ is connected. Thus, by Corollary 8.5, for a selfadjoint operator the spectral points outside the essential spectrum are eigenvalues of finite type.

Let $A: X \to X$ be a compact operator. Then $\sigma_{\mathrm{ess}}(A)$ is empty or consists of the point zero only (cf. Corollary 4.3). In particular, $\mathbf{C}\backslash\sigma_{\mathrm{ess}}(A)$ is connected, and hence we may apply Corollary 8.5 to A. This provides an alternative proof of the fact (Corollary II.3.2) that the non-zero part of the spectrum of a compact operator A consists of eigenvalues of finite type only.

Corollary 8.5 may also be used to show that $\sigma_{\mathrm{ess}}(A)$ is non-empty whenever X is infinite dimensional. Indeed, assume that $\sigma_{\mathrm{ess}}(A) = \emptyset$. Then, by Corollary 8.5, the spectrum of A consists of eigenvalues of finite type only. In particular, $\sigma(A)$ must consist of a finite number of points $\lambda_1, \ldots, \lambda_r$, say. But then

$$\dim X = \dim(\mathrm{Im}\, P_{\{\lambda_1\}}(A) \oplus \cdots \oplus \mathrm{Im}\, P_{\{\lambda_r\}}(A))$$
$$= \mathrm{rank}\, P_{\{\lambda_1\}}(A) + \cdots + \mathrm{rank}\, P_{\{\lambda_r\}}(A) < \infty.$$

XI.9 AN OPERATOR VERSION OF ROUCHÉ'S THEOREM

Throughout this section H is a separable Hilbert space, Ω is an open connected subset of $\mathbf{C}$, and $W\colon \Omega \to \mathcal{L}(H)$ is an operator function which is analytic on Ω. We say that $\lambda_0 \in \Omega$ is an *eigenvalue of finite type* of $W(\cdot)$ if $W(\lambda_0)$ is Fredholm, $W(\lambda_0)x = 0$ for some non-zero $x \in H$ and $W(\lambda)$ is invertible for all λ in some punctured disc $0 < |\lambda - \lambda_0| < \varepsilon$ around λ_0. In that case, $\mathrm{ind}\, W(\lambda_0) = 0$, and hence, by Theorem 8.1, the operator function W is equivalent at λ_0 to an operator function of the form

$$(1) \qquad D(\lambda) = P_0 + (\lambda - \lambda_0)^{\kappa_1} P_1 + \cdots + (\lambda - \lambda_0)^{\kappa_r} P_r,$$

where $P_0, P_1, \ldots, P_r$ are as in Theorem 8.1 and satisfy the additional condition that

$$(2) \qquad P_0 + P_1 + \cdots + P_r = I.$$

Note that (2) follows from the fact that in this case $D(\lambda)$ is invertible for $\lambda \neq \lambda_0$ and λ sufficiently close to λ_0. The sum $\kappa_1 + \cdots + \kappa_r$ of the indices in (1) is called the *algebraic multiplicity* of W at λ_0 and will be denoted by $m(\lambda_0; W(\cdot))$. We shall see from Theorem 9.1 below that the number $m(\lambda_0; W(\cdot))$ is well-defined and does not depend on the particular choice of the function $D(\cdot)$ in (1). If $H = \mathbf{C}^n$, then $\det D(\lambda) = (\lambda - \lambda_0)^m$, where $m = \kappa_1 + \cdots + \kappa_r$, and hence in that case, because of the equivalence between $W(\cdot)$ and $D(\cdot)$ at λ_0, the algebraic multiplicity $m(\lambda_0; W(\cdot))$ is precisely the order of λ_0 as a zero of $\det W(\cdot)$.

Let Γ be a Cauchy contour in Ω such that its inner domain Δ is a subset of Ω. The operator function W is said to be *normal with respect to* Γ if $W(\lambda)$ is invertible for all $\lambda \in \Gamma$ and $W(\lambda)$ is Fredholm for all λ in the inner domain Δ. Assume that W is such an operator function. Then Corollary 8.4 implies that $W(\lambda)$ is invertible for all $\lambda \in \Delta$, except for a finite number of points which are eigenvalues of finite type of W. This allows us to define the *algebraic multiplicity* $m(\Gamma; W(\cdot))$ *of* W *relative to the contour* Γ, namely

$$m(\Gamma; W(\cdot)) = m(\lambda_1; W(\cdot)) + \cdots + m(\lambda_p; W(\cdot)),$$

where $\lambda_1, \ldots, \lambda_p$ are the eigenvalues of finite type of W inside Γ.

In what follows $W'(\lambda)$ denotes the derivative of W at λ. We shall prove the following two theorems.

THEOREM 9.1. *Let* $W: \Omega \to \mathcal{L}(H)$ *be an analytic operator function, and assume that* W *is normal with respect to the contour* Γ. *Then*

$$(3) \qquad m(\Gamma; W(\cdot)) = \mathrm{tr}\left(\frac{1}{2\pi i} \int_\Gamma W'(\lambda)W(\lambda)^{-1}d\lambda\right).$$

THEOREM 9.2. *Let* $W, S: \Omega \to \mathcal{L}(H)$ *be analytic operator functions, and assume that* W *is normal with respect to* Γ. *If*

$$(4) \qquad \|W(\lambda)^{-1}S(\lambda)\| < 1, \qquad \lambda \in \Gamma,$$

then $V(\cdot) = W(\cdot) + S(\cdot)$ *is also normal with respect to* Γ *and*

$$(5) \qquad m(\Gamma; V(\cdot)) = m(\Gamma; W(\cdot)).$$

The second theorem may be viewed as an operator generalization of Rouché's theorem from complex function theory.

To prove Theorem 9.1 it will be convenient to use the following terminology. An $\mathcal{L}(H)$-valued operator function G is called *finitely meromorphic* at λ_0 if G has a pole at λ_0 and the coefficients of the principal part of its Laurent expansion at λ_0 are operators of finite rank, i.e., in some punctured neighbourhood of λ_0 we have an expansion

$$G(\lambda) = \sum_{\nu=-q}^{\infty} (\lambda - \lambda_0)^\nu G_\nu,$$

which converges in the operator norm on $\mathcal{L}(H)$, such that $G_{-1}, \ldots, G_{-q}$ are finite rank operators. In that case we write $\Xi G(\lambda)$ for the principal of G at λ_0. Thus

$$(6) \qquad \Xi G(\lambda) = \sum_{\nu=-q}^{-1} (\lambda - \lambda_0)^\nu G_\nu, \qquad \lambda \neq \lambda_0.$$

Note that ΞG is analytic on $\mathbb{C}\backslash\{\lambda_0\}$ and its values are finite rank operators. We need the following lemma.

LEMMA 9.3. *Let* G_1 *and* G_2 *be* $\mathcal{L}(H)$-*valued operator functions which are finitely meromorphic at* λ_0. *Then* $G_1 G_2$ *and* $G_2 G_1$ *are finitely meromorphic at* λ_0 *and*

$$(7) \qquad \mathrm{tr}\,\Xi(G_1 G_2)(\lambda) = \mathrm{tr}\,\Xi(G_2 G_1)(\lambda), \qquad \lambda \neq \lambda_0.$$

PROOF. Here $G_1 G_2$ is the operator function defined by $(G_1 G_2)(\lambda) = G_1(\lambda)G_2(\lambda)$. In what follows $\lambda \neq \lambda_0$ and λ is sufficiently close to λ_0. For $\nu = 1, 2$ put

$$H_\nu(\lambda) = G_\nu(\lambda) - \Xi G_\nu(\lambda).$$

Then H_1 and H_2 are analytic at λ_0 and

$$(G_1 G_2)(\lambda) = (H_1 H_2)(\lambda) + H_1(\lambda)(\Xi G_2(\lambda)) + (\Xi G_1(\lambda))H_2(\lambda) + (\Xi G_1(\lambda))(\Xi G_2(\lambda)).$$

The last three terms are finitely meromorphic at λ_0, and $H_1 H_2$ is analytic at λ_0. Thus $G_1 G_2$ is finitely meromorphic at λ_0 and

$$(8) \quad \Xi(G_1 G_2)(\lambda) = \Xi\{H_1(\lambda)\Xi G_2(\lambda)\} + \Xi\{(\Xi G_1(\lambda))H_2(\lambda)\} + \Xi\{(\Xi G_1(\lambda))(\Xi G_2(\lambda))\}.$$

All coefficients in the Laurent expansion of $H_1 \Xi G_2$ at λ_0 are operators of finite rank. Furthermore, the Laurent expansion of $H_1 \Xi G_2$ at λ_0 converges in the trace class norm. Since the trace is a continuous linear functional on the trace class operators with respect to the trace class norm (Theorem VII.2.1), we may conclude that

$$(9) \quad \operatorname{tr} \Xi\{H_1(\lambda)\Xi G_2(\lambda)\} = \Xi \operatorname{tr}\{H_1(\lambda)\Xi G_2(\lambda)\}.$$

Similarly,

$$(10) \quad \operatorname{tr} \Xi\{(\Xi G_1(\lambda))H_2(\lambda)\} = \Xi \operatorname{tr}\{(\Xi G_1(\lambda))H_2(\lambda)\}.$$

For the third term in the right hand side of (8) we also have

$$(11) \quad \operatorname{tr} \Xi\{(\Xi G_1(\lambda))(\Xi G_2(\lambda))\} = \Xi \operatorname{tr}\{(\Xi G_1(\lambda))(\Xi G_2(\lambda))\}.$$

Next, use Corollary VII.6.2(i) to derive the following identities:

$$(12) \quad \operatorname{tr} H_1(\lambda)\Xi G_2(\lambda) = \operatorname{tr}(\Xi G_2(\lambda))H_1(\lambda),$$

$$(13) \quad \operatorname{tr}(\Xi G_1(\lambda))H_2(\lambda) = \operatorname{tr} H_2(\lambda)\Xi G_1(\lambda),$$

$$(14) \quad \operatorname{tr}(\Xi G_1(\lambda))(\Xi G_2(\lambda)) = \operatorname{tr}(\Xi G_2(\lambda))(\Xi G_1(\lambda)).$$

Finally, write the analogues of (8), (9), (10) and (11) with G_1 and G_2 interchanged. Then the identities (12), (13) and (14) yield the desired result (7). $\square$

PROOF OF THEOREM 9.1. Let Δ be the inner domain of the Cauchy contour Γ. The operator function $W'(\cdot)W(\cdot)^{-1}$ is analytic on $\Delta \cup \Gamma$, except possibly at a finite number of points in Δ which are eigenvalues of finite type of W. Hence, by Cauchy's theorem for analytic functions, it suffices to prove the theorem in the case where Γ is a circle of sufficiently small radius ρ with center at an eigenvalue λ_0 of W. Recall that W is equivalent at λ_0 to the operator function D defined in (1). So there exists an open neighbourhood $\mathcal{U}$ of λ_0 such that

$$(15) \quad W(\lambda) = E(\lambda)D(\lambda)F(\lambda), \qquad \lambda \in \mathcal{U},$$

where $E(\lambda)$ and $F(\lambda)$ are invertible operators which depend analytically on λ in $\mathcal{U}$. We shall assume that the radius ρ of the circle Γ has been chosen in such a way that $\lambda \in \mathcal{U}$ whenever $|\lambda - \lambda_0| \leq \rho$. Omitting the variable λ we may write

$$(16) \quad \begin{aligned} W'W^{-1} &= (E'DF + ED'F + EDF')F^{-1}D^{-1}E^{-1} \\ &= E'E^{-1} + ED'D^{-1}E^{-1} + EDF'F^{-1}D^{-1}E^{-1}. \end{aligned}$$

From Corollary 8.4 we know that W^{-1} is finitely meromorphic at λ_0. Since W is analytic on $\mathcal{U}$, its derivative W' is also analytic on $\mathcal{U}$, and thus $W'W^{-1}$ is finitely meromorphic. It follows that

$$(17) \qquad K := \frac{1}{2\pi i}\int_\Gamma W'(\lambda)W(\lambda)^{-1}d\lambda$$

is a well-defined operator of finite rank. In (17) the integrand may be replaced by the principal part of $W'(\cdot)W(\cdot)^{-1}$ at λ_0. But then the integral exists in the trace class norm, and thus trace and integral may be interchanged. So we see that

$$(18) \qquad \begin{aligned} \operatorname{tr} K &= \operatorname{tr}\left(\frac{1}{2\pi i}\int_\Gamma \Xi\{W'(\lambda)W(\lambda)^{-1}\}d\lambda\right) \\ &= \frac{1}{2\pi i}\int_\Gamma \operatorname{tr}\Xi\{W'(\lambda)W(\lambda)^{-1}\}d\lambda. \end{aligned}$$

Next, we use (16). Note that

$$(19) \qquad D(\lambda)^{-1} = P_0 + (\lambda - \lambda_0)^{-\kappa_1}P_1 + \cdots + (\lambda - \lambda_0)^{-\kappa_r}P_r, \qquad \lambda \neq \lambda_0.$$

In particular, $D(\cdot)^{-1}$ is finitely meromorphic at λ_0. All other functions appearing in the right hand side of (16) are analytic at λ_0. It follows that for $\lambda_0 \neq \lambda \in \mathcal{U}$

$$(20) \qquad \begin{aligned} \operatorname{tr}\Xi\{W'(\lambda)W(\lambda)^{-1}\} =\ & \operatorname{tr}\Xi\{E(\lambda)D'(\lambda)D(\lambda)^{-1}E(\lambda)^{-1}\} \\ & + \operatorname{tr}\Xi\{E(\lambda)D(\lambda)F'(\lambda)F(\lambda)^{-1}D(\lambda)^{-1}E(\lambda)^{-1}\}. \end{aligned}$$

Now we use Lemma 9.3 to interchange factors in the two terms of the right hand side of (20). We have the following identities:

$$(21) \qquad \begin{aligned} \operatorname{tr}\Xi\{E(\lambda)D'(\lambda)D(\lambda)^{-1}E(\lambda)^{-1}\} &= \operatorname{tr}\Xi\{D'(\lambda)D(\lambda)^{-1}E(\lambda)^{-1}E(\lambda)\} \\ &= \operatorname{tr}\Xi\{D'(\lambda)D(\lambda)^{-1}\}, \end{aligned}$$

$$(22) \qquad \begin{aligned} \operatorname{tr}\Xi\{E(\lambda)D(\lambda)F'(\lambda)F(\lambda)^{-1}D(\lambda)^{-1}E(\lambda)^{-1}\} & \\ = \operatorname{tr}\Xi\{F'(\lambda)F(\lambda)^{-1}D(\lambda)^{-1}E(\lambda)^{-1}E(\lambda)D(\lambda)\} & \\ = \operatorname{tr}\Xi\{F'(\lambda)F(\lambda)^{-1}\} = 0. & \end{aligned}$$

From (1) and (19) we see that

$$D'(\lambda)D(\lambda)^{-1} = \kappa_1(\lambda - \lambda_0)^{-1}P_1 + \cdots + \kappa_r(\lambda - \lambda_0)^{-1}P_r, \qquad \lambda \neq \lambda_0,$$

and thus

$$(23) \qquad \operatorname{tr}\Xi\{D'(\lambda)D(\lambda)^{-1}\} = \left(\sum_{j=1}^r \kappa_j\right)(\lambda - \lambda_0)^{-1}, \qquad \lambda \neq \lambda_0.$$

Formulas (20)–(23) yield

$$(24) \qquad \operatorname{tr} \Xi \{ W'(\lambda) W(\lambda)^{-1} \} = \left(\sum_{j=1}^{r} \kappa_j \right) (\lambda - \lambda_0)^{-1}, \qquad \lambda \neq \lambda_0.$$

Our choice of Γ implies that $m\big(\Gamma; W(\cdot)\big)$ is equal to $\kappa_1 + \cdots + \kappa_r$. Thus, by using (24) in the second integral appearing in (18), we obtain the desired result (3). $\square$

Theorem 9.1 implies that the definition of the algebraic multiplicity of W at λ_0 (as given in the first paragraph of this section) does not depend on the particular choice of the function $D(\cdot)$ in (1). In fact (see also (24)), the residue of $\operatorname{tr} \Xi W'(\cdot) W(\cdot)^{-1}$ at λ_0 is equal to $\kappa_1 + \cdots + \kappa_r$, and hence the latter number is uniquely determined by W.

Before we prove Theorem 9.2, let us first consider the case when $W(\lambda) = W_A(\lambda) = \lambda I - A$, where A is a bounded linear operator on H (and $\Omega = \mathbf{C}$). From Corollary 8.5 it follows that λ_0 is an eigenvalue of finite type of $W_A(\cdot)$ if and only if λ_0 is an eigenvalue of finite type of A (see Section II.1). Theorem 9.1 implies that in that case $m\big(\lambda_0; W_A(\cdot)\big)$ is precisely the algebraic multiplicity of λ_0 as an eigenvalue of A. To see this, choose a positively oriented circle Γ_0 with center at λ_0 such that $\sigma(A) \cap \Gamma_0 = \emptyset$ and λ_0 is the only point in the spectrum of A inside Γ_0. Then W_A is normal with respect to Γ_0 and

$$(25) \qquad \frac{1}{2\pi i} \int_{\Gamma_0} W_A'(\lambda) W_A(\lambda)^{-1} d\lambda = \frac{1}{2\pi i} \int_{\Gamma_0} (\lambda I - A)^{-1} d\lambda.$$

The right hand side of (25) is the Riesz projection $P_{\{\lambda_0\}}$ of A corresponding to the point λ_0. The rank of $P_{\{\lambda_0\}}$ is finite and equal to $m(\lambda_0; A)$, the algebraic multiplicity of λ_0 as an eigenvalue of A (cf. Section II.1). By Theorem 9.1 the trace of the left hand side of (25) is equal to $m\big(\lambda_0; W_A(\cdot)\big)$. Since the trace of a finite rank projection is equal to the rank of the projection, it follows that

$$m\big(\lambda_0; W_A(\cdot)\big) = m(\lambda_0; A).$$

From the remarks in the previous paragraph one sees that Theorem 9.2 may be viewed as a generalization (and a further refinement) of Theorem II.4.2.

PROOF OF THEOREM 9.2. The proof is split into two parts. As before, Δ denotes the inner domain of Γ.

PART (i). First we show that V is normal with respect to Γ. Put $C(\cdot) = W(\cdot)^{-1} S(\cdot)$. For $\lambda \in \Gamma$ we have $\|C(\lambda)\| < 1$, and hence $I + C(\lambda)$ is invertible for $\lambda \in \Gamma$. By our hypotheses, the same holds true for $W(\lambda)$. Thus $V(\lambda) = W(\lambda)[I + C(\lambda)]$ is invertible for $\lambda \in \Gamma$. To prove that $V(\lambda)$ is Fredholm for $\lambda \in \Delta$, we study the behaviour of $C(\cdot)$ on Δ. From its definition we see that C is analytic at each point of Γ, and C is analytic on Δ, except possibly for a finite number of points inside Δ, which are eigenvalues of finite type of W. Let $\lambda_1, \ldots, \lambda_p$ be these exceptional points. From Corollary 8.4 we know that $W(\cdot)^{-1}$ is finitely meromorphic at the points $\lambda_1, \ldots, \lambda_p$. Since S is analytic on Ω, it follows that C is also finitely meromorphic at $\lambda_1, \ldots, \lambda_p$. So for λ close to λ_j we have

$$C(\lambda) = \sum_{\nu=-q_j}^{\infty} (\lambda - \lambda_j)^\nu C_{j,\nu},$$

where $C_{j,-1}, \ldots, C_{j,-q_j}$ are operators of finite rank. Put

$$N = \bigcap \{ \operatorname{Ker} C_{j,\nu} \mid \nu = -1, \ldots, -q_j, \ j = 1, \ldots, p \}.$$

The space N is a closed linear manifold of H and $\dim H/N < \infty$. Let $C_N(\lambda): N \to H$ be defined by $C_N(\lambda)x = C(\lambda)x$ for each $x \in N$. Then C_N is an $\mathcal{L}(N,H)$-valued operator function which is analytic at each point of $\Gamma \cup \Delta$. From (4) it follows that there exists $0 \le \gamma < 1$ such that $\|C(\lambda)\| \le \gamma$ for all $\lambda \in \Gamma$. In particular, for $x \in N$ and $y \in H$ we have

$$(26) \qquad\qquad\qquad |\langle C_N(\lambda)x, y\rangle| \le \gamma \|x\| \|y\|,$$

whenever $\lambda \in \Gamma$. But $\langle C_N(\cdot)x, y\rangle$ is analytic at each point of $\Gamma \cup \Delta$. So, by the maximum modulus principle, the inequality (26) holds for all $\lambda \in \Gamma \cup \Delta$, and thus

$$(27) \qquad\qquad\qquad \|C_N(\lambda)\| \le \gamma < 1, \qquad \lambda \in \Delta \cup \Gamma.$$

Now define $\widetilde{C}(\lambda): H \to H$ by taking $\widetilde{C}(\lambda) = C_N(\lambda)\Pi$, where Π is the orthogonal projection of H onto N. Then $\widetilde{C}$ is an $\mathcal{L}(H)$-valued operator function which is analytic at each point of $\Gamma \cup \Delta$ and $\|\widetilde{C}(\lambda)\| \le \gamma < 1$ for $\lambda \in \Gamma \cup \Delta$. It follows that $I + \widetilde{C}(\lambda)$ is invertible for each $\lambda \in \Gamma \cup \Delta$. We know that $W(\lambda)$ is Fredholm for each $\lambda \in \Delta$. So the same holds true for $W(\lambda)[I + \widetilde{C}(\lambda)]$. Note that $V(\lambda)$ and $W(\lambda)[I + \widetilde{C}(\lambda)]$ coincide on the space N. But N has finite codimension in H. So we may apply Lemma 3.1 twice to conclude that $V(\lambda)$ is Fredholm for each $\lambda \in \Delta$. Thus V is normal with respect to Γ.

PART (ii). In this part we prove formula (5) by using the method of linearization. Without loss of generality we may assume that 0 is in the inner domain of Γ. Choose a bounded Cauchy domain Θ such that $\Gamma \subset \Theta \subset \overline{\Theta} \subset \Omega$. Let K_0 be the space of all H-valued continuous functions on $\partial\Theta$ endowed with the inner product

$$(28) \qquad\qquad\qquad \langle f, g\rangle = \int_{\partial\Theta} \langle f(\zeta), g(\zeta)\rangle d\zeta,$$

and let K be a Hilbert space such that K_0 is a linear submanifold of K which is dense in K and for elements f and g in K_0 the inner product in K coincides with the one given by (28). Up to a linear isometry the Hilbert space K is uniquely determined by K_0. Since H is separable, the same is true for K.

For $0 \le t \le 1$ we let A_t be the operator on K defined by

$$(29) \qquad (A_t f)(z) = zf(z) - \frac{1}{2\pi i} \int_{\partial\Theta} [I - W(\zeta) - tS(\zeta)] f(\zeta) d\zeta, \qquad z \in \partial\Theta.$$

The operator A_t is bounded on K_0, and hence, by continuity, A_t extends to a bounded linear operator on K, which we also denote by A_t. From (29) it follows that the map

$$(30) \qquad t \mapsto A_t, \qquad [0,1] \to \mathcal{L}(K)$$

is continuous with respect to the operator norm on $\mathcal{L}(K)$.

Let $\omega \colon K \to H$ be the (unique) bounded linear operator such that

$$(31) \qquad \omega f = \frac{1}{2\pi i} \int_{\partial \Theta} \frac{1}{\zeta} f(\zeta) d\zeta, \qquad f \in K_0.$$

Note that ω is bounded on K_0, and hence, by continuity, it extends to all of K. Put $Z = \operatorname{Ker} \omega$. Then Z is a (separable) Hilbert space in its own right, and for $0 \le t \le 1$ the Z-extension of $W(\cdot) + tS(\cdot)$ is equivalent on Θ to $\lambda - A_t$. To prove this, one uses the same arguments as were used in the proof of Theorem III.2.1. In fact, all formulas in the proof of Theorem III.2.1 hold on K_0 and, by continuity, also on K.

Put $W_t(\lambda) = \lambda - A_t$. The above equivalence implies that W_t is normal with respect to Γ. By extension and equivalence the algebraic multiplicity does not change. Thus

$$(32) \qquad m\big(\Gamma; W(\cdot) + tS(\cdot)\big) = m\big(\Gamma; W_t(\cdot)\big).$$

By the remarks made in the second paragraph after the proof of Theorem 9.1, we have

$$(33) \qquad m\big(\Gamma; W_t(\cdot)\big) = \sum_{\lambda \text{ inside } \Gamma} m(\lambda; A_t).$$

Furthermore, by Theorem II.4.2, the quantity in the right hand side of (33) is a continuous function of t. It follows that the same holds true for the first quantity in (32). Since these functions are integer-valued, we conclude that $m\big(\Gamma; W(\cdot) + tS(\cdot)\big)$ does not depend on $t \in [0,1]$, and hence (5) is proved. $\square$

The results of this section also hold for operator functions whose values act on Banach spaces. To see this one needs the trace of a finite rank operator F acting on a Banach space X. This quantity is defined as follows:

$$\operatorname{tr} F = \sum_{j=1}^{n} g_j(x_j),$$

where $x_1, \ldots, x_n$ are vectors in X and $g_1, \ldots, g_n$ are continuous linear functionals on X such that

$$F = \sum_{j=1}^{n} g_j(\cdot) x_j.$$

One can prove (see, e.g., Gohberg-Krupnik [1]) that $\operatorname{tr} F$ is a linear functional on the finite rank operators F and $\operatorname{tr}(FG) = \operatorname{tr}(GF)$ for any bounded linear operator G. With this definition of the trace the proofs given above carry over to the Banach space case.

XI.10 SINGULAR VALUES FOR BOUNDED OPERATORS

In this section we introduce s-numbers (singular values) for arbitrary (not necessarily compact) operators acting on a Hilbert space. The starting point is Theorem VI.1.5 which identifies the s-numbers of a compact operator as certain approximation numbers. Let $A: H \to H$ be a bounded linear operator acting on the complex Hilbert space H. By definition the j-th singular value (s-number) of A is the number

$$(1) \qquad s_j(A) = \inf\{\|A - K\| \mid K \in \mathcal{L}(H), \operatorname{rank} K \leq j - 1\}.$$

Note that $s_1(A) = \|A\|$ and $s_1(A) \geq s_2(A) \geq \cdots$. Many properties of s-numbers of compact operators carry over to the non-compact case. For example, Proposition VI.1.3 and Corollaries VI.1.2, VI.1.4 and VI.1.6 hold for arbitrary bounded linear operators. The definition (1) also presents a way to introduce s-numbers for operators acting between Banach spaces.

Since $\left(s_j(A)\right)_{j=1}^{\infty}$ is a non-increasing sequence of non-negative numbers, we may define

$$(2) \qquad s_\infty(A) := \lim_{n \to \infty} s_n(A) = \inf_n s_n(A).$$

The next theorem (together with the fact that the set of compact operators on H is closed in $\mathcal{L}(H)$) shows that $s_\infty(A) = 0$ if and only if A is compact.

THEOREM 10.1. *For $A \in \mathcal{L}(H)$*

$$(3) \qquad s_\infty(A) = \inf\{\|A - K\| \mid K \text{ compact}\}.$$

PROOF. Let γ be the number defined by the right hand side of (3). Let $K \in \mathcal{L}(H)$ be of finite rank, $\operatorname{rank} K = n$, say. Then K is compact, and thus $\gamma \leq \|A - K\|$. It follows (see (1)) that $\gamma \leq s_{n+1}(A)$, and hence $\gamma \leq s_\infty(A)$. To prove the reverse inequality, take $\varepsilon > 0$. Choose a compact operator K such that $\|A - K\| < \gamma + \frac{1}{2}\varepsilon$. Next, choose a finite rank operator $F \in \mathcal{L}(H)$ such that $\|K - F\| < \frac{1}{2}\varepsilon$. Then $\|A - F\| < \gamma + \varepsilon$. According to the second identity in (2), we have $s_\infty(A) \leq \|A - F\|$. Hence $s_\infty(A) < \gamma + \varepsilon$. But $\varepsilon > 0$ is arbitrary, and therefore $s_\infty(A) \leq \gamma$. $\square$

THEOREM 10.2. *The number $s_\infty(A)$ is the square root of the maximum of the essential spectrum of A^*A.*

PROOF. Put $S = A^*A$. Since S is nonnegative, the essential spectrum of S is a compact subset of the nonnegative real line, and hence the maximum of the essential spectrum of S is a well-defined number μ, say. Let $\varepsilon > 0$. By Theorem 10.1 there exists a compact operator K such that $\|A - K\| < s_\infty(A) + \varepsilon$. Thus

$$\begin{aligned}
\|(A - K)^*(A - K)\| &= \sup_{\|x\|=1} \langle (A - K)^*(A - K)x, x \rangle \\
&= \sup_{\|x\|=1} \|(A - K)x\|^2 \\
&= \|A - K\|^2 < \left(s_\infty(A) + \varepsilon\right)^2.
\end{aligned}$$

Since $T := (A - K)^*(A - K) = S + C$, where C is a compact operator, T and S have the same essential spectrum. In particular, $\mu \in \sigma(T)$. Thus

$$\mu \leq \|T\| \leq \left(s_\infty(A) + \varepsilon\right)^2.$$

Since $\varepsilon > 0$ is arbitrary, this shows that $s_\infty(A)^2 \geq \mu$.

To prove the converse we consider the set $\Sigma = \{\lambda \in \sigma(S) | \lambda > \mu\}$. This set consists of eigenvalues of finite type only (apply Corollary 8.5). In particular, each $\lambda \in \Sigma$ is an isolated point of $\sigma(S)$. It follows that Σ is at most countable. Let $\mu_1 > \mu_2 > \cdots$ be the points in Σ. By H_0 we denote the smallest closed linear manifold containing all eigenvectors of S corresponding to the eigenvalues $\mu_1, \mu_2, \ldots$, and P will be the orthogonal projection onto the space $H_1 := H_0^\perp$. Put $S_0 = S(I - P)$. Note that the space H_0 has an orthonormal basis $\varphi_1, \varphi_2, \ldots$ consisting of eigenvectors of S. Thus we may write

$$(4) \qquad S_0 x = \sum_j \lambda_j \langle x, \varphi_j \rangle \varphi_j, \qquad x \in H,$$

where $\lambda_1 \geq \lambda_2 \geq \lambda_3 \geq \cdots$ is the sequence $\mu_1 > \mu_2 > \mu_3 > \cdots$ with each μ_j repeated as many times as the value of its algebraic multiplicity. The operator K, defined by

$$Kx = \sum_j (\lambda_j^{1/2} - \mu^{1/2}) \langle x, \varphi_j \rangle \varphi_j, \qquad x \in H,$$

is nonnegative and K has finite rank (if the sequence $\lambda_1, \lambda_2, \ldots$ is finite) or is compact.

Next, use the polar decomposition (see Section V.6) to write A as $A = UR$, where $R = (A^*A)^{1/2} = S^{1/2}$ and U is a partial isometry. Put $C = UK$. Then C is compact, and, by Theorem 10.1,

$$(5) \qquad s_\infty(A) \leq \|A - C\| = \|UR - UK\| \leq \|R - K\|.$$

Since S commutes with P, also its square root R commutes with P (Theorem V.6.1) and

$$R(I - P)x = S_0^{1/2} x = \sum_j \lambda_j^{1/2} \langle x, \varphi_j \rangle \varphi_j, \qquad x \in H.$$

It follows that $R - K = RP + \mu^{1/2}(I - P)$, and thus

$$(6) \qquad \|R - K\| \leq \max\{\|RP\|, \mu^{1/2}\}.$$

It remains to show that $\|RP\| \leq \mu^{1/2}$.

Let $\{E(t)\}_{t \in \mathbf{R}}$ be the resolution of the identity for the non-negative operator S. From Theorem V.5.3 we know that $E(\cdot)$ is constant on $(\mu, \infty) \setminus \{\lambda_1, \lambda_2, \ldots\}$. Furthermore, by Corollary V.5.2,

$$\operatorname{Im}\{E(\lambda_i) - E(\lambda_i - 0)\} = \operatorname{Ker}(\lambda_i - S) \subset H_0, \qquad i \geq 1.$$

It follows that $\mathrm{Im}\big(I - E(t)\big) \subset H_0$ for $t > \mu$, and hence $H_1 \subset \mathrm{Im}\, E(t)$ for $t > \mu$. This shows that

$$\mathrm{Im}\, E(\mu) = \cap\{\mathrm{Im}\, E(t) \mid t > \mu\} \supset H_1.$$

(With a little extra effort one shows that $E(\mu) = P$, but we don't need this equality.) Choose $\alpha < 0$ and $\beta > \|S\|$, and let

$$g(t) = \left\{ \begin{array}{ll} 0, & \alpha \leq t < 0, \\ t^{1/2}, & 0 \leq t \leq \beta. \end{array} \right.$$

Then for each $x \in H$,

$$\langle RPx, x\rangle = \langle RPx, Px\rangle = \int_{\alpha}^{\beta} g(t)d\langle E(t)Px, Px\rangle$$

$$= \int_{0}^{\mu} t^{1/2}d\langle E(t)Px, Px\rangle \leq \mu^{1/2}\|x\|^2,$$

because $\langle E(t)Px, Px\rangle = \|Px\|^2 \leq \|x\|^2$ for $t \geq \mu$. It follows that $\|RP\| \leq \mu^{1/2}$. $\square$

The proof of Theorem 10.2 shows that the infimum in (3) is actually a minimum. To see this, let C be the operator introduced in the proof of Theorem 10.2. Since $s_\infty(A) = \mu^{1/2}$ (the notation is as in the proof of Theorem 10.2), formulas (5) and (6) imply that $\|A - C\| = s_\infty(A)$, and hence the infimum in (3) is attained for $K = C$.

A further refinement of the arguments used in the proof of Theorem 10.2 yields the following alternative description of the s-numbers of $A \in \mathcal{L}(H)$. Let $\lambda_1 \geq \lambda_2 \geq \cdots$ be the eigenvalues (multiplicities taken into account) of A^*A strictly larger than the maximum μ of the essential spectrum of A^*A, and let N be the number of elements in the sequence $\lambda_1 \geq \lambda_2 \geq \cdots$. Then $s_j(A) = \lambda_j^{1/2}$, $j = 1, 2, \ldots$, if N is infinite, and otherwise

$$s_j(A) = \left\{ \begin{array}{ll} \lambda_j^{1/2}, & j = 1, \ldots, N, \\ \mu^{1/2}, & j = N+1, N+2, \ldots \,. \end{array} \right.$$

CHAPTER XII
WIENER-HOPF INTEGRAL OPERATORS

In this chapter we deal with integral operators of the following type:

$$g(t) - \int_0^\infty k(t-s)g(s)ds = f(t), \qquad 0 \le t < \infty.$$

Here g and f are $\mathbf{C}^m$-valued functions with components in $L_2([0,\infty])$ and the kernel function k is an $m \times m$ matrix function. The corresponding operators are called *Wiener-Hopf (integral) operators*. We shall restrict the attention to the case when the entries of k are integrable on the real line. Wiener-Hopf integral operators provide one of the main examples for the Fredholm theory. In this case the index may be expressed in terms of topological properties of the symbol. The chapter consists of three sections. The first has an introductory character and concerns the analogous equation on the full line. In Section 2 the first properties of Wiener-Hopf operators are derived. The last section is devoted to the Fredholm theory.

XII.1 CONVOLUTION OPERATORS

This section concerns operators of the form $A = I - L$, where

$$(1) \qquad (L\varphi)(t) = \int_{-\infty}^\infty k(t-s)\varphi(s)ds, \qquad -\infty < t < \infty.$$

The operators A and L will be considered on $L_2^m(\mathbf{R})$, the space of all $\mathbf{C}^m$-valued functions that are square integrable (relative to the Lebesgue measure) on the real line $\mathbf{R}$. The space $L_2^m(\mathbf{R})$ is a Hilbert space with inner product

$$(2) \qquad \langle \varphi, \psi \rangle = \sum_{j=1}^m \int_{-\infty}^\infty \varphi_j(t)\overline{\psi_j(t)}dt.$$

Here φ_j and ψ_j are the j-th components of φ and ψ, respectively. As usual, two functions that are equal almost everywhere are identified.

The function k in (1), which is called the *kernel function* of L, is assumed to be an $m \times m$ matrix function, $k = [k_{ij}]_{i,j=1}^m$, of which the entries k_{ij} are integrable on $\mathbf{R}$. One could take a more general class of kernel functions, as we shall do in the discrete case (in Volume II), but the present class is already rich enough to explain the main points of the theory. The integrability condition on the entries of k implies that L is a well-defined bounded linear operator on $L_2^m(\mathbf{R})$. In fact,

$$(3) \qquad \|L\| \le \left(\sum_{i=1}^m \sum_{j=1}^m \|k_{ij}\|_1^2 \right)^{1/2},$$

where $\|k_{ij}\|_1$ is the norm of k_{ij} as an element of $L_1(\mathbf{R})$. To prove (3), note that $L_2^m(\mathbf{R})$ is equal to the Cartesian product of m copies of $L_2(\mathbf{R})$, and hence L may be represented by an $m \times m$ matrix whose entries are operators on $L_2(\mathbf{R})$. In this case

$$
(4) \qquad L = \left[\begin{array}{ccc} L_{11} & \cdots & L_{1m} \\ \vdots & & \vdots \\ L_{m1} & \cdots & L_{mm} \end{array} \right] : L_2^m(\mathbf{R}) \to L_2^m(\mathbf{R}),
$$

where

$$
(5) \qquad L_{ij} : L_2(\mathbf{R}) \to L_2(\mathbf{R}), \qquad L_{ij}f = k_{ij} * f.
$$

Here the symbol $*$ denotes the usual convolution product. It is well-known (see, e.g., [R], Ch. 7, Exercise 4) that for $g \in L_1(\mathbf{R})$ and $f \in L_2(\mathbf{R})$ the convolution product $g*f \in L_2(\mathbf{R})$ and $\|g * f\| \le \|g\|_1 \|f\|$. Let $\varphi \in L_2^m(\mathbf{R})$, and let φ_j denote its j-th component. Then

$$
\begin{aligned}
\|L\varphi\|^2 &= \sum_{i=1}^m \|(L\varphi)_i\|^2 \\
&= \sum_{i=1}^m \left\| \sum_{j=1}^m k_{ij} * \varphi_j \right\|^2 \\
&\le \sum_{i=1}^m \left(\sum_{j=1}^m \|k_{ij}\|_1 \|\varphi_j\| \right)^2 \\
&\le \left(\sum_{i=1}^m \sum_{j=1}^m \|k_{ij}\|_1^2 \right) \|\varphi\|^2,
\end{aligned}
$$

which proves (3).

With the operator $A = I - L$ we associate the following $m \times m$ matrix function:

$$
(6) \qquad W(s) := I_m - \int_{-\infty}^{\infty} e^{ist} k(t)\,dt, \qquad s \in \mathbf{R}.
$$

Here I_m denotes the $m \times m$ identity matrix. Since the entries of k are in $L_1(\mathbf{R})$, we may conclude (cf., [R], Theorem 9.6) that the entries of W are continuous functions on $\mathbf{R}$ and

$$
(7) \qquad \lim_{|s| \to \infty} W(s) = I_m.
$$

The function W is called the *symbol* of A, and we shall refer to A as the *convolution operator with symbol W*.

To study A in terms of its symbol we need the Fourier transformation. Recall that for $f \in L_1(\mathbf{R}) \cap L_2(\mathbf{R})$ the Fourier transform $\widehat{f}(\lambda)$, which is defined by

$$
\widehat{f}(\lambda) = \frac{1}{\sqrt{2\pi}} \int_{-\infty}^{\infty} e^{-i\lambda t} f(t)\,dt, \qquad -\infty < \lambda < \infty,
$$

is an element of $L_2(\mathbf{R})$. By the Plancherel theorem ([R], Theorem 9.13) the map $f \mapsto \widehat{f}$ extends to an isometry U which maps all of $L_2(\mathbf{R})$ onto $L_2(\mathbf{R})$. Since $L_2^m(\mathbf{R})$ is equal to the Cartesian product of m copies of $L_2(\mathbf{R})$, the operator U induces in a natural way an isometry, also denoted by U, from $L_2^m(\mathbf{R})$ onto $L_2^m(\mathbf{R})$, namely

$$(8) \qquad U\varphi = U \begin{bmatrix} \varphi_1 \\ \vdots \\ \varphi_m \end{bmatrix} = \begin{bmatrix} U\varphi_1 \\ \vdots \\ U\varphi_m \end{bmatrix}.$$

The operator U is called the *Fourier transformation* on $L_2^m(\mathbf{R})$ and $U\varphi$ the *Fourier transform* of φ. Obviously, $U\colon L_2^m(\mathbf{R}) \to L_2^m(\mathbf{R})$ is unitary, i.e., $U^* = U^{-1}$, and hence

$$(9) \qquad (U^{-1}\varphi)(t) = (U\varphi)(-t), \quad \text{a.e.} \ .$$

THEOREM 1.1. *Let A be the convolution operator on $L_2^m(\mathbf{R})$ with symbol W, and let U be the Fourier transformation on $L_2^m(\mathbf{R})$. Then*

$$(10) \qquad U^{-1}AU\varphi = W(\cdot)\varphi(\cdot), \qquad \varphi \in L_2^m(\mathbf{R}).$$

PROOF. Let L_{ij} be as in (5). Take f in $L_1(\mathbf{R}) \cap L_2(\mathbf{R})$. By using (9) and the connections between Fourier transforms and convolution products ([R], Theorem 9.2) one proves that for almost all $s \in \mathbf{R}$

$$\begin{aligned} (U^{-1}L_{ij}f)(s) &= (UL_{ij}f)(-s) \\ &= [U(k_{ij} * f)](-s) \\ &= \left(\int_{-\infty}^{\infty} e^{ist}k_{ij}(t)dt \right)\widehat{f}(-s) \\ &= \left(\int_{-\infty}^{\infty} e^{ist}k_{ij}(t)dt \right)(U^{-1}f)(s). \end{aligned}$$

Let $M_{ij}\colon L_2(\mathbf{R}) \to L_2(\mathbf{R})$ be defined by

$$(M_{ij}g)(s) = \left(\int_{-\infty}^{\infty} e^{ist}k_{ij}(t)dt \right)g(s).$$

Since M_{ij} is an operator of multiplication by a bounded continuous function on $\mathbf{R}$, the operator M_{ij} is well-defined and bounded on $L_2(\mathbf{R})$. We have proved that

$$(11) \qquad U^{-1}L_{ij}Ug = M_{ij}g$$

for any $g \in U^{-1}[L_1(\mathbf{R}) \cap L_2(\mathbf{R})]$. Since the latter set is dense in $L_2(\mathbf{R})$, the continuity of the operators in (11) implies that (11) holds for any $g \in L_2(\mathbf{R})$. Thus $U^{-1}L_{ij}U = M_{ij}$.

Now use $U^{-1}AU = I - U^{-1}LU$, where L is given by (4), and the identity (8), to obtain formula (10). $\square$

The operator on $L_2^m(\mathbf{R})$ defined by the right hand side of (10) will be called the *operator of multiplication* by W and will be denoted by M_W.

COROLLARY 1.2. *Let A be the convolution operator on $L_2^m(\mathbf{R})$ with symbol* W. *Then*

$$(12) \qquad \|A\| = \sup\{\|W(t)\| \mid t \in \mathbf{R}\}.$$

PROOF. In (12) the term $\|W(t)\|$ is the norm of the $m \times m$ matrix $W(t)$ as an operator on $\mathbf{C}^m$; thus $\|W(t)\|$ is the largest singular value of the matrix $W(t)$. Let M_W be the operator of multiplication by W. By Theorem 1.1 it suffices to show that $\|M_W\|$ is given by the right hand side of (12). First take $\varphi \in L_2^m(\mathbf{R})$. Then

$$\|M_W\varphi\|^2 = \int_{-\infty}^{\infty} \|W(t)\varphi(t)\|^2 dt$$

$$\leq \int_{-\infty}^{\infty} \|W(t)\|^2 \|\varphi(t)\|^2 dt$$

$$\leq \left(\sup_{t\in\mathbf{R}} \|W(t)\|\right)^2 \cdot \|\varphi\|^2,$$

which shows that the right hand side of (12) is an upper bound for $\|M_W\|$.

Next, fix $t_0 \in \mathbf{R}$, and let x be an arbitrary vector in $\mathbf{C}^m$ with $\|x\| = 1$. Define a sequence in $L_2^m(\mathbf{R})$ by

$$\varphi_n(t) = \begin{cases} \sqrt{\frac{n}{2}}x & \text{for } t_0 - \frac{1}{n} \leq t \leq t_0 + \frac{1}{n}, \\ 0 & \text{otherwise.} \end{cases}$$

Then $\|\varphi_n\| = 1$ for $n = 1, 2, \ldots$, and hence

$$\|M_W\|^2 \geq \|M_W\varphi_n\|^2 = \int_{-\infty}^{\infty} \|W(t)\varphi_n(t)\|^2 dt$$

$$= \frac{n}{2} \int_{t_0-\frac{1}{n}}^{t_0+\frac{1}{n}} \|W(t)x\|^2 dt \to \|W(t_0)x\|^2.$$

Here we used the continuity in t of the integrand $\|W(t)x\|^2$. It follows that $\|M_W\| \geq \|W(t_0)x\|$. Since x is an arbitrary vector of norm one, we get $\|M_W\| \geq \|W(t_0)\|$. Also t_0 is arbitrary. Thus $\|M_W\|$ is an upper bound for the right hand side of (12). $\square$

COROLLARY 1.3. *Convolution operators on $L_2(\mathbf{R})$ commute with one another.*

PROOF. The statement follows from Theorem 1.1 and the fact that operators of multiplication by scalar functions on $L_2(\mathbf{R})$ commute with one another. $\square$

THEOREM 1.4. *Let A be the convolution operator on $L_2^m(\mathbf{R})$ with symbol W. Then A is invertible if and only if $\det W(s) \neq 0$ for each $s \in \mathbf{R}$. In that case*

$$(13) \qquad W(s)^{-1} = I_m - \int_{-\infty}^{\infty} e^{ist} k^{\times}(t)\,dt, \qquad s \in \mathbf{R}$$

for some $m \times m$ matrix function $k^{\times}$ with entries in $L_1(\mathbf{R})$ and

$$(14) \qquad (A^{-1}\psi)(t) = \psi(t) - \int_{-\infty}^{\infty} k^{\times}(t-s)\psi(s)\,ds, \qquad t \in \mathbf{R}.$$

PROOF. Since $L_2^m(\mathbf{R})$ is the Cartesian product of m copies of $L_2(\mathbf{R})$, the operator A may be represented as an $m \times m$ operator matrix,

$$(15) \qquad A = \begin{bmatrix} A_{11} & \cdots & A_{1m} \\ \vdots & & \vdots \\ A_{m1} & \cdots & A_{mm} \end{bmatrix},$$

of which the entries act on $L_2(\mathbf{R})$. Note that $A_{ij} = \delta_{ij}I - L_{ij}$, where L_{ij} is given by (5) and δ_{ij} is the Kronecker delta. Corollary 1.3 implies that the entries in (15) commute with another. Thus, by Proposition XI.7.2, the operator A is invertible if and only if $\det A$ is invertible.

Let U be the Fourier transformation. Then $U^{-1}A_{ij}U = \delta_{ij}I - M_{ij}$, where M_{ij} is as in (11). It follows that $U^{-1}A_{ij}U$ is the operator of multiplication by the (i,j)-th entry of the symbol W of A. But then

$$M := U^{-1}(\det A)U = \det(U^{-1}AU)$$

is the operator of multiplication by $\det W$.

Assume A is invertible. Then M is invertible and $\rho = \|M^{-1}\|^{-1} > 0$. Put

$$(16) \qquad E_n = \left\{ t \in \mathbf{R} \mid |\det W(t)| < \frac{1}{2}\rho, |t| \leq n \right\}.$$

Let χ_{E_n} be the characteristic function of the set E_n (i.e., $\chi_{E_n}(t) = 1$ if $t \in E_n$ and $\chi_{E_n}(t) = 0$ otherwise). Then

$$\rho^2 \|\chi_{E_n}\|^2 \leq \|M\chi_{E_n}\|^2$$
$$= \int_{-\infty}^{\infty} \left|(\det W(t))\chi_{E_n}(t)\right|^2 dt$$
$$\leq \frac{1}{4}\rho^2 \|\chi_{E_n}\|^2, \qquad n = 1, 2, \dots,$$

and hence $\chi_{E_n} = 0$ for $n = 1, 2, \ldots$. It follows that the set $E := \{t \in \mathbf{R} \mid |\det W(t)| < \frac{1}{2}\rho\}$ has Lebesgue measure zero. But $\det W(t)$ is a continuous function in t. Thus E is an open set of measure zero, which implies that $E = \emptyset$. In particular, $\det W(t) \neq 0$ for all $t \in \mathbf{R}$.

To prove the converse statement and the formula for A^{-1}, assume that $\det W(t) \neq 0$ for all $t \in \mathbf{R}$. In that case there exists an $m \times m$ matrix function $k^{\times}$ of which the entries are in $L_1(\mathbf{R})$ such that (13) holds. This result is the matrix version of a theorem of N. Wiener which will be proved in the Banach algebra part of this book (in Volume II). Note that $k^{\times}$ belongs to the class of kernel functions considered in this section. Let $A^{\times}$ be the operator on $L_2^m(\mathbf{R})$ defined by the right hand side of (14). Formula (13) implies that $A^{\times}$ is the convolution operator on $L_2^m(\mathbf{R})$ with symbol $W(\cdot)^{-1}$. But then we can use Theorem 1.1 to show that

$$U^{-1}AA^{\times}U\varphi = (U^{-1}AU)(U^{-1}A^{\times}U)\varphi$$
$$= W(\cdot)W(\cdot)^{-1}\varphi = \varphi,$$

and thus $AA^{\times} = I$. Similarly, $A^{\times}A = I$. Thus A is invertible and $A^{-1} = A^{\times}$. $\quad\square$

COROLLARY 1.5. *The convolution operator A is a Fredholm operator if and only if A is invertible.*

PROOF. Assume that A is Fredholm. Write A as an $m \times m$ operator matrix as in (15). Since the entries in (15) commute with one another, Theorem XI.7.3 implies that $det A$ is Fredholm. Let U be the Fourier transformation on $L_2(\mathbf{R})$, and put $M := U^{-1}(det A)U$. We have already seen that M is the operator of multiplication by $\det W$. Consider the set

$$F_n = \{t \in \mathbf{R} \mid \det W(t) = 0, |t| \leq n\},$$

and let $N(F_n)$ be the subspace of $L_2(\mathbf{R})$ consisting of all f that are zero almost everywhere on $\mathbf{R} \backslash F_n$. Note that $N(F_n) \subset \operatorname{Ker} M$. Since M is Fredholm, $\dim N(F_n) < \infty$, but then F_n must have measure zero. It follows that the set $\{t \in \mathbf{R} \mid \det W(t) = 0\}$ has measure zero, and thus M is injective. We also know that M has closed range. Thus there exists $\rho > 0$ such that $\|Mf\| \geq \rho\|f\|$ for each $f \in L_2(\mathbf{R})$. Now define E_n as in (16), and proceed as in the proof of Theorem 1.4. One obtains that $\det W(t) \neq 0$ for all $t \in \mathbf{R}$, and thus, by Theorem 1.4, the operator A is invertible. The converse statement is trivial. $\quad\square$

XII.2 WIENER-HOPF OPERATORS

This section concerns operators of the form $T = I - K$, where

$$(1) \qquad (K\varphi)(t) = \int_0^{\infty} k(t-s)\varphi(s)ds, \qquad 0 \leq t < \infty.$$

The operators T and K will be considered on $L_2^m([0, \infty])$, the space of all $\mathbf{C}^m$-valued functions that are square integrable (relative to the Lebesgue measure) on the real line, which we shall identify with the subspace of $L_2^m(\mathbf{R})$ consisting of all φ that are zero almost everywhere on $(-\infty, 0)$.

As in the previous section the function k in (1) is assumed to be an $m \times m$ matrix function, $k = [k_{ij}]_{i,j=1}^m$, of which the entries are in $L_1(\mathbf{R})$. This implies that K is a well-defined bounded operator on $L_2^m(\mathbf{R})$ and

$$(2) \qquad \|K\| \le \left(\sum_{i=1}^m \sum_{j=1}^m \|k_{ij}\|_1^2 \right)^{1/2},$$

where $\|k_{ij}\|_1$ is the norm of k_{ij} as an element of $L_1(\mathbf{R})$. To see this, note that

$$(3) \qquad L_2^m(\mathbf{R}) = L_2^m([0,\infty)) \oplus L_2^m((-\infty,0]).$$

Here $L_2^m((-\infty,0])$ is the subspace of $L_2^m(\mathbf{R})$ consisting of all functions φ that are zero almost everywhere on $[0,\infty)$. The decomposition (3) is an orthogonal one. Now, let L be defined by (1.1), and let Q be the projection of $L_2^m(\mathbf{R})$ onto $L_2^m([0,\infty))$ along $L_2^m((-\infty,0])$. Then

$$(4) \qquad K\varphi = QL\varphi, \qquad \varphi \in L_2^m([0,\infty)).$$

Since Q is orthogonal, we conclude that $\|K\| \le \|L\|$, and hence (2) follows from the inequality (3) in the previous section.

With the operator $T = I - K$ we associate its *symbol*, which is defined to be the $m \times m$ matrix function

$$(5) \qquad W(s) = I_m - \int_{-\infty}^{\infty} e^{ist} k(t)\,dt, \qquad s \in \mathbf{R}.$$

Note that k is uniquely determined by W. We shall refer to T as the *Wiener-Hopf operator with symbol W*.

Up to unitary equivalence a Wiener-Hopf operator is a compression of a multiplication operator. To see this, let U be the Fourier transformation, and put $H_2^m(\mathbf{R}) = UL_2^m((-\infty,0])$. Thus $\varphi \in H_2^m(\mathbf{R})$ if and only if

$$(6) \qquad \varphi(s) = \frac{1}{\sqrt{2\pi}} \int_{-\infty}^{0} e^{-ist} \psi(t)\,dt, \qquad s \in \mathbf{R},$$

for some $\psi \in L_2^m((-\infty,0])$. It follows that $\varphi \in H_2^m(\mathbf{R})$ has an analytic continuation to the upper half plane. Indeed, let φ be given by (6). Note that $e^{yt}\psi(t)$ is an integrable function and belongs to $L_2^m(\mathbf{R})$ if $y > 0$. Hence

$$(7) \qquad \varphi(x + iy) = \frac{1}{\sqrt{2\pi}} \int_{-\infty}^{0} e^{-i(x+iy)t} \psi(t)\,dt$$

is well-defined for $y > 0$ and φ is analytic in the open upper half plane. For fixed $y > 0$ the function $\varphi(\cdot + iy)$ is the Fourier transform of $e^{yt}\psi(t)$, and hence

$$\int_{-\infty}^{\infty} \|\varphi(x+iy)\|^2 dx = \int_{-\infty}^{0} \|e^{yt}\psi(t)\|^2 dt$$

$$\le \|\psi\|^2 = \|\varphi\|^2.$$

In particular,

$$(8) \qquad \rho := \sup_{y>0}\left(\int_{-\infty}^{\infty} \|\varphi(x+iy)\|^2 dx\right)^{1/2} < \infty.$$

Conversely, by the Paley-Wiener theorem (see [R], Theorem 19.2), if φ is a $\mathbf{C}^m$-valued function which is analytic in the open upper half plane and satisfies (8), then there exists a unique $\psi \in L_2^m((-\infty, 0])$ such that (7) holds and $\|\psi\| = \rho$. Thus via formulas (6) and (7) the space $H_2^m(\mathbf{R})$ may be identified with the space of all functions that are analytic on the open upper half plane, take values in $\mathbf{C}^m$ and satisfy the uniform square-integrability condition (8). In that case the norm of a function $\varphi \in H_2^m(\mathbf{R})$ is also given by the right hand side of (8). The space $H_2^m(\mathbf{R})$ is called the *Hardy space* of square-integrable $\mathbf{C}^m$-valued functions on $\mathbf{R}$.

THEOREM 2.1. *Let T be the Wiener-Hopf operator on $L_2^m(\mathbf{R})$ with symbol W, and let U be the Fourier transform. Then*

$$(9) \qquad U^{-1}TU\varphi = \mathbf{P}M_W\varphi, \qquad \varphi \in H_2^m(\mathbf{R}).$$

Here M_W is the operator of multiplication by W and $\mathbf{P}$ is the orthogonal projection of $L_2^m(\mathbf{R})$ onto $H_2^m(\mathbf{R})$.

PROOF. From formula (9) in the previous section we know that $U^{-1} = \mathcal{I}U = U\mathcal{I}$, where

$$(10) \qquad \mathcal{I}: L_2^m(\mathbf{R}) \to L_2^m(\mathbf{R}), \qquad (\mathcal{I}\varphi)(t) = \varphi(-t).$$

Take $\varphi \in H_2^m(\mathbf{R})$. Thus $\varphi = U\mathcal{I}\psi$ for some $\psi \in L_2^m([0,\infty))$. It follows that

$$U\varphi = \mathcal{I}U^{-1}\varphi = \mathcal{I}^2\psi = \psi \in L_2^m([0,\infty)).$$

In this way one proves that $UH_2^m(\mathbf{R}) = L_2^m([0,\infty))$. Since U is unitary, we see that $\mathbf{P} = U^{-1}QU$, where Q is the (orthogonal) projection of $L_2^m(\mathbf{R})$ onto $L_2^m([0,\infty))$ along $L_2^m((-\infty, 0])$.

Again, take $\varphi \in H_2^m(\mathbf{R})$. Then

$$U^{-1}TU\varphi = U^{-1}QAU\varphi = \mathbf{P}U^{-1}AU\varphi,$$

where $A = I - L$ with L as in (4). From Theorem 1.1 we know that $U^{-1}AU = M_W$, which proves (9). $\square$

THEOREM 2.2. *Let T be the Wiener-Hopf operator on $L_2^m([0,\infty))$ with symbol W. Then*

$$(11) \qquad \|T\| = \sup\{\|W(t)\| \mid t \in \mathbf{R}\}.$$

PROOF. Let A be the convolution operator on $L_2^m(\mathbf{R})$ with symbol W. By Corollary 1.2 it is sufficient to show that $\|T\| = \|A\|$. We know (see formula (4)) that

$T\varphi = QA\varphi$ for each $\varphi \in L_2^m([0, \infty))$. Here Q is the orthogonal projection of $L_2^m(\mathbf{R})$ onto $L_2^m([0, \infty))$. Since $\|Q\| = 1$, it follows that $\|T\| \leq \|A\|$.

To prove the reverse inequality, let S_τ be the operator defined by

$$(12) \qquad (S_\tau\varphi)(t) = \varphi(t - \tau), \qquad -\infty < t < \infty.$$

For each $\tau \in \mathbf{R}$ the operator S_τ is a well-defined bounded linear operator on $L_2^m(\mathbf{R})$,

$$(13) \qquad \|S_\tau\varphi\| = \|\varphi\|, \qquad \varphi \in L_2^m(\mathbf{R}),$$

and S_τ is invertible. Furthermore, $S_\tau^* = S_{-\tau} = S_\tau^{-1}$ for all $\tau \in \mathbf{R}$. Write $A = I - L$ and $T = I - K$, where L is defined by (1.1) and K by (1). A simple computation shows that

$$(14) \qquad S_{-\tau}AS_\tau = I - S_{-\tau}LS_\tau = I - L = A, \qquad \tau \in \mathbf{R}.$$

Now, let $\mathcal{M}$ be the subset of $L_2^m(\mathbf{R})$ consisting of all $f \in L_2^m(\mathbf{R})$ with compact support. Thus, if $\varphi \in \mathcal{M}$, then there exists $a_\varphi \geq 0$ such that $\varphi(t) = 0$ for almost all $|t| \geq a_\varphi$. In that case $S_\tau\varphi$ has support in $[0, \infty)$ whenever $\tau > a_\varphi$, and thus $QS_\tau\varphi = S_\tau\varphi$ for $\tau > a_\varphi$. Take $\varphi, \psi \in \mathcal{M}$, and choose $\tau > 0$ sufficiently large. Then

$$\begin{aligned}
|\langle A\varphi, \psi \rangle| &= |\langle S_{-\tau}AS_\tau\psi, \varphi \rangle| \\
&= |\langle AS_\tau\varphi, S_\tau\psi \rangle| \\
&= |\langle TQS_\tau\varphi, QS_\tau\psi \rangle| \\
&\leq \|T\|\|QS_\tau\varphi\|\|QS_\tau\psi\| \leq \|T\|\|\varphi\|\|\psi\|.
\end{aligned}$$

Since $\mathcal{M}$ is dense in $L_2^m(\mathbf{R})$, this implies that

$$|\langle A\varphi, \psi \rangle| \leq \|T\|\|\varphi\|\|\psi\|, \qquad \varphi, \psi \in L_2^m(\mathbf{R}),$$

and hence $\|A\| \leq \|T\|$. $\quad\square$

The class of symbols defined by (5) is closed under the usual product of matrix functions. Thus, if W_1 and W_2 are symbols and $W(s) = W_1(s)W_2(s)$ for all $s \in \mathbf{R}$, then there exists an $m \times m$ matrix function k of which the entries are in $L_1(\mathbf{R})$ such that (5) holds. This statement follows from the fact that $L_1(\mathbf{R})$ is closed under the convolution product and the fact that (modulo a suitable normalization) the Fourier transform of a convolution product is the product of the Fourier transforms.

THEOREM 2.3. *Let T_1 and T_2 be the Wiener-Hopf operators with symbols W_1 and W_2, respectively, and let T be the Wiener-Hopf operator with symbol W, where $W(s) = W_1(s)W_2(s)$ for all $s \in \mathbf{R}$. Then $T - T_1T_2$ is a compact operator. Furthermore, if W_1 or W_2 is a rational matrix function, then $T - T_1T_2$ has finite rank.*

For the proof of Theorem 2.3 we need the following two lemmas.

LEMMA 2.4. *Let $k = [k_{ij}]_{i,j=1}^m$ be an $m \times m$ matrix function of which the entries are integrable on $[0, \infty)$, and let H be the operator on $L_2^m(\mathbf{R})$ defined by*

$$(15) \qquad (H\varphi)(t) = \int_0^\infty k(t + s)\varphi(s)\,ds, \qquad t \geq 0.$$

Then H is compact and

$$(16) \qquad \|H\| \le \left(\sum_{i=1}^{m} \sum_{j=1}^{m} \|k_{ij}\|_1^2 \right)^{1/2},$$

where $\|k_{ij}\|_1$ is the norm of k_{ij} as an element of $L_1([0,\infty))$.

PROOF. Let L on $L_2^m(\mathbf{R})$ be defined by (1.1), let Q be the projection of $L_2^m(\mathbf{R})$ onto $L_2^m([0,\infty))$ along $L_2^m((-\infty,0])$, and let $\mathcal{I}$ on $L_2^m(\mathbf{R})$ be as in formula (10). Then

$$(17) \qquad H\varphi = QL(I - Q)(\mathcal{I}\varphi), \qquad \varphi \in L_2^m([0,\infty)).$$

It follows that H is a well-defined bounded linear operator on $L_2^m([0,\infty))$, and the right hand side of the inequality (1.3) is an upper bound for $\|H\|$, which implies (16).

Since $L_2^m([0,\infty))$ is a Cartesian product of m copies of $L_2([0,\infty))$, the operator H may be represented by the $m \times m$ operator matrix

$$(18) \qquad H = \left[\begin{array}{ccc} H_{11} & \cdots & H_{1m} \\ \vdots & & \vdots \\ H_{m1} & \cdots & H_{mm} \end{array} \right],$$

where H_{ij} is the operator on $L_2([0,\infty))$ defined by

$$(H_{ij}f)(t) = \int_0^\infty k_{ij}(t+s)f(s)ds, \qquad t \ge 0.$$

To prove that H is compact, it suffices to prove that each of the entries H_{ij} is compact. In other words to prove the compactness we may, without loss of generality, assume that $m = 1$.

Take $k(t) = e^{-t}p(t)$, where $p(t) = a_0 + a_1 t + \cdots + a_n t^n$ is a polynomial with complex coefficients. Then

$$(H_p f)(t) := \int_0^\infty e^{-(t+s)}p(t+s)f(s)dt$$

$$= \sum_{j=0}^{n} \sum_{\nu=0}^{n} \binom{n}{\nu} a_j e^{-t} t^{n-\nu} \int_0^\infty e^{-s} s^\nu f(s)ds, \qquad t \ge 0.$$

It follows that rank H_p is finite. Since the functions $e^{-t}p(t)$, p arbitrary complex polynomial, are dense in $L_1([0,\infty))$ (see B. Sz-Nagy [3], items 7.3.2 and 7.3.3), we can find a sequence of complex polynomials $p_1, p_2, \ldots$ such that $e^{-t}p_n(t) \to k(t)$ in the norm of $L_1([0,\infty))$. But then we may apply inequality (16) to show that $\|H - H_{p_n}\| \to 0$ if

$n \to \infty$. It follows that H is the limit in the operator norm of a sequence of finite rank operators. Thus H is compact. $\square$

The operator H defined by (15) is called the *Hankel operator* on $L_2^m([0,\infty))$ with kernel function k. Assume that the entries of k are functions of the form

$$(19) \qquad t^n e^{-i\alpha t} \qquad (\Im\alpha < 0).$$

Then the argument used in the last part of the proof of Lemma 2.4 shows that each of the entries H_{ij} in (18) is an operator of finite rank, and thus H is an operator of finite rank.

LEMMA 2.5. *Let A be the convolution operator with symbol W. Write A as a 2×2 operator matrix relative to the decomposition* (3):

$$(20) \qquad A = \begin{bmatrix} A_{11} & A_{12} \\ A_{21} & A_{22} \end{bmatrix}.$$

Then A_{11} is the Wiener-Hopf operator with symbol W and the operators A_{12} and A_{21} are compact. If, in addition, W is a rational matrix function, then A_{12} and A_{21} have finite rank.

PROOF. Formula (4) implies that $A_{11} = T$ is the Wiener-Hopf operator with symbol W. Lemma 2.3 and formula (17) imply that A_{12} is compact. By changing t to $-t$ one proves in a similar way that A_{21} is compact.

Recall that a matrix function is rational if each of its entries is a rational function, i.e., the quotient of two polynomials. Note that

$$(21) \qquad (\lambda - \alpha)^{-(n+1)} = \frac{(-i)^{n+1}}{n!} \int_0^\infty t^n e^{-i\alpha t} e^{i\lambda t} dt \qquad (\Im\alpha < 0),$$

$$(22) \qquad (\lambda - \alpha)^{-(n+1)} = -\frac{(-i)^{n+1}}{n!} \int_{-\infty}^0 t^n e^{-i\alpha t} e^{i\lambda t} dt \qquad (\Im\alpha > 0).$$

Since a symbol is continuous on the real line, the entries of a rational symbol have no poles on $\mathbf{R}$. Furthermore, by the continuous analogue of the Riemann-Lebesgue lemma (see [R], Theorem 9.6) $W(\lambda) \to I_m$ if $\lambda \to \infty$. It follows, by the method of partial fractional expansion, that the entries of $W(\lambda)$ are functions of the form

$$c + \sum_{j=1}^r \sum_{n=1}^{q_j} c_{ij}(\lambda - \alpha_j)^{-(n+1)},$$

where c and c_{jn} are constants and $\Im\alpha_j \neq 0$ for $j = 1,\ldots,r$. So the equalities (21) and (22) imply that a symbol W is rational if and only if the entries of its kernel function k are finite linear combinations of functions of the form

$$(23) \qquad h(t) = \begin{cases} a_1 t^{n_1} e^{-i\alpha_1 t}, & t > 0, \\ a_2 t^{n_2} e^{-i\alpha_2 t}, & t < 0, \end{cases}$$

where a_1, a_2 are arbitrary complex numbers, $\Im \alpha_1 < 0$ and $\Im \alpha_2 > 0$. Thus, if W is rational, then the result stated in the paragraph preceding the present lemma can be used to show that A_{12} and A_{21} are operators of finite rank. $\square$

PROOF OF THEOREM 2.3. Let A, A_1 and A_2 be the convolution operators on $L_2^m(\mathbf{R})$ with symbols W, W_1 and W_2, respectively. Write A, A_1 and A_2 as 2×2 operator matrices relative to the decomposition (3):

$$A = \begin{bmatrix} A_{11} & A_{12} \\ A_{21} & A_{22} \end{bmatrix}, \qquad A_\nu = \begin{bmatrix} A_{11}^{(\nu)} & A_{12}^{(\nu)} \\ A_{21}^{(\nu)} & A_{22}^{(\nu)} \end{bmatrix}, \qquad \nu = 1, 2.$$

Then (Lemma 2.5) we have $T = A_{11}$, $T_1 = A_{11}^{(1)}$ and $T_2 = A_{11}^{(2)}$. From Theorem 1.1 we know that $A = A_1 A_2$, and hence

$$(24) \qquad\qquad\qquad\qquad T = T_1 T_2 + A_{12}^{(1)} A_{21}^{(2)}.$$

By Lemma 2.5 the operators $A_{12}^{(1)}$ and $A_{21}^{(2)}$ are compact, and hence $T - T_1 T_2$ is compact.

Next, assume that W_1 or W_2 is rational. Then we know from Lemma 2.5 that $A_{12}^{(1)}$ or $A_{21}^{(2)}$ is finite rank. In either case the product $A_{12}^{(1)} A_{21}^{(2)}$ has finite rank, which proves the theorem. $\square$

COROLLARY 2.6. *Let T_1 and T_2 be Wiener-Hopf operators on $L_2([0, \infty))$ with (scalar) symbols W_1 and W_2. Then $T_1 T_2 - T_2 T_1$ is a compact operator. If, in addition, W_1 or W_2 is rational, then $T_1 T_2 - T_2 T_1$ has finite rank.*

PROOF. Let T be the Wiener-Hopf operator with symbol $W(\cdot) = W_1(\cdot)W_2(\cdot)$. Since $m = 1$, we have $W_1 W_2 = W_2 W_1$, and thus T is also the Wiener-Hopf operator with symbol $W_2(\cdot)W_1(\cdot)$. It follows that $T - T_1 T_2$ and $T - T_2 T_1$ are both compact (by Theorem 2.3), and hence the difference $T_1 T_2 - T_2 T_1$ is also compact. The second statement is proved in a similar way. $\square$

XII.3 THE FREDHOLM INDEX

In this section we derive the first Fredholm theorem for Wiener-Hopf operators.

THEOREM 3.1. *Let T be the Wiener-Hopf operator on $L_2^m([0, \infty))$ with symbol W. Then T is a Fredholm operator if and only if $\det W(s) \neq 0$ for all $s \in \mathbf{R}$, and in that case the index of T is equal to the negative of the winding number relative to the origin of the curve parametrized by the function*

$$(1) \qquad\qquad\qquad t \mapsto \det W(t), \qquad -\infty \leq t \leq \infty.$$

Before we prove the theorem, first a few remarks about the winding number. Let α be a continuous function on $\mathbf{R}$ such that $\lim_{|t| \to \infty} \alpha(t)$ exists, and assume that the closed curve Γ_α parametrized by $t \mapsto \alpha(t)$, $-\infty \leq t \leq \infty$, does not pass through the origin. The *winding number* $\kappa(\Gamma_\alpha; 0)$ of Γ_α relative to the origin is, by definition, equal to $\frac{1}{2\pi}$ times the total variation of the argument function $\arg(\alpha(t))$ when the variable t varies over $-\infty \leq t \leq \infty$, i.e.,

$$(2) \qquad \kappa(\Gamma_\alpha; 0) := \frac{1}{2\pi}[\arg\alpha(t)]_{t=-\infty}^{\infty}.$$

If, in addition, α is piecewise continuously differentiable, then the winding number is also given by

$$(3) \qquad \kappa(\Gamma_\alpha; 0) = \frac{1}{2\pi i} \int_{\Gamma_\alpha} \frac{1}{\lambda} d\lambda.$$

For example, if $\tau(t) = (t - a)(t - b)^{-1}$ with a and b non-real, then

$$(4) \qquad \kappa(\Gamma_\tau; 0) = \begin{cases} 0 & \text{for} \quad (\Im a)(\Im b) > 0, \\ 1 & \text{for} \quad \Im a > 0, \ \Im b < 0, \\ -1 & \text{for} \quad \Im a < 0, \ \Im b > 0. \end{cases}$$

Let β be another continuous function on $\mathbf{R}$ such that $\lim_{|t|\to\infty} \beta(t)$ exists. If

$$|\beta(t) - \alpha(t)| < |\alpha(t)|, \qquad -\infty \leq t \leq \infty,$$

then the curve Γ_β parametrized by β does not meet the origin and $\kappa(\Gamma_\beta; 0) = \kappa(\Gamma_\alpha; 0)$. For piecewise continuously differentiable functions α and β this follows from [R], Theorem 10.35, and for arbitrary continuous functions it is proved by an approximation argument.

PROOF OF THEOREM 3.1. We split the proof into seven parts. The first part contains a general remark.

PART (i). Assume $m = 1$, and let A be the convolution operator on $L_2(\mathbf{R})$ with symbol W. Let

$$(5) \qquad A = \begin{bmatrix} A_{11} & A_{12} \\ A_{21} & A_{22} \end{bmatrix}$$

be the 2×2 operator matrix representation of A relative to the decomposition

$$(6) \qquad L_2(\mathbf{R}) = L_2([0, \infty)) \oplus L_2((-\infty, 0]).$$

We claim that the Banach conjugate A_{11}' of A_{11} is similar to A_{22}. To prove this, take φ in $L_2((-\infty, 0])$, and let $V\varphi$ be the continuous linear functional on $L_2([0, \infty))$ defined by

$$(V\varphi)(\psi) = \int_0^\infty \psi(t)\varphi(-t)dt, \qquad \psi \in L_2([0, \infty)).$$

Obviously, V is an isometry from $L_2((-\infty, 0])$ onto $L_2([0, \infty))'$. For $\varphi \in L_2((-\infty, 0])$

and $\psi \in L_2([0,\infty))$ we have

$$
\begin{aligned}
(A'_{11}V\varphi)(\psi) &= (V\varphi)(A_{11}\psi) \\
&= \int_0^\infty (A_{11}\psi)(t)\varphi(-t)dt \\
&= \int_0^\infty \psi(t)\varphi(-t)dt - \int_0^\infty \left(\int_0^\infty k(t-s)\psi(s)ds \right) \varphi(-t)dt \\
&= \int_0^\infty \psi(s)\varphi(-s)ds - \int_0^\infty \left(\int_0^\infty k(t-s)\varphi(-t)dt \right) \psi(s)ds \\
&= \int_0^\infty \psi(s)\left\{ \varphi(-s) - \int_{-\infty}^0 k(-s-t)\varphi(t)dt \right\} ds \\
&= \int_0^\infty \psi(s)(A_{22}\varphi)(-s)ds = (VA_{22}\varphi)\psi.
\end{aligned}
$$

The change in the order of the integrals is justified by Fubini's theorem. The above calculation shows that $A'_{11}V = VA_{22}$, and hence A_{22} is similar to A'_{11}.

PART (ii). Assume that $\det W(t) \neq 0$ for all $t \in \mathbf{R}$. We shall prove that T is a Fredholm operator. Note that Theorem 1.4 implies that $W(\cdot)^{-1}$ is of the form

$$
W(s)^{-1} = I_m - \int_0^\infty e^{ist}k^\times(t)dt, \qquad s \in \mathbf{R},
$$

where $k^\times$ is an $m \times m$ matrix function with entries in $L_1(\mathbf{R})$. Let $T^\times$ be the Wiener-Hopf operator with symbol $W(\cdot)^{-1}$, i.e.,

$$
(7) \qquad (T^\times\varphi)(t) = \varphi(t) - \int_0^\infty k^\times(t-s)\varphi(s)ds, \qquad t \geq 0.
$$

Since $W(\cdot)W(\cdot)^{-1} = W(\cdot)^{-1}W(\cdot) = I$, Theorem 2.3 implies that $I - TT^\times$ and $I - T^\times T$ are compact. Hence T is invertible modulo the compact operators, and thus T is a Fredholm operator.

PART (iii). Take $m = 1$, and assume that T is Fredholm. We shall prove that $\det W(t) \neq 0$ for all $t \in \mathbf{R}$. Consider the representation (5). We know (see Lemma 2.5) that $A_{11} = T$ and thus A_{11} is Fredholm. By duality, the same is true for A'_{11} (apply Theorem XI.5.1 and use that the conjugate of a finite rank operator has again finite rank). Since $m = 1$, the operators A'_{11} and A_{22} are similar (see Part (i)). It follows that

$$
\begin{bmatrix} A_{11} & 0 \\ 0 & A_{22} \end{bmatrix}
$$

is a Fredholm operator. According to Lemma 2.5 the operators A_{12} and A_{21} are compact, and hence A is the sum of a Fredholm operator and a compact operator. So A is also Fredholm, and we can apply Corollary 1.5 to show that A is invertible, which implies $\det W(t) \neq 0$ for all $t \in \mathbf{R}$ (by Theorem 1.4).

PART (iv). Assume that T is Fredholm. We shall prove that $\det W(t) \neq 0$ for all $t \in \mathbf{R}$. Since $L_2^m([0,\infty))$ is the Cartesian product of m copies of $L_2([0,\infty))$ we may represent T as an $m \times m$ operator matrix:

$$
(8) \qquad T = \begin{bmatrix} T_{11} & \cdots & T_{1m} \\ \vdots & & \vdots \\ T_{m1} & \cdots & T_{mm} \end{bmatrix}.
$$

Here $T_{ij} = \delta_{ij} I - K_{ij}$, where δ_{ij} is the Kronecker delta and

$$
(9) \qquad (K_{ij}f)(t) = \int_0^\infty k_{ij}(t-s)f(s)ds, \qquad t \geq 0.
$$

The function k_{ij} is the (i,j)-th entry of the kernel function k of T. Note that $I - K_{ij}$ are Wiener-Hopf operators on $L_2([0,\infty))$ with (scalar) symbols, and hence we can apply Corollary 2.6 to show that the entries T_{ij} in (8) commute with one another modulo the compact operators. So Theorem XI.7.3 implies that $\det T$ is a Fredholm operator.

Since our class of symbols is closed under the product of (matrix) functions, $\det W$ is again a symbol. Let $T_{\det W}$ be the corresponding Wiener-Hopf operator (acting on $L_2([0,\infty))$). A repeated application of Theorem 2.3 implies that $\det T - T_{\det W}$ is a compact operator. It follows that $T_{\det W}$ is a perturbation of $\det T$ by a compact operator, and hence $T_{\det W}$ is a Fredholm operator. By Part (iii) of the proof this implies that its symbol does not vanish on the real line, that is, $\det W(t) \neq 0$ for all $t \in \mathbf{R}$.

PART (v). It remains to prove the formula for the index. To do this we first show that, without loss of generality, we may assume that the symbol W is rational. Consider functions of the form

$$
(10) \qquad h(t) = \begin{cases} a_1 t^{n_1} e^{-i\alpha_1 t}, & t > 0, \\ a_2 t^{n_2} e^{-i\alpha_2 t}, & t < 0, \end{cases}
$$

where a_1, a_2 are arbitrary complex numbers, $\Im \alpha_1 < 0$ and $\Im \alpha_2 > 0$. The linear span of functions h of the form (10) is dense in $L_1(\mathbf{R})$. So we can find a sequence $k_1, k_2, \ldots$ of $m \times m$ matrix functions, $k_n = [(k_n)_{ij}]_{i,j=1}^m$, such that

$$
(11) \qquad \sum_{i=1}^m \sum_{j=1}^m \| k_{ij} - (k_n)_{ij} \|_1^2 \to 0 \qquad (n \to \infty),
$$

and for each n the (i,j)-th entry of k_n is a finite linear combination of functions h of the form (10). In (11) the function k_{ij} is the (i,j)-th entry of the kernel function k of T. Put

$$
(T_n \varphi)(t) = \varphi(t) - \int_0^\infty k_n(t-s)\varphi(s)ds, \qquad t \geq 0,
$$

$$W_n(s) = I_m - \int_{-\infty}^{\infty} e^{ist} k_n(t) dt, \qquad s \in \mathbf{R}.$$

From (11) it follows (cf. formula (2.2)) that $\|T - T_n\| \to 0$ if $n \to \infty$, and hence, by Theorem 2.2,

$$(12) \qquad \sup_{t \in \mathbf{R}} \|W(t) - W_n(t)\| \to 0 \qquad (n \to \infty).$$

The special form of the entries of k_n imply (see the paragraph preceding Lemma 2.5) that the symbols $W_1(\lambda), W_2(\lambda), \ldots$ are rational matrix functions.

Now assume that T is Fredholm (and thus $\det W(t) \neq 0$ for all $t \in \mathbf{R}$). Since $T_n \to T$ in the operator norm, there exists a positive integer n_0 such that for $n \geq n_0$ the operator T_n is Fredholm and $\operatorname{ind} T_n = \operatorname{ind} T$. From (12) it follows that there exists $n_1 \geq n_0$ such that

$$(13) \qquad |\det W_n(t) - \det W(t)| < |\det W(t)|, \qquad t \in \mathbf{R},$$

for each $n \geq n_1$. Note that $\det W(\pm\infty) = \det W_n(\pm\infty) = 1$ for all n. Thus (13) implies (see the paragraph preceding the present proof) that for $n \geq n_1$ the curve parametrized by

$$(14) \qquad t \mapsto \det W_n(t), \qquad -\infty \leq t \leq \infty,$$

does not pass through the origin and the winding number relative to the origin of this curve is equal to the winding number relative to the origin of the curve parametrized by (1). We conclude that it suffices to derive the formula for the index for the case when the symbol is rational.

PART (vi). Assume that T is Fredholm, and let its symbol W be a rational matrix function. In this part we shall show that it suffices to derive the formula of the index for the case when $m = 1$. Again consider the representation (8). Since the symbol is rational, Corollary 2.6 implies that the entries T_{ij} in (8) commute with one another modulo the operators of finite rank. But then we can use Theorem XI.7.6 to show that $detT$ is Fredholm and $\operatorname{ind} T = \operatorname{ind}(detT)$.

Let $T_{\det W}$ be the Wiener-Hopf operator with symbol $\det W$. We already proved that $detT - T_{\det W}$ is a compact operator. Thus $T_{\det W}$ is a Fredholm operator and

$$\operatorname{ind} T = \operatorname{ind}(detT) = \operatorname{ind} T_{\det W}.$$

Note that $\det W$ is a rational function. So it suffices to derive the formula of the index for the case when $m = 1$ and W rational.

PART (vii). Now take $m = 1$, and let T be a Wiener-Hopf operator on $L_2([0, \infty))$ with a rational symbol ω. We assume that T is a Fredholm operator (and thus $\omega(t) \neq 0$ for all $t \in \mathbf{R}$), and we shall compute the index of T. Write $\omega(s) = p(s)/q(s)$, where p and q are scalar polynomials with no common zeros. By the continuous analogue of the Riemann-Lebesgue lemma, $\omega(\lambda) \to 1$ if $|\lambda| \to \infty$, and hence the degree of p is equal to the degree of q. Furthermore, p and q have no zeros on the real line, because ω has no poles on $\mathbf{R}$ and $\omega(t) \neq 0$ for $t \in \mathbf{R}$. Let n_+ (resp. k_+) be the number of zeros of p (resp. q) in the open upper half plane, and let n_- (resp. k_-) be the number of zeros of p (resp. q) in the open lower half plane. Then $n_+ + n_- = k_+ + k_-$, and hence we may put

$$(15) \qquad r := n_+ - k_+ = -(n_- - k_-).$$

It follows that ω may be factored as

$$(16) \qquad \omega(s) = \left(\frac{\lambda - i}{\lambda + i}\right)^r \cdot \prod_{j=1}^{\ell_+} \frac{(\lambda - a_j^+)}{(\lambda - b_j^+)} \cdot \prod_{j=1}^{\ell_-} \frac{(\lambda - a_j^-)}{(\lambda - b_j^-)},$$

where a_j^+, b_j^+ are in open upper half plane and a_j^-, b_j^- are in the open lower half plane. Put $\tau(s) = (s - a)(s - b)^{-1}$ with a, b non-real, and let S_τ be the Wiener-Hopf operator on $L_2([0, \infty))$ with symbol τ. From formulas (2.21) and (2.22) we see that

$$(17) \qquad (S_\tau f)(t) = \begin{cases} f(t) - i(b - a) \int_0^t e^{-ib(t-s)} f(s)ds, & \text{if } \Im b < 0, \\[2mm] f(t) + i(b - a) \int_t^\infty e^{-ib(t-s)} f(s)ds, & \text{if } \Im b > 0. \end{cases}$$

Let $\sigma(\lambda) = (\lambda - b)(\lambda - a)^{-1} = \tau(\lambda)^{-1}$, and let S_σ be the Wiener-Hopf operator on $L_2([0, \infty))$ with symbol σ. One obtains $S_\sigma f$ by interchanging a and b in the right hand side of (17). A direct computation shows that

$$(18) \qquad S_\sigma S_\tau = I, \qquad \Im b < 0,$$

$$(19) \qquad \operatorname{Ker} S_\tau = \{ce^{-iat} \mid c \in \mathbf{C}\}, \qquad \Im b > 0.$$

$$(20) \qquad (S_\tau)^* = S_{\bar\tau}, \qquad (S_\sigma)^* = S_{\bar\sigma}.$$

It follows that

$$n(S_\tau) = d(S_\tau) = 0, \qquad (\Im a)(\Im b) > 0,$$

$$n(S_\tau) = 0 \quad d(S_\tau) = 1, \qquad \Im a > 0, \quad \Im b < 0,$$

$$n(S_\tau) = 1, \quad d(S_\tau) = 0, \qquad \Im a < 0, \quad \Im b > 0.$$

Let κ_τ be the winding number relative to the origin of the curve parametrized by $t \to \tau(t)$. By comparing the above formulas for $n(S_\tau)$ and $d(S_\tau)$ with the expression (4) we see that $\operatorname{ind} S_\tau = -\kappa_\tau$. From the definition of the winding number in (2) it follows that the winding number corresponding to a product of functions is equal to the sum of the winding numbers corresponding to the factors. So, by repeatedly using (4), we obtain from (16) that the winding number (relative to the origin) of the curve parametrized by $t \mapsto \omega(t)$ is equal to r. Furthermore, by repeatedly applying Corollary 2.6, we see that the index of T is the index of a product of Fredholm operators of the form S_τ. Since the index of a product is the sum of the indices of the separate factors, we conclude that $\operatorname{ind} T = -r$. $\quad\square$

CHAPTER XIII
WIENER-HOPF INTEGRAL OPERATORS WITH RATIONAL SYMBOLS

In this chapter we study in more detail Wiener-Hopf integral operators with a rational matrix symbol. The technique of Wiener-Hopf factorization is introduced. The fact that the symbols are rational allows us to represent them in a special way. We use this representation to construct explicitly the factors in a canonical Wiener-Hopf factorization. In this way explicit formulas for the inverse and the Fredholm characteristics are obtained. Also convolution operators on a finite interval are analyzed in terms of the special representation of the symbol. An example from linear transport theory illustrates the general theory.

XIII.1 PRELIMINARIES AND SCALAR CASE

Let W be a rational matrix function. Thus each of the entries of W is the quotient of two polynomials. The expression "W has no poles in the set Σ" will mean that none of the entries of W has a pole in Σ. In that case W is analytic on Σ. In what follows $\mathcal{R}_\infty^{m \times m}(\mathbf{R})$ denotes the set of all rational $m \times m$ matrix functions that have no poles on the real line and at infinity. The latter means that

$$(1) \qquad W(\infty) := \lim_{\lambda \to \infty} W(\lambda)$$

exists.

Let T be a Wiener-Hopf operator on $L_2^m([0, \infty))$ with symbol W. Thus $T = I - K$, where

$$(2) \qquad (K\varphi)(t) = \int_0^\infty k(t - s)\varphi(s)ds, \qquad 0 \leq t < \infty,$$

$$(3) \qquad W(\lambda) = I_m - \int_{-\infty}^\infty e^{i\lambda t}k(t)dt.$$

Here k is an $m \times m$ matrix function whose entries belong to $L_1(\mathbf{R})$. Throughout this chapter we assume that W is rational. This implies that $W \in \mathcal{R}_\infty^{m \times m}(\mathbf{R})$ and $W(\infty) = I_m$. Indeed, (3) implies that W is continuous on $\mathbf{R}$ (and thus W has no poles on $\mathbf{R}$), and, by the continuous analogue of the Riemann-Lebesgue lemma, the limit (1) exists and is equal to I_m.

We shall use the fact that the converse statement is also true, i.e, if $W \in \mathcal{R}_\infty^{m \times m}(\mathbf{R})$ and $W(\infty) = I_m$, then there exists an $m \times m$ matrix function k with entries in $L_1(\mathbf{R})$ such that (3) holds. Recall that this fact was established in the proof of Lemma XII.2.5 by using the partial fraction expansion of the entries of W.

THEOREM 1.1. *Let T be the Wiener-Hopf operator on $L_2^m([0,\infty))$ with rational symbol W. Assume that W has no poles in $\Im\lambda \geq 0$ (resp. $\Im\lambda \leq 0$) and $\det W(\lambda) \neq 0$ for $\Im\lambda \geq 0$ (resp. $\Im\lambda \leq 0$). Then T is invertible and T^{-1} is the Wiener-Hopf operator with symbol $W(\cdot)^{-1}$.*

To prove Theorem 1.1 we need the following further refinement of Theorem XII.2.3.

PROPOSITION 1.2. *Let T_1 and T_2 be Wiener-Hopf operators on $L_2^m([0,\infty))$ with rational symbols W_1 and W_2, respectively, and let T be the Wiener-Hopf operator with symbol W, where $W(s) = W_1(s)W_2(s)$ for all $s \in \mathbf{R}$. Then $T = T_1T_2$, whenever W_1 has no poles in $\Im\lambda < 0$ or W_2 has no poles in $\Im\lambda > 0$.*

PROOF. We use the notation introduced in the proof of Theorem XII.2.3. Assume that $W_1(\lambda)$ has no poles in $\Im\lambda < 0$. We know that $W_1(\lambda)$ has no poles on the real line and at infinity. Thus the partial fraction expansions of the entries of $W_1(\lambda)$ and formula (22) in Section XII.2 tells us that

$$(4) \qquad W_1(\lambda) = I_m - \int\limits_{-\infty}^{0} e^{i\lambda t}k_1(t)dt,$$

where k_1 is an $m \times m$ matrix function with entries in $L_1(\mathbf{R})$ and $k_1(t) = 0$ for $t > 0$. By definition the operator $A_{12}^{(1)}$ appearing in formula (24) of Section XII.2 is given by

$$(A_{12}^{(1)}\varphi)(t) = -\int\limits_{-\infty}^{0} k_1(t-s)\varphi(s)ds, \qquad t \geq 0,$$

and hence $A_{12}^{(1)} = 0$. But then formula (24) of Section XII.2 implies that $T = T_1T_2$. The same result holds true when $W_2(\lambda)$ has no poles in $\Im\lambda > 0$, because then the operator $A_{21}^{(2)}$ in formula (24) of Section XII.2 is the zero operator. $\square$

PROOF OF THEOREM 1.1. Our hypotheses imply that $W(\cdot)^{-1} \in \mathcal{R}_\infty^{m\times m}(\mathbf{R})$ and $W(\infty)^{-1} = I_m$. Thus $W(\cdot)^{-1}$ is the symbol of a Wiener-Hopf operator $T^\times$, say. Note that either $W(\cdot)^{-1}$ has no poles in $\Im\lambda \geq 0$ or $W(\cdot)^{-1}$ has no poles in $\Im\lambda \leq 0$. Now apply Proposition 1.2, first with $W_1 = W$ and $W_2 = W(\cdot)^{-1}$, and next with $W_1 = W(\cdot)^{-1}$ and $W_2 = W$. It follows that $TT^\times$ and $T^\times T$ are equal to the identity operator on $L_2^m([0,\infty))$, and hence $T^\times = T^{-1}$. $\square$

To show the importance of Theorem 1.1, let us restrict the attention to the scalar case. So let T be a Wiener-Hopf operator on $L_2([0,\infty))$ with a scalar rational symbol ω. Assume $\omega(\lambda) \neq 0$ for $\lambda \in \mathbf{R}$. In Part (vii) of the proof of Theorem XII.3.1 we have shown (see formula (16) in Section XII.3) that ω may be represented in the form

$$(5) \qquad \omega(\lambda) = \omega_-(\lambda)\left(\frac{\lambda-i}{\lambda+i}\right)^{\kappa}\omega_+(\lambda), \qquad \lambda \in \mathbf{R},$$

where ω_+ (resp. ω_-) is a scalar rational function with no poles and zeros in $\Im\lambda \geq 0$ (resp. $\Im\lambda \leq 0$),

$$\omega_\pm(\infty) = \lim_{\lambda\to\infty} \omega_\pm(\lambda) = 1,$$

and κ is the winding number relative to the origin of the curve parametrized by $\omega(t)$ for $-\infty \leq t \leq \infty$. The representation (5) is called the *Wiener-Hopf factorization* of ω relative to the real line. In the next theorem we write T_φ for the Wiener-Hopf operator on $L_2([0,\infty))$ with symbol φ.

THEOREM 1.3. *Let T be the Wiener-Hopf operator on $L_2([0,\infty))$ with scalar rational symbol ω. Assume $\omega(t) \neq 0$ for $t \in \mathbf{R}$, and let*

$$\omega(\lambda) = \omega_-(\lambda)\left(\frac{\lambda-i}{\lambda+i}\right)^\kappa \omega_+(\lambda), \qquad \lambda \in \mathbf{R},$$

be the Wiener-Hopf factorization of ω relative to the real line. Then T is invertible if and only if $\kappa = 0$, and in that case

$$T^{-1} = T_{\omega_+^{-1}} \circ T_{\omega_-^{-1}}.$$

Furthermore, if $\kappa > 0$, then T is right invertible, $d(T) = \kappa$ and a right inverse of T is given by

$$T^+ = T_{\omega_+^{-1}} S_1^\kappa T_{\omega_-^{-1}}.$$

If $\kappa < 0$, then T is left invertible, $n(T) = -\kappa$ and a left inverse of T is given by

$$T^+ = T_{\omega_+^{-1}} (S_2)^{-\kappa} T_{\omega_-^{-1}}.$$

Here S_2 (resp. S_1) is the Wiener-Hopf operator on $L_2([0,\infty))$ with symbol $(\lambda-i)(\lambda+i)^{-1}$ (resp. $(\lambda+i)(\lambda-i)^{-1}$).

PROOF. Because of Proposition 1.2 we may write

$$(6) \qquad\qquad T = T_{\omega_-} T_\delta T_{\omega_+},$$

where T_δ is the Wiener-Hopf operator on $L_2([0,\infty))$ with symbol $\delta(\lambda) = [(\lambda-i)(\lambda+i)^{-1}]^\kappa$. By Theorem 1.1 the factors T_{ω_-} and T_{ω_+} are invertible operators with inverses $T_{\omega_-^{-1}}$ and $T_{\omega_+^{-1}}$, respectively. Proposition 1.2 yields

$$(7) \qquad\qquad T_\delta = \begin{cases} (S_1)^{-\kappa} & \text{for} \quad \kappa \leq 0, \\ S_2^\kappa & \text{for} \quad \kappa \geq 0. \end{cases}$$

From the formulas (18), (19) and (20) in Section XII.3 we see that

$$(8) \qquad\qquad S_1 S_2 = I, \qquad n(S_1) = 1, \qquad d(S_2) = 1.$$

Thus S_1 and S_2 are right and left invertible, respectively, and

$$(9a) \qquad\qquad n(S_1^k) = \mathrm{ind}(S_1^k) = k\,\mathrm{ind}(S_1) = k,$$

$$(9b) \qquad\qquad d(S_2^k) = -\operatorname{ind}(S_2^k) = -k\operatorname{ind}(S_2) = k,$$

for $k = 1, 2, \dots$. The equivalence relation (6) implies that T has the desired properties.
$\square$

Theorem 1.3 provides an effective method to find the inverse of a Wiener-Hopf operator on $L_2([0, \infty))$ with a scalar rational symbol. The aim of the next sections is to apply this method (if possible) to matrix symbols. The first step is to extend the notion of Wiener-Hopf factorization to rational matrix functions.

XIII.2 WIENER-HOPF FACTORIZATION

Recall that $\mathcal{R}_\infty^{m \times m}(\mathbf{R})$ is the set of all rational $m \times m$ matrix functions that have no poles on the real line and at infinity. The following theorem is the main result of this section.

THEOREM 2.1. *Let* $W \in \mathcal{R}_\infty^{m \times m}(\mathbf{R})$, *and assume that* $\det W(\lambda) \neq 0$ *for* $\lambda \in \mathbf{R} \cup \{\infty\}$. *Then there exist* W_+, W_- *in* $\mathcal{R}_\infty^{m \times m}(\mathbf{R})$ *and integers* $\kappa_1 \leq \kappa_2 \leq \cdots \leq \kappa_m$ *such that*

$$(1) \qquad W(\lambda) = W_-(\lambda)
\begin{bmatrix}
\left(\frac{\lambda-i}{\lambda+i}\right)^{\kappa_1} & & & \\
& \left(\frac{\lambda-i}{\lambda+i}\right)^{\kappa_2} & & \\
& & \ddots & \\
& & & \left(\frac{\lambda-i}{\lambda+i}\right)^{\kappa_m}
\end{bmatrix}
W_+(\lambda), \qquad \lambda \in \mathbf{R},$$

and

(j) W_+ *has no poles in* $\Im\lambda \geq 0$ *and* $\det W_+(\lambda) \neq 0$ *for* $\Im\lambda \geq 0$,

(jj) W_- *has no poles in* $\Im\lambda \leq 0$ *and* $\det W_+(\lambda) \neq 0$ *for* $\Im\lambda \leq 0$,

(jjj) $\det W_\pm(\infty) \neq 0$.

The factorization (1) is called a *right Wiener-Hopf factorization of* W *relative to the real line*. One obtains a *left Wiener-Hopf factorization* if in (2) the positions of the functions $W_+(\lambda)$ and $W_-(\lambda)$ are interchanged. In what follows we shall often omit the word "right". We shall show in the next section that the integers $\kappa_1, \dots, \kappa_m$ in (1) are uniquely determined by W; they are called the *(right) factorization indices*. If in (1) all indices $\kappa_1, \dots, \kappa_m$ are equal to zero, then (1) is said to be a *(right) canonical factorization*. In general (except for the scalar case), the factors W_+ and W_- are not uniquely determined by W.

For a scalar rational function an explicit construction of a Wiener-Hopf factorization has been given in the previous section. For the matrix case the standard construction of the Wiener-Hopf factorization (which we shall give below) does not yield explicit formulas for the factors W_+ and W_- nor for the indices, but only an algorithm which yields the factors and the indices in a finite number of steps. We shall come back to this in Section 6.

PROOF OF THEOREM 2.1. It will be convenient to pass from the real line to the unit circle by using the Möbius transformation

$$\eta(\zeta) = i\frac{1+\zeta}{1-\zeta}. \tag{2}$$

Put $\Phi(\zeta) = W\big(\eta(\zeta)\big)$. Then Φ is a rational $m \times m$ matrix function. Since η maps the unit circle T onto $\mathsf{R} \cup \{\infty\}$, the function Φ has no poles on T and $\det \Phi(\zeta) \neq 0$ for all $\zeta \in \mathsf{T}$. We shall show that Φ can be factored as

$$\Phi(\zeta) = \Phi_-(\zeta)\begin{bmatrix} \zeta^{\kappa_1} & & & \\ & \zeta^{\kappa_2} & & \\ & & \ddots & \\ & & & \zeta^{\kappa_m} \end{bmatrix}\Phi_+(\zeta), \qquad |\zeta| = 1, \tag{3}$$

where $\kappa_1 \leq \kappa_2 \leq \cdots \leq \kappa_m$ are integers and the functions Φ_+ and Φ_- are rational $m \times m$ matrix functions with no poles on T which have the following properties:

    (i) Φ_+ has no poles on $|\zeta| \leq 1$ and $\det \Phi_+(\zeta) \neq 0$ for $|\zeta| \leq 1$,

    (ii) Φ_- has no poles on $|\zeta| \geq 1$ and $\det \Phi_-(\zeta) \neq 0$ for $|\zeta| \geq 1$,

    (iii) Φ_- has no pole at infinity and $\det \Phi_-(\infty) \neq 0$.

Assume that the factorization has been established. As a map of the Riemann sphere $\mathsf{C} \cup \{\infty\}$ into itself the Möbius transformation η is bijective and the inverse map η^{-1} is given by $\eta^{-1}(\lambda) = (\lambda - i)(\lambda + i)^{-1}$. Put

$$W_+(\lambda) = \Phi_+\big(\eta^{-1}(\lambda)\big), \qquad W_-(\lambda) = \Phi_-\big(\eta^{-1}(\lambda)\big). \tag{4}$$

The functions W_+ and W_- have the desired properties. Indeed, η^{-1} maps the real line into the unit circle and $\eta^{-1}(\infty) = 1$. Thus, W_+, W_- belong to $\mathcal{R}_\infty^{m \times m}(\mathsf{R})$ and

$$\det W_\pm(\lambda) = \det \Phi_\pm\big(\eta^{-1}(\lambda)\big) \neq 0, \qquad \lambda \in \mathsf{R} \cup \{\infty\}.$$

Furthermore, since η^{-1} maps the open upper half plane onto the open unit disc $|\zeta| < 1$ and the open lower half plane onto the set $|\zeta| > 1$, the conditions (j) and (jj) are also fulfilled. The factorization (1) follows from (3) by replacing ζ by $(\lambda - i)(\lambda + i)^{-1}$. So it suffices to establish the factorization (3).

We shall refer to (3) as a *(right) Wiener-Hopf factorization relative to the circle.* For a scalar rational function φ which has no poles and zeros on T, the factorization (3) may be constructed by applying the inverse Möbius transformation η^{-1} and the Wiener-Hopf factorization result for scalar rational functions relative to the real line. A more direct argument goes as follows. Write $\varphi(\zeta)$ as a quotient $q_1(\zeta)/q_2(\zeta)$ of two polynomials which have no common zeros. Since φ has no poles and zeros on T, the polynomials q_1 and q_2 have no zeros on T. Thus we may write

$$q_1(\lambda) = c_1 \prod_{j=1}^{k^+}(\lambda - t_j^+) \prod_{j=1}^{k^-}(\lambda - t_j^-),$$

$$q_2(\lambda) = c_2 \prod_{j=1}^{\ell^+}(\lambda - \tau_j^+) \prod_{j=1}^{\ell^-}(\lambda - \tau_j^-),$$

where t_j^+ and τ_j^+ are inside the unit circle and the points t_j^- and τ_j^- are outside T. Put $\kappa = k^+ - \ell^+$ and

$$(5) \qquad \varphi_+(\zeta) = d\frac{\prod_{j=1}^{k^-}(\zeta - t_j^-)}{\prod_{j=1}^{\ell^-}(\zeta - \tau_j^-)}, \qquad \varphi_-(\zeta) = \frac{\prod_{j=1}^{k^+}(1 - \zeta^{-1}t_j^+)}{\prod_{j=1}^{\ell^+}(1 - \zeta^{-1}\tau_j^+)},$$

with $d = c_1/c_2$. Then

$$(6) \qquad \varphi(\zeta) = \varphi_-(\zeta)\zeta^k\varphi_+(\zeta), \qquad \zeta \in \mathsf{T},$$

and the factors φ_- and φ_+ have the properties described by (i), (ii) and (iii). Thus (6) is a Wiener-Hopf factorization of φ relative to the unit circle, and we are finished with the scalar case.

Next, we consider the general case. Let Φ be a rational $m \times m$ matrix function which has no poles on T, and assume that $\det \Phi(\zeta) \neq 0$ for $|\zeta| = 1$. Let q be a common multiple of all the denominators of Φ. Then $\Phi(\zeta) = \frac{1}{q(\zeta)}P(\zeta)$, where $P(\zeta)$ is an $m \times m$ matrix whose entries are polynomials in ζ. (In particular, P has no poles in $|\zeta| \leq 1$.) Since the entries of Φ have no poles on T, we may assume that q has no zeros on T, and hence we know (see the previous paragraph) that the scalar rational function $\varphi(\zeta) = q(\zeta)^{-1}$ admits a Wiener-Hopf factorization relative to T:

$$\varphi(\zeta) = \frac{1}{q(\zeta)} = \varphi_-(\zeta)\zeta^\kappa\varphi_+(\zeta), \qquad \zeta \in \mathsf{T}.$$

Now assume that P admits a Wiener-Hopf factorization relative to T:

$$(7) \qquad P(\zeta) = P_-(\zeta)\big([\zeta^{n_j}\delta_{ij}]_{i,j=1}^m\big)P_+(\zeta), \qquad \zeta \in \mathsf{T}.$$

Then

$$\Phi(\zeta) = \big(\varphi_-(\zeta)P_-(\zeta)\big)\big([\zeta^{\kappa+n_j}\delta_{ij}]_{i,j=1}^m\big)\big(\varphi_+(\zeta)P_+(\zeta)\big), \qquad \zeta \in \mathsf{T},$$

is a Wiener-Hopf factorization of Φ relative to T. So we have to prove (7). In other words, it suffices to prove the factorization for the case when the entries of $\Phi(\zeta)$ are polynomials in ζ.

To simplify the proof, let us say that two rational $m \times m$ matrix functions Φ_1 and Φ_2 are *(left) strictly equivalent* if $\Phi_1(\zeta) = E(\zeta)\Phi_2(\zeta)$, where E is a rational $m \times m$ matrix function such that

(α) E has no poles on $|\zeta| \geq 1$ and $\det E(\zeta) \neq 0$ for $|\zeta| \geq 1$,

(β) E has no pole at infinity and $\det E(\infty) \neq 0$.

We have to prove that Φ is strictly equivalent to a function of the form

$$\big([\zeta^{\kappa_j}\delta_{ij}]_{i,j=1}^m\big)\Phi_+(\zeta),$$

where Φ_+ is as in (i).

Let $\Phi(\zeta)$ be an $m \times m$ matrix whose entries are polynomials in ζ, and assume $\det \Phi(\zeta) \neq 0$ for $|\zeta| = 1$. Let $\tau_1, \ldots, \tau_q$ be the zeros of $\det \Phi(\zeta)$ in $|\zeta| < 1$, and let $\ell_1, \ldots, \ell_q$ be the corresponding multiplicities. We assume that $\tau_q = 0$, and put $\ell_q = 0$ if τ_q is not a zero for $\det \Phi(\cdot)$. Consider the j-th row of $\Phi(\zeta)$:

$$f_j(\zeta) = [\varphi_{j1}(\zeta) \cdots \varphi_{jm}(\zeta)].$$

Let p_j be the minimum of the multiplicities of τ_1 as a zero of the polynomials $\varphi_{j1}(\zeta), \ldots, \varphi_{jm}(\zeta)$. We call p_j the *j-th row multiplicity* of τ_1. Obviously, $p_1 + \cdots + p_m \leq \ell_1$. If $\sum_{j=1}^{m} p_j < \ell_1$, then

$$\det[(\zeta - \tau_1)^{-p_j} \varphi_{jk}(\zeta)]_{j,k=1}^{m}$$

is zero in $\zeta = \tau_1$. So there exists complex numbers $c_1, \ldots, c_m$, not all equal to zero, such that

$$f(\zeta) = \sum_{j=1}^{m} c_j (\zeta - \tau_1)^{-p_j} f_j(\zeta)$$

has a zero in τ_1. Choose $c_r \neq 0$ such that $p_r \leq p_j$ when $c_j \neq 0$. Consider

$$E_r(\zeta) = \begin{bmatrix} 1 & & & & & & & \\ & \ddots & & & & & & \\ & & 1 & & & & & \\ \frac{c_1}{(\zeta-\tau_1)^{p_1-p_r}} & \cdots & & c_r & \cdots & & \frac{c_m}{(\zeta-\tau_1)^{p_m-p_r}} & \\ & & & & 1 & & & \\ & & & & & \ddots & & \\ & & & & & & 1 \end{bmatrix}.$$

Note that $\det E_r(\zeta) = c_r$ for $\zeta \neq \tau_1$. Since $c_r \neq 0$, it follows that for $E = E_r$ the conditions (α) and (β) are fulfilled. Thus multiplication on the left by E_r is a strict equivalence operation. Observe that $E_r(\zeta)\Phi(\zeta)$ is obtained from $\Phi(\zeta)$ by replacing the r-th row in $\Phi(\zeta)$ by $(\zeta - \tau_1)^{p_r} f(\zeta)$. This implies that the entries of $E_r(\zeta)\Phi(\zeta)$ are again polynomials. From

$$\det\big(E_r(\zeta)\Phi(\zeta)\big) = c_r \det \Phi(\zeta), \qquad \zeta \neq \tau_1,$$

we may conclude that in $|\zeta| < 1$ the zeros of $\det\big(E_r(\zeta)\Phi(\zeta)\big)$ coincide with those of $\det \Phi(\zeta)$, multiplicities taken into account. Now, let $\widehat{p}_1, \ldots, \widehat{p}_m$ be the row multiplicities of $E_r(\zeta)\Phi(\zeta)$ for τ_1. Then $\widehat{p}_j = p_j$ for $j \neq r$ and $\widehat{p}_r \geq p_r + 1$. Thus (apply a number of strict equivalence operations if necessary) we may assume that the sum of the row multiplicities for τ_1 is equal to ℓ_1.

So, assume that $\sum_{j=1}^{m} p_j = \ell_1$. Consider

$$F(\zeta) = \begin{bmatrix} \left(\frac{\zeta}{\zeta-\tau_1}\right)^{p_1} & & 0 \\ & \ddots & \\ 0 & & \left(\frac{\zeta}{\zeta-\tau_1}\right)^{p_m} \end{bmatrix}.$$

Multiplication on the left by $F(\zeta)$ is a strict equivalence operation. Put $\Phi_1(\zeta) = F(\zeta)\Phi(\zeta)$. Then the entries $\Phi_1(\zeta)$ are polynomials. Moreover, in $|\zeta| < 1$ the function $\det \Phi_1(\zeta)$ has zeros in $\tau_2, \ldots, \tau_q$ with multiplicities $\ell_2, \ldots, \ell_{q-1}, \ell_q + \ell_1$. Now apply an induction argument. It follows that after a finite number of strict equivalence operations we may reduce the factorization problem to the case where the entries of $\Phi(\zeta)$ are polynomials, $\det \Phi(\zeta)$ has only a zero in the origin and the sum of the row multiplicities for 0 is equal to the multiplicity of 0 as a zero of $\det \Phi(\zeta)$. But this case is easy to treat. Indeed, put

$$\Phi_+(\zeta) = \begin{bmatrix} \zeta^{-\kappa_1} & & & 0 \\ & \zeta^{-\kappa_2} & & \\ & & \ddots & \\ 0 & & & \zeta^{-\kappa_m} \end{bmatrix} \Phi(\zeta),$$

where $\kappa_1, \ldots, \kappa_m$ are the row multiplicities for 0 of the rows in $\Phi(\zeta)$. Then the entries of $\Phi_+(\zeta)$ are polynomials and $\det \Phi_+(\zeta) = \zeta^{-\ell} \det \Phi(\zeta)$, where $\ell = \sum_{j=1}^m \kappa_j$. Thus $\det \Phi_+(\zeta) \neq 0$ for $|\zeta| \leq 1$ and

$$\Phi(\zeta) = \left([\zeta^{\kappa_j} \delta_{ij}]_{i,j=1}^m \right) \Phi_+(\zeta), \qquad \zeta \in \mathsf{T},$$

is a Wiener-Hopf factorization relative to T. Note that by reordering the rows in the diagonal term (which is a strict equivalence operation), we may assume that $\kappa_1 \leq \kappa_2 \leq \cdots \leq \kappa_m$. $\square$

From the properties of the factors in (1) it is clear that the condition "$\det W(\lambda) \neq 0$ for $\lambda \in \mathsf{R} \cup \{\infty\}$" in Theorem 2.1 is a necessary condition for the existence of a Wiener-Hopf factorization relative to the real line.

XIII.3 INVERSION AND FREDHOLM CHARACTERISTICS

The results of this section may be viewed as the matrix analogue of Theorem 1.3. We shall prove the following two theorems.

THEOREM 3.1. *Let T be a Wiener-Hopf operator on $L_2^m([0,\infty))$ with a rational symbol W. Then T is invertible if and only if*

(i) $\det W(\lambda) \neq 0$ *for all $\lambda \in \mathsf{R}$,*

(ii) *W admits a (right) canonical factorization relative to the real line.*

In that case the inverse of T is obtained in the following way. Construct a right canonical factorization $W(\lambda) = W_-(\lambda)W_+(\lambda)$, $\lambda \in \mathsf{R}$, and choose $m \times m$ matrix function γ_- and γ_+ with entries in $L_1(\mathsf{R})$ such that

$$(1a) \qquad W_-(\lambda)^{-1} = I_m + \int_{-\infty}^{0} e^{i\lambda t} \gamma_-(t)dt,$$

$$(1b) \qquad W_+(\lambda)^{-1} = I_m + \int_{0}^{\infty} e^{i\lambda t} \gamma_+(t)dt.$$

Then

$$(2) \qquad (T^{-1}\psi)(t) = \psi(t) + \int_0^\infty \gamma(t,s)\psi(s)ds, \qquad t \geq 0,$$

where

$$(3) \qquad \gamma(t,s) = \begin{cases} \gamma_+(t-s) + \int_0^s \gamma_+(t-\alpha)\gamma_-(\alpha-s)d\alpha, & 0 \leq s < t < \infty, \\[2mm] \gamma_-(t-s) + \int_0^t \gamma_+(t-\alpha)\gamma_-(\alpha-s)d\alpha, & 0 \leq t < s < \infty. \end{cases}$$

THEOREM 3.2. *Let T be a Wiener-Hopf operator on $L_2^m([0,\infty))$ with a rational symbol. Assume that $\det W(\lambda) \neq 0$ for all $\lambda \in \mathbf{R}$, and let*

$$(4) \qquad W(\lambda) = W_-(\lambda)\left(\left[\delta_{kj}\left(\frac{\lambda-i}{\lambda+i}\right)^{\kappa_j}\right]_{k,j=1}^m\right)W_+(\lambda), \qquad \lambda \in \mathbf{R},$$

be a Wiener-Hopf factorization of W relative to the real line. Then T is a Fredholm operator

$$(5) \qquad n(T) = \sum_{\kappa_j \leq 0} -\kappa_j, \qquad d(T) = \sum_{\kappa_j \geq 0} \kappa_j,$$

and a generalized inverse of T is given by the operator

$$(6) \qquad T^+ = V_1 \begin{bmatrix} S_2^{-\kappa_1} & & & & & & & \\ & \ddots & & & & & & \\ & & S_2^{-\kappa_r} & & & & & \\ & & & I & & & & \\ & & & & \ddots & & & \\ & & & & & I & & \\ & & & & & & S_1^{\kappa_{s+1}} & \\ & & & & & & & \ddots & \\ & & & & & & & & S_1^{\kappa_m} \end{bmatrix} V_2,$$

where V_1 and V_2 are the Wiener-Hopf operators on $L_2^m([0,\infty))$ with symbols $W_+(\cdot)^{-1}$ and $W_-(\cdot)^{-1}$, respectively, S_1 and S_2 are the Wiener-Hopf operators on $L_2([0,\infty))$ defined by

$$(7) \qquad (S_1 f)(t) = f(t) - 2\int_t^\infty e^{t-s}f(s)ds, \qquad t \geq 0,$$

$$(8) \qquad (S_2 f)(t) = f(t) - 2\int_0^t e^{s-t}f(s)ds, \qquad t \geq 0,$$

the numbers $\kappa_1, \ldots, \kappa_r$ are the negative factorization indices, $\kappa_{s+1}, \ldots, \kappa_m$ are the positive factorization indices, and the unspecified entries of the $m \times m$ operator matrix in (6) are zero.

Theorem 3.1 will appear as a corollary of Theorem 3.2. Therefore we begin with the proof of the latter theorem.

PROOF OF THEOREM 3.2. Let T_1 and T_2 be the Wiener-Hopf operators with symbols W_+ and W_-, respectively, and let T_D be the Wiener-Hopf operator with symbol $D(\lambda) = \left[\delta_{kj} \left(\frac{\lambda-i}{\lambda+i} \right)^{\kappa_j} \right]_{k,j=1}^m$. By Proposition 1.2

$$(9) \qquad T = T_2 T_D T_1.$$

From Theorem 1.1 we know that T_1 and T_2 are invertible operators, $T_1^{-1} = V_1$ and $T_2^{-1} = V_2$, where V_1 and V_2 are as in the theorem. So to find the Fredholm properties of T we have to analyze the operator T_D.

The symbol of the Wiener-Hopf operator S_2, defined in (8), is equal to $\tau(\lambda) = (\lambda - i)(\lambda + i)^{-1}$, and S_1 is the Wiener-Hopf operator with symbol $\sigma(\lambda) = \tau(\lambda)^{-1}$. So we may apply formulas (8) and (9) in Section 1 to show that S_1 and S_2 are Fredholm operators and

$$(10a) \qquad n(S_1^k) = k, \qquad d(S_1^k) = 0, \qquad \operatorname{ind} S_1^k = k,$$

$$(10b) \qquad n(S_2^k) = 0, \qquad d(S_2^k) = k, \qquad \operatorname{ind} S_2^k = -k,$$

for $k = 1, 2, \ldots$. Now T_D is the $m \times m$ diagonal operator matrix

$$\operatorname{diag}\left(S_1^{-\kappa_1}, \ldots, S_1^{-\kappa_r}, I, \ldots, I, S_2^{\kappa_{s+1}}, \ldots, S_2^{\kappa_m} \right).$$

From (10a) and (10b) it follows that T_D is a Fredholm operator and

$$(11) \qquad n(T_D) = \sum_{j=1}^{r} -\kappa_j, \qquad d(T_D) = \sum_{j=s+1}^{m} \kappa_j.$$

The equivalence relation (9) implies that T is a Fredholm operator, $n(T) = n(T_D)$ and $d(T) = d(T_D)$, which yields (8). From $S_1 S_2 = I$ it follows that operator defined by the diagonal matrix in the right hand side of (6) is a generalized inverse of T_D. Since $V_1 = T_1^{-1}$ and $V_2 = T_2^{-1}$, we conclude that (6) defines a generalized inverse of T. $\square$

PROOF OF THEOREM 3.1. Assume that T is invertible. In particular, T is Fredholm, and thus Theorem XII.3.1 implies that condition (i) is fulfilled, and hence, by Theorem 2.1, the symbol W admits a Wiener-Hopf factorization as in (4). Since T is invertible, $n(T) = d(T) = 0$. But then formula (5) implies that all the indices κ_j in (4) must be zero, and the factorization is a canonical one, which proves (ii).

Conversely, assume (i) and (ii) hold. Then W admits a factorization as in (4) with $\kappa_j = 0$ for $j = 1, \ldots, m$. Thus Theorem 3.2 implies that T is invertible and

$T^{-1} = V_1 V_2$, where V_1 and V_2 are the Wiener-Hopf operators with symbols $W_+(\cdot)^{-1}$ and $W_-(\cdot)^{-1}$, respectively. Since $W_+(\lambda)^{-1}$ has no poles in $\Im\lambda \geq 0$, we know (see the proof of Proposition 1.2) that (1b) holds for some $m \times m$ matrix function γ_+ with entries in $L_1(\mathbf{R})$. It follows that

$$(V_1\psi)(t) = \psi(t) + \int_0^t \gamma_+(t - s)\psi(s)ds, \qquad t \geq 0.$$

Similarly,

$$(V_2\psi)(t) = \psi(t) + \int_t^\infty \gamma_-(t - s)\psi(s)ds, \qquad t \geq 0,$$

where γ_- is as in (1a). We may assume that $\gamma_+(t) = 0$ for $t < 0$ and $\gamma_-(t) = 0$ for $t > 0$. It follows that

$$(V_1 V_2\psi)(t) = (V_2\psi)(t) + \int_0^\infty \gamma_+(t - \alpha)(V_2\psi)(\alpha)d\alpha$$

$$= \psi(t) + \int_0^\infty \gamma_-(t - s)\psi(s)ds + \int_0^\infty \gamma_+(t - \alpha)\psi(\alpha)d\alpha$$

$$+ \int_0^\infty \gamma_+(t - \alpha)\left(\int_0^\infty \gamma_-(\alpha - s)\psi(s)ds\right)d\alpha.$$

By Fubini's theorem the order of the integrals in the last term may be interchanged. This yields

$$(T^{-1}\psi)(t) = (V_1 V_2\psi)(t)$$

$$= \psi(t) + \int_t^\infty \gamma_-(t - s)\psi(s)ds + \int_0^t \gamma_+(t - s)\psi(s)ds$$

$$+ \int_0^\infty \left(\int_0^{\min(t,s)} \gamma_+(t - \alpha)\gamma_-(\alpha - s)d\alpha\right)\psi(s)ds,$$

which proves (2) and (3). $\square$

Formula (5) can be used to prove that the factorization indices are uniquely determined by W. Indeed, let

$$W(\lambda) = W_-(\lambda)\left(\left[\delta_{kj}\left(\frac{\lambda - i}{\lambda + i}\right)^{\kappa_j}\right]_{k,j=1}^m\right)W_+(\lambda), \qquad \lambda \in \mathbf{R},$$

be a Wiener-Hopf factorization of W relative to the real line. Put

$$W_\nu(\lambda) = \left(\frac{\lambda - i}{\lambda + i}\right)^{-\nu} W(\lambda).$$

Note that $\widehat{\kappa}_j = \kappa_j - \nu$, $j = 1, \ldots, m$, are the factorization indices for W_ν. Let T_ν be the Wiener-Hopf operator on $L_2^m([0, \infty))$ with symbol W_ν, and apply the second identity in (5) to T_ν and $T_{\nu+1}$. One obtains that

$$d(T_\nu) - d(T_{\nu+1}) = \sum_{\kappa_j - \nu \geq 0} (\kappa_j - \nu) - \sum_{\kappa_j - \nu - 1 \geq 0} (\kappa_j - \nu - 1)$$

$$= \sum_{\kappa_j - \nu \geq 1} (\kappa_j - \nu) - \sum_{\kappa_j - \nu \geq 1} (\kappa_j - \nu - 1)$$

$$= \sum_{\kappa_j - \nu \geq 1} 1 = \#\{j \mid \kappa_j \geq \nu + 1\}.$$

It follows that

$$\#\{j \mid \kappa_j = \nu\} = d(T_{\nu-1}) - 2d(T_\nu) + d(T_{\nu+1}),$$

which shows that the factorization indices are uniquely determined by W.

To make Theorem 3.1 an effective tool for the inversion of Wiener-Hopf operators one has to compute explicitly the inverse Fourier transforms in (1a) and (1b). The special representation that will be introduced in the next section, provides a solution for this problem.

XIII.4 INTERMEZZO ABOUT REALIZATION

In this section we show that the matrix functions in $\mathcal{R}_\infty^{m \times m}(\mathbf{R})$ admit a special representation.

THEOREM 4.1. *A rational $m \times m$ matrix function W with no poles on the real line and at infinity admits the following representation*

(1) $$W(\lambda) = D + C(\lambda - A)^{-1}B, \qquad \lambda \in \mathbf{R}.$$

Here A is a square matrix of size $n \times n$, say, and A has no real eigenvalue, B and C are matrices of size $n \times m$ and $m \times n$, respectively, and $D = W(\infty)$.

PROOF. Let $\lambda_1, \ldots, \lambda_p$ be the poles (of the entries) of W. Fix $1 \leq j \leq p$, and consider the Laurent series expansion of W in a punctured neighbourhood of λ_j:

$$W(\lambda) = \sum_{\nu=-q_j}^{\infty} (\lambda - \lambda_j)^\nu A_{j,\nu}.$$

Introduce the following block matrices

$$N_j = \begin{bmatrix} \lambda_j I & I & & & \\ & \lambda_j I & I & & \\ & & \ddots & \ddots & \\ & & & \lambda_j I & I \\ & & & & \lambda_j I \end{bmatrix}, \qquad Q_j = \begin{bmatrix} A_{j,-1} \\ A_{j,-2} \\ \vdots \\ A_{j,-q_j} \end{bmatrix},$$

$$R_j = \begin{bmatrix} I & 0 & \cdots & 0 \end{bmatrix}.$$

Here I denotes the $m \times m$ identity matrix, the blanks in N_j stand for zero entries, and N_j has (block) size $q_j \times q_j$. The matrix $\lambda - N_j$ is invertible for $\lambda \neq \lambda_j$, and $W(\lambda) - R_j(\lambda - N_j)^{-1}Q_j$ is analytic in λ_j. We carry out this construction for each j, and define

$$A = \begin{bmatrix} N_1 & & & \\ & N_2 & & \\ & & \ddots & \\ & & & N_p \end{bmatrix}, \qquad B = \begin{bmatrix} Q_1 \\ Q_2 \\ \vdots \\ Q_p \end{bmatrix},$$

$$C = [\, R_1 \quad R_2 \quad \cdots \quad R_p \,].$$

Note that A is a block diagonal matrix with diagonal elements $N_1, N_2, \ldots, N_p$. So the eigenvalues of A are precisely the poles $\lambda_1, \ldots, \lambda_p$. In particular, A has no eigenvalues on the real line. From

$$C(\lambda - A)^{-1}B = \sum_{j=1}^{p} R_j(\lambda - N_j)^{-1}Q_j,$$

we conclude that $V(\lambda) := W(\lambda) - C(\lambda - A)^{-1}B$ is analytic in λ on the entire complex plane. Furthermore,

$$\lim_{\lambda \to \infty} V(\lambda) = W(\infty).$$

Liouville's theorem for entire functions implies that $V(\lambda)$ is identically equal to $D = W(\infty)$, which yields the representation (1). $\square$

      If W is as in (1), then we shall say that W is in *realized form*, and we shall call the right hand side of (1) a *realization* of W. Note that for $W \in \mathcal{R}_{\infty}^{m \times m}(\mathbf{R})$ the matrix A in a realization of W has no real eigenvalue by definition.

      In what follows we identify a $p \times q$ matrix with the linear transformation from $\mathbf{C}^q$ into $\mathbf{C}^p$ defined by the canonical action of the matrix relative to the standard bases in $\mathbf{C}^q$ and $\mathbf{C}^p$. In particular, the matrix A appearing in (1) will be viewed as an operator on $\mathbf{C}^n$.

      The next theorem shows how realizations may be used to construct inverse Fourier transforms.

      THEOREM 4.2. *Let* $W \in \mathcal{R}_{\infty}^{m \times m}(\mathbf{R})$, *and let* $W(\lambda) = D + C(\lambda - A)^{-1}B$, $\lambda \in \mathbf{R}$, *be a realization. Then*

$$(2a) \qquad\qquad W(\lambda) = D - \int_{-\infty}^{\infty} e^{i\lambda t} k(t)\, dt$$

with

$$(2b) \qquad\qquad k(t) = \begin{cases} iCe^{-itA}(I - P)B, & t > 0, \\ -iCe^{-itA}PB, & t < 0, \end{cases}$$

where P is the Riesz projection of A corresponding to the eigenvalues in the upper half plane.

PROOF. It suffices to show that for λ not an eigenvalue of A the following identity holds:

$$(3) \qquad (\lambda - A)^{-1}x = \begin{cases} -i \int\limits_0^\infty e^{i\lambda t}e^{-itA}x\,dt, & x \in \operatorname{Ker} P, \\[2mm] i \int\limits_{-\infty}^0 e^{i\lambda t}e^{-itA}x\,dt, & x \in \operatorname{Im} P. \end{cases}$$

To prove (3) take $x \in \operatorname{Ker} P$. Put $A_1 = A|\operatorname{Ker} P$. Note $-iA_1$ has all its eigenvalues in the open left half plane. Thus $e^{-itA}x = e^{-itA_1}x$ is exponentially decaying (see Section I.5). Furthermore,

$$\frac{d}{dt}\big(e^{i(\lambda-A)t}x\big) = i(\lambda - A)e^{i(\lambda-A)t}x.$$

It follows that

$$(\lambda - A)\bigg(i\int\limits_0^\infty e^{i\lambda t}e^{-itA}x\,dt\bigg) = \int\limits_0^\infty \big(e^{i(\lambda-A)t}x\big)'dt = -x,$$

which proves the part of (3) connected with $\operatorname{Ker} P$. For $x \in \operatorname{Im} P$ a similar argument can be used. $\square$

XIII.5 INVERSION OF CONVOLUTION OPERATORS

In this section $I - L$ is a convolution operator on $L_2^m(\mathbf{R})$ with a rational symbol W. Recall that W has no poles on the real line and

$$W(\infty) := \lim_{\lambda \to \infty} W(\lambda) = I_m,$$

and so, by Theorem 4.1, the symbol W admits a realization

$$(1) \qquad W(\lambda) = I + C(\lambda - A)^{-1}B, \qquad \lambda \in \mathbf{R}.$$

Our aim is to describe the inversion of $I - L$ in terms of the matrices A, B and C.

THEOREM 5.1. *Let $I-L$ be a convolution operator on $L_2^m(\mathbf{R})$ with a rational symbol W, and let (1) be a given realization of W. Then $I - L$ is invertible if and only if the matrix $A^\times := A - BC$ has no real eigenvalue. In that case*

$$(2) \qquad \big((I - L)^{-1}\varphi\big)(t) = \varphi(t) - \int\limits_{-\infty}^\infty k^\times(t - s)\varphi(s)\,ds, \qquad t \in \mathbf{R},$$

with

$$(3) \qquad k^\times(t) = \begin{cases} -iCe^{-itA^\times}(I - P^\times)B, & t > 0, \\[2mm] iCe^{-itA^\times}P^\times B, & t < 0, \end{cases}$$

where $P^\times$ is the Riesz projection of $A^\times$ corresponding to the eigenvalues in the upper half plane.

For the proof of Theorem 4.1 we need the following lemma.

LEMMA 5.2. *Let W be as in* (1). *Then* $\det W(\lambda) \neq 0$ *for all* $\lambda \in \mathbf{R}$ *if and only if* $A^\times := A - BC$ *has no real eigenvalue, and in that case*

$$(4) \qquad W(\lambda)^{-1} = I - C(\lambda - A^\times)^{-1}B, \qquad \lambda \in \mathbf{R}.$$

PROOF. By Corollary VII.6.2

$$\begin{aligned}
\det W(\lambda) &= \det[I + C(\lambda - A)^{-1}B] \\
&= \det[I + (\lambda - A)^{-1}BC] \\
&= \det[(\lambda - A)^{-1}(\lambda - A^\times)] \\
&= \frac{\det(\lambda - A^\times)}{\det(\lambda - A)}, \qquad \lambda \in \mathbf{R}.
\end{aligned}$$

Thus if $\lambda \in \mathbf{R}$, then $\det W(\lambda) \neq 0$ if and only if λ is not an eigenvalue of $A^\times$.

Next, assume that $\det(\lambda - A^\times) \neq 0$, and let us solve the equation $W(\lambda)x = y$. Introduce a new unknown by setting $z = (\lambda - A)^{-1}Bx$. Then given y we have to compute x from

$$(5) \qquad \begin{cases} \lambda z = Az + Bz, \\ y = Cz + x. \end{cases}$$

This is easy. Apply B to the second equation in (5) and subtract the result from the first equation in (5). This yields the following equivalent system:

$$\begin{cases} \lambda z = A^\times z + By, \\ x = -Cz + y. \end{cases}$$

Hence $z = (\lambda - A^\times)^{-1}By$, and

$$W(\lambda)^{-1}y = x = y - C(\lambda - A^\times)^{-1}By,$$

which proves (4). $\quad\square$

PROOF OF THEOREM 5.1. By Theorem XII.1.4 the operator $I - L$ is invertible if and only if $\det W(\lambda) \neq 0$ for all $\lambda \in \mathbf{R}$, and, by the previous lemma, the latter condition is equivalent to the requirement that $A^\times := A - BC$ has no real eigenvalue. From Theorem XII.1.4 we also know that in that case the inverse of $I - L$ is given by (2) with $k^\times$ uniquely determined by

$$W(\lambda)^{-1} = I_m - \int_{-\infty}^{\infty} e^{i\lambda t}k^\times(t)dt, \qquad \lambda \in \mathbf{R}.$$

Since $W(\lambda)^{-1}$ has the realization (4) we can apply Theorem 4.2 to get the expression (3) for $k^{\times}$. $\square$

The following lemma will be used in the next section.

LEMMA 5.3. *Let W be as in* (1), *and assume that* $\det W(\lambda) \neq 0$ *for all* $\lambda \in \mathbf{R}$. *Put* $A^{\times} = A - BC$. *Then for* $\lambda \in \mathbf{R}$

$$C(\lambda - A^{\times})^{-1} = W(\lambda)^{-1}C(\lambda - A)^{-1},$$
$$(\lambda - A^{\times})^{-1}B = (\lambda - A)^{-1}BW(\lambda)^{-1},$$
$$(\lambda - A^{\times})^{-1} = (\lambda - A)^{-1} - (\lambda - A)^{-1}BW(\lambda)^{-1}C(\lambda - A)^{-1}.$$

PROOF. From Lemma 5.2 we know that $\lambda - A^{\times}$ is invertible for each $\lambda \in \mathbf{R}$. The desired formulas follow by a direct computation, using (4) and the identity

$$(6) \qquad\qquad BC = A - A^{\times} = (\lambda - A^{\times}) - (\lambda - A). \ \square$$

XIII.6 EXPLICIT CANONICAL FACTORIZATION

This section concerns explicit factorization in terms of a given realization. We begin with canonical factorization.

THEOREM 6.1. *Let W be a rational $m \times m$ matrix function without poles on the real line and at infinity, and let W be given in realized form:*

$$(1) \qquad\qquad W(\lambda) = D + C(\lambda - A)^{-1}B, \qquad \lambda \in \mathbf{R}.$$

Let n be order of the matrix A. Then W admits a right canonical factorization relative to the real line if and only if

(i) *D is invertible and $A^{\times}:= A - BD^{-1}C$ has no real eigenvalues,*

(ii) $\mathbf{C}^n = \operatorname{Im} P \oplus \operatorname{Ker} P^{\times}$.

Here P and $P^{\times}$ are the Riesz projections of A and $A^{\times}$, respectively, corresponding to the eigenvalues in the upper half plane. If conditions (i) *and* (ii) *are fulfilled, then a right canonical factorization $W(\lambda) = W_-(\lambda)W_+(\lambda)$, $\lambda \in \mathbf{R}$, is obtained by taking*

$$(2) \qquad\qquad W_-(\lambda) = D + C(\lambda - A)^{-1}(I - \Pi)B, \qquad \lambda \in \mathbf{R},$$

$$(3) \qquad\qquad W_+(\lambda) = I + D^{-1}C\Pi(\lambda - A)^{-1}B, \qquad \lambda \in \mathbf{R},$$

$$(4) \qquad W_-(\lambda)^{-1} = D^{-1} - D^{-1}C(I - \Pi)(\lambda - A^{\times})^{-1}BD^{-1}, \qquad \lambda \in \mathbf{R},$$

$$(5) \qquad\qquad W_+(\lambda)^{-1} = I - D^{-1}C(\lambda - A^{\times})^{-1}\Pi B, \qquad \lambda \in \mathbf{R},$$

where Π is the projection of $\mathbf{C}^n$ along $\operatorname{Im} P$ onto $\operatorname{Ker} P^{\times}$.

PROOF. We know (see the last paragraph of Section 2) that the condition "$\det W(\lambda) \neq 0$ for all $\lambda \in \mathbf{R} \cup \{\infty\}$" is a necessary condition for the existence of a canonical factorization. Thus in order that W admits such a factorization, the matrix $D = W(\infty)$ must be invertible and, by applying Lemma 5.2 to

$$(6) \qquad V(\lambda) := D^{-1}W(\lambda) = I_m + D^{-1}C(\lambda - A)^{-1}B, \qquad \lambda \in \mathbf{R},$$

one sees that $A^{\times} := A - BD^{-1}C$ cannot have a real eigenvalue. Thus condition (i) is fulfilled whenever W admits a canonical factorization.

Assume that conditions (i) and (ii) hold true. Write A, B, C and $A^{\times}$ as operator matrices relative to the decomposition in (ii):

$$(7) \qquad A = \begin{bmatrix} A_{11} & A_{12} \\ 0 & A_{22} \end{bmatrix} : \operatorname{Im} P \oplus \operatorname{Ker} P^{\times} \to \operatorname{Im} P \oplus \operatorname{Ker} P^{\times},$$

$$(8) \qquad B = \begin{bmatrix} B_1 \\ B_2 \end{bmatrix} : \mathbf{C}^m \to \operatorname{Im} P \oplus \operatorname{Ker} P^{\times},$$

$$(9) \qquad C = [\, C_1 \quad C_2 \,] : \operatorname{Im} P \oplus \operatorname{Ker} P^{\times} \to \mathbf{C}^m,$$

$$(10) \qquad A^{\times} = \begin{bmatrix} A_{11}^{\times} & 0 \\ A_{21}^{\times} & A_{22}^{\times} \end{bmatrix} : \operatorname{Im} P \oplus \operatorname{Ker} P^{\times} \to \operatorname{Im} P \oplus \operatorname{Ker} P^{\times}.$$

The zeros in the left lower corner of the matrix in (7) and in the right upper corner of the matrix in (10) are justified by the fact that $\operatorname{Im} P$ is invariant under A and $\operatorname{Ker} P^{\times}$ is invariant under $A^{\times}$. From $A^{\times} = A - BD^{-1}C$ it follows that

$$(11) \qquad A_{12} = B_1 D^{-1} C_2, \qquad A_{21}^{\times} = -B_2 D^{-1} C_1,$$

$$(12) \qquad A_{11}^{\times} = A_{11} - B_1 D^{-1} C_1, \qquad A_{22}^{\times} = A_{22} - B_2 D^{-1} C_2.$$

Let W_- and W_+ be the matrix functions defined by (2) and (3), respectively. By using the block matrix representations (7)–(9) we may rewrite W_- and W_+ in the following form:

$$(13) \qquad W_-(\lambda) = D + C_1(\lambda - A_{11})^{-1} B_1, \qquad \lambda \in \mathbf{R},$$

$$(14) \qquad W_+(\lambda) = I + D^{-1} C_2(\lambda - A_{22})^{-1} B_2, \qquad \lambda \in \mathbf{R}.$$

From (7) and the first identity in (11) one sees that

$$W_-(\lambda)W_+(\lambda) = D + [\, C_1 \quad C_2 \,] \begin{bmatrix} \lambda - A_{11} & -B_1 D^{-1} C_2 \\ 0 & \lambda - A_{22} \end{bmatrix}^{-1} \begin{bmatrix} B_1 \\ B_2 \end{bmatrix}$$

$$= D + C(\lambda - A)^{-1}B = W(\lambda), \qquad \lambda \in \mathbf{R},$$

which gives us the desired factorization.

Recall that A has no eigenvalues on the real line. Since $\operatorname{Im} P$ is the maximal A-invariant subspace M of $\mathbf{C}^n$ such that $A|M$ has all its eigenvalues in the open upper half plane, the matrix A_{11} has all its eigenvalues in the open upper half plane and A_{22} has all its eigenvalues in the open lower half plane. A similar argument applied to $A^\times$ shows that $A_{22}^\times$ has all its eigenvalues in the open lower half plane and $A_{11}^\times$ has all its eigenvalues in the open upper half plane. In particular, $A_{11}^\times$ and $A_{22}^\times$ have no real eigenvalues. But then we can use the identities in (12) and Lemma 5.2 to conclude that

$$(15) \qquad W_-(\lambda)^{-1} = D^{-1} - D^{-1}C_1(\lambda - A_{11}^\times)^{-1}B_1 D^{-1}, \qquad \lambda \in \mathbf{R},$$

$$(16) \qquad W_+(\lambda)^{-1} = I - D^{-1}C_2(\lambda - A_{22}^\times)^{-1}B_2, \qquad \lambda \in \mathbf{R}.$$

To get (15) from (13) one applies Lemma 5.2 to $D^{-1}W_-(\lambda)$. From the location of the eigenvalues of A_{11}, A_{22}, $A_{11}^\times$ and $A_{22}^\times$ it is now clear that the conditions (j)–(jjj) in Theorem 2.1 are fulfilled. Thus our factorization $W(\lambda) = W_-(\lambda)W_+(\lambda)$, $\lambda \in \mathbf{R}$, is a right canonical factorization. The formulas (4) and (5) follow from (15) and (16), the block matrix representations (8)–(10), and the definition of the projection Π.

Next, we consider the converse. Assume that W admits a right canonical factorization relative to the real line:

$$(17) \qquad W(\lambda) = W_-(\lambda)W_+(\lambda), \qquad \lambda \in \mathbf{R}.$$

We already know that this implies (i). So we have to prove (ii). Take $x \in \operatorname{Im} P \cap \operatorname{Ker} P^\times$, and put

$$\varphi_-(\lambda) = C(\lambda - A)^{-1}x, \quad \varphi_+(\lambda) = D^{-1}C(\lambda - A^\times)^{-1}x \qquad (\lambda \in \mathbf{R}).$$

Since $x \in \operatorname{Im} P$, the function φ_- has an analytic continuation to the closed lower half plane $\Im \lambda \leq 0$, and $x \in \operatorname{Ker} P^\times$ implies that φ_+ has an analytic continuation to $\Im \lambda \geq 0$. Note that $W(\lambda)\varphi_+(\lambda) = \varphi_-(\lambda)$, $\lambda \in \mathbf{R}$, because of Lemma 5.3 (applied to $D^{-1}W(\lambda)$). It follows (use the factorization (17)) that

$$(18) \qquad W_+(\lambda)\varphi_+(\lambda) = W_-(\lambda)^{-1}\varphi_-(\lambda), \qquad \lambda \in \mathbf{R}.$$

Now employ the properties of the factors W_- and W_+ and Liouville's theorem. We obtain that both terms in (18) are identically zero, and thus $\varphi_-(\lambda) = 0$ for all $\lambda \in \mathbf{R}$. But then we can apply the third identity in Lemma 5.3 to show that

$$(19) \qquad (\lambda - A^\times)^{-1}x = (\lambda - A)^{-1}x$$

holds for all $\lambda \in \mathbf{R}$ and hence also for all λ outside the eigenvalue sets of A and $A^\times$. Take a contour Γ in the open upper half plane such that the eigenvalues of A and $A^\times$ in the open upper half plane are inside Γ. Integrating (19) over Γ yields $0 = P^\times x = Px = x$. Thus $\operatorname{Im} P \cap \operatorname{Ker} P^\times = (0)$.

We proceed by showing that $\mathbf{C}^n = \operatorname{Im} P + \operatorname{Ker} P^\times$. Take $y \in \mathbf{C}^n$ such that $y \perp (\operatorname{Im} P + \operatorname{Ker} P^\times)$. Let y^* be the row vector of which the j-th entry is equal to the complex conjugate of the j-th entry of y ($j = 1, \ldots, m$). Put

$$\psi_-(\lambda) = y^*(\lambda - A^\times)^{-1}BD^{-1}, \quad \psi_+(\lambda) = y^*(\lambda - A)^{-1}B \qquad (\lambda \in \mathbf{R}).$$

Since $y^*P = 0$, we have $\psi_+(\lambda) = y^*(\lambda - A)^{-1}(I - P)B$, and hence ψ_+ has an analytic continuation to $\Im \lambda \geq 0$. From $y^*(I - P^\times) = 0$ it follows that ψ_- has an analytic continuation to $\Im \lambda \leq 0$. Lemma 5.3 (applied to $D^{-1}W(\lambda)$) implies that $\psi_-(\lambda)W(\lambda) = \psi_+(\lambda)$, $\lambda \in \mathbf{R}$, and hence we can use the same arguments as in the preceding paragraph to show that $y = 0$. The equality in (ii) is now proved. $\square$

Let W be as in (1). From the remark made in the first paragraph of the proof of Theorem 6.1 one may deduce that W admits a Wiener-Hopf factorization relative to the real line if and only if D is invertible and $A^\times := A - BD^{-1}C$ has no real eigenvalue. Let us assume that these conditions are fulfilled. Then, as for canonical factorization, one can describe explicitly the factors in a Wiener-Hopf factorization of W in terms of the matrices A, B, C and D appearing in the realization (1). Also the factorization indices may be expressed in terms of these matrices. The formulas for the factors are more complex than in the case of canonical factorization and we shall not give them here. To obtain the factorization indices in terms of the realization (1) we proceed as follows. Put $M = \operatorname{Im} P$ and $M^\times = \operatorname{Ker} P^\times$, where P and $P^\times$ are the Riesz projections of A and $A^\times$, respectively, corresponding to the eigenvalues in the upper half plane. Then there are precisely t negative factorization indices $\kappa_1, \ldots, \kappa_t$ and precisely s positive factorization indices $\kappa_{t+1}, \ldots, \kappa_{t+s}$, where

$$t = \dim \frac{M \cap M^\times}{M \cap M^\times \cap \operatorname{Ker} C}, \qquad s = \dim \frac{M + M^\times + \operatorname{Im} B}{M + M^\times}.$$

Furthermore, these factorization indices are given by

$$\kappa_j = -\#\{\nu \geq 1 \mid \dim(K_{\nu-1}/K_\nu) \geq j\}, \qquad j = 1, \ldots, t,$$

$$\kappa_j = \#\{\nu \geq 1 \mid \dim(H_\nu/H_{\nu-1}) \geq s + t - j + 1\}, \qquad j = t+1, \ldots, t+s,$$

where

$$K_\nu = M \cap M^\times \cap \operatorname{Ker} C \cap \cdots \cap \operatorname{Ker} CA^{\nu-1}, \qquad \nu \geq 1$$

$$H_\nu = M + M^\times + \operatorname{Im} B + \cdots + \operatorname{Im} A^{\nu-1}B, \qquad \nu \geq 1$$

$$K_0 = M \cap M^\times, \qquad H_0 = M + M^\times.$$

For these and related results we refer to Bart-Gohberg-Kaashoek [5].

XIII.7 EXPLICIT INVERSION

In this section the factorization formulas derived in the previous section are applied to construct explicitly the inverse of a Wiener-Hopf operator on $L_2^m([0, \infty))$ with a rational (matrix) symbol. The necessary and sufficient conditions for invertibility and the formula for the inverse are expressed explicitly in terms of the data appearing in a realization of the symbol. Recall that a rational symbol W belongs to the class $\mathcal{R}_\infty^{m\times m}(\mathbf{R})$ and $W(\infty) = I_m$, the $m \times m$ identity matrix.

THEOREM 7.1. *Let T be a Wiener-Hopf operator on $L_2^m([0,\infty))$ with a rational symbol W given in realized form:*

$$(1) \qquad W(\lambda) = I_m + C(\lambda - A)^{-1}B, \qquad \lambda \in \mathbf{R}.$$

Let n be the order of the matrix A, and put $A^\times = A - BC$. Then T is invertible if and only if $A^\times$ has no real eigenvalue and

$$(2) \qquad \mathbf{C}^n = \operatorname{Im} P \oplus \operatorname{Ker} P^\times,$$

where P and $P^\times$ are the Riesz projections of A and $A^\times$, respectively, corresponding to the eigenvalues in the upper half plane. In that case,

$$(3) \qquad (T^{-1}\psi)(t) = \psi(t) + \int\limits_0^\infty \gamma(t,s)\psi(s)ds, \qquad t \geq 0,$$

with

$$(4) \qquad \gamma(t,s) = \begin{cases} iCe^{-itA^\times}\Pi e^{isA^\times}B, & 0 \leq s < t < \infty, \\ -iCe^{-itA^\times}(I - \Pi)e^{isA^\times}B, & 0 \leq t < s < \infty. \end{cases}$$

Here Π is the projection of $\mathbf{C}^n$ along $\operatorname{Im} P$ onto $\operatorname{Ker} P^\times$.

PROOF. Theorems 3.1 and 6.1 imply that T is invertible if and only if $A^\times$ has no real eigenvalue and (2) holds. Assume that the latter two conditions are satisfied. We have to construct the inverse of T. Again we apply Theorems 3.1 and 6.1. By Theorem 6.1 (which we apply with $D = I$) the symbol W admits a right canonical factorization $W(\lambda) = W_-(\lambda)W_+(\lambda)$, $\lambda \in \mathbf{R}$, with

$$W_-(\lambda)^{-1} = I_m - C(I - \Pi)(\lambda - A^\times)^{-1}B, \qquad \lambda \in \mathbf{R},$$

$$W_+(\lambda)^{-1} = I_m - C(\lambda - A^\times)^{-1}\Pi B, \qquad \lambda \in \mathbf{R}.$$

To construct T^{-1} we have to compute the inverse Fourier transforms of $W_-(\cdot)^{-1}$ and $W_+(\cdot)^{-1}$. To do this we use Theorem 4.2. Note that $(I - \Pi)P^\times = I - \Pi$ and $P^\times\Pi = 0$. It follows that

$$W_-(\lambda)^{-1} = I_m + \int\limits_{-\infty}^0 e^{i\lambda t}\gamma_-(t)dt,$$

$$W_+(\lambda)^{-1} = I_m + \int\limits_0^\infty e^{i\lambda t}\gamma_+(t)dt,$$

where

$$\gamma_-(t) = -iC(I - \Pi)e^{-itA^\times}B, \qquad t \le 0, \tag{5}$$

$$\gamma_+(t) = iCe^{-itA^\times}\Pi B, \qquad t \ge 0. \tag{6}$$

To compute the kernel γ in (3) we shall apply formula (3.3). Since $\operatorname{Im}P$ is invariant under A and $\operatorname{Ker}P^\times$ under $A^\times$, we have $\Pi A(I - \Pi) = 0$ and $\Pi A^\times(I - \Pi) = \Pi A^\times - A^\times\Pi$, and thus

$$\Pi BC(I - \Pi) = \Pi(A - A^\times)(I - \Pi) = A^\times\Pi - \Pi A^\times. \tag{7}$$

It follows that

$$\frac{d}{d\alpha}(e^{i\alpha A^\times}\Pi e^{-i\alpha A^\times}) = ie^{i\alpha A^\times}(A^\times\Pi - \Pi A^\times)e^{-i\alpha A^\times}$$
$$= ie^{i\alpha A^\times}\Pi BC(I - \Pi)e^{-i\alpha A^\times},$$

and hence for $0 \le s < t < \infty$

$$\gamma(t,s) = \gamma_+(t - s) + \int_0^s \gamma_+(t - \alpha)\gamma_-(\alpha - s)d\alpha$$
$$= \gamma_+(t - s) - iCe^{-itA^\times}\left(\int_0^s ie^{i\alpha A^\times}\Pi BC(I - \Pi)e^{-i\alpha A^\times}\,d\alpha\right)e^{isA^\times}B$$
$$= iCe^{-itA^\times}\Pi e^{isA^\times}B.$$

A similar calculation yields the desired expression for $\gamma(t,s)$ when $0 \le t < s < \infty$. $\square$

XIII.8 KERNEL, IMAGE AND GENERALIZED INVERSE

In this section we use the realization approach to derive explicit formulas for the Fredholm characteristics and a generalized inverse of a Wiener-Hopf operator.

THEOREM 8.1. *Let T be a Wiener-Hopf operator on $L_2^m([0,\infty))$ with a rational symbol W given in realized form:*

$$W(\lambda) = I_m + C(\lambda - A)^{-1}B, \qquad \lambda \in \mathbf{R}. \tag{1}$$

Let n be the order of the matrix A, and put $A^\times = A - BC$. Then T is a Fredholm operator if and only if $A^\times$ has no real eigenvalue. Next assume that the latter condition holds, and let P and $P^\times$ be the Riesz projection of A and $A^\times$, respectively, corresponding to the eigenvalues in the upper half plane. Then

$$\operatorname{Ker}T = \{\varphi \in L_2^m([0,\infty)) \mid \varphi(t) = Ce^{-itA^\times}x, x \in \operatorname{Im}P \cap \operatorname{Ker}P^\times\}, \tag{2}$$

$$(3) \qquad \operatorname{Im} T = \left\{ \psi \in L_2^m([0,\infty)) \int_0^\infty P^\times e^{isA^\times} B\psi(s)ds \in \operatorname{Im} P + \operatorname{Ker} P^\times \right\},$$

$$(4) \qquad n(T) = \dim(\operatorname{Im} P \cap \operatorname{Ker} P^\times), \qquad d(T) = \dim \frac{\mathbf{C}^n}{\operatorname{Im} P + \operatorname{Ker} P^\times},$$

$$(5) \qquad \operatorname{ind}(T) = \operatorname{rank} P - \operatorname{rank} P^\times,$$

and a generalized inverse of T in the weak sense is given by the operator

$$(6) \qquad (T^+\psi)(t) = \psi(t) + \int_0^\infty \widetilde{\gamma}(t,s)\psi(s)ds, \qquad t \geq 0,$$

with

$$(7) \qquad \widetilde{\gamma}(t,s) = \begin{cases} iCe^{-itA^\times}(I - \widetilde{\widetilde{\Pi}})e^{isA^\times}B, & 0 \leq s < t < \infty, \\ -iCe^{-itA^\times}\widetilde{\widetilde{\Pi}}e^{isA^\times}B, & 0 \leq t < s < \infty. \end{cases}$$

Here $\widetilde{\widetilde{\Pi}} = P^\times + (I - P^\times)S^+P^\times$, where S^+ is a generalized inverse of the operator

$$(8) \qquad S: \operatorname{Im} P \to \operatorname{Im} P^\times, \qquad Sx = P^\times x \quad (x \in \operatorname{Im} P).$$

PROOF. From Theorem XII.3.1 we know that T is Fredholm if and only if $\det W(\lambda) \neq 0$ for each $\lambda \in \mathbf{R}$. According to Lemma 5.2 the latter condition is equivalent to the requirement that $A^\times$ has no real eigenvalue. This proves the first part of the theorem.

In what follows we assume that $A^\times$ has no real eigenvalue. Formulas (2)–(6) can be proved by using Theorem 3.2 and explicit formulas for the factors in the Wiener-Hopf factorization (3.4). In particular, one may derive (4) and (5) from formula (3.5) and the formulas for the factorization indices κ_j given at the end of Section 6. We shall follow a different route and employ matricial coupling (see Section III.4). We shall show that the Wiener-Hopf operator T and the finite dimensional operator S defined by (8) are matricially coupled. More precisely, the following coupling relation holds:

$$(9) \qquad \begin{bmatrix} T & U \\ R & Q \end{bmatrix}^{-1} = \begin{bmatrix} T^\times & U^\times \\ R^\times & S \end{bmatrix}.$$

Here

$$U: \operatorname{Im} P^\times \to L_2^m([0,\infty)), \qquad (Ux)(t) = iCe^{-itA}(I - P)x,$$

$$U^\times: \operatorname{Im} P \to L_2^m([0,\infty)), \qquad (U^\times x)(t) = iCe^{-itA^\times}(I - P^\times)x,$$

$$R: L_p^m([0,\infty)) \to \operatorname{Im} P, \qquad R\varphi = -\int_0^\infty Pe^{isA}B\varphi(s)ds,$$

$$R^\times: L_p^m([0,\infty)) \to \operatorname{Im} P^\times, \qquad R^\times\varphi = \int_0^\infty P^\times e^{isA^\times}B\varphi(s)ds,$$

$$Q: \operatorname{Im} P^\times \to \operatorname{Im} P, \qquad Qx = Px,$$

and $T^\times$ is the Wiener-Hopf operator on $L_2^m([0,\infty))$ with symbol $W(\cdot)^{-1}$.

To prove (9) we have to verify (8) identities. Here we shall establish four of them, namely

$$(10) \qquad TT^\times + UR^\times = I_{L_2^m([0,\infty))}$$

$$(11) \qquad RT^\times + QR^\times = 0,$$

$$(12) \qquad TU^\times + US = 0,$$

$$(13) \qquad RU^\times + QS = I_{\operatorname{Im}P}.$$

The other four identities can be obtained by interchanging the roles of $W(\cdot)$ and $W(\cdot)^{-1}$. It will be convenient to consider the convolution operators $I-L$ and $I-L^\times$ on $L_2^m(\mathbf{R})$ with symbols $W(\cdot)$ and $W(\cdot)^{-1}$, respectively. With respect to the decomposition $L_2^m(\mathbf{R}) = L_2^m([0,\infty)) \oplus L_2^m((-\infty,0])$, we may write $I-L$ and $I-L^\times$ in the following form

$$I - L = \begin{bmatrix} T & H \\ * & * \end{bmatrix}, \qquad I - L^\times = \begin{bmatrix} T^\times & * \\ H^\times & * \end{bmatrix}.$$

From Theorem XII.1.4 we know that $(I-L)^{-1} = I - L^\times$. Thus to prove (10) it suffices to show that $HH^\times = UR^\times$. Take $\varphi \in L_2^m([0,\infty))$. According to Theorem 5.1

$$(H^\times \varphi)(t) = -iCe^{-itA^\times} \int_0^\infty P^\times e^{isA^\times} B\varphi(s)ds$$

$$= -iCe^{-itA^\times} R^\times \varphi, \qquad t \le 0.$$

It follows (cf. Theorem 4.2) that

$$(HH^\times \varphi)(t) = -iCe^{-itA} \int_{-\infty}^0 (I - P)e^{isA} B(H^\times \varphi)(s)ds$$

$$= i\left[U\left(\int_{-\infty}^0 (I - P)e^{isA} BC e^{-isA^\times} P^\times ds \right) R^\times \varphi \right](t), \qquad t \ge 0.$$

From

$$(14) \qquad \frac{d}{ds}(e^{isA} e^{-isA^\times}) = ie^{isA} BC e^{-isA^\times},$$

we conclude that

$$i \int_{-\infty}^0 (I - P)e^{isA} BC e^{-isA^\times} P^\times ds = (I - P)P^\times,$$

and thus $HH^\times \varphi = U(I - P)P^\times R^\times = UR^\times$, which proves (10).

To prove (11), note that all operators involved are bounded linear operators. Hence it suffices to establish (11) on a dense subset of $L_2^m([0,\infty))$. We take as dense subset the set of all $\mathbf{C}^m$-valued continuous functions on $[0,\infty)$ with compact support. Let φ be such a function. We have

$$(T^\times \varphi)(t) = \varphi(t) + iCe^{-itA^\times}(I - P^\times)\varphi_1(t) - iCe^{-itA^\times}P^\times\varphi_2(t), \qquad t \geq 0,$$

where

$$\varphi_1(t) = \int_0^t (I - P^\times)e^{isA^\times}B\varphi(s)ds, \qquad t \geq 0,$$

$$\varphi_2(t) = \int_t^\infty P^\times e^{isA^\times}B\varphi(s)ds, \qquad t \geq 0.$$

Note that φ_1 and φ_2 are differentiable. This allows us to use partial integration and to show that

$$-i\int_0^\infty Pe^{isA}BCe^{-isA^\times}(I - P^\times)\varphi_1(s)ds$$

$$= -\int_0^\infty \left(Pe^{isA}e^{-isA^\times}(I - P^\times)\right)'\varphi_1(s)ds$$

$$= -Pe^{isA}e^{-isA^\times}(I - P^\times)\varphi_1(s)\Big|_0^\infty + \int_0^\infty Pe^{isA}e^{-isA^\times}(I - P^\times)\varphi_1'(s)ds$$

$$= \int_0^\infty Pe^{isA}(I - P^\times)B\varphi(s)ds.$$

A similar calculation yields

$$i\int_0^\infty Pe^{isA}BCe^{-isA^\times}P^\times\varphi_2(s)ds$$

$$= Pe^{isA}e^{-isA^\times}P^\times\varphi_2(s)\Big|_0^\infty + \int_0^\infty Pe^{isA}P^\times B\varphi(s)ds$$

$$= -P\int_0^\infty P^\times e^{isA^\times}B\varphi(s)ds + \int_0^\infty Pe^{isA}P^\times B\varphi(s)ds$$

$$= -QR^\times\varphi + \int_0^\infty Pe^{isA}P^\times B\varphi(s).$$

It follows that

$$RT^\times \varphi = R\varphi + \int_0^\infty Pe^{isA}(I - P^\times)B\varphi(s)ds - QR^\times \varphi$$

$$+ \int_0^\infty Pe^{isA}P^\times B\varphi(s)ds = -QR^\times \varphi,$$

which proves (11).

Next, observe that

$$i\int_0^t (I - P)e^{isA}BCe^{-isA^\times}(I - P^\times)ds = (I - P)e^{itA}e^{-itA^\times}(I - P^\times) - (I - P)(I - P^\times),$$

$$i\int_t^\infty Pe^{isA}BCe^{-isA^\times}(I - P^\times)ds = -Pe^{itA}e^{-itA^\times}(I - P^\times).$$

Take $x \in \operatorname{Im} P$. Then $(I - P)(I - P^\times)x = -(I - P)P^\times x$ and the two preceding identities yield:

$$\begin{aligned}
(TU^\times x)(t) &= (U^\times x)(t) - iC(I - P)e^{-itA^\times}(I - P^\times)x \\
&\quad + iCe^{-itA}(I - P)(I - P^\times)x - iCPe^{-itA^\times}(I - P^\times)x \\
&= -iCe^{-itA}(I - P)P^\times x \\
&= -(USx)(t), \qquad t \geq 0,
\end{aligned}$$

and (12) is proved. For $x \in \operatorname{Im} P$ we also have

$$\begin{aligned}
RU^\times x &= -i\int_0^\infty Pe^{isA}BCe^{-isA^\times}(I - P^\times)x\,ds \\
&= -Pe^{isA}e^{-isA^\times}(I - P^\times)\Big|_0^\infty \\
&= P(I - P^\times)x = -QSx + x,
\end{aligned}$$

which proves (13).

The coupling relation (9) is now proved, and thus we can apply the results of Section III.4. By Corollary III.4.3

$$(15) \qquad \operatorname{Ker} T = \{\varphi \mid \varphi = U^\times x, x \in \operatorname{Ker} S\},$$

$$(16) \qquad \operatorname{Im} T = \{\psi \mid R^\times \psi \in \operatorname{Im} S\}.$$

$$(17) \qquad n(T) = n(S), \qquad d(T) = d(S).$$

Note that $\operatorname{Ker} S = \operatorname{Im} P \cap \operatorname{Ker} P^\times$ and

$$(18) \qquad\qquad P^\times z \in \operatorname{Im} S \Leftrightarrow z \in \operatorname{Im} P + \operatorname{Ker} P^\times.$$

With these remarks we have proved (2), (3) and the first identity in (4). To prove the second identity in (4) one uses that

$$(19) \qquad\qquad d(S) = \dim[\operatorname{Im} P^\times / \operatorname{Im} S] = \dim \frac{\mathbf{C}^n}{\operatorname{Im} P + \operatorname{Ker} P^\times}.$$

From (17) and the definition of S it follows that

$$n(T) = \operatorname{rank} P - \operatorname{rank} S, \qquad d(T) = \operatorname{rank} P^\times - \operatorname{rank} S,$$

and thus $\operatorname{ind}(T)$ is given by (5). Finally, let S^+ be a generalized inverse of S. Put

$$T^+ = T^\times - U^\times S^+ R^\times.$$

Then, by using (9),

$$\begin{aligned}
TT^+T &= TT^\times T - (TU^\times)S^+(R^\times T) \\
&= TT^\times T - USS^+SR \\
&= TT^\times T - USR \\
&= TT^\times T + UR^\times T = T,
\end{aligned}$$

which shows that T^+ is a generalized inverse of T in the weak sense. We know that the action of $T^\times$ is given by

$$(T^\times \psi)(t) = \psi(t) - \int_0^\infty k^\times(t - s)\psi(s)\,ds, \qquad t \geq 0,$$

where $k^\times$ is the kernel given by formula (5.3). Since

$$(U^\times S^+ R^\times \varphi)(t) = \int_0^\infty iCe^{-itA^\times}(I - P^\times)S^+ P^\times e^{isA^\times}B\varphi(s)\,ds, \qquad t \geq 0,$$

we see that the action of T^+ is given by (6) and (7). $\square$

From $\operatorname{Ker} S = \operatorname{Im} P \cap \operatorname{Ker} P^\times$ and (19) it follows that S is invertible if and only if $\mathbf{C}^n = \operatorname{Im} P \oplus \operatorname{Ker} P^\times$. In that case the operator T^+, given by (6) and (7), is the inverse of T and $\widetilde{\Pi}$ is the projection of $\mathbf{C}^n$ onto $\operatorname{Im} P$ along $\operatorname{Ker} P^\times$. It follows that Theorem 8.1 contains Theorem 7.1 as a special case.

XIII.9 AN EXAMPLE FROM TRANSPORT THEORY (1)

In this section we apply the theory of Wiener-Hopf operators developed in the present chapter to solve a finite dimensional version of a linear transport equation.

Transport theory concerns the mathematical analysis of equations that describe transport phenomena in matter, e.g., a flow of electrons through a metal strip or radiative transfer in a stellar atmosphere. Always these phenomena concern the migration of particles in a medium. Collision of the particles may result in absorption or production of new particles. For a homogeneous medium and without interaction between the particles, the mathematical equation describing a stationary transport problem is an integro-differential equation of the following form

$$(1) \qquad \mu \frac{\partial \psi}{\partial t}(t,\mu) + \psi(t,\mu) = \int_{-1}^{1} k(\mu,\mu')\psi(t,\mu')d\mu', \qquad -1 \leq \mu \leq 1.$$

This equation is a balance equation. The unknown function ψ is a density function related to the (expected) number of particles in an infinitesimal volume element. The right hand side of (1) describes the effect of the collisions. The function k, which is called the *scattering function*, is assumed to be real symmetric. The variable μ is the cosine of the scattering angle and therefore $-1 \leq \mu \leq 1$. The variable t is not a time variable, but a position variable (sometimes referred to as the optical depth).

In this section we assume that the medium is semi-infinite, and hence the position variable runs over the interval $0 \leq t < \infty$. Since the density of the incoming particles is known, the values of $\psi(0,\mu)$ are known for $0 \leq \mu \leq 1$. It follows that equation (1) appears with the following boundary condition:

$$(2) \qquad \psi(0,\mu) = \varphi_+(\mu), \qquad 0 < \mu \leq 1,$$

where φ_+ is a given function on $[0,1]$. There is also a boundary condition at infinity which is often stated as an integrability condition on the solution ψ. Equation (1) with $0 \leq t < \infty$ and the boundary condition (2) is called the *half range problem*.

In this section we consider a finite dimensional version of the half range problem. We assume that scattering occurs in a finite number of directions only. This assumption reduces the equation (1) and the boundary condition (2) to

$$(3) \qquad \mu_j \frac{\partial \psi}{\partial t}(t,\mu_j) + \psi(t,\mu_j) = \sum_{\nu=1}^{n} k(\mu_j,\mu_\nu)\psi(t,\mu_\nu),$$
$$j = 1,\ldots,n, \ 0 \leq t < \infty,$$

$$(4) \qquad \psi(t,\mu_j) = \varphi_+(\mu_j), \qquad \mu_j > 0.$$

To deal with the latter version of the half range problem, introduce the $\mathbf{C}^n$-valued vector function

$$\psi(t) = \begin{bmatrix} \psi(t,\mu_1) \\ \vdots \\ \psi(t,\mu_n) \end{bmatrix}, \qquad 0 \leq t < \infty,$$

and the matrices

$$T = (\mu_j \delta_{jk})_{j,k=1}^n, \qquad F = \big(k(\mu_j, \mu_k)\big)_{j,k=1}^n.$$

This allows us to rewrite (3) and (4) in the following form:

$$(5) \qquad \begin{cases} T\psi'(t) = -\psi(t) + F\psi(t), & 0 \le t < \infty, \\ P_+\psi(0) = x_+. \end{cases}$$

Here T and F are selfadjoint $n \times n$ matrices, P_+ is the spectral projection of T corresponding to the positive eigenvalues and x_+ is a given vector in $\operatorname{Im} P_+$. In what follows we assume that T is invertible (which corresponds to the requirement that all μ_j in (3) are different from 0). We shall look for solutions ψ of (5) in the space $L_2^n([0,\infty))$.

First we show that the problem (5) is equivalent to a Wiener-Hopf integral equation with a rational matrix symbol. Introduce the following matrix function:

$$(6) \qquad h(t) = \begin{cases} e^{-tT^{-1}} P_+ T^{-1}, & t > 0, \\ -e^{-tT^{-1}} P_- T^{-1}, & t < 0. \end{cases}$$

Here $P_- = I - P_+$. Since P_+ (resp. P_-) is the spectral projection of T corresponding to the positive (resp. negative) eigenvalues, the function $e^{-tT^{-1}} P_+$ (resp. $e^{-tT^{-1}} P_-$) is exponentially decaying on $0 \le t < \infty$ (resp. $-\infty < t \le 0$). It follows that h is a $n \times n$ matrix function of which the entries are integrable on $\mathbf{R}$.

THEOREM 9.1. *Let $\psi \in L_2^n([0,\infty))$. Then ψ is a solution of the equation* (5) *if and only if ψ is a solution of the Wiener-Hopf integral equation*

$$(7) \qquad \psi(t) - \int_0^\infty h(t-s)F\psi(s)ds = f(t), \qquad 0 \le t < \infty,$$

where f is the function in $L_2^n([0,\infty))$ defined by

$$(8) \qquad f(t) = e^{-tT^{-1}} x_+, \qquad 0 \le t < \infty.$$

PROOF. Assume that ψ is a solution of (5). Applying T^{-1} to the first identity in (5) and solving the resulting equation yields

$$(9) \qquad \psi(t) = e^{-tT^{-1}}\psi(0) + e^{-tT^{-1}} \int_0^t e^{sT^{-1}} T^{-1} F\psi(s)ds, \qquad 0 \le t < \infty.$$

Next, apply $e^{tT^{-1}} P_-$ to both sides of (9) and use that $e^{tT^{-1}}$ and P_- commute. Since $e^{tT^{-1}} P_-$ is exponentially decaying on $0 \le t < \infty$, the function $e^{tT^{-1}} P_- F\psi(t)$ is integrable on $0 \le t < \infty$, and thus

$$(10) \qquad \lim_{t\to\infty} e^{tT^{-1}} P_- \psi(t) = P_-\psi(0) + \int_0^\infty e^{sT^{-1}} P_- T^{-1} F\psi(s)ds.$$

Also the function $e^{tT^{-1}}P_-\psi(t)$ is integrable on $0 \le t < \infty$, and thus the left hand side of (10) is equal to zero, which proves that

$$(11) \qquad P_-\psi(0) = -\int_0^\infty e^{sT^{-1}}P_-T^{-1}F\psi(s)ds.$$

Now, replace $\psi(0)$ in (9) by $P_+\psi(0) + P_-\psi(0)$, use the boundary condition in (5) and apply (11). We conclude that

$$\psi(t) = e^{-tT^{-1}}x_+ - \int_0^\infty e^{-(t-s)T^{-1}}P_-T^{-1}F\psi(s)ds + \int_0^t e^{-(t-s)T^{-1}}T^{-1}F\psi(s)ds$$

$$= e^{-tT^{-1}}x_+ + \int_0^\infty h(t-s)F\psi(s)ds, \qquad 0 \le t < \infty.$$

Thus ψ is a solution of (7) with f given by (8).

To prove the converse statement, assume that ψ is a solution of (7) with f given by (8). Thus

$$(12) \qquad \begin{aligned} \psi(t) &= e^{-tT^{-1}}x_+ + e^{-tT^{-1}}\int_0^t e^{sT^{-1}}P_+T^{-1}F\psi(s)ds \\ &\quad - e^{-tT^{-1}}\int_t^\infty e^{sT^{-1}}P_-T^{-1}F\psi(s)ds, \qquad 0 \le t < \infty. \end{aligned}$$

It follows that ψ is absolutely continuous on each compact interval of $[0,\infty)$, and hence the integrands in the right hand side of (12) are continuous functions of the variable s. But then ψ is differentiable on $[0,\infty)$, and we see that for $0 \le t < \infty$

$$\begin{aligned} \psi'(t) &= -T^{-1}\psi(t) + P_+T^{-1}F\psi(t) + P_-T^{-1}F\psi(t) \\ &= -T^{-1}\psi(t) + T^{-1}F\psi(t), \end{aligned}$$

and hence ψ satisfies the first equation in (5). From (12) it also follows that

$$\psi(0) = x_+ - \int_0^\infty e^{sT^{-1}}P_-T^{-1}F\psi(s)ds, \qquad 0 \le t < \infty,$$

which implies that $P_+\psi(0) = P_+x_+ = x_+$. We conclude that ψ is a solution of the problem (5). $\square$

Let us compute the symbol of equation (7):

$$\int_{-\infty}^{\infty} e^{i\lambda t} h(t)dt = \int_{0}^{\infty} e^{i(\lambda+iT^{-1})t} P_+ T^{-1} dt - \int_{-\infty}^{0} e^{i(\lambda+iT^{-1})t} P_- T^{-1} dt$$

$$= -i(\lambda+iT^{-1})^{-1} e^{i(\lambda+iT^{-1})t} P_+ T^{-1} \Big|_{0}^{\infty}$$

$$+ i(\lambda+iT^{-1})^{-1} e^{i(\lambda+iT^{-1})t} P_- T^{-1} \Big|_{-\infty}^{0}$$

$$= i(\lambda+iT^{-1})^{-1} P_+ T^{-1} + i(\lambda+iT^{-1})^{-1} P_- T^{-1}$$

$$= i(\lambda+iT^{-1})^{-1} T^{-1}.$$

It follows that the symbol W of the Wiener-Hopf equation (7) has the following realization:

$$(13) \qquad\qquad W(\lambda) = I - iI(\lambda+iT^{-1})^{-1}T^{-1}F.$$

This allows us to apply Theorem 7.1 with

$$(14) \qquad\qquad A = -iT^{-1}, \qquad B = T^{-1}F, \qquad C = -iI.$$

THEOREM 9.2. *Assume that $I - F$ is positive definite. Then for each* $f \in L_2^n([0,\infty))$ *the equation*

$$(15) \qquad\qquad \psi(t) - \int_{0}^{\infty} h(t-s)F\psi(s)ds = f(t), \qquad 0 \le t < \infty,$$

has a unique solution $\psi \in L_2^n([0,\infty))$. *In particular, if the right hand side f is given by* (8), *then the solution of* (15) *is*

$$(16) \qquad\qquad \psi(t) = e^{-tT^{-1}(I-F)}\Pi x_+, \qquad 0 \le t < \infty,$$

where Π is the projection of $\mathbf{C}^n$ along $\operatorname{Ker} P_+$ onto the spectral subspace of $(I - F)^{-1}T$ corresponding to eigenvalues in the open right half plane.

A priori it is not clear that a projection Π as in (16) may be defined; it will be part of the proof to show the existence of Π.

PROOF OF THEOREM 9.2. We apply Theorem 7.1 with A, B and C as in (14). Note that

$$(17) \qquad\qquad A^\times = A - BC = -iT^{-1}(I - F).$$

To prove that (15) is uniquely solvable in $L_2^n([0,\infty))$, we have to show that $A^\times$ has no real eigenvalue and $\mathbf{C}^n = M \oplus M^\times$, where M is the spectral subspace of A corresponding to the eigenvalues in $\Im\lambda > 0$ and $M^\times$ is the spectral subspace of $A^\times$ corresponding to the eigenvalues in $\Im\lambda < 0$. Put $S = (I - F)^{-1}T$, and consider the sesquilinear form:

$$[x, y] = \langle(I - F)x, y\rangle. \tag{18}$$

Since $I - F$ is positive definite, S is well-defined and $[\cdot, \cdot]$ is an inner product on $\mathbf{C}^n$. From

$$[Sx, y] = \langle(I - F)Sx, y\rangle = \langle Tx, y\rangle,$$

it follows that S is selfadjoint with respect to the inner product $[\cdot, \cdot]$. But then the same is true for the operator $iA^\times = S^{-1}$. Thus, $A^\times$ is invertible and has its eigenvalues on the imaginary axis. In particular, $A^\times$ has no real eigenvalue.

Let P_+ and Q_+ be the spectral projections of T and S, respectively, corresponding to the eigenvalues in $\Re\lambda > 0$. The equality $iA = T^{-1}$ implies that $M = \operatorname{Ker} P_+$. Similarly, $iA^\times = S^{-1}$ yields $M^\times = \operatorname{Im} Q_+$. Since T and S are invertible, $T|\operatorname{Ker} P_+$ is negative definite and $S|\operatorname{Im} Q_+$ is positive definite. Thus

$$0 \neq x \in \operatorname{Ker} P_+ \Rightarrow \langle Tx, x\rangle < 0,$$
$$0 \neq x \in \operatorname{Im} Q_+ \Rightarrow [Sx, x] > 0.$$

But $[Sx, x] = \langle Tx, x\rangle$ for all $x \in \mathbf{C}^n$. It follows that $\operatorname{Ker} P_+ \cap \operatorname{Im} Q_+ = \{0\}$. In particular, $\operatorname{rank} P_+ \geq \operatorname{rank} Q_+$. By repeating the argument with the roles of P_+ and Q_+ interchanged, we see that $\operatorname{rank} Q_+ \geq \operatorname{rank} P_+$, and hence $\operatorname{rank} Q_+ = \operatorname{rank} P_+$. But then we may conclude that $\mathbf{C}^n = \operatorname{Ker} P_+ \oplus \operatorname{Im} Q_+$.

We have now proved that (15) is uniquely solvable in $L_2^n([0, \infty))$. Also, we have shown that the projection Π in (16) is well-defined. In fact, Π is the projection of $\mathbf{C}^n$ along M onto $M^\times$. It remains to prove (16). Let f be given by (8). From Theorem 7.1 we know that the solution of (15) is given by

$$\psi(t) = f(t) + \int_0^\infty \gamma(t, s)f(s)ds, \qquad 0 \leq t < \infty,$$

where the resolvent kernel γ is defined by formula (4) in Section 7. Note that f may be rewritten as

$$f(t) = iCe^{-itA}x_+, \qquad 0 \leq t < \infty.$$

Since

$$\frac{d}{ds}(e^{isA^\times}e^{-isA}) = -e^{isA^\times}(iBC)e^{-isA},$$

it follows that

$$\psi(t) = iCe^{-itA}x_+ + iCe^{-itA^\times} \int_0^t \Pi e^{isA^\times}(iBC)e^{-isA}x_+ ds$$

$$- iCe^{-itA^\times} \int_t^\infty (I - \Pi)e^{isA^\times}(iBC)e^{-isA}x_+ ds$$

$$= iCe^{-itA}x_+ - iCe^{-itA^\times}\left(\Pi e^{isA^\times}e^{-isA}x_+ \Big|_{s=0}^t\right)$$

$$+ iCe^{-itA^\times}\left((I - \Pi)e^{isA^\times}e^{-isA}x_+ \Big|_{s=t}^\infty\right).$$

Next use that

$$e^{-itA}x_+ = e^{-tT^{-1}}P_+x_+ \to 0 \qquad (t \to \infty)$$

$$(I - \Pi)e^{itA^\times} = (I - \Pi)e^{tS^{-1}}(I - Q_+) \to 0 \qquad (t \to \infty).$$

In the latter limit we employ the fact that $(I - \Pi)Q_+ = 0$. It is now clear that

$$\psi(t) = iCe^{-itA}x_+ - iCe^{-itA^\times}\Pi e^{itA^\times}e^{-itA}x_+$$

$$+ iCe^{-itA^\times}\Pi x_+ - iCe^{-itA^\times}(I - \Pi)e^{itA^\times}e^{-itA}x_+$$

$$= iCe^{-itA^\times}\Pi x_+, \qquad 0 \le t < \infty.$$

Since $iA^\times = T^{-1}(I - F)$, the above calculations prove (16). $\square$.

COROLLARY 9.3. *If* $I - F$ *is positive definite, then the problem* (5) *has a unique solution in* $L_2^n([0\infty))$, *namely*

$$(19) \qquad \psi(t) = e^{-tT^{-1}(I-F)}\Pi x_+, \qquad 0 \le t < \infty,$$

where Π *is the projection of* $\mathbf{C}^n$ *along* $\operatorname{Ker} P_+$ *onto the spectral subspace of* $(I - F)^{-1}T$ *corresponding to the eigenvalues in the open right half plane.*

PROOF. Apply Theorems 9.1 and 9.2. $\square$

To solve (5) effectively requires one to compute the projection Π in (19) and to analyse further the matrix $R = T^{-1}(I - F)$. From the proof of Theorem 9.2 we know that R is selfadjoint with respect to the inner product $[\cdot, \cdot]$ defined in (18), and hence R admits a diagonal representation. In the last part of this section we show how one may compute Π and find a diagonal representation of R for a concrete class of examples.

In what follows the underlying space in (5) assumed to be $\mathbf{C}^{2n}$, and we take

$$(20) \qquad T = \operatorname{diag}(\alpha_1, \ldots, \alpha_n, -\alpha_1, \ldots, -\alpha_n),$$

$$(21) \qquad F = \langle \cdot, g\rangle g,$$

where

$$g = (m_1, \ldots, m_n, m_1, \ldots, m_n)^T.$$

Thus T is a $2n \times 2n$ diagonal matrix whose diagonal elements are specified by the right hand side of (20). We shall assume that

$$(22) \qquad \alpha_1 > \cdots > \alpha_n > 0, \qquad m_j \neq 0 \quad (j = 1, \ldots, n),$$

$$(23) \qquad \sum_{j=1}^{n} |m_j|^2 < \frac{1}{2}.$$

It follows that T is selfadjoint and the spectral projection P_+ is given by

$$(24) \qquad P_+ = \mathrm{diag}(1, \ldots, 1, 0, \ldots, 0).$$

Condition (23) implies that $I - F$ is positive definite, and hence with T and F as in (20) and (21), equation (5) is uniquely solvable in $L_2^{2n}([0, \infty))$. We shall prove that its solution is given by

$$(25) \qquad \psi(t) = \left[\begin{array}{c} \Lambda_1 e^{-t\Gamma} \Lambda_1^{-1} x_0 \\ \Lambda_2 e^{-t\Gamma} \Lambda_1^{-1} x_0 \end{array} \right], \qquad t \geq 0,$$

where x_0 is the vector in $\mathbf{C}^n$ whose coordinates are the first n coordinates of x_+,

$$(26) \qquad \Lambda_1 = \left[\frac{c_j m_j}{c_j - \gamma_k} \right]_{j,k=1}^n, \qquad \Lambda_2 = \left[\frac{c_j m_j}{c_j + \gamma_k} \right]_{j,k=1}^n,$$

$$(27) \qquad \Gamma = \mathrm{diag}(\gamma_1, \ldots, \gamma_n).$$

Here $c_j = 1/\alpha_j$ for $j = 1, \ldots, n$ and $\gamma_1, \ldots, \gamma_n$ are the n different zeros in the open right half plane of the function

$$(28) \qquad \omega(\lambda) = 1 + 2 \sum_{j=1}^{n} \frac{c_j^2 |m_j|^2}{\lambda^2 - c_j^2}.$$

The matrix Λ_1 is of so-called Hilbert type and its inverse can be computed explicitly. In fact (see Knuth [1])

$$\Lambda_1^{-1} = D_1 \left(\left[\frac{c_k^{-1} m_k^{-1}}{c_k - \gamma_j} \right]_{j,k=1}^n \right) D_2.$$

Here D_1 and D_2 are $n \times n$ diagonal matrices. Their ν-th diagonal elements are given by

$$\frac{\prod_{j=1}^{n}(c_j - \gamma_\nu)}{\prod_{j \neq \nu}(\gamma_j - \gamma_\nu)}, \qquad \frac{\prod_{j=1}^{n}(c_\nu - \gamma_j)}{\prod_{j \neq \nu}(c_\nu - c_j)},$$

respectively. Thus for the class of examples considered here the problem to solve the equation (5) is reduced to finding the zeros in the open right half plane of the scalar function (28).

To establish formula (25) we have to analyse the spectral properties of the matrix $R = T^{-1}(I - F)$ with T and F as in (20) and (21). This is done in the next two lemmas.

LEMMA 9.4. *The $2n \times 2n$ matrix R has $2n$ different eigenvalues. They are of the form $\pm\gamma_j$, $j = 1,\ldots,n$, where*

$$(29) \qquad\qquad 0 < \gamma_1 < c_1 < \gamma_2 < c_2 < \cdots < \gamma_n < c_n,$$

and they coincide with the zeros of the function (28).

Recall that $c_j = 1/\alpha_j$. Put $\gamma_{-j} = -\gamma_j$ for $j = 1,\ldots,n$. Thus $\gamma_{-1},\ldots,\gamma_{-n},\gamma_1,\ldots,\gamma_n$ are the eigenvalues of $R = T^{-1}(I - F)$.

LEMMA 9.5. *For $|k| = 1,\ldots,n$ the vector*

$$(30) \qquad f_k = \left(\frac{c_1 m_1}{c_1 - \gamma_k}, \ldots, \frac{c_n m_n}{c_n - \gamma_k}, \frac{c_1 m_1}{c_1 + \gamma_k}, \ldots, \frac{c_n m_n}{c_n + \gamma_k} \right)^T$$

is an eigenvector of R corresponding to the eigenvalue γ_k.

PROOF OF LEMMA 9.4. Let W be the symbol of the Wiener-Hopf equation associated with (5). Thus $W(\lambda) = I + C(\lambda - A)^{-1}B$, where $A = -iT^{-1}$, $B = T^{-1}F$ and $C = -iI$. In what follows T and F are as in (20) and (21), and hence

$$(31) \qquad\qquad A = \text{diag}(-ic_1,\ldots,-ic_n,ic_1,\ldots,ic_n).$$

Put $e = \|g\|^{-1}g$. Then $Fe = \|g\|g$ and

$$\langle W(\lambda)e, e \rangle = 1 + \langle (\lambda - A)^{-1}Ag, g \rangle$$

$$= 1 + \sum_{j=1}^{n} \frac{-ic_j|m_j|^2}{\lambda + ic_j} + \sum_{j=1}^{n} \frac{ic_j|m_j|^2}{\lambda - ic_j}$$

$$= 1 - 2\sum_{j=1}^{n} \frac{c_j^2|m_j|^2}{\lambda^2 + c_j^2}$$

$$= \omega(i\lambda),$$

where $\omega(\cdot)$ is defined by (28). The previous calculation implies that there exists an invertible matrix S such that

$$(32) \qquad\qquad W(\lambda) = S \begin{bmatrix} \omega(i\lambda) & 0 \\ V(\lambda) & I_{\mathbf{C}^{2n-1}} \end{bmatrix} S^{-1}.$$

Recall that $W(\lambda)^{-1} = I - C(\lambda - A^{\times})^{-1}B$ (see Lemma 5.2). We know that $iA^{\times} = T^{-1}(I - F)$ is selfadjoint relative to an equivalent inner product. It follows that the eigenvalues of $A^{\times}$ are on the imaginary axis, and hence the same holds true for the poles of $W(\cdot)^{-1}$. But then we can use (32) to show that the function ω has its zeros on the real line. Since $\omega(\lambda) = \omega(-\lambda)$, a simple sketch of the function $t \mapsto \omega(t)$, $0 \le t < \infty$, shows that ω has precisely $2n$ zeros, which are of the form $\pm\gamma_j$, where $\gamma_1, \ldots, \gamma_n$ satisfy (29).

It remains to show that each zero of ω is an eigenvalue of $R = iA^{\times}$. So assume that $\omega(\gamma) = 0$. Then

$$0 = \omega(\gamma) = \langle W(-i\gamma)e, e\rangle = 1 + \langle(-i\gamma - A)^{-1}Ag, g\rangle.$$

Thus $\langle f, g\rangle = 1$ for $f = (i\gamma + A)^{-1}Ag$. We claim that f is an eigenvector for R. Indeed,

$$\begin{aligned}
Rf &= iAf - iBCf = iAf - iAFf \\
&= iA(i\gamma + A)^{-1}Ag - i\langle f, g\rangle Ag \\
&= iAg + \gamma(i\gamma + A)^{-1}Ag - iAg = \gamma f.
\end{aligned}$$

Since $f \neq 0$, we conclude that γ is an eigenvalue of R and f is a corresponding eigenvector. $\square$

PROOF OF LEMMA 9.5. From the last part of the proof of Lemma 9.4 we know that $f_k := (i\gamma_k + A)^{-1}Ag$ is an eigenvector of R corresponding to the eigenvalue γ_k. Since A is given by (31), we see that f_k is equal to the right hand side of (30). $\square$

To prove (25) requires one to specify further the formula of ψ appearing in Corollary 9.3. In particular, we have to compute the matrix $R = T^{-1}(I - F)$ and the projection Π for the case considered here. To do this, introduce the $2n \times 2n$ matrix

$$\Lambda = [f_1 \cdots f_n \ f_{-1} \cdots f_{-n}]$$

with $f_{\pm k}$, $k = 1, \ldots, n$, defined by (30). The matrix Λ is invertible and diagonalizes R, i.e., $R = \Lambda D \Lambda^{-1}$, where

$$D = \operatorname{diag}(\gamma_1, \ldots, \gamma_n, -\gamma_1, \ldots, -\gamma_n).$$

If follows that

$$e^{-tT^{-1}(I-F)} = \Lambda e^{-tD}\Lambda^{-1} = \Lambda \begin{bmatrix} e^{-t\Gamma} & 0 \\ 0 & e^{t\Gamma} \end{bmatrix} \Lambda^{-1},$$

where Γ is given by (27). Next, we partition Λ as a 2×2 block matrix whose entries are $n \times n$ matrices. This yields

$$\Lambda = \begin{bmatrix} \Lambda_1 & \Lambda_2 \\ \Lambda_2 & \Lambda_1 \end{bmatrix},$$

where Λ_1 and Λ_2 are as in (26). Put

$$M = \operatorname{Im}\begin{bmatrix} 0 \\ I_{\mathbb{C}^n} \end{bmatrix}, \qquad M^{\times} = \operatorname{Im}\begin{bmatrix} \Lambda_1 \\ \Lambda_2 \end{bmatrix}.$$

Then $M = \operatorname{Ker} P_+$ and $M^\times$ is the spectral subspace of $T^{-1}(I - F)$ (and hence of $(I - F)^{-1}T$) corresponding to the eigenvalues in the open right half plane. Thus $\mathbf{C}^{2n} = M \oplus M^\times$, and the projection Π of $\mathbf{C}^{2n}$ along M onto $M^\times$ is given by

$$\Pi = \begin{bmatrix} \Lambda_1 & 0 \\ \Lambda_2 & I \end{bmatrix} \begin{bmatrix} I & 0 \\ 0 & 0 \end{bmatrix} \begin{bmatrix} \Lambda_1 & 0 \\ \Lambda_2 & I \end{bmatrix}^{-1}$$

$$= \begin{bmatrix} \Lambda_1 & 0 \\ \Lambda_2 & 0 \end{bmatrix} \begin{bmatrix} \Lambda_1^{-1} & 0 \\ -\Lambda_2\Lambda_1^{-1} & I \end{bmatrix} = \begin{bmatrix} I & 0 \\ \Lambda_2\Lambda_1^{-1} & 0 \end{bmatrix}.$$

Finally, partition the given vector x_+ as $\begin{bmatrix} x_0 \\ 0 \end{bmatrix}$, where $x_0 \in \mathbf{C}^n$ and 0 is the zero vector in $\mathbf{C}^n$. Then

$$\psi(t) = e^{-tT^{-1}(I-F)}\Pi x_+$$

$$= \Lambda \begin{bmatrix} e^{-t\Gamma} & 0 \\ 0 & e^{t\Gamma} \end{bmatrix} \Lambda^{-1} \begin{bmatrix} I & 0 \\ \Lambda_2\Lambda_1^{-1} & 0 \end{bmatrix} \begin{bmatrix} x_0 \\ 0 \end{bmatrix}$$

$$= \Lambda \begin{bmatrix} e^{-t\Gamma} & 0 \\ 0 & e^{t\Gamma} \end{bmatrix} \Lambda^{-1} \begin{bmatrix} \Lambda_1 \\ \Lambda_2 \end{bmatrix} \Lambda_1^{-1}x_0$$

$$= \Lambda \begin{bmatrix} e^{-t\Gamma} & 0 \\ 0 & e^{t\Gamma} \end{bmatrix} \begin{bmatrix} \Lambda_1^{-1}x_0 \\ 0 \end{bmatrix}$$

$$= \begin{bmatrix} \Lambda_1 e^{-t\Gamma}\Lambda_1^{-1}x_0 \\ \Lambda_2 e^{-t\Gamma}\Lambda_1^{-1}x_0 \end{bmatrix},$$

which proves formula (25).

XIII.10 CONVOLUTION OPERATORS ON A FINITE INTERVAL

The results discussed in the last three sections carry over to convolution operators on a finite interval $[0, \tau]$. Such operators are of the form $A = I - K$, where

$$(1) \qquad (K\varphi)(t) = \int_0^\tau k(t - s)\varphi(s)ds, \qquad 0 \le t \le \tau.$$

The kernel function k will be an $m \times m$ matrix function whose entries are integrable on $[0, \tau]$, and the operator K will be considered on $L_2^m([0, \tau])$. To make more precise the type of kernel functions we shall deal with, let us first assume that k is the restriction to $[-\tau, \tau]$ of an $m \times m$ matrix function ℓ with entries in $L_1(\mathbf{R})$ whose Fourier transform $\widehat{\ell}$ is rational. This implies, by Theorem 4.1, that $\widehat{\ell}$ admits a realization, and hence, by Theorem 4.2, the kernel function k can be written in the form:

$$(2) \qquad k(t) = \begin{cases} iCe^{-itA}(I - P)B, & 0 \le t \le \tau, \\ -iCe^{-itA}PB, & -\tau \le t < 0, \end{cases}$$

where A is a square matrix of size $n \times n$, say, which has no real eigenvalue, B and C are matrices of sizes $n \times m$ and $m \times n$, respectively, and P is the Riesz projection of A

corresponding to the eigenvalues in the upper half plane. The actual value of k at zero is not relevant; for simplicity we assume here that $k(0) = k(0+)$.

In what follows we shall work with a slightly more general representation of k. We shall assume that the kernel function k of (1) admits a representation (2), where A is an arbitrary $n \times n$ matrix and P is a projection commuting with A. Thus A is allowed to have real eigenvalues and P is not required to be a Riesz projection. We shall refer to (2) as an *exponential representation* of k.

One of our aims is to find the condition of invertibility of $I - K$ and to determine its inverse in terms of the representation (2).

The representation (2) implies that the operator K in (1) has a semi-separable kernel function (cf. Section IX.1). Indeed, let $f_1, \ldots, f_{n_1}$, and $g_1, \ldots, g_{n_2}$ be bases of $\operatorname{Ker} P$ and $\operatorname{Im} P$, respectively, and let E be the $n \times n$ matrix whose columns are the vectors $f_1, \ldots, f_{n_1}, g_1, \ldots, g_{n_2}$. Then E is invertible and

$$(3) \qquad P = E \begin{bmatrix} 0 & 0 \\ 0 & I_{n_2} \end{bmatrix} E^{-1}.$$

Here, as usual, the symbol I_k denotes the $k \times k$ identity matrix. Since P commutes with A, it follows that

$$(4) \qquad k(t - s) = \begin{cases} F_1(t)G_1(s), & 0 \leq s \leq t \leq \tau, \\ F_2(t)G_2(s), & 0 \leq t < s \leq \tau, \end{cases}$$

where

$$(5a) \qquad F_1(t) = iCe^{-itA}E \begin{bmatrix} I_{n_1} \\ 0 \end{bmatrix}, \qquad G_1(t) = [\, I_{n_1} \quad 0\,]E^{-1}e^{itA}B,$$

$$(5b) \qquad F_2(t) = iCe^{-itA}E \begin{bmatrix} 0 \\ I_{n_2} \end{bmatrix}, \qquad G_2(t) = -[\, 0 \quad I_{n_2}\,]E^{-1}e^{itA}B.$$

The semi-separable representation (4) allows us to apply the results of Sections IX.2 and IX.3, which yields the following theorems.

THEOREM 10.1. *Let K be the integral operator on $L_2^m([0, \tau])$ defined by (1), and assume that k has the exponential representation (2). Let n be the order of A, and put $A^\times = A - BC$. Then $I - K$ is invertible if and only if the map*

$$(6) \qquad S_\tau = Pe^{i\tau A}e^{-i\tau A^\times}P \colon \operatorname{Im} P \to \operatorname{Im} P$$

is invertible. In that case

$$((I - K)^{-1}f)(t) = f(t) + \int_0^\tau \gamma(t, s)f(s)\,ds, \qquad 0 \leq t \leq \tau,$$

with

$$(7) \qquad \gamma(t,s) = \begin{cases} iCe^{-itA^\times}\Pi_\tau e^{isA^\times}B, & 0 \le s \le t \le \tau, \\ -iCe^{-itA^\times}(I - \Pi_\tau)e^{isA^\times}B, & 0 \le t < s \le \tau. \end{cases}$$

Here Π_τ is the projection of $\mathbf{C}^n$ along $\operatorname{Im}P$ defined by

$$(8) \qquad \Pi_\tau x = x - S_\tau^{-1}Pe^{i\tau A}e^{-i\tau A^\times}x, \qquad x \in \mathbf{C}^n.$$

THEOREM 10.2. *Let K be the integral operator on $L_2^m([0,\tau])$ defined by* (1), *and assume that k has the exponential representation* (2). *Let n be the order of A, and put*

$$(9) \qquad S_\tau = Pe^{i\tau A}e^{-i\tau A^\times}P\colon \operatorname{Im}P \to \operatorname{Im}P,$$

where $A^\times = A - BC$. Then

$$\operatorname{Ker}(I - K) = \{\varphi \mid \varphi(t) = Ce^{-itA^\times}Py, \qquad y \in \operatorname{Ker}S_\tau\},$$

$$\operatorname{Im}(I - K) = \left\{f \in L_2^m([0,\tau]) \mid Pe^{i\tau A}e^{-i\tau A^\times}\int_0^\tau e^{isA^\times}Bf(s)ds \in \operatorname{Im}S_\tau\right\},$$

and a generalized inverse of $I - K$ in the weak sense is given by the operator

$$((I - K)^+f)(t) = f(t) + \int_0^\tau \widetilde{\gamma}(t,s)f(s)ds, \qquad 0 \le t \le \tau,$$

with

$$(10) \qquad \widetilde{\gamma}(t,s) = \begin{cases} iCe^{-itA^\times}(I - \widetilde{\Pi}_\tau)e^{isA^\times}B, & 0 \le s \le t \le \tau, \\ -iCe^{-itA^\times}\widetilde{\Pi}_\tau e^{isA^\times}B, & 0 \le t < s \le \tau. \end{cases}$$

Here $\widetilde{\Pi}_\tau x = S_\tau^+ Pe^{i\tau A}e^{-i\tau A^\times}x$ for $x \in \mathbf{C}^n$, where S_τ^+ is a generalized inverse of S_τ.

We shall refer to the operator S_τ in (6) and (9) as the *indicator* of $I - K$ associated with the representation (2). Note that S_τ acts on a finite dimensional space, and hence it always has a generalized inverse.

PROOF OF THEOREMS 10.1 AND 10.2. We shall employ the semiseparable representation (4) and we shall apply Theorem IX.2.3. Let us compute the indicator V of $I - K$ associated with the representation (4) (cf. Section IX.2). Put

$$A(t) = \begin{bmatrix} G_1(t)F_1(t) & G_1(t)F_2(t) \\ -G_2(t)F_2(t) & -G_2(t)F_2(t) \end{bmatrix}, \qquad 0 \le t \le \tau,$$

where F_1, F_2 and G_1, G_2 are as in (5a) and (5b). Furthermore, let $U(t) = E^{-1}e^{itA}e^{-itA^\times}E$ for $0 \le t \le \tau$. Since

$$A(t) = E^{-1}e^{itA}(iBC)e^{-itA}E, \qquad 0 \le t \le \tau,$$

we have

$$\int_0^t A(s)U(s)ds = \int_0^t E^{-1}e^{isA}(iBC)e^{-isA^\times}E\,ds$$

$$= E^{-1}\left\{\int_0^t \frac{d}{ds}(e^{isA}e^{-isA^\times})ds\right\}E$$

$$= E^{-1}\{e^{itA}e^{-itA^\times} - I_n\}E$$

$$= U(t) - I_n, \qquad 0 \le t \le \tau.$$

It follows that $U(t)$ is the fundamental matrix (normalized to I_n at $t = 0$) of the differential equation $x'(t) = A(t)x(t)$, $0 \le t \le \tau$, and therefore

$$V = [\,0 \quad I_{n_2}\,]U(\tau)\begin{bmatrix} 0 \\ I_{n_2} \end{bmatrix}$$

$$= [\,0 \quad I_{n_2}\,]E^{-1}e^{i\tau A}e^{-i\tau A^\times}E\begin{bmatrix} 0 \\ I_{n_2} \end{bmatrix}$$

$$= [\,0 \quad I_{n_2}\,]E^{-1}Pe^{i\tau A}e^{-i\tau A^\times}PE\begin{bmatrix} 0 \\ I_{n_2} \end{bmatrix}$$

$$= [\,0 \quad I_{n_2}\,]E^{-1}S_\tau E\begin{bmatrix} 0 \\ I_{n_2} \end{bmatrix},$$

where S_τ is the indicator of $I - K$ associated with the representation (2). In the above calculation we used that P satisfies (3). Note that the linear transformation

$$E_0 = E\begin{bmatrix} 0 \\ I_{n_2} \end{bmatrix} : \mathbf{C}^{n_2} \to \operatorname{Im} P$$

is invertible and $E_0^{-1} = [\,0 \quad I_{n_2}\,]E^{-1}|\operatorname{Im} P$. Thus we have shown that $V = E_0^{-1}S_\tau E_0$.

It follows that the operators $I - K$ and S_τ are matricially coupled. To find the coupling relation we refer to Theorem IX.3.2 for the case considered here. Introduce the following operators

(11) $$H: L_2^m([0,\tau]) \to L_2^m([0,\tau]),$$

$$(H\varphi)(t) = iCe^{-itA}\int_0^t e^{isA}B\varphi(s)ds, \qquad 0 \le t \le \tau;$$

(12) $$R: \operatorname{Im} P \to L_2^m([0,\tau]), \qquad (Rx)(t) = iCe^{-itA}x, \qquad 0 \le t \le \tau,$$

(13) $$Q: L_2^m([0,\tau]) \to \operatorname{Im} P, \qquad Q\varphi = -\int_0^\tau Pe^{isA}B\varphi(s)ds.$$

The final result of the previous paragraph and Theorem IX.2.3 imply that the operators $I - H$ and

$$\begin{bmatrix} I - K & R \\ Q & I_{\mathrm{Im}P} \end{bmatrix} : L_2^m([0,\tau]) \oplus \mathrm{Im}\,P \to L_2^m([0,\tau]) \oplus \mathrm{Im}\,P$$

are invertible and

$$(14) \qquad \begin{bmatrix} I - K & R \\ Q & I_{\mathrm{Im}P} \end{bmatrix}^{-1} = \begin{bmatrix} (I - H)^{-1} & (I - H)^{-1}R \\ -Q(I - H)^{-1} & S_\tau \end{bmatrix},$$

which is the desired coupling relation. Furthermore (see formula (19) in Section IX.2 for the case considered here),

$$(15) \qquad ((I - H)^{-1}f)(t) = f(t) + iCe^{-itA^\times} \int_0^t e^{isA^\times} Bf(s)\,ds, \qquad 0 \le t \le \tau.$$

It follows (see Corollary III.4.3) that $I - K$ is invertible if and only if S_τ is invertible, and in that case

$$(16) \qquad (I - K)^{-1} = (I - H)^{-1} + (I - H)^{-1}RS_\tau^{-1}Q(I - H)^{-1}.$$

Since (cf. formulas (20) and (23) in Section IX.2)

$$(17) \qquad ((I - H)^{-1}Rx)(t) = iCe^{-itA^\times} x, \qquad 0 \le t \le \tau,$$

$$(18) \qquad Q(I - H)^{-1}f = -Pe^{i\tau A}e^{-i\tau A^\times} \int_0^\tau e^{isA^\times} Bf(s)\,ds,$$

formulas (15) and (16) yield the expression for $(I - K)^{-1}$ appearing in Theorem 10.1. To complete the proof of Theorem 10.1, it remains to show that the map Π_τ defined by (8) is a projection of $\mathbf{C}^n$ along $\mathrm{Im}\,P$. To do this, note that $\Pi_\tau P = 0$ according to the definition of S_τ. On the other hand $\mathrm{Im}(I - \Pi_\tau) \subset \mathrm{Im}\,P$, and hence $\Pi_\tau(I - \Pi_\tau) = 0$. Thus Π_τ is a projection and $\mathrm{Ker}\,\Pi_\tau = \mathrm{Im}\,P$.

Next, we prove Theorem 10.2. The descriptions of $\mathrm{Ker}(I - K)$ and $\mathrm{Im}(I - K)$ follow directly from Corollary III.4.3 by using the coupling relation (14) and the formulas (17) and (18). To prove the second part of the theorem, assume that S_τ^+ is a generalized inverse of S_τ. Put

$$(19) \qquad (I - K)^+ = (I - H)^{-1} + (I - H)^{-1}RS_\tau^+Q(I - H)^{-1}.$$

We want to show that $(I - K)^+$ is a generalized inverse of $I - K$ in the weak sense. To do this, note that the coupling relation (14) implies that

$$(I - K)(I - H)^{-1} = I + RQ(I - H)^{-1}$$
$$(I - K)(I - H)^{-1}R = -RS_\tau,$$
$$Q(I - H)^{-1}(I - K) = S_\tau Q.$$

Since $S_\tau S_\tau^+ S_\tau = S_\tau$, we see that

$$
\begin{aligned}
(I - K)(I - K)^+(I - K) &= (I - K)(I - H)^{-1}(I - K) \\
&\quad + (I - K)(I - H)^{-1} R S_\tau^+ Q(I - H)^{-1}(I - K) \\
&= I - K + RQ(I - H)^{-1}(I - K) - R S_\tau S_\tau^+ S_\tau Q \\
&= I - K + R S_\tau Q - R S_\tau Q = I - K,
\end{aligned}
$$

and hence $(I - K)^+$ is the desired generalized inverse. Inserting (15), (17) and (18) into (19) yield the expression for $(I - K)^+$ appearing in Theorem 10.2. $\square$

The operators K considered in this section are Hilbert-Schmidt operators. The following theorem provides a simple criterion for K to be a trace class operator.

THEOREM 10.3. *Let K be the integral operator on $L_2^m([0,\tau])$ defined by (1), and assume that k has the exponential representation (2). Then K is a trace class operator if and only if $CB = 0$, and in that case*

$$
(20) \qquad \operatorname{tr} K = -i\tau \operatorname{tr} CPB, \qquad \det(I - K) = \det S_\tau,
$$

where S_τ is the indicator of $I - K$ associated with the representation (2).

PROOF. Let H, R and Q be defined by (11), (12) and (13), respectively. Since $K = H + RQ$ and RQ has finite rank, we have to show that H is a trace class operator if and only if $CB = 0$.

Let M_A be the multiplication operator on $L_2^n([0,\tau])$ induced by A, i.e.,

$$
(M_A\varphi)(t) = A\varphi(t), \qquad 0 \le t \le \tau.
$$

Then M_A is a bounded operator which commutes with the operator $J \colon L_2^n([0,\tau]) \to L_2^n([0,\tau])$ defined by

$$
(21) \qquad (J\varphi)(t) = -i \int_0^t \varphi(s)\,ds, \qquad 0 \le t \le \tau.
$$

Note that

$$
(JM_A\varphi)(t) = -i \int_0^t A\varphi(s)\,ds, \qquad 0 \le t \le \tau.
$$

Next apply Theorem 10.1 to $K = JM_A$ (and thus with $A = 0$, $B = -A$, $C = I$ and $P = 0$). It follows that $I - JM_A$ is invertible and

$$
[(I - JM_A)^{-1}\varphi](t) = \varphi(t) - i \int_0^t e^{-i(t-s)A} A\varphi(s)\,ds, \qquad 0 \le t \le \tau.
$$

But then

$$[J(I - JM_A)^{-1}\varphi](t) = (J\varphi)(t) - \int_0^t \left(\int_0^s e^{-i(s-\alpha)A} A\varphi(\alpha)d\alpha \right) ds$$

$$= (J\varphi)(t) - \int_0^t \left(\int_\alpha^t e^{-i(s-\alpha)A} A\varphi(\alpha)ds \right) d\alpha$$

$$= (J\varphi)(t) - i \int_0^t \left\{ \int_\alpha^t \left(\frac{d}{ds} e^{-i(s-\alpha)A} \right) ds \right\} \varphi(\alpha)d\alpha$$

$$= -i \int_0^t e^{-i(t-\alpha)A} \varphi(\alpha)d\alpha, \qquad 0 \leq t \leq \tau.$$

So we have proved the identity

(22) $$H = -M_C J(I - JM_A)^{-1} M_B,$$

where $M_B \colon L_2^m([0,\tau]) \to L_2^n([0,\tau])$ and $M_C \colon L_2^n([0,\tau]) \to L_2^m([0,\tau])$ are the multiplication operators induced by B and C, respectively.

Formula (22) implies that

$$H = -M_C J M_B - M_C J^2 M_A (I - JM_A)^{-1} M_B.$$

Since J is a Hilbert-Schmidt operator, J^2 is a trace class operator (Lemma VIII.2.2). Hence H is a trace class operator if and only if $M_C J M_B$ is trace class. Let b_{jk} and c_{jk} be the (j,k)-th entries of the matrices of B and C, respectively. Identifying $L_2^m([0,\tau])$ with a product of m copies of $L_2([0,\tau])$, one can write the operator

$$M_C J M_B \colon L_2^m([0,\tau]) \to L_2^m([0,\tau])$$

as an $m \times m$ matrix whose entries are operators on $L_2([0,\tau])$. The (j,k)-th entry of this operator matrix is equal to

$$\left(\sum_{\nu=1}^n c_{j\nu} b_{\nu k} \right) J_0,$$

where J_0 is the operator on $L_2([0,\tau])$ defined by the right hand side of (21). Now $M_C J M_B$ is trace class if and only if each entry of its operator matrix is trace class. Since J_0 is not a trace class operator (cf. formula (10) in Section VI.1), the latter happens if and only if

$$\sum_{\nu=1}^n c_{j\nu} b_{\nu k} = 0, \qquad j,k = 1,\ldots,m,$$

which is equivalent to $CB = 0$. This proves the first part of the theorem.

Next, assume that K is a trace class operator. It remains to prove the two identities in (20). This we do by applying Theorem IX.3.2 to the semi-separable representation (4). According to formula (8) in Section IX.3, we have

$$
\operatorname{tr} K = \int_0^\tau \operatorname{tr} G_2(s)F_2(s)ds = \int_0^\tau \operatorname{tr} F_2(s)G_2(s)ds
$$

$$
= -i \int_0^\tau \operatorname{tr} Ce^{-isA}E \begin{bmatrix} 0 & 0 \\ 0 & I_{n_2} \end{bmatrix} E^{-1}e^{isA}B\,ds
$$

$$
= -i \int_0^\tau \operatorname{tr} Ce^{-isA}Pe^{isA}B\,ds
$$

$$
= -i \int_0^\tau \operatorname{tr} CPB\,ds = -i\tau \operatorname{tr} CPB.
$$

Here we used (5a) and (5b), the identity (3) and the fact that A and P commute.

Let V be the indicator of $I - K$ associated with the representation (4). From the proof of Theorem 10.1 and 10.2 we know that V and S_τ are similar. Thus $\det V = \det S_\tau$, and we can use formula (9) in Section IX.3 to prove the second identity in (20). $\square$

In the proof of Theorems 10.1 and 10.2 we have only used the semi-separable representation (4) and not the fact that A commutes with P. It follows that Theorems 10.1 and 10.2 also hold true if K is an integral operator on $L_2^m([0,\tau])$ with the property that its kernel function k admits a representation of the following form:

$$
k(t,s) = \begin{cases} iCe^{-itA}(I - P)e^{isA}B, & 0 \le s \le t \le \tau, \\ -iCe^{-itA}Pe^{isA}B, & 0 \le t < s \le \tau, \end{cases}
$$

where P is a projection (which does not have to commute with A). With the exception of the first identity in (20), also Theorem 10.3 is valid for such an operator.

To illustrate the results of this section we consider the *finite slab* version of the transport equation. In this version the medium is like a strip and the densities of the incoming particles at one side and of the outgoing particles at the other side are known. As in the previous section we assume that scattering occurs only in a finite number of directions. In that case the finite slab problem reduces to an equation of the following form:

$$
(23) \qquad \begin{cases} T\psi'(t) = -\psi(t) + F\psi(t), & 0 \le t \le \tau, \\ P_+\psi(0) = x_+, \quad P_-\psi(\tau) = x_-. \end{cases}
$$

Here T and F are selfadjoint matrices, T is assumed to be invertible, P_+ and P_- are the spectral projections of T corresponding to the positive and negative eigenvalues, respectively. The vector x_+ (resp. x_-) is a given vector in $\operatorname{Im} P_+$ (resp. $\operatorname{Im} P_-$). We shall look for solutions of (23) in the space $L_2^m([0,\tau])$. It is of particular interest to consider the case when equation (23) has a non-trivial solution with x_+ and x_- both equal to the zero vector. (In Physics this case is meaningful; it concerns a so-called multiplying medium.)

As for the half range problem we first show that the finite slab problem (23) is equivalent to an integral equation of convolution type.

THEOREM 10.4. *Let $\psi \in L_2^n([0,\tau])$. Then ψ is a solution of (23) if and only if ψ is a solution of the integral equation*

$$(24) \qquad \psi(t) - \int_0^\tau h(t-s)F\psi(s)ds = f(t), \qquad 0 \le t \le \tau,$$

where

$$(25) \qquad h(t) = \begin{cases} e^{-tT^{-1}}P_+T^{-1}, & t \ge 0, \\ -e^{-tT^{-1}}P_-T^{-1}, & t < 0, \end{cases}$$

$$(26) \qquad f(t) = e^{-tT^{-1}}x_+ + e^{(\tau-t)T^{-1}}x_-, \qquad 0 \le t \le \tau.$$

PROOF. Assume that ψ is a solution of (23). Then

$$(27) \qquad \psi(t) = e^{-tT^{-1}}\psi(0) + e^{-tT^{-1}}\int_0^t e^{sT^{-1}}T^{-1}F\psi(s)ds, \qquad 0 \le t \le \tau.$$

Apply $e^{tT^{-1}}P_-$ to both sides of (27), evaluate the result at $t = \tau$, and use that $e^{tT^{-1}}$ and P_- commute. It follows that

$$(28) \qquad P_-\psi(0) = e^{\tau T^{-1}}P_-\psi(\tau) - \int_0^\tau e^{sT^{-1}}P_-T^{-1}F\psi(s)ds.$$

Now, replace $\psi(0)$ in (27) by $P_+\psi(0)+P_-\psi(0)$, and use (28) and the boundary conditions in (23). We conclude that

$$\psi(t) = e^{-tT^{-1}}x_+ + e^{(\tau-t)T^{-1}}x_-$$
$$- \int_0^\tau e^{-(t-s)T^{-1}}P_-T^{-1}F\psi(s)ds + \int_0^t e^{-(t-s)T^{-1}}T^{-1}F\psi(s)ds$$
$$= f(t) + \int_0^\tau h(t-s)F\psi(s)ds, \qquad 0 \le t \le \tau,$$

where h and f are given by (25) and (26), respectively. Thus ψ is a solution of (24).

The reverse implication is proved by the same type of arguments as used in the second part of the proof of Theorem 9.1. $\square$

From (25) it is clear that the kernel function $k(\cdot) = h(\cdot)F$ of the integral operator associated with (24) admits an exponential representation. In fact, to get the representation (2) one has to take

$$(29) \qquad A = -iT^{-1}, \quad B = T^{-1}F, \quad C = -iI, \quad P = P_-.$$

It follows that we may apply Theorems 10.1 and 10.2 to obtain solutions of (24).

THEOREM 10.5. *Assume that $I - F$ is nonnegative definite. Then for each $f \in L_2^n([0,\tau])$ the equation*

$$(30) \qquad \psi(t) - \int_0^\tau h(t-s)F\psi(s)ds = f(t), \qquad 0 \le t \le \tau,$$

has a unique solution $\psi \in L_2^n([0,\tau])$. Moreover, if the right hand side f is given by (26), then the solution of (30) is

$$(31) \quad \psi(t) = e^{-tT^{-1}(I-F)}\{x_+ + M(\tau)^{-1}(x_- - P_- e^{-\tau T^{-1}(I-F)}x_+)\}, \qquad 0 \le t \le \tau,$$

where $M(\tau)\colon \operatorname{Im} P_- \to \operatorname{Im} P_-$ is defined by

$$(32) \qquad M(\tau)x = P_- e^{-\tau T^{-1}(I-F)}x, \qquad x \in \operatorname{Im} P_-.$$

PROOF. The first step is to show that $M(\tau)$ is invertible. Since $M(\tau)$ acts on a finite dimensional space, it suffices to show that $M(\tau)$ is injective. Take $0 \ne y \in \operatorname{Im} P_-$, and consider the expression

$$(33) \qquad \rho_y(\tau) = \langle TM(\tau)y, y \rangle.$$

We shall prove that $\rho_y(\tau)$ as a function of τ is monotonely decreasing on $0 \le \tau < \infty$. Assume for the moment that this has been proved. Then

$$(34) \qquad \rho_y(\tau) \le \rho_y(0) = \langle Ty, y \rangle.$$

Since P_- is the spectral projection of T corresponding to the negative eigenvalues, $0 \ne y \in \operatorname{Im} P_-$ implies that the third term in (34) is strictly negative. But then $\rho_y(\tau) \ne 0$, and hence $M(\tau)y$ must be different zero. Thus $M(\tau)$ is injective.

To prove that ρ_y is monotonely decreasing, we first assume that $I - F$ is positive definite. Put $R = T^{-1}(I - F)$. The operator R is selfadjoint with respect to the inner product

$$[x, z] := \langle (I - F)x, z \rangle.$$

Indeed, because of the selfadjointness of T and F, we have

$$[Rx, z] = \langle (I - F)T^{-1}(I - F)x, z\rangle$$
$$= \langle (I - F), T^{-1}(I - F)z\rangle$$
$$= [x, Rz]$$

for x and z in $\mathbf{C}^n$. It follows that $e^{-\tau R}$ is nonnegative definite relative to the inner product $[\cdot, \cdot]$. Now, observe that ρ_y is differentiable and

$$\rho_y'(\tau) = \langle -(I - F)e^{-\tau T^{-1}(I-F)}y, y\rangle$$
$$= -[e^{-\tau R}y, y] \leq 0, \qquad 0 \leq \tau < \infty.$$

Thus ρ_y is monotonely decreasing if $I - F$ is positive definite.

To prove the latter result for the nonnegative definite case, we choose a sequence $F_1, F_2, \ldots$ of selfadjoint $n \times n$ matrices such that $F_j \to F$ as $j \to \infty$ and $I - F_j$ is positive definite for each j. Let $M_j(\tau)$ be the operator defined by (32) with F replaced by F_j, and consider

$$\rho_{y,j}(\tau) = \langle TM_j(\tau)y, y\rangle, \qquad 0 \leq \tau < \infty.$$

Obviously, $M_j(\tau) \to M(\tau)$ as $j \to \infty$, and hence

(35) $$\rho_y(\tau) = \lim_{j \to \infty} \rho_{y,j}(\tau), \qquad 0 \leq \tau < \infty.$$

Since $I - F_j$ is positive definite, $\rho_{y,j}$ is monotonely decreasing (see the preceding paragraph). But then, by (35), the same must be true for ρ_y. We have now shown that $M(\tau)$ is invertible.

Next, we show that the equation (30) is uniquely solvable. Consider the kernel function $k(\cdot) = h(\cdot)F$ of the integral operator associated with (30). We know that $k(\cdot)$ can be written in the form (2) by taking A, B, C and P as in (29). Since

(36) $$A^\times = A - BC = -iT^{-1}(I - F),$$

the indicator S_τ associated with this exponential representation of $k(\cdot)$ is given by

(37) $$S_\tau = Pe^{i\tau A}PM(\tau): \operatorname{Im} P \to \operatorname{Im} P.$$

The map $Pe^{i\tau A}P: \operatorname{Im} P \to \operatorname{Im} P$ is invertible; its inverse being $Pe^{-i\tau A}P$. Also $M(\tau)$ is invertible. It follows that S_τ is invertible, and hence, by Theorem 10.1, the equation (30) is uniquely solvable in $L_2^n([0, \tau])$.

Finally, let us compute the solution of (30) for f given by (26). To do this we use formula (7) for the resolvent kernel. Note that f may be written in the form

$$f(t) = iCe^{-itA}x_+ + iCe^{-i(t-\tau)A}x_-, \qquad 0 \leq t \leq \tau.$$

First, assume that $x_- = 0$. Then, according to Theorem 10.1, the solution ψ of (30) has the following representation:

$$\psi(t) = f(t) + iCe^{-itA^\times}\Pi_\tau \int_0^t e^{isA^\times}(iBC)e^{-isA}x_+ds$$

$$- iCe^{-itA^\times}(I - \Pi_\tau)\int_t^\tau e^{isA^\times}(iBC)e^{-isA}x_+ds.$$

Recall that

$$e^{isA^\times}(iBC)e^{-isA} = -\frac{d}{ds}(e^{isA^\times}e^{-isA}).$$

Thus the two integrals in the formula for ψ can be computed. This yields:

$$\psi(t) = f(t) - iCe^{-itA^\times}\Pi_\tau e^{itA^\times}e^{-itA}x_+ + iCe^{-itA^\times}\Pi_\tau x_+$$

$$+ iCe^{-itA^\times}(I - \Pi_\tau)e^{i\tau A^\times}e^{-i\tau A}x_+ - iCe^{-itA^\times}(I - \Pi_\tau)e^{itA^\times}e^{-itA}x_+$$

$$= f(t) - iCe^{-itA}x_+ + iCe^{-itA^\times}\Pi_\tau x_+ + iCe^{-itA^\times}(I - \Pi_\tau)e^{i\tau A^\times}e^{-i\tau A}x_+.$$

From the definition of Π_τ (see formula (8)) we conclude that

$$(I - \Pi_\tau)e^{i\tau A^\times}e^{-i\tau A}x_+ = S_\tau^{-1}Px_+ = 0,$$

because $P = P_- = (I - P_+)$. Thus, if $x_- = 0$, then

$$\psi(t) = iCe^{-itA^\times}\Pi_\tau x_+, \qquad 0 \le t \le \tau.$$

In a similar way one shows that for $x_+ = 0$ the solution ψ is given by

$$\psi(t) = iCe^{-itA^\times}(I - \Pi_\tau)e^{i\tau A^\times}x_-, \qquad 0 \le t \le \tau.$$

It follows that for f given by (26) the solution ψ of (30) is equal to

$$(38) \qquad \psi(t) = iCe^{-itA^\times}\{\Pi_\tau x_+ + (I - \Pi_\tau)e^{i\tau A^\times}x_-\}, \qquad 0 \le t \le \tau.$$

By (8) and (37)

$$\Pi_\tau x_+ + (I - \Pi_\tau)e^{i\tau A^\times}x_- = x_+ - S_\tau^{-1}Pe^{i\tau A}e^{-i\tau A^\times}x_+ + S_\tau^{-1}Pe^{i\tau A}x_-$$

$$= x_+ - M(\tau)^{-1}Pe^{-i\tau A^\times}x_+ + M(\tau)^{-1}x_-$$

$$= x_+ + M(\tau)^{-1}\{x_- - Pe^{-i\tau A^\times}x_+\}.$$

Here we used that $x_- \in \operatorname{Im} P$ and that P commutes with $e^{i\tau A}$. Since $iA^\times = T^{-1}(I - F)$, $iC = I$ and $P = P_-$, formula (38) and the preceding calculation yield the desired expression (31). $\square$

COROLLARY 10.6. *If $I - F$ is nonnegative definite, then the finite slab problem (23) has a unique solution ψ, namely*

$$\psi(t) = e^{-tT^{-1}(I-F)}\{x_+ + M(\tau)^{-1}(x_- - P_- e^{-\tau T^{-1}(I-F)}x_+)\}, \qquad 0 \leq t \leq \tau,$$

where $M(\tau)\colon \operatorname{Im} P_- \to \operatorname{Im} P_-$ is defined by (32).

PROOF. Note that a solution of (23) is continuous on $[0, \tau]$ and thus belongs to $L_2^n([0, \tau])$. So we can apply Theorems 10.4 and 10.5 to get the desired result. $\square$

The next theorem deals with the case of a multiplying medium. It concerns the finite slab problem (23) with both x_- and x_+ equal to the zero vector, i.e., the equation

$$(39) \qquad \begin{cases} T\psi(t) = -\psi(t) + F\psi(t), & 0 \leq t \leq \tau, \\ P_+\psi(0) = 0, & P_-\psi(0) = 0. \end{cases}$$

As before $M(\tau)\colon \operatorname{Im} P_- \to \operatorname{Im} P_-$ is defined by (32).

THEOREM 10.7. *The problem (39) has a nontrivial solution if and only if $\det M(\tau) \neq 0$, and in that case the general solution of (39) is given by*

$$(40) \qquad \psi(t) = e^{-tT^{-1}(I-F)}y, \qquad 0 \leq t \leq x,$$

where y is any vector in $\operatorname{Im} P_-$ such that $M(\tau)y = 0$. Furthermore, if all the eigenvalues of $T^{-1}(I - F)$ are real, then there are only finitely many τ's for which (39) has a nontrivial solution.

PROOF. Note that any solution of (39) is a function belonging to $L_2^n([0, \tau])$. Thus, by Theorem 10.4, the solutions of the problem (39) coincide with the solutions $\psi \in L_2^n([0, \tau])$ of the equation

$$(41) \qquad \psi(t) - \int_0^\tau h(t - s)F\psi(s)ds = 0, \qquad 0 \leq t \leq \tau.$$

Here $h(\cdot)$ is as in (25). Now apply Theorems 10.1 and 10.2 with A, B, C and P as in (29), and use (37). It follows that (39) has a nontrivial solution if and only if $\det M(\tau) \neq 0$, and in that case the general solution of (39) is given by (40) with $y \in \operatorname{Ker} M(\tau)$. For the proof of the latter statement we also use that

$$iCe^{-itA^\times} = e^{-tT^{-1}(I-F)}, \qquad 0 \leq t \leq \tau.$$

Next, assume that all the eigenvalues of $T^{-1}(I - F)$ are real. Note that $\det M(\tau)$ depends analytically on τ and $\det M(0) \neq 0$. It follows that the number of zeros of $\det M(\cdot)$ on $[0, \infty)$ is finite or countable, and in the latter case the only accumulation point of these zeros is the point infinity. Since $T^{-1}(I - F)$ has only real eigenvalues, $\det M(\tau)$ is a linear combination of functions (in τ) of the form $\tau^j e^{\lambda \tau}$ with λ real. Let λ_0 be the largest λ in the exponent of functions $\tau^j e^{\lambda \tau}$ which appear in the linear combination with a nonzero coefficient. Then

$$e^{-\lambda_0 \tau} \det M(\tau) = p(\tau) + q(\tau),$$

where $p(\tau)$ is a nonzero polynomial and $q(\tau)$ is analytic in τ and tends to zero if $\tau \to \infty$. In particular, $|p(\tau)| \to c$, $0 < c \leq +\infty$ if $\tau \to \infty$, and it follows that $\det M(\tau) \neq 0$ for τ sufficiently large. Hence $M(\tau)$ has only a finite number of zeros. $\square$

We conclude this section with the following simple example. Consider the equation

$$(42) \quad \begin{cases} \dot{\psi}_1(t) = (c-1)\psi_1(t) + c\psi_2(t), & 0 \leq t \leq \tau, \\ -\dot{\psi}_2(t) = c\psi_1(t) + (c-1)\psi_2(t), & 0 \leq t \leq \tau, \\ \psi_1(0) = 0, \quad \psi_2(0) = 0. \end{cases}$$

The problem is to find scalar functions ψ_1 and ψ_2 that satisfy (42). Hence c is a positive constant. Equation (42) may be rewritten in the form (39) by taking

$$T = \begin{bmatrix} 1 & 0 \\ 0 & -1 \end{bmatrix}, \qquad F = c \begin{bmatrix} 1 & 1 \\ 1 & 1 \end{bmatrix},$$

$$P_+ = \begin{bmatrix} 1 & 0 \\ 0 & 0 \end{bmatrix}, \qquad P_- = \begin{bmatrix} 0 & 0 \\ 0 & -1 \end{bmatrix}.$$

Note that $I - F$ is nonnegative definite if and only if $c \leq \frac{1}{2}$. Thus for $c \leq \frac{1}{2}$ equation (42) has only a trivial solution, namely $\psi_1 = \psi_2 = 0$ (cf. Corollary 10.6). In what follows we assume that $c > \frac{1}{2}$.

We have to analyse

$$R := T^{-1}(I - F) = \begin{bmatrix} 1 - c & c \\ c & -(1-c) \end{bmatrix}.$$

This matrix has two different eigenvalues, namely $\pm i\sqrt{2c-1}$. (Here we use that $c > \frac{1}{2}$). Put $\alpha = i\sqrt{2c-1}$, and consider the matrix

$$\Lambda = \begin{bmatrix} c & c \\ 1 - c - \alpha & 1 - c + \alpha \end{bmatrix}.$$

The matrix Λ is invertible and

$$\Lambda^{-1} = \frac{1}{2c\alpha} \begin{bmatrix} \alpha + 1 - c & -c \\ \alpha - 1 + c & c \end{bmatrix}, \qquad R = \Lambda \begin{bmatrix} \alpha & 0 \\ 0 & -\alpha \end{bmatrix} \Lambda^{-1}.$$

Now consider the scalar function

$$m(t) = \begin{bmatrix} 0 & 1 \end{bmatrix} e^{-tT^{-1}(I-F)} \begin{bmatrix} 0 \\ 1 \end{bmatrix}$$

$$= \begin{bmatrix} 0 & 1 \end{bmatrix} e^{-tR} \begin{bmatrix} 0 \\ 1 \end{bmatrix}$$

$$= \begin{bmatrix} 0 & 1 \end{bmatrix} \Lambda \begin{bmatrix} e^{-t\alpha} & 0 \\ 0 & e^{t\alpha} \end{bmatrix} \Lambda^{-1} \begin{bmatrix} 0 \\ 1 \end{bmatrix}$$

$$= \cos(t\sqrt{2c-1}) + (1-c)\frac{\sin(t\sqrt{2c-1})}{\sqrt{2c-1}}, \qquad t \geq 0.$$

Theorem 10.7 tells us that (42) has a nontrivial solution if and only if $m(\tau) = 0$, and in that case the general solution of (42) is given by

$$\psi_1(t) = \gamma [\, 1 \quad 0\,] e^{-tT^{-1}(I-F)} \begin{bmatrix} 0 \\ 1 \end{bmatrix} = \gamma c \frac{\sin(\tau\sqrt{2c-1})}{\sqrt{2c-1}},$$

$$\psi_2(t) = \gamma m(t), \qquad 0 \le t \le \tau,$$

where γ is an arbitrary complex parameter. Note that for each $c > \frac{1}{2}$ there exist infinitely many τ's such that equation (42) has nontrivial solutions.

COMMENTS ON PART III

The first six sections of Chapter XI contain the standard elements of the abstract Fredholm theory which may also be found in other books (see, e.g., Kato [1]). Sections 7, 8 and 9 in this chapter concern more recent material. Theorem 7.1 is taken from Calderon [1] (see also Fedosov [1]) and Theorem 7.6 from Markus-Feldman [1]. The results of Sections 8 and 9 are based on the papers Markus-Sigal [1], Gohberg-Sigal [1], and Gohberg-Kaashoek-Lay [1], [2]. Theorems 9.1 and 9.2 may also be proved for finitely meromorphic operator functions that are of so-called Fredholm type. For these and other generalizations we refer to Gohberg-Sigal [1], Sigal [1] and Bart-Kaashoek-Lay [1]. The results in the last section of Chapter XI come from the book Gohberg-Krein [3]. The approach followed in this section allows one to introduce and to develop a theory of singular values for operators on Banach spaces. In this wider framework the singular values are approximation numbers which are extensively studied in the books Pietsch [1] and König [1]. Chapter XII contains standard material about Wiener-Hopf integral operators, which is taken from the paper Gohberg-Krein [2] (see also the book Gohberg-Feldman [1]). Chapter XIII, which contains more recent material, is based on Section 4.5 of the book Bart-Gohberg-Kaashoek [1] and the papers Bart-Gohberg-Kaashoek [2], [3]. The notion of a realization has its origin in mathematical system theory (see, e.g., Kalman-Falb-Arbib [1], Kailath [1]). The examples from linear transport theory in the last two sections of Chapter XIII are inspired by Chapter 6 in Bart-Gohberg-Kaashoek [1], Van der Mee [1] and Ran-Rodman [1]. We shall return to the study of concrete classes of Fredholm operators, like Toeplitz and Wiener-Hopf integral operators, in the second volume. Unbounded Fredholm operators will be treated in Part IV.

EXERCISES TO PART III

In what follows, X and Y are complex Banach spaces and I denotes the identity operator on X or Y. In the first six exercises, $A \in \mathcal{L}(X, Y)$ and $B \in \mathcal{L}(Y, X)$.

1. Assume that the product AB is a Fredholm operator. Which of the following statements is true?

    (a) The operator A is Fredholm.

    (b) The operator B is Fredholm.

    (c) Both A and B are Fredholm.

2. Assume that AB is Fredholm. Show that

    (a) $\operatorname{Im} A$ is closed and $d(A) < \infty$,

    (b) $\operatorname{Im} B$ is closed and $n(B) < \infty$.

3. If AB is Fredholm, does it follow that BA is Fredholm?

4. If $I - AB$ is Fredholm, does it follow that $I - BA$ is Fredholm?

5. Assume that AB is Fredholm. Prove that A is Fredholm if and only if B is Fredholm.

6. If A is a Fredholm operator and T is a generalized inverse of A, then T is a Fredholm operator and $\operatorname{ind} T = -\operatorname{ind} A$. Prove this.

7. Let $A, B \in \mathcal{L}(X, Y)$. The operator A is Fredholm and T is a generalized inverse of A. Assume that the operator $C = I + TB$ is invertible. Prove that $A + B$ is a Fredholm operator and

    (a) $n(A + B) \le n(A)$,

    (b) $d(A + B) \le d(A)$,

    (c) $\operatorname{ind}(A + B) = \operatorname{ind} A$.

Hint: use Exercises 5 and 6.

8. Let A, B and C be as in the previous exercise, and assume now that C is Fredholm. Prove that B is a Fredholm operator and $\operatorname{ind}(A + B) = \operatorname{ind} A + \operatorname{ind} C$.

9. Use Exercise 7 to give a new proof of Theorem XI.4.1 (with, perhaps, a different bound for $\|B\|$).

10. Use Exercise 8 to give a new proof of Theorem XI.4.2.

11. Assume that the operator

$$T = \begin{bmatrix} A & 0 \\ 0 & B \end{bmatrix} : X \oplus Y \to X \oplus Y$$

is Fredholm. Which of the following statements is true?

    (a) The operator A is Fredholm.

(b) The operator B is Fredholm.

(c) Both A and B are Fredholm.

12. Assume that the operator

$$T = \begin{bmatrix} A & C \\ 0 & B \end{bmatrix} : X \oplus Y \to X \oplus Y$$

is Fredholm. Answer for this case the question posed in the previous exercise.

13. Consider the operator

$$T = \begin{bmatrix} A & C \\ 0 & B \end{bmatrix} : X \oplus Y \to X \oplus Y,$$

and assume that A and B are Fredholm operators. Prove that T is a Fredholm operator and

$$\operatorname{ind} T = \operatorname{ind} A + \operatorname{ind} B.$$

In the next three exercises A is a bounded linear operator on the Banach space X.

14. Let N be a positive integer, and assume that $A^N - I$ is a compact operator. Prove that the essential spectrum of A is a subset of the set of N-th roots of unity.

15. Let p be a non-zero polynomial, and assume that $p(A)$ is compact. Prove that the essential spectrum of A is a subset of the set of zeros of p.

16. For each positive integer N, put

$$\rho_N(A) = \inf\{\|A^N - K\| \mid K \text{ compact}\}.$$

Prove that $\sigma_{\text{ess}}(A) = \{0\}$ if and only if $\rho_N(A)^{1/N} \to 0$ as $N \to \infty$ (cf., Exercise 18 to Part I).

17. Let $A \in \mathcal{L}(X, Z)$ and $B \in \mathcal{L}(Y, Z)$ be operators acting between complex Banach spaces, and assume that $Z = \operatorname{Im} A \oplus \operatorname{Im} B$. Show that A and B have closed range.

18. Let $A \in \mathcal{L}(X, Y)$, and assume that $\operatorname{Im} A$ is closed. Show that the conjugate A' has closed range and $\operatorname{Im} A' = (\operatorname{Ker} A)^\perp$. (Hint: first consider the case when $\operatorname{Ker} A = \{0\}$.)

19. Let $A \in \mathcal{L}(X, Y)$, and assume that $\operatorname{Im} A' = (\operatorname{Ker} A)^\perp$. Prove that A has closed range and $\operatorname{Im} A = {}^\perp(\operatorname{Ker} A')$.

An operator $A \in \mathcal{L}(X, Y)$ is called *semi-Fredholm* if $\operatorname{Im} A$ is closed and $n(A)$ or $d(A)$ is finite. The *index* of such an operator A is defined by

$$\operatorname{ind} A = \begin{cases} +\infty, & \text{if } n(A) = \infty, \\ -\infty, & \text{if } d(A) = \infty, \\ n(A) - d(A), & \text{otherwise.} \end{cases}$$

20. Prove that $A \in \mathcal{L}(X, Y)$ is semi-Fredholm if and only if the conjugate operator A' is semi-Fredholm, and show that in that case $\operatorname{ind} A = -\operatorname{ind} A'$.

21. Extend the first perturbation theorem in Section XI.4 to semi-Fredholm operators. (Hint: see Section V.1 of Goldberg [1].)

22. Extend the second perturbation theorem in Section XI.4 to semi-Fredholm operators. (Hint: see Section V.2 of Goldberg [1].)

23. Let $A \in \mathcal{L}(X, Y)$ and $B \in \mathcal{L}(Y, X)$, and assume that the product AB is semi-Fredholm. Show that one of the factors is semi-Fredholm, but not necessarily both.

24. If $A \in \mathcal{L}(X, Y)$ and $B \in \mathcal{L}(Y, X)$ are semi-Fredholm, does it follow that AB is semi-Fredholm? If yes, prove this statement; if no, give an additional condition under which the statement is true.

25. Assume that the operator

$$T = \begin{bmatrix} A & C \\ 0 & B \end{bmatrix} : X \oplus Y \to X \oplus Y$$

is Fredholm. Which of the following statements is true?

    (a) The operator A is semi-Fredholm.

    (b) The operator B is semi-Fredholm.

    (c) Both A and B are semi-Fredholm.

Answer the same question assuming only that T is semi-Fredholm.

26. The analogue of the statement in Exercise 13 for semi-Fredholm operators holds true. Prove or disprove this.

27. Determine the kernel function of the Wiener-Hopf operator T on $L_2([0, \infty))$ with symbol $\left(\frac{\lambda - i}{\lambda + i}\right)^n$. Show that T is Fredholm and compute $n(T)$ and $d(T)$.

28. Let T be the Wiener-Hopf operator on $L_2([0, \infty))$ with symbol

$$\omega(\lambda) = -\frac{3}{2}\left(\frac{\lambda - i}{\lambda + i}\right) + \frac{7}{2} - \left(\frac{\lambda + i}{\lambda - i}\right).$$

Compute the kernel function of T. Show that T is invertible and determine its inverse.

29. Let T_α be the Wiener-Hopf operator on $L_2([0, \infty))$ with symbol

$$\omega_\alpha(\lambda) = 2\alpha\left(\frac{\lambda - i}{\lambda + i}\right) + 1 - 3\alpha + \alpha\left(\frac{\lambda + i}{\lambda - i}\right).$$

For which $\alpha \in \mathbb{C}$ is the operator T_α left, right or two-sided invertible? Construct the inverse of T_α if it exists.

30. For $a \in \mathbb{R}$ let T_a be the Wiener-Hopf operator on $L_2([0, \infty))$ with symbol

$$\omega(\lambda) = a\left(\frac{\lambda - i}{\lambda + i}\right) + 1 - a\left(\frac{\lambda + i}{\lambda - i}\right).$$

Determine the spectrum of T_a.

The remaining exercises deal with convolution operators with kernel function

$$k(t) = \sum_{j=1}^{n} d_j e^{-|t|c_j}, \qquad -\infty < t < \infty.$$

Here $0 < c_1 < c_2 < \cdots < c_n$ and $d_j > 0$ $(j = 1, \ldots, n)$.

31. Show that k may be represented in the form

$$k(t) = \begin{cases} iCe^{-itA}(I - P)B, & t > 0, \\ -iCe^{-itA}PB, & t < 0, \end{cases}$$

by taking

$$\begin{aligned}
A &= \operatorname{diag}(-ic_1, \ldots, -ic_n, ic_1, \ldots, ic_n) \\
P &= \operatorname{diag}(0, \ldots, 0, 1, \ldots, 1) \\
B &= Ag, \qquad C = g^T,
\end{aligned}$$

where g is the $2n \times 1$ matrix whose transpose is given by

$$g^T = \left[\sqrt{\frac{d_1}{c_1}} \cdots \sqrt{\frac{d_n}{c_n}} \ \sqrt{\frac{d_1}{c_1}} \cdots \sqrt{\frac{d_n}{c_n}} \right].$$

32. Let L be the convolution operator on $L_2(\mathbf{R})$ defined by

$$(Lf)(t) = \int_{-\infty}^{\infty} k(t - s)f(s)ds, \qquad -\infty < t < \infty.$$

Prove that $I - L$ is invertible if and only if

$$(*) \qquad \qquad \sum_{j=1}^{n} \frac{d_j}{c_j} < \frac{1}{2}.$$

Determine the inverse of $I - L$ if it exists. Hint: use Theorem XIII.5.1 and Lemmas XIII.9.4 and 9.5.

33. Let T be the Wiener-Hopf operator on $L_2([0, \infty))$ given by

$$(Tf)(t) = f(t) - \int_{0}^{\infty} k(t - s)f(s)ds, \qquad 0 \le t < \infty.$$

Show that T is invertible if and only if condition $(*)$ in the previous exercise is fulfilled. Determine the inverse of T if it exists.

34. Let $\omega(\cdot)$ be the symbol of the Wiener-Hopf operator T defined in the previous exercise. Determine a right canonical factorization of $\omega(\cdot)$ whenever it exists.

35. Let K be the convolution operator on $L_2([0,\tau])$ defined by

$$(Kf)(t) = \int_0^\tau k(t-s)f(s)ds, \qquad 0 \le t \le \tau.$$

Show that K is a trace class operator and compute $\operatorname{tr} K$. Prove that $I - K$ is invertible if

$$\sum_{j=1}^n \frac{d_j}{c_j} \le \frac{1}{2},$$

and in that case give a formula for the resolvent kernel.

PART IV

CLASSES OF UNBOUNDED LINEAR OPERATORS

One of the most important classes of operators is the class of unbounded operators. This includes differential operators which play a vital role in applications. In this part an introduction to the theory of unbounded operators is presented. We start with the study of their adjoints and conjugates, and then concentrate on certain types of ordinary and partial differential operators (Chapter XIV). In the next three chapters (XV-XVII) results proved earlier for bounded operators are extended to unbounded operators. The main topics are functional calculus, the spectral theorem, and the theory of Fredholm operators. The last chapter contains an introduction to the theory of strongly continuous semigroups with applications to the abstract Cauchy problem and linear transport theory.

CHAPTER XIV
UNBOUNDED LINEAR OPERATORS

This chapter gives an introduction to the theory of unbounded linear operators between Banach spaces. The important notions of closed and closable operators and their conjugates are analyzed with much attention paid to ordinary and partial differential operators. In particular, maximal and minimal operators and the properties of their inverses are studied. The chapter is divided into 6 sections. The first two sections are devoted to the general theory, and the other four sections deal mainly with differential operators.

XIV.1 PRELIMINARIES

Throughout this chapter X and Y will denote complex Banach spaces. A linear operator A with domain $\mathcal{D}(A) \subset X$ and range $\operatorname{Im} A \subset Y$ is denoted by $A(X \to Y)$. For $X = Y$ the *resolvent set* $\rho(A)$ is the set of all complex numbers λ for which $\lambda - A$ is invertible and the inverse $R(\lambda) = (\lambda - A)^{-1}$ is bounded on X. The operator valued function $R(\cdot)$ is called the *resolvent* of A.

PROPOSITION 1.1. *The resolvent set $\rho(A)$ of the operator $A(X \to X)$ is open. If $\rho(A) \neq \emptyset$, then the resolvent $R(\cdot)$ is analytic on $\rho(A)$. Moreover, if $\lambda_0 \in \rho(A)$ and $|\lambda - \lambda_0| < \|R(\lambda_0)\|^{-1}$, then λ is in $\rho(A)$ and*

$$
(1) \qquad R(\lambda) = \sum_{k=0}^{\infty} (-1)^k (\lambda - \lambda_0)^k R(\lambda_0)^{k+1};
$$

the series converges in the norm on $\mathcal{L}(X)$. If λ and μ are in $\rho(A)$, then

$$
(2) \qquad R(\lambda) - R(\mu) = (\mu - \lambda)R(\lambda)R(\mu).
$$

PROOF. The proof is the same as for the bounded case; see, e.g., [GG], Sections X.6 and X.8. $\square$

The identity (2) will be called the *resolvent identity* (or *resolvent equation*). As for bounded operators, the *spectrum $\sigma(A)$* of A is defined to be the complement in $\mathbb{C}$ of $\rho(A)$. However, while the spectrum of a bounded operator is a non-empty compact set, the spectrum of an unbounded operator may be empty or all of $\mathbb{C}$.

For example, let $X = Y = C([0,1])$, the space of complex valued continuous functions endowed with the supremum norm, and define A_1 and A_2 by

$$
(3) \qquad \mathcal{D}(A_1) = \{f \in C([0,1]) \mid f' \in C[0,1]\}, \qquad A_1 f = f',
$$

$$
(4) \qquad \mathcal{D}(A_2) = \{f \in \mathcal{D}(A_1) \mid f(0) = 0\}, \qquad A_2 f = f'.
$$

Since $e^{\lambda t} \in \text{Ker}(\lambda - A_1)$, the spectrum $\sigma(A_1) = \mathbb{C}$. On the other hand, $\sigma(A_2) = \emptyset$. To see this, note that $\text{Ker}(\lambda - A_2) = \{0\}$. Also, for any $g \in Y$, the equation $(\lambda - A_2)y = g$ has a solution $y \in \mathcal{D}(A_2)$, namely

$$y(t) = -\int_0^t e^{\lambda(t-s)} g(s)\,ds, \qquad 0 \le t \le 1.$$

Hence $\lambda - A_2$ is invertible and

$$\|(\lambda - A_2)^{-1}g\| = \|y\| \le e^{|\lambda|}\|g\|.$$

Recall (see [GG], Section X.3) that an operator is *closed* if its graph $G(A) = \{(x, Ax) \mid x \in \mathcal{D}(A)\}$ is a closed linear manifold in the Banach space $X \times Y$ with norm $\|(x, y)\| = \|x\| + \|y\|$. It is clear that A is closed if and only if $x_n \in \mathcal{D}(A)$, $n = 1, 2, \ldots$, $x_n \to x$ and $Ax_n \to y$ imply $x \in \mathcal{D}(A)$ and $Ax = y$. The differential operators defined by (3) and (4) are closed. More general differential operators are shown to be closed in Section XIV.3.

PROPOSITION 1.2. *If the operator $A(X \to X)$ has a non-empty resolvent set $\rho(A)$, then A is closed.*

PROOF. If $\lambda_0 \in \rho(A)$, then $(\lambda_0 - A)^{-1}$ is a bounded operator on X. Hence $(\lambda_0 - A)^{-1}$ is closed which implies that $\lambda_0 - A$ is closed. But then A is closed. $\square$

An operator $A(X \to Y)$ is called *closable* if it can be extended to a closed linear operator $C(X \to Y)$, i.e., $\mathcal{D}(A) \subset \mathcal{D}(C)$ and $Cx = Ax$ for $x \in \mathcal{D}(A)$. In that case C is called a *closed linear extension* of A.

PROPOSITION 1.3. *An operator $A(X \to Y)$ is closable if and only if $(0, y) \in \overline{G(A)}$ implies $y = 0$.*

PROOF. Suppose that C is a closed linear extension of A and $(0, y) \in \overline{G(A)}$. Then $(0, y) \in G(C)$ and therefore $y = C0 = 0$. On the other hand, suppose that $(0, y) \notin \overline{G(A)}$ whenever $y \ne 0$. Put

$$\mathcal{D} = \{x \in X \mid \exists y_x \in Y \text{ such that } (x, y_x) \in \overline{G(A)}\}.$$

The set $\mathcal{D}$ is a linear manifold in X. For each $x \in \mathcal{D}$ there is precisely one $y_x \in Y$ such that $(x, y_x) \in \overline{G(A)}$. Indeed, assume (x, y_x) and (x, y'_x) are in $\overline{G(A)}$. Then $(0, y_x - y'_x) \in \overline{G(A)}$, and by our hypothesis $y_x = y'_x$. Define $B(X \to Y)$ by setting

$$\mathcal{D}(B) = \mathcal{D}, \qquad Bx = y_x.$$

Then B is a well-defined linear operator and $G(B) = \overline{G(A)}$. Thus B is closed. If $x \in \mathcal{D}(A)$, then $(x, Ax) \in G(A)$. This implies that $x \in \mathcal{D}(B)$ and $Bx = Ax$. Hence B is a closed linear extension of A. $\square$

Let A be closable. The operator B constructed in the proof of Proposition 1.3 is called the *minimal closed linear extension* of A and is denoted by $\overline{A}$. Since $G(\overline{A}) = \overline{G(A)}$, any other closed linear extension of A is also an extension of $\overline{A}$.

PROPOSITION 1.4. *If $A(X \to Y)$ is closable and has finite rank, then A is bounded on its domain.*

PROOF. First we prove that $\operatorname{Im} A = \operatorname{Im} \overline{A}$. Obviously, $\operatorname{Im} A \subset \operatorname{Im} \overline{A}$. Take $y \in \operatorname{Im} \overline{A}$, say $y = \overline{A}x$. Then $(x, y) \in G(\overline{A}) = \overline{G(A)}$, and hence there exists a sequence $x_1, x_2, \ldots$ in $\mathcal{D}(A)$ such that $x_n \to x$ and $Ax_n \to y$. In particular, $y \in \overline{\operatorname{Im} A}$. Since $\operatorname{Im} A$ is closed, $y \in \operatorname{Im} A$. Thus $\operatorname{Im} A = \operatorname{Im} \overline{A}$.

It follows that $\overline{A}$ also has finite rank. So we may assume without loss of generality that A is closed. Then $\operatorname{Ker} A$ is closed. Indeed, let $z \in \overline{\operatorname{Ker} A}$, and let $x_1, x_2, \ldots$ be a sequence in $\operatorname{Ker} A$ such that $x_n \to z$. Since $Ax_n = 0$, it follows that $(z, 0) \in \overline{G(A)} = G(A)$, and hence $Az = 0$. Thus $z \in \operatorname{Ker} A$.

Since the map $[x] \to Ax$ is an isomorphism from $\mathcal{D}(A)/\operatorname{Ker} A$ onto $\operatorname{Im} A$, we have $\dim \mathcal{D}(A)/\operatorname{Ker} A = \dim \operatorname{Im} A < \infty$. Hence $\mathcal{D}(A) = \operatorname{Ker} A \oplus Z$ for some finite dimensional subspace Z. It follows (see [GG], Theorem IX.2.5) that $\mathcal{D}(A)$ is a closed subspace of X. Thus $A \colon \mathcal{D}(A) \to Y$ is a closed operator acting between Banach spaces and thus, by the closed graph theorem, the operator A is bounded on $\mathcal{D}(A)$. $\square$

For later purposes we introduce the following terminology. Let $A(X \to Y)$ and $B(X \to Y)$ be operators. We say that A is a *restriction* of B or, alternatively, B is an *extension* of A if

$$(5) \qquad \mathcal{D}(A) \subset \mathcal{D}(B), \qquad Ax = Bx \qquad (x \in \mathcal{D}(A)).$$

In that case we write $A \subset B$. Note that (5) is equivalent to $G(A) \subset G(B)$.

XIV.2 ADJOINT AND CONJUGATE OPERATORS

Recall that given a bounded operator A from a Hilbert space H_1, into a Hilbert space H_2, the adjoint A^* of A is defined as follows. For each $y \in H_2$ the linear functional $F_y(x) = \langle Ax, y \rangle$ is bounded on H_1. Hence, by the Riesz representation theorem, there exists a unique $z \in H_1$ such that

$$\langle Ax, y \rangle = F_y(x) = \langle x, z \rangle, \qquad x \in H_1.$$

Define $A^*y = z$. Then A^* is a bounded linear operator from H_2 into H_1, and $\langle Ax, y \rangle = \langle x, A^*y \rangle$ for all $x \in H_1$ and $y \in H_2$.

Suppose now that A is densely defined but not necessarily bounded. Then given $y \in H_2$, the functional F_y defined on $\mathcal{D}(A)$ by $F_y(x) = \langle Ax, y \rangle$ may be unbounded and the Riesz representation theorem does not apply. So we must modify the above definition in order to define A^*. To do this we put

$$(1) \qquad \mathcal{D}(A^*) = \left\{ y \in H_2 \,\Big|\, \sup_{0 \neq x \in \mathcal{D}(A)} \frac{|\langle Ax, y \rangle|}{\|x\|} < \infty \right\}.$$

Take $y \in \mathcal{D}(A^*)$. Since $\overline{\mathcal{D}(A)} = H_1$, the functional F_y has a unique bounded linear extension $\overline{F}_y$ to all of H_1. Therefore the Riesz representation theorem ensures the existence of a unique $z \in H_1$ such that $\overline{F}_y(x) = \langle x, z \rangle$, $x \in \mathcal{D}(A)$. Define $A^*y = z$. Hence

$$\langle Ax, y \rangle = \langle x, A^*y \rangle, \qquad x \in \mathcal{D}(A), \qquad y \in \mathcal{D}(A^*).$$

The operator A^* $(H_2 \to H_1)$ is linear and is called the *adjoint* of A.

If $\mathcal{D}(A)$ is not dense in X, then the set $\mathcal{D}(A^*)$ defined by (1) is well-defined. But given $y \in \mathcal{D}(A^*)$, the linear functional $F_y(x) = \langle Ax, y \rangle$, $x \in \mathcal{D}(A)$, can be extended in many different ways to a bounded linear functional on X, i.e., there are many different $z \in X$ such that $F_y(x) = \langle x, z \rangle$, $x \in \mathcal{D}(A)$, and hence A^* has to be defined as a multi-valued map. In what follows we shall not consider this case.

It can happen that $\mathcal{D}(A^*) = \{0\}$, as is seen in the following example. Take $H_1 = H_2 = \ell_2$. Let $e_k = (\delta_{kj})_{j=1}^\infty$, δ_{kj} the Kronecker delta, $k = 1, 2, \ldots$, and let $\{e_{ij}\}_{i,j=1}^\infty$ be any double indexing of $\{e_k\}_{k=1}^\infty$. For each i, j define $Ae_{ij} = e_i$ and extend A linearly to the span of $\{e_k\}_{k=1}^\infty$. Obviously, A is densely defined in ℓ_2. Suppose $y = (\alpha_1, \alpha_2, \ldots)$ is in $\mathcal{D}(A^*)$. Then for each positive integer n and i,

$$n|\alpha_i|^2 \le \sum_{j=1}^\infty |\langle Ae_{ij}, y \rangle|^2 = \sum_{j=1}^\infty |\langle e_{ij}, A^*y \rangle|^2 \le \|A^*y\|^2.$$

Hence $y = (\alpha_i)_{i=1}^\infty = 0$.

If H_i is a Hilbert space with inner product $\langle \cdot, \cdot \rangle_i$, $i = 1, 2$, then $H_1 \times H_2$ is a Hilbert space with inner product $[\cdot, \cdot]$ defined by

$$[(x_1, x_2), (y_1, y_2)] = \langle x_1, y_1 \rangle_1 + \langle x_2, y_2 \rangle_2.$$

For any $M \subset H_1 \times H_2$ the orthogonal complement $M^\perp$ of M is to be taken with respect to the inner product $[\cdot, \cdot]$.

PROPOSITION 2.1. *Let H_1, H_2 be Hilbert spaces, and let the operator A $(H_1 \to H_2)$ have dense domain in H_1. Then*

$$(2) \qquad\qquad G(A)^\perp = \{(-A^*y, y) \mid y \in \mathcal{D}(A^*)\},$$

and the operator $A^*(H_2 \to H_1)$ is closed.

PROOF. Let $G'(A^*)$ denote the right hand side of (2). From

$$[(x, Ax), (-A^*y, y)] = \langle x, -A^*y \rangle + \langle Ax, y \rangle = 0,$$

it follows that $G'(A^*) \subset G(A)^\perp$. Given $(u, v) \in G(A)^\perp$,

$$0 = [(x, Ax), (u, v)] = \langle x, u \rangle + \langle Ax, v \rangle, \qquad x \in \mathcal{D}(A).$$

Hence $v \in \mathcal{D}(A^*)$ and $A^*v = -u$, which shows that $G(A)^\perp \subset G'(A^*)$.

Let $J : H_1 \times H_2 \to H_2 \times H_1$ be defined by $J(x, y) = (y, -x)$. Then J is a homeomorphism and $JG'(A^*) = G(A^*)$. From (2) it follows that $G'(A^*)$ is closed, and hence $G(A^*)$ is closed. $\square$

The concept of the adjoint of a densely defined operator has a natural extension to operators between Banach spaces as follows. Given the operator $A(X \to Y)$ with domain $\mathcal{D}(A)$ dense in X, define the *conjugate* A' between the conjugate spaces Y' and X' by

$$\mathcal{D}(A') = \{g \in Y' \mid g \circ A \text{ is bounded on } \mathcal{D}(A)\},$$
$$A'g = \overline{g \circ A},$$

where $\overline{g \circ A}$ is the unique bounded linear extension of $g \circ A$ to all of X. The operator $A'(Y' \to X')$ is linear.

The next example exhibits the conjugate of the operator of differentiation on $L_p([0,1])$, $1 \le p < \infty$. To do this we make use of the following result. For a subset S of a Banach space X and a subset N of the conjugate space X', we define

$$S^{\perp} = \{F \in X' \mid F(s) = 0 \text{ for all } s \in S\},$$
$$^{\perp}N = \{x \in X \mid G(x) = 0 \text{ for all } G \in N\}.$$

Obviously, $S \subset^{\perp}(S^{\perp})$. The Hahn-Banach theorem implies that $^{\perp}(M^{\perp}) = M$ if M is a subspace (= closed linear manifold) of X. If N is a subspace of X, then, in general, it does not follow that $(^{\perp}N)^{\perp} = N$. However, the latter equality holds true if $\dim N < \infty$. To see this, let $G_1, \ldots, G_n$ be a basis of N, and assume that $F \in (^{\perp}N)^{\perp}$. Then

$$\bigcap_{j=1}^{n} \operatorname{Ker} G_j = {}^{\perp}N \subset \operatorname{Ker} F.$$

Hence F is a linear combination of $G_1, \ldots, G_n$ (see Lemma XVI.1.2), and thus $F \in N$. Since always $N \subset (^{\perp}N)^{\perp}$, we have proved that $(^{\perp}N)^{\perp} = N$ for any finite dimensional subspace N of X'.

Now let us consider the conjugate of the operator of differentiation. More precisely, we shall determine the conjugate of the operator $A(X \to X)$, where $X = L_p([0,1])$, $1 \le p < \infty$,

$$\mathcal{D}(A) = \{f \in X \mid f \text{ absolutely continuous on } [0,1], f' \in X, f(0) = f(1)\},$$
$$Af = f'.$$

The domain $\mathcal{D}(A)$ is dense in X. Given a bounded linear functional F on $L_p([0,1])$, $1 \le p < \infty$, there exists a unique $g \in L_q([0,1])$, $p^{-1} + q^{-1} = 1$, with the property that

$$F(f) = [f,g] := \int_0^1 f(s)g(s)\,ds, \qquad f \in L_p([0,1]).$$

This allows us to identify $L_p([0,1])'$ with $L_q([0,1])$. Thus A' acts on $L_q([0,1])$. We shall show that

$$\begin{aligned}
\text{(3)} \qquad \mathcal{D}(A') &= \{g \in L_q([0,1]) \mid g \text{ absolutely continuous on } [0,1], \\
&\qquad\qquad g' \in L_q([0,1]), g(0) = g(1)\}, \\
A'g &= -g'.
\end{aligned}$$

Let M be the linear manifold described on the right hand side of (3). If $g \in M$, then for every $f \in \mathcal{D}(A)$, integration by parts yields

$$[Af, g] = \int_0^1 f'(s)g(s)ds = -\int_0^1 f(s)g'(s)ds.$$

By Hölders inequality, $|[Af, g]| \leq \|f\| \|g'\|$. Hence $g \in \mathcal{D}(A')$ and $A'g = -g'$. It remains to prove that $\mathcal{D}(A') \subset M$. Suppose $h \in \mathcal{D}(A')$, $A'h = h^*$ and $f \in \mathcal{D}(A)$. Then integration by parts gives

$$
\begin{aligned}
(4) \qquad \int_0^1 f'(s)h(s)ds = [Af, h] = [f, A'h] &= \int_0^1 f(s)h^*(s)ds \\
&= f(0)(H(1) - H(0)) - \int_0^1 f'(s)H(s)ds,
\end{aligned}
$$

where $H(s) = \int_0^s h^*(t)dt$. We shall now show that $h + H$ is constant a.e. To do this we take $Q \in {}^{\perp}\{1\}$, i.e., the function $Q \in L_p([0,1])$ and $\int_0^1 Q(s)ds = 0$. Clearly, $f(s) = \int_0^s Q(t)dt$ is in $\mathcal{D}(A)$. Therefore by (4),

$$\int_0^1 Q(x)[h(x) + H(x)]dx = 0,$$

which shows that $h + H \in ({}^{\perp}\mathrm{span}\{1\})^{\perp} = \mathrm{span}\{1\}$, i.e., $h + H$ is a constant c a.e. Thus we may redefine h on a set of measure zero so that $h = -H + c$. Hence h is absolutely continuous on $[0, 1]$ and $h' = -h^* \in L_q([0, 1])$. Also for $g \in \mathcal{D}(A)$, (4) shows that

$$g(0)(H(1) - H(0)) = \int_0^1 cg'(x)dx = c(g(1) - g(0)) = 0.$$

So if we take $g(0) \neq 0$, then $0 = H(1) - H(0) = h(0) - h(1)$. Hence $h \in \mathcal{D}(A)$.

Other examples of adjoints of differential operators appear in Section XIV.4. Since bounded linear functionals on a Hilbert space H correspond to elements in H via the Riesz representation theorem, we shall only speak of the adjoint of an operator on H rather than its conjugate.

PROPOSITION 2.2. *The conjugate of a densely defined operator is closed.*

PROOF. Given $A(X \to Y)$, suppose $g_n \to g$ and $A'g_n \to f$. Then for all $x \in \mathcal{D}(A)$,

$$f(x) = \lim(A'g_n)(x) = \lim g_n(Ax) = g(Ax).$$

Hence $g \in \mathcal{D}(A')$ and $A'g = f$. $\square$

PROPOSITION 2.3. *Let the domain of the operator $A(X \to X)$ be dense in X. Then A is closable if and only if for every $y \neq 0$ in Y there exists $g \in \mathcal{D}(A')$ such that $g(y) \neq 0$.*

PROOF. Suppose A is closable and $0 \neq y \in Y$. Let C be a closed linear extension of A. Then $(0, y) \notin G(C) \supset \overline{G(A)}$. Hence there exists $w \in (X \times Y)'$ such that $w(0, y) \neq 0$ and $w\big(G(A)\big) = 0$. Define $f \in X'$ and $g \in Y'$ by

$$f(x) = w(x, 0), \qquad g(v) = w(0, v).$$

Then

$$f(x) + g(Ax) = w(x, Ax) = 0, \qquad x \in \mathcal{D}(A).$$

Hence $g \in \mathcal{D}(A')$ and $g(y) = w(0, y) \neq 0$.

Suppose for each $u \neq 0$ in Y there exists $g \in \mathcal{D}(A')$ so that $g(u) \neq 0$. If $(0, y) \in \overline{G(A)}$, then there exists a sequence (x_n) in $\mathcal{D}(A)$ so that $x_n \to 0$ and $Ax_n \to y$. For each $g \in \mathcal{D}(A')$,

$$0 = \lim A'g(x_n) = \lim g(Ax_n) = g(y).$$

Hence $y = 0$ and A is closable by Proposition 1.3. $\square$

Proposition 2.3 is equivalent to the statement that A is closable if and only if $^{\perp}\mathcal{D}(A') = \{0\}$.

PROPOSITION 2.4. *Let H_1 and H_2 be Hilbert spaces, and let $A(H_1 \to H_2)$ have domain dense in H_1. Then A is closable if and only if $\mathcal{D}(A^*)$ is dense in H_2. In this case,*

$$\overline{A} = A^{**}, \qquad A^* = (\overline{A})^*,$$

where $\overline{A}$ is the minimal closed linear extension of A.

PROOF. Suppose A is closable and $w \perp \mathcal{D}(A^*)$. Then $w = 0$ by Proposition 2.3. Hence $\overline{\mathcal{D}(A^*)} = H_2$. Conversely, if $\overline{\mathcal{D}(A^*)} = H_2$ and $\langle v, y \rangle = 0$ for all $y \in \mathcal{D}(A^*)$, then $v = 0$ and therefore A is closable by Proposition 2.3. Let U be defined on $H_1 \times H_2$ by $U(x, y) = (y, x)$. Now by Proposition 2.1,

$$G(\overline{A}) = \overline{G(A)} = G(A)^{\perp\perp} = \big(UG(-A^*)\big)^{\perp} = UG(-A^*)^{\perp} = U^2 G(A^{**}) = G(A^{**}).$$

Hence $\overline{A} = A^{**}$. If we replace A by A^* and use the fact that A^* is closed (Proposition 2.2), we get $A^* = A^{***} = (\overline{A})^*$. $\square$

PROPOSITION 2.5. *Let H_1 and H_2 be Hilbert spaces, and let $A\ (H_1 \to H_2)$ be densely defined. Then*

(i) $\operatorname{Im} A^{\perp} = \operatorname{Ker} A^*,$

(ii) $\overline{\operatorname{Im} A} = \operatorname{Ker} A^{*\perp}.$

If, in addition, A is closable, then

(iii) $\operatorname{Ker} A = \operatorname{Im} A^{*\perp} \cap \mathcal{D}(A)$.

PROOF. (i). If $q \in \operatorname{Im} A^{\perp}$, then $\langle Ax, g \rangle = 0$ for all $x \in \mathcal{D}(A)$. Hence $g \in \mathcal{D}(A^*)$ and $A^* g = 0$. Thus $\operatorname{Im} A^{\perp} \subset \operatorname{Ker} A^*$. The reverse inclusion is proved as in the bounded case (see [GG], Theorem II.11.4). Statement (ii) is an immediate corollary of (i) and the fact that $(M^{\perp})^{\perp} = \overline{M}$ for a linear manifold M. The proof of (iii) is analogous to that of (i) (and uses that $\mathcal{D}(A^*)$ is dense in H_2). $\square$

PROPOSITION 2.6. *A closed densely defined operator $A(H_1 \to H_2)$ is invertible if and only if its adjoint A^* is invertible, in which case $(A^*)^{-1} = (A^{-1})^*$.*

PROOF. Assume A is invertible. By Proposition 2.5 the operator A^* is injective. Since A^{-1} is bounded from H_2 into H_1, the adjoint $(A^{-1})^*$ has domain H_1. Hence for $z \in H_1$ and $x \in \mathcal{D}(A)$

$$\langle Ax, (A^{-1})^* z \rangle = \langle A^{-1} Ax, z \rangle = \langle x, z \rangle.$$

Therefore,

$$(5) \qquad (A^{-1})^* z \in \mathcal{D}(A^*), \qquad A^*(A^{-1})^* z = z.$$

We have shown that A^* is both injective and surjective. Since A^* is closed, it is invertible (by the closed graph theorem) and $(A^*)^{-1} = (A^{-1})^*$ because of (5). If A^* is invertible, then by Proposition 2.4 and the above result applied to A^* in place of A, we have $A = A^{**}$ is invertible. $\square$

XIV.3 ORDINARY DIFFERENTIAL OPERATORS

Let τ be the differential expression of the form

$$\tau = D^n + a_{n-1}(t) D^{n-1} + \cdots + a_1(t) D + a_0(t),$$

where $D = \frac{d}{dt}$ and each a_k is locally integrable on an interval J. Let $AC_n(J)$ denote the set of complex valued functions g on J with the property that $g^{(n-1)}$ exists and is absolutely continuous on each compact subinterval of J. Thus $g^{(n)}$ exists a.e. on J. Define the linear differential operator $T_{\max, \tau, J}\big(L_2(J) \to L_2(J)\big)$ by

$$\mathcal{D}(T_{\max, \tau, J}) = \{ g \in AC_n(J) \cap L_2(J) \mid \tau g \in L_2(J) \},$$
$$T_{\max, \tau, J} g = \tau g.$$

This operator is called the *maximal operator* corresponding to τ and J. Where it is clear which τ and J we are dealing with, we write $T_{\max}$ instead of $T_{\max, \tau, J}$. We shall now show that certain restrictions of $T_{\max}$ are invertible when J is compact.

THEOREM 3.1. *Let $T_{\max}$ $(L_2([a, b]) \to L_2([a, b]))$ be the maximal operator corresponding to τ and the compact interval $[a, b]$. Let T be the restriction of $T_{\max}$ to those $g \in \mathcal{D}(T_{\max})$ which satisfy the boundary conditions*

$$(1) \qquad \sum_{j=1}^{n} \alpha_{ij}^{(1)} g^{(j-1)}(a) + \sum_{j=1}^{n} \alpha_{ij}^{(2)} g^{(j-1)}(b) = 0, \qquad 1 \le i \le n,$$

where each $\alpha_{ij}^{(k)}$ is a given constant. Then T is invertible if and only if

$$(2) \qquad\qquad \det\big(N_1 + N_2 U(b)\big) \neq 0,$$

with N_k the $n \times n$ matrix $[\alpha_{ij}^{(k)}]_{i,j=1}^n$, $k = 1,2$, and $U(t)$ the unique continuous $n \times n$ matrix such that

$$U(t) = I_n + \int_a^t A(s)U(s)ds, \qquad a \le t \le b,$$

where

$$A(t) = \begin{bmatrix} 0 & 1 & 0 & \cdots & 0 \\ 0 & 0 & 1 & \cdots & 0 \\ \vdots & \vdots & \vdots & \ddots & \vdots \\ 0 & 0 & 0 & & 1 \\ -a_0(t) & -a_1(t) & -a_2(t) & \cdots & -a_{n-1}(t) \end{bmatrix}.$$

If (2) holds, then T^{-1} is the Hilbert-Schmidt operator on $L_2([a,b])$ given by

$$(T^{-1}f)(t) = \int_a^b G(t,s)f(s)ds, \qquad a \le t \le b,$$

with

$$G(t,s) = \begin{cases} CU(t)(I - P)U(s)^{-1}B, & a \le s < t \le b, \\ -CU(t)PU(s)^{-1}B, & a \le t < s \le b. \end{cases}$$

Here

$$P = \big(N_1 + N_2 U(b)\big)^{-1} N_2 U(b),$$

$$B = \begin{bmatrix} 0 \\ \vdots \\ 0 \\ 1 \end{bmatrix}, \qquad C = [\, 1 \quad 0 \quad \cdots \quad 0 \,].$$

PROOF. Since the entries of $A(t)$ are integrable on $[a,b]$, we know from Lemma IX.2.2 that $U(t)$ and $U(t)^{-1}$ are well-defined. From the integral expression for $U(t)$ it follows that the entries of $U(t)$ are absolutely continuous functions on $a \le t \le b$ and $U(t)' = A(t)U(t)$ a.e. on $[a,b]$. We shall first show that

$$(3) \qquad \operatorname{Im} T = \Big\{ f \in L_2([a,b]) \mid N_2 U(b) \int_a^b U(s)^{-1} B f(s)ds \in \operatorname{Im}\big(N_1 + N_2 U(b)\big) \Big\}$$

and

$$(4) \qquad \operatorname{Ker} T = \big\{ g \in L_2([a,b]) \mid g(t) = CU(t)x, x \in \operatorname{Ker}(N_1 + N_2 U(b)) \big\}.$$

Assume $f \in \operatorname{Im} T$. Then

$$(5) \qquad \begin{cases} x'(t) = A(t)x(t) + Bf(t), & a \le t \le b, \\ N_1 x(a) + N_2 x(b) = 0, \end{cases}$$

has a solution. In fact, we may take

$$x(t) = \begin{bmatrix} g(t) \\ g'(t) \\ \vdots \\ g^{(n-1)}(t) \end{bmatrix}, \qquad a \le t \le b,$$

where $Tg = f$. The solution of (5) is of the form

$$(6) \qquad x(t) = U(t)x_0 + U(t)\int_a^t U(s)^{-1} Bf(s)\,ds.$$

Since $x(\cdot)$ satisfies the boundary conditions, we see that

$$0 = N_1 x(a) + N_2 x(b)$$
$$= N_1 x_0 + N_2 U(b)x_0 + N_2 U(b)\int_a^b U(s)^{-1} Bf(s)\,ds.$$

So

$$N_2 U(b)\int_a^b U(s)^{-1} Bf(s)\,ds = -\bigl(N_1 + N_2 U(b)\bigr)x_0 \in \operatorname{Im}(N_1 + N_2 U(b)).$$

Conversely, assume that

$$N_2 U(b)\int_a^b U(s)^{-1} Bf(s)\,ds \in \operatorname{Im}(N_1 + N_2 U(b)).$$

Then there is $x_0 \in \mathbf{C}^n$ such that

$$N_2 U(b)\int_a^b U(s)^{-1} Bf(s)\,ds = -\bigl(N_1 + N_2 U(b)\bigr)x_0.$$

It follows that

$$(7) \qquad x(t) = U(t)x_0 + U(t)\int_a^t U(s)^{-1} Bf(s)\,ds, \qquad a \le t \le b,$$

satisfies (5), and therefore $g = Cx \in \mathcal{D}(T)$ and $Tg = f$. Hence $f \in \operatorname{Im} T$.

To prove (4), we set $f = 0$ in (5) and (6) to obtain

$$
(8) \qquad \begin{cases} x'(t) = A(t)x(t), & a \le t \le b, \\ N_1 x(a) + N_2 x(b) = 0. \end{cases}
$$

The general solution of the first equation in (8) is of the form $x(t) = U(t)x(a)$. For $x(t)$ to satisfy the second equation in (8),

$$
\big(N_1 + N_2 U(b)\big)x(a) = 0
$$

must hold. Also, $g(t) = CU(t)x(a) \in \operatorname{Ker} T$. Thus (4) follows.

It is clear from (3) and (4) that T is bijective if $N_1 + N_2 U(b)$ is invertible, or equivalently, $\det\big(N_1 + N_2 U(b)\big) \ne 0$. In this case, it follows from (6) and the above discussion that if $f \in L_2([a,b])$ and $Tg = f$, then $g(t) = Cx(t)$ with $x(t)$ given by (7). Using the boundary conditions we see that

$$
x(a) = -[N_1 + N_2 U(b)]^{-1} N_2 U(b) \int_a^b U(s)^{-1} B f(s)\,ds.
$$

It follows that for $P = [N_1 + N_2 U(b)]^{-1} N_2 U(b)$,

$$
x(t) = -U(t)P \int_a^b U(s)^{-1} B f(s)\,ds + U(t) \int_a^t U(s)^{-1} B f(s)\,ds
$$

$$
= U(t)(I - P) \int_a^t U(s)^{-1} B f(s)\,ds - U(t)P \int_t^b U(s)^{-1} B f(s)\,ds
$$

for each $t \in [a,b]$. The formula for $G(t,s)$ as stated in the theorem is now clear from $T^{-1}f = Cx(t)$. In particular, T^{-1} is a bounded linear operator, and thus T is invertible.

Next we shall show that

$$
(9) \qquad \dim \operatorname{Ker} T = \dim \operatorname{Ker}\big(N_1 + N_2 U(b)\big).
$$

By (4) it suffices to show that $CU(t)x = 0$ for all $t \in [a,b]$ implies $x = 0$. Let $U_j(t)$ be the j-th row of $U(t)$. Since

$$
U'(t) = A(t)U(t), \qquad a \le t \le b, \qquad \text{a.e.,}
$$

we see that for $1 \le j \le n - 1$,

$$
(10) \qquad U'_j(t) = U_{j+1}(t), \qquad a \le t \le b, \qquad \text{a.e. .}
$$

From $CU(t)x = 0$ it follows that $U_1(t)x = 0$. Apply (10) repeatedly and obtain $U_j(t)x = 0$, $1 \le j \le n$. But then $\det U(t) \ne 0$ implies $x = 0$.

From (9) we see that T is invertible implies that $N_1 + N_2 U(b)$ has full column rank. Since $N_1 + N_2 U(b)$ is a square matrix, we conclude that $\det\big(N_1 + N_2 U(b)\big) \ne 0$ whenever T is invertible. $\square$

It is clear that with $N_1 = I$ and $N_2 = 0$, the operator T is surjective. Since T is a restriction of $T_{\max}$, it follows that $T_{\max}$ is also surjective. Setting $N_1 = N_2 = 0$ in (9) yields $T = T_{\max}$ and $\dim \operatorname{Ker} T_{\max} = n$. Hence we have the following result.

COROLLARY 3.2. *The maximal operator corresponding to τ and a compact interval is surjective and its kernel is n-dimensional.*

PROPOSITION 3.3. *The maximal operator corresponding to τ and any interval is closed.*

PROOF. Let us first assume that the interval J is compact. If we take $N_1 = I$ and $N_2 = 0$ in Theorem 3.1, then the corresponding operator T has a bounded inverse on $L_2(J)$. Hence T is closed. Since $T_{\max}$ is an extension of the invertible operator T, it follows that

$$G(T_{\max}) = G(T) \oplus \big(\operatorname{Ker} T_{\max} \times \{0\}\big).$$

Therefore $G(T_{\max})$ is closed since $G(T)$ is closed and $\operatorname{Ker} T_{\max} \times \{0\}$ is finite dimensional.

Now let J be any interval, and let $T_1 = T_{\max,\tau,J}$. Suppose $f_1, f_2, \ldots$ is a sequence in $\mathcal{D}(T_1)$ such that

$$f_n \to f \in L_2(J), \qquad T_1 f_n \to g \in L_2(J).$$

For any compact subinterval J_0 of J, let $T_0 = T_{\max,\tau,J_0}$. If we consider f_n, f and g as elements of $L_2(J_0)$, then $f_n \to f$ and $T_0 f_n \to g$ in $L_2(J_0)$. Since T_0 is closed by what we have just proved, $f \in \mathcal{D}(T_0)$ and $T_0 f = g$ on $L_2(J_0)$. But J_0 was an arbitrary subinterval of J. Therefore $f \in \mathcal{D}(T_1)$ and $T_1 f = g$. $\square$

PROPOSITION 3.4. *If J is a compact interval, then any closed operator A which is a restriction of $T_{\max,\tau,J}$ has a closed range. If, in addition, A is injective, then $A^{-1}\big(\operatorname{Im} A \to L_2(J)\big)$ is compact.*

PROOF. Let us first assume that A is injective. By Theorem 3.1 there is an invertible operator T which is a restriction of $T_{\max}$ and has a compact inverse. Define

$$\mathcal{D}(F) = \operatorname{Im} A, \qquad F = A^{-1} - (T^{-1}|\operatorname{Im} A).$$

Since A^{-1} is closed and T^{-1} is bounded, the operator F is closed. Also $\operatorname{Im} F \subset \operatorname{Ker} T_{\max}$ and the latter space is finite dimensional. Hence F is bounded by Proposition 1.4. So $\mathcal{D}(F)$ is closed. Since F has finite rank, it follows that F is compact. From $A^{-1} = F + (T^{-1}|\operatorname{Im} A)$ and the compactness of F and T^{-1} we have that A^{-1} is compact. In particular, A^{-1} is bounded and thus $\operatorname{Im} A$ is closed.

If A is not injective, take A_1 to be the restriction of A to $\mathcal{D}(A) \cap \operatorname{Ker} A^{\perp}$. Then A_1 is closed, injective and a restriction of $T_{\max}$. Hence $\operatorname{Im} A = \operatorname{Im} A_1$ is closed. $\square$

Let $T_{R,\tau,J}$ be the restriction of the maximal operator to those $f \in \mathcal{D}(T_{\max,\tau,J})$ that have compact support in the interior of J. By the *support* of a function f we mean the closure of the set $\{x \mid f(x) \neq 0\}$. Thus $f \in \mathcal{D}(T_{R,\tau,J})$ if and only if $f \in \mathcal{D}(T_{\max,\tau,J})$ and there exists a compact set C (depending on f) such that

$$\{x \in J \mid f(x) \neq 0\} \subset C \subset \mathrm{int} J.$$

Since $T_{\max}$ is closed, $T_R = T_{R,\tau,J}$ is closable. The minimal closed linear extension of T_R is called the *minimal operator* corresponding to τ and J and is denoted by $T_{\min,\tau,J}$. When it is clear which τ and J we are dealing with, we write $T_{\min}$ instead of $T_{\min,\tau,J}$.

PROPOSITION 3.5. *The minimal operator $T_{\min}$ corresponding to τ and the compact interval $[a,b]$ is injective, has closed range and $T_{\min}^{-1}$ is compact on $\mathrm{Im}\, T_{\min}$. If $g \in \mathcal{D}(T_{\min})$, then $g^{(k)}(a) = g^{(k)}(b) = 0, 0 \leq k \leq n-1$.*

PROOF. Since $T_{\min}$ is a closed operator and a restriction of $T_{\max}$, Proposition 3.4 implies that $\mathrm{Im}(T_{\min})$ is closed. Let T_1 be the invertible operator in Theorem 3.1 corresponding to $N_1 = I$, $N_2 = 0$. Then T_1 is a closed linear extension of T_R, because

$$\mathcal{D}(T_R) \subset \{g \in \mathcal{D}(T_{\max}) \mid g^{(k)}(a) = 0, 0 \leq k \leq n-1\} = \mathcal{D}(T_1).$$

Hence T_1 is an extension of $T_{\min}$, and thus $T_{\min}$ is injective. But then Proposition 3.4 implies that $T_{\min}^{-1}$ is compact on $\mathrm{Im}\, T_{\min}$. Finally, let T_2 be the invertible operator in Theorem 3.1 corresponding to $N_1 = 0$, $N_2 = I$. Then

$$\mathcal{D}(T_R) \subset \{h \in \mathcal{D}(T_{\max}) \mid h^{(k)}(b) = 0, 0 \leq k \leq n-1\} = \mathcal{D}(T_2).$$

It follows that T_2 is a closed linear extension of T_R and hence also of $T_{\min}$. It follows that $\mathcal{D}(T_{\min}) \subset \mathcal{D}(T_1) \cap \mathcal{D}(T_2)$, which proves the last statement of the proposition. $\square$

XIV.4   ADJOINTS OF ORDINARY DIFFERENTIAL OPERATORS

Throughout this section, $\tau = D^n + \sum_{k=1}^{n-1} a_k(t) D^k$, with $a_k \in C^k(J)$, the space of complex valued functions which have continuous k-th order derivatives on the interval J (not necessarily compact). Let $C_0^\infty(J)$ be the space of infinitely differentiable functions with compact support in the interior of J. The extra smoothness conditions on the coefficients of τ imply that $C_0^\infty(J)$ is contained in $\mathcal{D}(T_{R,\tau,J})$. Since $C_0^\infty(J)$ is dense in $L_2(J)$, it follows that the domains of $T_{R,\tau,J}$, $T_{\min,\tau,J}$ and $T_{\max,\tau,J}$ are also dense in $L_2(J)$.

Our aim now is to determine the adjoints $T_{\max}^*$ and $T_{\min}^*$. First let us remark that Proposition 2.4 shows that $T_R^* = T_{\min}^*$. In order to have a clue to the domains of the adjoints, take f, g in $C_0^\infty(J)$. Then

$$\langle \tau f, g \rangle = \sum_{k=0}^{n} \int_J a_k(t) f^{(k)}(t) \overline{g(t)} dt, \qquad a_n(t) = 1.$$

Note that the integrals are well-defined since f and its derivatives have compact support. Successive integration by parts yields

$$\int_J a_k(t) f^{(k)}(t) \overline{g(t)} dt = \int_J (-1)^k \left(a_k(t)\overline{g(t)}\right)^{(k)} f(t) dt,$$

and therefore

$$(1) \qquad \langle \tau f, g \rangle = \langle f, \tau^* g \rangle, \qquad f, g \in C_0^\infty(J),$$

where

$$\tau^* g = (-1)^n g^{(n)} + \sum_{k=0}^{n-1} (-1)^k (\overline{a}_k g)^{(k)}.$$

By Leibnitz's rule,

$$(\overline{a}_j g)^{(j)} = \sum_{\nu=0}^{j} \binom{j}{\nu} \overline{a}_j^{(j-\nu)} g^{(\nu)}.$$

Hence we can rewrite τ^* in the form

$$\tau^* = \sum_{k=0}^{n} b_k(t) D^k,$$

where

$$b_k = \sum_{j=k}^{n} (-1)^j \binom{j}{k} \overline{a}_j^{(j-k)}, \qquad 0 \le k \le n,$$

which belongs to $C^k(J)$ because $a_j \in C^j(J)$. Thus τ^* is a differential expression of the same type as τ. One calls τ^* the *Lagrange adjoint* of τ.

THEOREM 4.1. *Let $T_{\max,\tau}$ be the maximal operator corresponding to τ and an arbitrary interval J. Then*

$$(2) \qquad T_{\min,\tau}^* = T_{R,\tau}^* = T_{\max,\tau^*}, \qquad T_{\max,\tau}^* = T_{\min,\tau^*}.$$

The proof relies on the following lemma.

LEMMA 4.2. *If τ^* is the Lagrange adjoint of τ, then $\tau^{**} = \tau$.*

PROOF. Let $J_0 = [a, b]$ be any compact interval in the interior of J. Take f in $C_0^\infty(J_0)$ and g in $C^n(J_0)$. One shows (in the same way as (1) is proved) that

$$\langle f, \tau g \rangle = \langle \tau^* f, g \rangle = \langle f, \tau^{**} g \rangle.$$

Since $C_0^\infty(J_0)$ is dense in $L_2(J_0)$ (cf. Lemma 5.1 in the next section), we have $\tau g = \tau^{**} g$ a.e. on J_0. But then $\tau g = \tau^{**} g$ on J_0 since these two functions are continuous. Now

$$\tau^{**} h = h^{(n)} + \sum_{k=0}^{n-1} c_k(t) h^{(k)},$$

where $c_k \in C^k(J)$. Now let $g(t) = t^p$ for $0 \le p \le n - 1$. Then $\tau g = \tau^{**} g$ on J_0 yields $a_k = c_k$ on J_0. But J_0 was an arbitrary compact interval in the interior of J. Hence $a_k = c_k$ on the interior of J, and thus, by continuity, $a_k = c_k$. $\square$

PROOF OF THEOREM 4.1. Take $f \in \mathcal{D}(T_{R,\tau})$ and $g \in \mathcal{D}(T_{\max,\tau^*})$. Since f has compact support in the interior of J, one shows (cf. (1)) that

$$\langle T_{R,\tau}f, g \rangle = \langle \tau f, g \rangle = \langle f, \tau^* g \rangle.$$

From these identities it is clear that

$$\mathcal{D}(T_{\max,\tau^*}) \subset \mathcal{D}(T_{R,\tau}^*), \qquad T_{\max,\tau^*} g = T_{R,\tau}^* g$$

for $g \in \mathcal{D}(T_{\max,\tau^*})$. Suppose $h \in \mathcal{D}(T_{R,\tau}^*)$. To prove the first two identities in (2), it remains to show that $h \in \mathcal{D}(T_{\max,\tau^*})$. Let $J_0 = [a, b]$ be a compact subinterval of J. Define

$$V\colon L_2([a, b]) \to L_2([a, b]), \qquad (Vf)(t) = \int_a^t f(s)ds.$$

Straightforward computations verify that for $f \in L_2([a, b])$ and $k \geq 1$ the following identities hold:

$$(3) \qquad (V^k f)(t) = \int_a^t \frac{(t-s)^{k-1}}{(k-1)!} f(s)ds, \qquad a \leq t \leq b,$$

$$(4) \qquad ((V^k)^* f)(t) = \int_t^b \frac{(s-t)^{k-1}}{(k-1)!} f(s)ds, \qquad a \leq t \leq b,$$

$$(5) \qquad D^k (V^*)^k f = (-1)^k f \quad \text{a.e.,} \; k \geq 1,$$

$$(6) \qquad V^k D^k f = f, \quad f \in AC_k([a, b]), \quad f^{(j)}(a) = 0, \qquad 1 \leq j \leq k-1.$$

Consider the space M consisting of all $f \in \mathcal{D}(T_{R,\tau})$ such that the support of f is contained in $[a, b]$. Thus for $f \in M$ the support of f is a compact set in the interior of J, which is also contained in $[a, b]$. It follows that for each $f \in M$

$$f^{(j)}(a) = f^{(j)}(b) = 0, \qquad j = 0, \ldots, n-1.$$

Now put $h^* = T_{R,\tau}^* h$, and take $f \in M$. Formula (6) yields $f^{(k)} = V^{n-k} f^{(n)}$ for $0 \leq k \leq n$, and hence

$$\langle V^n f^{(n)}, h^* \rangle = \langle f, h^* \rangle = \langle \tau f, h \rangle$$

$$= \left\langle \sum_{k=0}^n a_k f^{(k)}, h \right\rangle = \left\langle \sum_{k=0}^n a_k V^{n-k} f^{(n)}, h \right\rangle$$

$$= \left\langle f^{(n)}, \sum_{k=0}^n (V^{n-k})^* \bar{a}_k h \right\rangle.$$

Here $a_n = 1$. Therefore $f^{(n)}$ is orthogonal to

$$(7) \qquad k := -(V^*)^n h^* + \sum_{k=0}^{n} (V^{n-k})^* \overline{a}_k h$$

with respect to the inner product in $L_2([a,b])$. Our aim is to show that k is equal a.e. to a polynomial of degree $\leq n - 1$. To do this let $[c,d]$ be an arbitrary compact interval in the interior of $[a,b]$. Identify $L_2([c,d])$ with the subspace of $L_2([a,b])$ consisting of all functions that are equal to zero a.e. on $[a,b]\backslash[c,d]$. Let $\mathcal{P}_0$ denote the set of polynomials of degree $\leq n - 1$ restricted to $[c,d]$. Take $u \in \mathcal{P}_0^\perp$, the orthogonal complement of $\mathcal{P}_0$ in $L_2([c,d])$, and put

$$g(t) = \begin{cases} \displaystyle\int_c^t \frac{(t-s)^{n-1}}{(n-1)!} u(s)\,ds, & c \leq t \leq d, \\[4mm] \qquad\quad 0 & t \in J\backslash[c,d]. \end{cases}$$

Then $g \in M$, and by the above result with g in place of f, we have $g^{(n)} \perp k$. Now $g^{(n)} = u$ a.e. on $[c,d]$ and $g^{(n)}(t) = 0$ for $t \notin [c,d]$. It follows that

$$\int_c^d u(s)\overline{k(s)}\,ds = \int_a^b g^{(n)}(s)\overline{k(s)}\,ds = 0.$$

Let k_0 be the restriction of k to $[c,d]$. It follows that $k_0 \in (\mathcal{P}_0^\perp)^\perp$, and hence there exists a polynomial p_0 of degree at most $n - 1$ such that $k_0 = p_0$ a.e. on $[c,d]$. This holds for any interval $[c,d]$ in the interior of $[a,b]$. But then there exists a polynomial p of degree at most $n - 1$ such that $k = p$ a.e. on $[a,b]$. Therefore by (7)

$$(8) \qquad h = p + (V^n)^* h^* - \sum_{k=0}^{n-1} (V^{n-k})^* \overline{a}_k h$$

almost everywhere on $[a,b]$. If we define h_1 to be the right hand side of (8), then $h = h_1$ considered as elements of $L_2([a,b])$. Clearly, h_1 is absolutely continuous and $h_1' = h_2$ almost everywhere on $[a,b]$, where

$$h_2 = Dp - (V^{n-1})^* h^* + \overline{a}_{n-1} h_1 + \sum_{k=0}^{n-2} (V^{n-1-k})^* \overline{a}_k h_1.$$

The function h_2 is absolutely continuous on $[a,b]$. In particular, h_2 is continuous on $[a,b]$. But then

$$h_1(t) - h_1(a) = \int_a^t h_1'(s)\,ds = \int_a^t h_2(s)\,ds, \qquad a \leq t \leq b,$$

implies that h_1 is differentiable and $h_1' = h_2$ on all of $[a, b]$. Next, note that $h_2' = h_3$ almost everywhere on $[a, b]$, where

$$h_3 = D^2 p + (-1)^2 (V^{n-2})^* h^* - (-1)(\bar{a}_{n-1} h_1)'$$
$$- (-1)^2 (\bar{a}_{n-2} h_1) - \sum_{k=0}^{n-3} (-1)^2 (V^{n-2-k})^* \bar{a}_k h_1.$$

Again h_3 is continuous on $[a, b]$. Hence h_1 is twice differentiable and $h_1^{(2)} = h_3$. Proceeding in this way one finds that h_1 is in $AC_n([a, b])$. Since $[a, b]$ is an arbitrary interval of J, we may conclude that $h \in AC_n(J) \cap L_2(J)$. From formulas (8) and (5) and the fact that p is a polynomial of degree $\leq n - 1$, we see that $\tau^*(h) = h^*$ on $[a, b]$. Hence $\tau^*(h) = h^* \in L_2(J)$, and we have proved that $h \in \mathcal{D}(T_{\max,\tau^*})$.

To prove the third identity in (2), we replace τ by τ^* and apply Proposition 2.4 to $T_{\min,\tau^*}$. This yields

$$T_{\max,\tau}^* = T_{\min,\tau^*}^{**} = T_{\min,\tau^*}. \qquad \square$$

COROLLARY 4.3. *If $T_{\min,\tau}$ is the minimal operator corresponding to τ and a compact interval J, then* $\operatorname{codim} \operatorname{Im} T_{\min,\tau} = n$.

PROOF. By Proposition 2.5 and Theorem 4.1,

$$\operatorname{codim} \operatorname{Im} T_{\min,\tau} = \dim(\operatorname{Im} T_{\min,\tau})^{\perp}$$
$$= \dim \operatorname{Ker} T_{\min,\tau}^* = \dim \operatorname{Ker} T_{\max,\tau^*}.$$

Since the interval J is compact, $\dim \operatorname{Ker} T_{\max,\tau^*} = n$ by Corollary 3.2. $\square$

Next we shall describe the adjoint of a differential operator of the type considered in Theorem 3.1. In order to do this we need the following preliminary results.

Successive integration by parts shows that for all f and g in $AC_n([a, b])$, the following *Green's formula* holds:

$$(9) \qquad \langle \tau f, g \rangle - \langle f, \tau^* g \rangle = F_b(f, g) - F_a(f, g),$$

where

$$(10) \qquad F_t(f, g) = \sum_{i=1}^{n} \sum_{k=1}^{i} (-1)^{k-1} f^{(i-k)}(t) \left(a_i(t) \bar{g}(t) \right)^{(k-1)}.$$

Applying Leibnitz's rule to $F_t(f, g)$, we get

$$F_t(f, g) = \sum_{i=1}^{n} \sum_{k=1}^{i} \sum_{j=0}^{k-1} (-1)^{k-1} \binom{k-1}{j} a_i^{(k-j-1)}(t) f^{(i-k)}(t) \bar{g}^{(j)}(t)$$

$$= \sum_{i=1}^{n} \sum_{m=0}^{i-1} \sum_{j=0}^{i-m-1} (-1)^{i-m-1} \binom{i-m-1}{j} a_i^{(i-m-j-1)}(t) f^{(m)}(t) \overline{g}^{(j)}(t)$$

$$= \sum_{m=0}^{n-1} \sum_{j=0}^{n-m-1} \sum_{i=m+j+1}^{n} (-1)^{i-m-1} \binom{i-m-1}{j} a_i^{(i-m-j-1)}(t) f^{(m)}(t) \overline{g}^{(j)}(t)$$

$$= \sum_{m=0}^{n-1} \sum_{j=0}^{n-m-1} \sum_{k=j}^{n-m-1} (-1)^{k} \binom{k}{j} a_{m+k+1}^{(k-j)}(t) f^{(m)}(t) \overline{g}^{(j)}(t).$$

The latter identity allows us to rewrite $F_t(f,g)$ in the following form:

$$(11) \qquad\qquad F_t(f,g) = W_t(g)^* F(t)^* W_t(f).$$

Here $F(t) = [F_{mj}(t)]_{m,j=0}^{n-1}$ is the $n \times n$ matrix with entries

$$F_{mj}(t) = \begin{cases} \displaystyle\sum_{k=j}^{n-m-1} (-1)^{k} \binom{k}{j} \overline{a}_{m+k+1}^{(k-j)}(t), & \text{if } m+j \le n-1, \\[2mm] 0, & \text{otherwise.} \end{cases}$$

Furthermore,

$$(12) \qquad W_t(f) = \begin{bmatrix} f(t) \\ f'(t) \\ \vdots \\ f^{(n-1)}(t) \end{bmatrix}, \qquad W_t(g) = \begin{bmatrix} g(t) \\ g'(t) \\ \vdots \\ g^{(n-1)}(t) \end{bmatrix},$$

and the symbol $*$ denotes the usual matrix adjoint. The character W in (12) stands for Wronskian. Note that $F_{mj}(t) = (-1)^{j}\overline{a}_n(t)$, $j + m = n - 1$. Since $a_n(t) \equiv 1$ and $F_{mj}(t) = 0$ for $j + m > n - 1$, we conclude that

$$(13) \qquad\qquad |\det F(t)| = 1, \qquad a \le t \le b.$$

In particular, $F(t)$ is invertible for $a \le t \le b$.

As we are interested in finding the adjoint of a differential operator subject to the boundary conditions

$$\sum_{j=0}^{n-1} \alpha_{ij} g^{(j)}(a) + \sum_{j=0}^{n-1} \beta_{ij} g^{(j)}(b) = 0, \qquad 1 \le i \le k,$$

it will be necessary to define the corresponding adjoint boundary conditions. We start with the $k \times n$ matrices

$$(14) \qquad\qquad N_1 = [\alpha_{ij}]_{i=1,\ j=0}^{k,\ n-1}, \qquad N_2 = [\beta_{ij}]_{i=1,\ j=0}^{k,\ n-1}.$$

Let ℓ be the rank of the matrix $N = [\ N_1 \quad N_2\ ]$. Then $\dim \operatorname{Ker} N = 2n - \ell$, and we can construct matrices G_1 and G_2 of size $n \times (2n - \ell)$ such that

$$(15) \qquad \operatorname{Ker} N = \operatorname{Ker}[\ N_1 \quad N_2\ ] = \operatorname{Im} \begin{bmatrix} G_1 \\ G_2 \end{bmatrix}.$$

Next we introduce

$$(16a) \qquad [\alpha_{ij}^{\#}]_{i=1,\ j=0}^{2n-\ell\ \ n-1} := -G_1^* F(a),$$

$$(16b) \qquad [\beta_{ij}^{\#}]_{i=1,\ j=0}^{2n-\ell\ \ n-1} := G_2^* F(b).$$

Here $F(t)$ is the $n \times n$ matrix appearing in (11). We shall refer to the system of equations

$$\sum_{j=0}^{n-1} \alpha_{ij}^{\#} g^{(j)}(a) + \sum_{j=0}^{n-1} \beta_{ij}^{\#} g^{(j)}(b) = 0, \qquad 1 \le i \le 2n - \ell,$$

as the *adjoint boundary conditions*. We are now prepared to prove the following duality theorem.

> THEOREM 4.4. *Let* $T\bigl(L_2([a,b]) \to L_2([a,b])\bigr)$ *be the restriction of* $T_{\max,\tau}$ *to those* $g \in \mathcal{D}(T_{\max,\tau})$ *which satisfy the boundary conditions*
>
> $$(17) \qquad B_i(g) = \sum_{j=0}^{n-1} \alpha_{ij} g^{(j)}(a) + \sum_{j=0}^{n-1} \beta_{ij} g^{(j)}(b) = 0, \qquad 1 \le i \le k.$$
>
> *Assume that the rank of the* $k \times 2n$ *matrix* $[[\alpha_{ij}][\beta_{ij}]]$ *is* ℓ. *Then the adjoint* T^* *is the restriction of* $T_{\max,\tau^*}$ *to those* $f \in \mathcal{D}(T_{\max,\tau^*})$ *which satisfy the adjoint boundary conditions*
>
> $$(18) \qquad B_i^{\#}(f) = \sum_{j=0}^{n-1} \alpha_{ij}^{\#} f^{(j)}(a) + \sum_{j=0}^{n-1} \beta_{ij}^{\#} f^{(j)}(b) = 0, \qquad 1 \le i \le 2n - \ell,$$
>
> *where* $\alpha_{ij}^{\#}$ *and* $\beta_{ij}^{\#}$ *are defined in* (16a) *and* (16b).

PROOF. Define $A\bigl(L_2([a,b]) \to L_2([a,b])\bigr)$ by

$$\mathcal{D}(A) = \{ f \in \mathcal{D}(T_{\max,\tau^*}) \mid B_p^{\#}(f) = 0, 1 \le p \le 2n - \ell \},$$
$$Af = \tau^* f.$$

We must show that $A = T^*$. Let $f \in \mathcal{D}(A)$ and $g \in \mathcal{D}(T)$ be given. Consider the vectors $W_a(g)$ and $W_b(g)$ defined by formula (12). Formula (17) implies that

$$\begin{bmatrix} W_a(g) \\ W_b(g) \end{bmatrix} \in \operatorname{Ker} N,$$

and hence, according to (15), there exists a vector d in $\mathbf{C}^{2n-\ell}$ such that $W_a(g) = G_1 d$ and $W_b(g) = G_2 d$. From (9) and (11) we have

$$
\begin{aligned}
\langle Tg, f \rangle - \langle g, Af \rangle &= \langle \tau g, f \rangle - \langle g, \tau^* f \rangle \\
&= W_b(f)^* F(b)^* W_b(g) - W_a(f)^* F(a)^* W_a(g) \\
&= \langle W_b(g), F(b) W_b(f) \rangle - \langle W_a(g), F(a) W_a(f) \rangle \\
&= \langle G_2 d, F(b) W_b(f) \rangle - \langle G_1 d, F(a) W_a(f) \rangle \\
&= \langle d, G_2^* F(b) W_b(f) \rangle - \langle d, G_1^* F(a) W_a(f) \rangle \\
&= \langle d, N_1^{\#} W_a(f) + N_2^{\#} W_b(f) \rangle,
\end{aligned}
$$

where $N_1^{\#}$ (resp. $N_2^{\#}$) is the matrix in the left hand side of (16a) (resp. (16b)). But then we can use (18) to show that $\langle Tg, f \rangle = \langle g, Af \rangle$. Hence $f \in \mathcal{D}(T^*)$ and $T^* f = Af$ which proves that A is a restriction of T^*.

It remains to prove that $\mathcal{D}(T^*) \subset \mathcal{D}(A)$. Take $h \in \mathcal{D}(T^*)$. From the second part of Proposition 3.5 it follows that $T_{\min,\tau}$ is a restriction of T. Hence T^* is a restriction of $T_{\min,\tau}^*$. By Theorem 4.1, we have $T_{\min,\tau}^* = T_{\max,\tau^*}$. Thus $h \in \mathcal{D}(T_{\max,\tau^*})$ and $T^* h = \tau^* h$. There exist polynomials $q_1, \ldots, q_{2n-\ell}$ such that

$$
(19) \qquad G_1 = W_a(q_1, \ldots, q_{2n-\ell}), \qquad G_2 = W_b(q_1, \ldots, q_{2n-\ell}),
$$

where

$$
W_t(h_1, \ldots, h_r) := \begin{bmatrix} h_1(t) & \cdots & h_r(t) \\ h_1'(t) & \cdots & h_r'(t) \\ \vdots & & \vdots \\ h_1^{(n-1)}(t) & \cdots & h_r^{(n-1)}(t) \end{bmatrix}.
$$

To see this, put

$$
(20) \qquad p_{aj}(t) = \sum_{i=0}^{n-j-1} \frac{(-1)^n}{j!} \binom{n+i-1}{i} \frac{(t-a)^{i+j}(t-b)^n}{(b-a)^{n+i}}, \qquad 0 \le j \le n-1,
$$

$$
(21) \qquad p_{bj}(t) = \sum_{i=0}^{n-j-1} \frac{(-1)^n}{j!} \binom{n+i-1}{i} \frac{(t-b)^{i+j}(t-a)^n}{(a-b)^{n+i}}, \qquad 0 \le j \le n-1.
$$

Then

$$
(22) \qquad W_a(p_{a0}, \ldots, p_{a\,n-1}) = I, \qquad W_b(p_{a0}, \ldots, p_{a\,n-1}) = 0,
$$

$$
(23) \qquad W_a(p_{b0}, \ldots, p_{b\,n-1}) = 0, \qquad W_b(p_{b0}, \ldots, p_{b\,n-1}) = I.
$$

Now define $q_1, \ldots, q_{2n-\ell}$ by setting

$$
[q_1(t) \cdots q_{2n-\ell}(t)] = [p_{a0}(t) \cdots p_{a\,n-1}(t)] G_1 + [p_{b0}(t) \cdots p_{b\,n-1}(t)] G_2.
$$

From this definition and the formulas (22) and (23) it is clear that (19) is fulfilled. Since $N_1 G_1 + N_2 G_2 = 0$, formula (19) implies that $q_1, \ldots, q_{2n-\ell}$ belong to the domain of T. Hence, because of (9) and (11),

$$0 = \langle Tq_\nu, h \rangle - \langle q_\nu, T^*h \rangle = \langle \tau q_\nu, h \rangle - \langle q_\nu, \tau^*h \rangle$$
$$= W_b(h)^* F(b)^* W_b(q_\nu) - W_a(h)^* F(b)^* W_a(q_\nu)$$

for $\nu = 1, \ldots, 2n - \ell$. Again using (19), we see that

$$W_b(h)^* F(b)^* G_2 - W_a(h)^* F(a)^* G_1 = 0.$$

Take matrix adjoints in the latter identity and use (16a) and (16b). This shows that $B_\nu^{\#}(h) = 0$ for $\nu = 1, \ldots, 2n - \ell$. Thus $h \in \mathcal{D}(A)$. $\square$

The linear functionals $B_1, \ldots, B_k$ appearing in (17) are linearly independent on $\mathcal{D}(T_{\max,\tau})$ if and only if the rank ℓ of the $k \times 2n$ matrix $N = [\, N_1 \quad N_2 \,]$ is precisely equal to k. Here N_1 and N_2 are as in (14). To see this note that for $\gamma_1, \ldots, \gamma_k$ and g in $\mathcal{D}(T_{\max,\tau})$

$$(24) \qquad \sum_{i=1}^{k} \gamma_i B_i(g) = [\gamma_1 \cdots \gamma_k](N_1 W_a(g) + N_2 W_b(g)).$$

Now assume that $\ell = \text{rank}[\, N_1 \quad N_2 \,] < k$. Then there exist $\gamma_1, \ldots, \gamma_k$, not all zero, such that

$$(25) \qquad [\gamma_1 \cdots \gamma_k][\, N_1 \quad N_2 \,] = 0,$$

and hence (24) shows that $B_1, \ldots, B_k$ are linearly dependent on $\mathcal{D}(T_{\max,\tau})$. To prove the converse statement, assume that there are $\gamma_1, \ldots, \gamma_k$, not all zero, such that the left hand side of (24) is zero for each $g \in \mathcal{D}(T_{\max,\tau})$. Since the interval we are working with is compact, all polynomials are in $\mathcal{D}(T_{\max,\tau})$. In particular, we may take $g = p_{aj}$ or $g = p_{bj}$, where p_{aj} and p_{bj}, $0 \le j \le n - 1$, are the polynomials defined by (20) and (21). But then (24) yields

$$0 = [\gamma_1 \cdots \gamma_k](N_1 W_a(p_{aj}) + N_2 W_b(p_{aj}))$$
$$0 = [\gamma_1 \cdots \gamma_k](N_1 W_a(p_{bj}) + N_2 W_b(p_{bj}))$$

for $j = 0, \ldots, n - 1$. So we can use (22) and (23) to show that (25) holds, and hence rank $N < k$. Note that in the above discussion the actual form of τ is irrelevant, because for a compact interval, $\mathcal{D}(T_{\max,\tau})$ is independent of τ for the class of τ's considered in this section.

The linear functionals $B_1^{\#}, \ldots, B_{2n-\ell}^{\#}$ defined by (18) are linearly independent on $\mathcal{D}(T_{\max,\tau^*})$. To prove this, it is sufficient to show (use the result of the previous paragraph) that the rank of the matrix $[\, N_1^{\#} \quad N_2^{\#} \,]$ is precisely equal to $2n - \ell$. Here $N_1^{\#}$ and $N_2^{\#}$ are the matrices defined by the left hand sides of (16a) and (16b), respectively. Thus

$$(26) \qquad [\, N_1^{\#} \quad N_2^{\#} \,] = [\, G_1^* \quad G_2^* \,] \begin{bmatrix} -F(a) & 0 \\ 0 & F(b) \end{bmatrix}.$$

From the definitions of G_1 and G_2 we know that $\mathrm{rank}[\ G_1^*\ \ G_2^*\ ] = 2n - \ell$. Since $F(a)$ and $F(b)$ are invertible (see formula (13)), we conclude from (26) that $[\ N_1^{\#}\ \ N_2^{\#}\ ]$ has the desired rank. The results of this and the previous paragraph will be used in Chapter XVI.

For $J = [a, b]$, a compact interval, Theorems 4.1 and 4.4 remain true if the leading coefficient (i.e., the coefficient of D^n) in the differential expression τ is a function in $C^n([a, b])$ which does not vanish on $[a, b]$. To be more precise, let $\rho = \sum_{j=0}^{n} c_j(t)D^j$, where $c_j \in C^j([a, b])$ for $0 \le j \le n$. By definition the *Lagrange adjoint* of ρ is the differential expression

$$\rho^* g = \sum_{j=0}^{n}(-1)^j (\bar{c}_j g)^{(j)}.$$

Now assume that $c_n(t) \ne 0$ for all $a \le t \le b$. Put $a_j = c_j/c_n$ for $0 \le j \le n - 1$, and let

$$\tau g = D^n g + \sum_{j=0}^{n-1} a_j D^j g.$$

Note that $a_j \in C^j([a, b])$ for $0 \le j \le n-1$, and hence τ belongs to the class of differential expressions to which Theorems 4.1 and 4.4 apply. Obviously,

$$(27) \qquad\qquad \rho g = c_n(\tau g), \qquad \rho^* g = \tau^*(\bar{c}_n g).$$

The definitions of $T_{\mathrm{max},\rho}$ and $T_{\mathrm{min},\rho}$ are analogous to those of $T_{\mathrm{max},\tau}$ and $T_{\mathrm{min},\tau}$. From (27) it follows that

$$T_{\mathrm{max},\rho} = M_{c_n} T_{\mathrm{max},\tau}, \qquad T_{\mathrm{min},\rho} = M_{c_n} T_{\mathrm{min},\tau}$$
$$T_{\mathrm{max},\rho^*} = T_{\mathrm{max},\tau^*} M_{c_n}^*, \qquad T_{\mathrm{min},\rho^*} = T_{\mathrm{min},\tau^*} M_{c_n}^*.$$

Here M_{c_n} is the operator of multiplication by c_n on $L_2([a, b])$, i.e.,

$$M_{c_n} : L_2([a, b]) \to L_2([a, b]), \qquad M_{c_n} f = c_n f.$$

Since c_n is continuous on $[a, b]$ and $c_n(t) \ne 0$ for $a \le t \le b$, the operator M_{c_n} is invertible and $(M_{c_n})^{-1} = M_{1/c_n}$. From these remarks it is now clear that Theorems 4.1 and 4.4 also hold for ρ in place of τ.

As an example of the above results, let ρ be the differential expression given by

$$\rho f = D(pf') + qf = pf'' + p'f' + qf,$$

where $p \in C^2([a, b])$ and $q \in C([a, b])$ are real-valued functions. Assume that $p(t) \ne 0$ for all $a \le t \le b$. Define $T\big(L_2([a, b]) \to L_2([a, b])\big)$ by

$$\mathcal{D}(T) = \{f \in AC_2([a, b]) \mid \rho f \in L_2([a, b]), f(a) = f'(b) = 0\},$$
$$Tf = \rho f.$$

Let us compute T^*. A simple calculation gives $\rho^* = \rho$. To find the adjoint boundary conditions, note that the boundary conditions for T are given by

$$B_1 f = f(a) = 0, \qquad B_2 f = f'(b) = 0.$$

Therefore

$$N_1 = \begin{bmatrix} 1 & 0 \\ 0 & 0 \end{bmatrix}, \qquad N_2 = \begin{bmatrix} 0 & 0 \\ 0 & 1 \end{bmatrix}, \qquad N = [\, N_1 \ \ N_2 \,] = \begin{bmatrix} 1 & 0 & 0 & 0 \\ 0 & 0 & 0 & 1 \end{bmatrix}.$$

A basis for $\operatorname{Ker} N$ is $\{[\,0\ 1\ 0\ 0\,]^T, [\,0\ 0\ 1\ 0\,]^T\}$. Thus we may take

$$G_1 = \begin{bmatrix} 0 & 0 \\ 1 & 0 \end{bmatrix}, \qquad G_2 = \begin{bmatrix} 0 & 1 \\ 0 & 0 \end{bmatrix}.$$

Furthermore,

$$\begin{aligned} F_{00}(t) &= p'(t) - p'(t) = 0, & F_{01}(t) &= -p(t), \\ F_{10}(t) &= p(t), & F_{11}(t) &= 0. \end{aligned}$$

So by (26),

$$\begin{aligned} \begin{bmatrix} \alpha_{10}^{\#} & \alpha_{11}^{\#} & \beta_{10}^{\#} & \beta_{11}^{\#} \\ \alpha_{20}^{\#} & \alpha_{21}^{\#} & \beta_{20}^{\#} & \beta_{21}^{\#} \end{bmatrix} &= \begin{bmatrix} 0 & 1 & 0 & 0 \\ 0 & 0 & 1 & 0 \end{bmatrix} \begin{bmatrix} 0 & p(a) & 0 & 0 \\ -p(a) & 0 & 0 & 0 \\ 0 & 0 & 0 & -p(b) \\ 0 & 0 & p(b) & 0 \end{bmatrix} \\ &= \begin{bmatrix} -p(a) & 0 & 0 & 0 \\ 0 & 0 & 0 & -p(b) \end{bmatrix}. \end{aligned}$$

Thus

$$B_1^{\#} f = -p(a) f(a), \qquad B_2^{\#} f = -p(b) f'(b).$$

Since $p(a) \neq 0$ and $p(b) \neq 0$, the operators T and T^* have the same domain. Also $\rho = \rho^*$. So Theorem 4.4 yields $T = T^*$.

XIV.5 INTERMEZZO ABOUT SOBOLEV SPACES

We now introduce the notions of weak derivatives and Sobolev spaces which are needed to study the Dirichlet problem in the next section. We shall assume that Ω is a *bounded open* set in $\mathbf{R}^n$. The points of $\mathbf{R}^n$ are denoted by $x = (x_1, \ldots, x_n)$. For $D_j = \frac{\partial}{\partial x_j}$ and $\alpha = (\alpha_1, \ldots, \alpha_n)$, an n-tuple of non-negative integers, we define $D^\alpha = D_1^{\alpha_1} D_2^{\alpha_2} \cdots D_n^{\alpha_n}$ and $D_i^0 = I$. The *order* of D^α is $|\alpha| = \sum_{j=1}^n \alpha_j$. For each non-negative integer k, we denote by $C^k(\Omega)$ the set of continuous complex valued functions on Ω whose partial derivatives up to order k exist and are continuous on Ω. The set $C^\infty(\Omega) = \bigcap_{k=0}^\infty C^k(\Omega)$ and $C_0^\infty(\Omega)$ is the set of functions in $C^\infty(\Omega)$ which have compact support in Ω. Recall that the *support* of a complex valued function f defined on Ω is the closure in Ω of the set $\{x \in \Omega \mid f(x) \neq 0\}$. We denote the support of f by $\operatorname{supp} f$. We write $f \overset{\prime}{\in} C^k(\overline{\Omega})$ if $f \in C^k(\mathcal{O})$, where $\mathcal{O}$ is some open set containing the closure $\overline{\Omega}$ of Ω.

By $L_2(\Omega)$ we denote the Hilbert space of all Lebesgue square integrable functions on Ω endowed with the usual inner product.

To define Sobolev spaces it is necessary to start with the concept of a weak derivative which relies on the following lemma.

LEMMA 5.1. *The space $C_0^\infty(\Omega)$ is dense in $L_2(\Omega)$.*

PROOF. We write $C(\Omega)$ for the space of all complex valued continuous functions on Ω and $C_0(\Omega)$ for the space consisting of the functions in $C(\Omega)$ that have compact support in Ω. Since $C_0(\Omega)$ is dense in $L_2(\Omega)$ ([R], Theorem 3.14), it suffices to prove that $C_0^\infty(\Omega)$ is dense in $C_0(\Omega)$ with respect to the $L_2(\Omega)$-norm. Let φ be a real-valued function on $\mathbf{R}^n$ with the following properties:

(a) $\varphi \in C^\infty(\mathbf{R}^n)$,

(b) $\varphi(x) > 0$, $\|x\| < 1$,

(c) $\varphi(x) = 0$, $\|x\| \geq 1$,

(d) $\int\limits_{\mathbf{R}^n} \varphi(x)dx = 1$.

An example of such a function is

$$\varphi(x) = \begin{cases} b\exp[(\|x\|^2 - 1)^{-1}], & \|x\| < 1 \\ 0, & \|x\| \geq 1 \end{cases}$$

with the constant b chosen so that (d) holds. Suppose $u \in C_0(\Omega)$ is given with compact support K. Extend u to all of $\mathbf{R}^n$ by setting $u(x) = 0$ for $x \notin \Omega$. Let ε be any positive number which is strictly less than the distance from K to $\mathbf{R}^n \backslash \Omega$. Define for all $x \in \mathbf{R}^n$,

$$u_\varepsilon = \frac{1}{\varepsilon^n} \int\limits_{\mathbf{R}^n} \varphi\left(\frac{x-y}{\varepsilon}\right) u(y)dy.$$

Since $\varphi \in C^\infty(\mathbf{R}^n)$, differentiation under the integral sign shows that u_ε is in $C^\infty(\mathbf{R}^n)$. Also, $\operatorname{supp} u_\varepsilon$ is a compact subset of Ω. For suppose $u(x) \neq 0$. Then by (c) and the assumption that $\operatorname{supp} u = K \subset \Omega$, there exists $y \in K$ such that $\|x - y\| < \varepsilon$. Hence $\operatorname{supp} u_\varepsilon$ is a closed subset of the compact set of points whose distance from K is at most ε. It follows from the definition of ε that this latter set is contained in Ω. Hence $u_\varepsilon \in C_0^\infty(\Omega)$. It remains to prove that $\lim_{\varepsilon \to 0} \|u_\varepsilon - u\| = 0$. By (d) we may write

$$u(x) = \frac{1}{\varepsilon^n} \int\limits_{\mathbf{R}^n} \varphi\left(\frac{x-y}{\varepsilon}\right) u(x)dy.$$

Therefore

$$|u_\varepsilon(x) - u(x)| \leq \frac{1}{\varepsilon^n} \int\limits_{\mathbf{R}^n} \varphi\left(\frac{x-y}{\varepsilon}\right) |u(y) - u(x)|dy$$

$$\leq \sup_{\|x-y\|<\varepsilon} |u(y) - u(x)|.$$

Since u is uniformly continuous on $\mathbf{R}^n$, it follows that as $\varepsilon \to 0$, u_ε converges uniformly to u on Ω. Hence $\lim_{\varepsilon \to 0} \|u_\varepsilon - u\| = 0$. $\square$

By $L_1(\Omega)$ we denote the space of Lebesgue integrable functions on Ω. Let $\alpha = (\alpha_1, \ldots, \alpha_n)$ be an n-tuple of non-negative integers. A function $f \in L_1(\Omega)$ is said to have a *weak α-derivative* $g \in L_1(\Omega)$ if for every $\varphi \in C_0^\infty(\Omega)$,

$$\int_\Omega f(x) D^\alpha \varphi(x) dx = (-1)^{|\alpha|} \int_\Omega g(x)\varphi(x) dx.$$

In this case we write $D^\alpha f = g$. Here integration is with respect to Lebesgue measure. Since $C_0^\infty(\Omega)$ is dense in $L_2(\Omega)$, it follows that if g_1 and g_2 are weak α-derivatives of f, then $g_1 = g_2$ a.e. If $f \in C^{|\alpha|}(\Omega)$, then integration by parts shows that $D^\alpha f$ is the usual α-derivative of f.

To illustrate the notion of a weak derivative, let $\Omega = (0, 2)$ and define

$$f(x) = \begin{cases} x, & 0 < x \le 1, \\ 1, & 1 < x < 2. \end{cases}$$

Then for every $\varphi \in C_0^\infty(\Omega)$,

$$\int_0^2 f(x)\varphi'(x) dx = \int_0^1 x\varphi'(x) dx + \int_1^2 \varphi'(x) dx$$

$$= \varphi(1) - \int_0^1 \varphi(x) dx - \varphi(1)$$

$$= -\int_0^2 g(x)\varphi(x) dx,$$

where

$$g(x) = \begin{cases} 1, & 0 < x \le 1, \\ 0, & 1 < x < 2. \end{cases}$$

Hence g is the weak derivative of f in $L_1(\Omega)$. Now g is differentiable at $x \ne 1$, yet g does not have a weak derivative in $L_1(\Omega)$. Indeed, suppose $Dg = h$ for some $h \in L_1(\Omega)$. Then for every $\varphi \in C_0^\infty(\Omega)$,

$$-\int_0^2 h(x)\varphi(x) dx = \int_0^2 g(x)\varphi'(x) dx = \int_0^1 \varphi'(x) dx = \varphi(1).$$

But this is impossible since there exist $\varphi_n \in C_0^\infty(\Omega)$, $n = 1, 2, \ldots$, such that $0 \le \varphi_n \le 1$, $\varphi_n(1) = 1$ and φ_n has support in the interval $\left(1 - \frac{1}{n}, 1 + \frac{1}{n}\right)$, whence

$$1 = \varphi_n(1) = -\int_0^2 h(x)\varphi_n(x) dx \to 0.$$

In the larger framework of the theory of distributions, g has a derivative, namely the Dirac δ functional supported at 1, i.e., $\delta(\varphi) = \varphi(1)$ for all $\varphi \in C_0^\infty(\Omega)$. The argument used above shows that this distribution δ is not defined by a function in $L_1(\Omega)$.

For each non-negative integer m, let $H_m(\Omega)$ denote the set of those $f \in L_2(\Omega)$ which have weak α-derivatives in $L_2(\Omega)$ for $0 \le |\alpha| \le m$. Since $D^\alpha = I$ for $|\alpha| = 0$, we set $H_0(\Omega) = L_2(\Omega)$. Define an inner product $\langle \cdot, \cdot \rangle_m$ on $H_m(\Omega)$ by

$$\langle f, g \rangle_m = \sum_{|\alpha| \le m} \int_\Omega (D^\alpha f)(x)\overline{(D^\alpha g)(x)}dx.$$

The space $H_m(\Omega)$ with this inner product is called the *Sobolev space of order* m. Note that $\| \cdot \|_0 \le \| \cdot \|_m$, where $\| \ \|_0$ is the usual norm on $L_2(\Omega)$ and $\|f\|_m = \langle f, f \rangle_m^{1/2}$. Clearly, $C_0^\infty(\Omega) \subset H_m(\Omega)$. Define $H_m^0(\Omega)$ to be the closure in $H_m(\Omega)$ of $C_0^\infty(\Omega)$. Note that $H_0^0(\Omega) = H_0(\Omega)$ by Lemma 5.1. For $m \ge 1$ the space $H_m^0(\Omega)$ is a proper subspace of $H_m(\Omega)$. For example, it can be shown (cf. Friedman [1], Section 10.2) that the constant function 1 belongs to $H_m(\Omega)$ but not to $H_m^0(\Omega)$ for $m \ge 1$.

THEOREM 5.2. *The Sobolev space* $H_m(\Omega)$ *is a Hilbert space.*

PROOF. Let (f_j) be a Cauchy sequence in $H_m(\Omega)$. Since

$$\|D^\alpha f_j - D^\alpha f_k\|_0 \le \|f_j - f_k\|_m, \qquad |\alpha| \le m,$$

the sequence $(D^\alpha f_j)$ converges in $L_2(\Omega)$ to some g_α. Therefore for each $\varphi \in C_0^\infty(\Omega)$ and $|\alpha| \le m$,

$$\int_\Omega g_0(x)D^\alpha \varphi(x)dx = \lim \int_\Omega f_j(x)D^\alpha \varphi(x)dx$$

$$= \lim(-1)^{|\alpha|} \int_\Omega D^\alpha f_j(x)\varphi(x)dx$$

$$= (-1)^{|\alpha|} \int_\Omega g_\alpha(x)\varphi(x)dx.$$

Hence $g_0 \in H_m(\Omega)$, $D^\alpha g_0 = g_\alpha$ and

$$\|g_0 - f_j\|_m^2 = \sum_{|\alpha| \le m} \|g_\alpha - D^\alpha f_j\|_0^2 \to 0. \quad \square$$

The next theorem (Theorem 5.4 below) is fundamental for the study of the generalized Dirichlet problem. To prove it we need the following result.

PROPOSITION 5.3. *Let* S *denote the cube* $\prod_{i=1}^N [a_i, b_i]$ *in* $\mathbf{R}^N$, *with* $b_i - a_i = \ell > 0$, $1 \le i \le N$. *For each* $\varphi \in C^1(S)$ *we have the inequality:*

$$(1) \qquad \int_S |\varphi(x)|^2 dx \le \frac{\ell^2 N}{2} \int_S \sum_{i=1}^N |D_i \varphi(x)|^2 dx + \frac{1}{\ell^N}\left| \int_S \varphi(x)dx \right|^2.$$

PROOF. For $x = (x_1, x_2, \ldots, x_N)$ and $y = (y_1, y_2, \ldots, y_N)$ in S,

$$
\varphi(x) - \varphi(y) = \int_{y_1}^{x_1} D_1\varphi(\xi_1, y_2, \ldots, y_N)d\xi_1 + \int_{y_2}^{x_2} D_2\varphi(x_1, \xi_2, y_3, \ldots, y_N)d\xi_2
$$

$$
(2) \qquad + \cdots + \int_{y_N}^{x_N} D_N\varphi(x_1, x_2, \ldots, x_{N-1}, \xi_N)d\xi_N.
$$

The Cauchy-Schwarz inequality and (2) imply

$$
|\varphi(x) - \varphi(y)|^2 = |\varphi(x)|^2 - \varphi(x)\overline{\varphi}(y) - \overline{\varphi}(x)\varphi(y) + |\varphi(y)|^2
$$

$$
\leq \ell N \left(\int_{a_1}^{b_1} |D_1\varphi(\xi_1, y_2, \ldots, y_N)|^2 d\xi_1 \right.
$$

$$
(3) \qquad + \int_{a_2}^{b_2} |D_2\varphi(x_1, \xi_2, y_3, \ldots, y_N)|^2 d\xi_2
$$

$$
\left. + \cdots + \int_{a_N}^{b_N} |D_N\varphi(x_1, x_2, \ldots, x_{N-1}, \xi_N)|^2 d\xi_N \right).
$$

Successive integration of each side of (3) with respect to $x_1, x_2, \ldots, x_N, y_1, \ldots, y_N$ (i.e., integration on $S \times S$) yields

$$
2\ell^N \int_S |\varphi(x)|^2 dx - 2\left| \int_S \varphi(x)dx \right|^2 \leq \ell N \ell^{N+1} \sum_{i=1}^N \int_S |D_i\varphi(x)|^2 dx,
$$

which is equivalent to (1). $\square$

THEOREM 5.4. *The embedding from $H_1^0(\Omega)$ into $L_2(\Omega)$ is compact.*

PROOF. The closure of a set U in a Banach space is compact if and only if it is *totally bounded*. That is to say (see [W], 24B), for every $\varepsilon > 0$ there exists a finite set $u_1, \ldots, u_m$ in U such that for each $u \in U$,

$$
\min_k \|u - u_k\| < \varepsilon.
$$

Since $C_0^\infty(\Omega)$ is dense in $H_1^0(\Omega)$, the theorem will be proved once we show that the set

$$
U = \{\varphi \mid \varphi \in C_0^\infty(\Omega), \|\varphi\|_1 = 1\}
$$

is totally bounded in $L_2(\Omega)$. Given $\varepsilon > 0$, let Q be a cube in $\mathbf{R}^n$ which contains Ω, and let $Q = \cup_{k=1}^p S_k$, where $S_1, S_2, \ldots, S_p$ are cubes whose edges have length $\ell \leq \varepsilon/\sqrt{3n}$ such that the interiors of $S_1, \ldots, S_p$ do not intersect. Take $\varphi \in U$, and put $\varphi = 0$ on $Q \backslash \Omega$. Since $\varphi \in C_0^\infty(\Omega)$, we have $\varphi \in C^1(S_j)$ for each j. Thus by (1),

(4)
$$
\begin{aligned}
\int_\Omega |\varphi(x)|^2 dx &= \sum_{k=1}^p \int_{S_k} |\varphi(x)|^2 dx \\
&\leq \frac{\varepsilon^2}{6} \int_Q \sum_{i=1}^n |D_i\varphi(x)|^2 + \left(\frac{1}{\ell}\right)^n \sum_{k=1}^p \left| \int_{S_k} \varphi(x) dx \right|^2 \\
&\leq \frac{\varepsilon^2}{6} \|\varphi\|_1^2 + \left(\frac{1}{\ell}\right)^n \sum_{k=1}^p \left| \int_{S_k} \varphi(x) dx \right|^2 .
\end{aligned}
$$

Consider the map $K\varphi = (\alpha_1(\varphi), \ldots, \alpha_p(\varphi))$, where $\alpha_k(\varphi) = \int_{S_k} \varphi(x) dx$. Obviously, $K\varphi \in \mathbf{C}^p$ and

$$
\|K\varphi\|_{\mathbf{C}^p}^2 \leq \ell^n \|\varphi\|_0^2 \leq \ell^n \|\varphi\|_1^2 .
$$

Hence $K(U)$ is bounded and therefore totally bounded in $\mathbf{C}^p$. So one may find $\varphi_1, \ldots, \varphi_m$ in U such that for each $\psi \in U$ there exists j (depending on ψ) with

$$
\left(\frac{1}{\ell}\right)^n \sum_{k=1}^p \left| \int_{S_k} (\psi(x) - \varphi_j(x)) dx \right|^2 = \left(\frac{1}{\ell}\right)^n \|K\psi - K\varphi_j\|_{\mathbf{C}^p}^2 < \frac{\varepsilon^2}{4} .
$$

Replacing φ by $\psi - \varphi_j$ in (4) gives

$$
\|\psi - \varphi_j\|_0^2 \leq \frac{\varepsilon^2}{6} \|\psi - \varphi_j\|_1^2 + \frac{\varepsilon^2}{4} \leq \frac{2}{3}\varepsilon^2 + \frac{1}{4}\varepsilon^2 < \varepsilon^2 .
$$

Hence U is totally bounded in $L_2(\Omega)$. $\square$

XIV.6   THE OPERATOR DEFINED BY THE DIRICHLET PROBLEM

In this section we introduce and study a closed unbounded operator associated with the generalized Dirichlet problem.

Let Ω be a bounded open set in $\mathbf{R}^n$. It is assumed throughout this section that the *boundary $\partial\Omega$ of Ω is of class C^2*. By this we mean that given any point $x \in \partial\Omega$, there exists an open neighborhood U of x and a homeomorphism h from $\overline{U}$ onto $E = \{y \in \mathbf{R}^n \mid \|y\| \leq 1\}$, with the following properties:

(a) $h(x) = 0$;

(b) h and h^{-1} are in $C^2(\overline{U})$ and $C^2(E)$, respectively;

(c) $h(U \cap \partial\Omega) = \{y \in \mathbf{R}^n \mid y = (y_1, \ldots, y_{n-1}, 0), \|y\| < 1\}$.

Let L be a partial differential expression of the form

(1)
$$
L = \sum_{i,j=1}^n a_{ij}(x) D_i D_j + \sum_{i=0}^n b_i(x) D_i ,
$$

where each a_{ij} and b_i are in $C^\infty(\overline{\Omega})$ and $D_0 f = f$. At times it will be convenient to rewrite L in the so-called *divergence form*:

$$(2) \qquad L = \sum_{i,j=1}^{n} D_i\big(a_{ij}(x)D_j\big) + \sum_{i=0}^{n} a_i(x)D_i,$$

with $a_i \in C^\infty(\overline{\Omega})$ for $i = 0,\dots,n$.

A classical Dirichlet problem is to determine if for every $f \in C(\Omega)$ there exists a unique $u \in C(\overline{\Omega})$, the space of continuous complex valued functions on $\overline{\Omega}$, such that

$$(3) \qquad \begin{cases} Lu = f & \text{on} \quad \Omega, \\ u = 0 & \text{on} \quad \partial\Omega. \end{cases}$$

If now f is in $L_2(\Omega)$, then the equation $Lu = f$ makes sense if $u \in H_2(\Omega)$ and each D_i is a weak derivative. Furthermore, if $u \in C(\overline{\Omega}) \cap H_1^0(\Omega)$, then one can prove (see, e.g., Friedman [1], Section 10.2) that the boundary condition in (3) is automatically fulfilled. This leads to a *generalized Dirichlet problem*, which is concerned with the existence and uniqueness of the solution of

$$(4) \qquad \begin{cases} Lu = f, \\ u \in H_1^0(\Omega) \cap H_2(\Omega), \end{cases}$$

where f is arbitrary in $L_2(\Omega)$. The operator $A\big(L_2(\Omega) \to L_2(\Omega)\big)$ defined by

$$(5) \qquad \mathcal{D}(A) = H_1^0(\Omega) \cap H_2(\Omega), \qquad Au = Lu,$$

will be referred to as the *Dirichlet operator* on $L_2(\Omega)$ associated with L.

In this section we study the Dirichlet operator under the additional assumption that L is *uniformly elliptic*, which means that there exists a constant $c > 0$ such that

$$(6) \qquad \Re\left(\sum_{i,j=1}^{n} a_{ij}(x)\overline{z}_i z_j \right) \le -c\left(\sum_{i=1}^{n} |z_i|^2 \right)$$

for every $x \in \Omega$ and $(z_1, z_2,\dots, z_n) \in \mathbf{C}^n$. For example, if Δ is the Laplacian $\sum_{i=1}^{n} D_i^2$, then $-\Delta$ is uniformly elliptic.

THEOREM 6.1. *Let $A\big(L_2(\Omega) \to L_2(\Omega)\big)$ be the Dirichlet operator associated with the differential expression L. Assume that L is uniformly elliptic. Then A is a closed densely defined operator. Furthermore, there exists $\lambda_0 \in \mathbf{R}$ such that $\lambda + A$ is invertible and $(\lambda + A)^{-1}$ is compact whenever $\lambda \ge \lambda_0$.*

Note that the above theorem implies that for $f \in L_2(\Omega)$ the generalized Dirichlet problem

$$u \in H_1^0(\Omega) \cap H_2(\Omega), \qquad (\lambda + L)u = f$$

is uniquely solvable if L is uniformly elliptic and $\lambda \in \mathbf{R}$ is sufficiently large.

The proof of Theorem 6.1 depends on the following preliminary results.

LEMMA 6.2. *If L is uniformly elliptic, then there exist constants $\alpha > 0$ and $\beta \geq 0$ such that for every $\varphi \in C_0^\infty(\Omega)$,*

$$(7) \qquad \Re\langle L\varphi, \varphi\rangle_0 \geq \alpha\|\varphi\|_1^2 - \beta\|\varphi\|_0^2.$$

PROOF. We use the divergence form (2) for L. For each $\varphi \in C_0^\infty(\Omega)$, integration by parts yields

$$(8) \qquad \begin{aligned} \langle L\varphi, \varphi\rangle_0 = &- \sum_{i,j=1}^n \int_\Omega a_{ij}(x)D_j\varphi(x)D_i\overline{\varphi}(x)dx \\ &+ \sum_{i=0}^n \int_\Omega a_i(x)(D_i\varphi(x))\overline{\varphi}(x)dx. \end{aligned}$$

Since each a_i is bounded on Ω, uniform ellipticity and (8) imply that there exist constants $M \geq 0$ and $c > 0$ such that for all $\varphi \in C_0^\infty(\Omega)$

$$(9) \qquad \Re\langle L\varphi, \varphi\rangle_0 \geq c\int_\Omega \sum_{i=1}^n |D_i\varphi(x)|^2 dx - 2M\int_\Omega \sum_{i=0}^n |D_i\varphi(x)||\varphi(x)|dx.$$

Now for any real numbers s, t and any $\varepsilon > 0$,

$$(10) \qquad 2st = 2(s\sqrt{\varepsilon})\left(\frac{1}{\sqrt{\varepsilon}}t\right) \leq (s\sqrt{\varepsilon})^2 + \left(\frac{1}{\sqrt{\varepsilon}}t\right)^2 = \varepsilon s^2 + \frac{1}{\varepsilon}t^2.$$

Hence (9) and (10) yield

$$\begin{aligned} \Re\langle L\varphi, \varphi\rangle_0 &\geq c\int_\Omega \sum_{i=1}^n |D_i\varphi(x)|^2 dx - M\int_\Omega \sum_{i=0}^n \varepsilon|D_i\varphi(x)|^2 + \frac{1}{\varepsilon}|\varphi(x)|^2 dx \\ &= (c - M\varepsilon)\|\varphi\|_1^2 - \left(c + \frac{M}{\varepsilon}(n+1)\right)\|\varphi\|_0^2. \end{aligned}$$

For ε sufficiently small,

$$\alpha = c - M\varepsilon > 0, \qquad \beta = c + \frac{M}{\varepsilon}(n+1) > 0,$$

and

$$\Re\langle L\varphi, \varphi\rangle_0 \geq \alpha\|\varphi\|_1^2 - \beta\|\varphi\|_0^2, \qquad \varphi \in C_0^\infty(\Omega). \quad \square$$

We continue to use the divergence form (2) for L. If $u \in H_2(\Omega)$, then for all

$\varphi \in C_0^\infty(\Omega),$

$$
\begin{aligned}
\langle Lu, \varphi \rangle_0 &= \sum_{i,j=1}^{n} - \int_\Omega a_{ij}(x) D_j u(x) D_i \overline{\varphi}(x) dx \\
&\quad + \sum_{i=0}^{n} \int_\Omega a_i(x)\big(D_i u(x)\big)\overline{\varphi}(x) dx \\
&= \sum_{i,j=1}^{n} \int_\Omega u(x) D_j\big(a_{ij}(x) D_i \overline{\varphi}(x)\big) dx - \sum_{i=1}^{n} \int_\Omega u(x) D_i\big(a(x)\overline{\varphi}(x)\big) dx \\
&\quad + \int_\Omega u(x) a_0(x)\overline{\varphi}(x) dx = \langle u, L^*\varphi \rangle_0,
\end{aligned}
$$

(11)

where

$$
(12) \qquad L^*\varphi = \sum_{i,j=1}^{n} D_j(\overline{a}_{ij} D_i \varphi) - \sum_{i=1}^{n} D_i(\overline{a}_i \varphi) + \overline{a}_0 \varphi.
$$

The expression L^* is called the *Lagrange adjoint* of L.

For $u \in H_2(\Omega)$ the linear functional $F_u(\varphi) := \langle u, L^*\varphi \rangle$ is $\| \; \|_0$-bounded on $C_0^\infty(\Omega)$, because of (11). Conversely, if $u \in L_2(\Omega)$ and F_u is a $\| \; \|_0$-bounded linear functional on $C_0^\infty(\Omega)$, then since $C_0^\infty(\Omega)$ is dense in $L_2(\Omega)$ (by Lemma 5.1), the functional F_u has a unique bounded linear extension to $L_2(\Omega)$. Hence, by the Riesz representation theorem, there exists a unique $f \in L_2(\Omega)$ such that

$$
\langle u, L^*\varphi \rangle_0 = \langle f, \varphi \rangle_0, \qquad \varphi \in C_0^\infty(\Omega).
$$

In this case we write $Lu \overset{w}{=} f$. If $u \in H_2(\Omega)$, then $Lu = f$, because of (11).

For the proof of Theorem 6.1 we need conditions which guarantee that $Lu \overset{w}{=} f$ implies $Lu = f$, that is, the weak solution u is in fact a strong solution. It turns out that the latter implication holds if $u \in H_1^0(\Omega)$ (which is contained in $L_2(\Omega)$) and L is uniformly elliptic. More precisely, the following theorem is true.

THEOREM 6.3. *If the differential expression L is uniformly elliptic, $u \in H_1^0(\Omega)$ and $Lu \overset{w}{=} f$ for some $f \in L_2(\Omega)$, then $u \in H_2(\Omega)$ and $Lu = f$.*

The proof of Theorem 6.3 requires a considerable amount of work which belongs to the theory of partial differential equations, and therefore it is omitted here. However, to give an impression of the proof we review briefly its main steps. The first major step is to establish interior regularity, that is to say, if V is an open set with $\overline{V} \subset \Omega$, then $u \in H_2(V)$ and there exists a constant C depending on V such that

$$
\|u\|_{H_2(V)} \le C\big\{\|f\|_{L_2(\Omega)} + \|u\|_{L_2(\Omega)}\big\}.
$$

Next one proves that if now Ω is the r-half-ball

$$
\Omega = \big\{x = (x_1, \ldots, x_n) \in \mathbf{R}^n \mid \|x\| < r, \; x_n > 0\big\},
$$

and

$$V = \left\{ x \in \Omega \mid \|x\| < \frac{1}{2}r \right\},$$

then $u \in H_2(V)$ and there exists a constant C such that

$$\|u\|_{H_2(\Omega)} \leq C\{\|f\|_{L_2(\Omega)} + \|u\|_{H_1(\Omega)}\}.$$

Finally, one covers $\partial\Omega$ by a finite number of open sets Ω_i, $i = 1,\ldots,N$, such that each open set $\Omega_i \cap \Omega$ can be mapped by a 1-1 C^2-map onto a half-ball. By applying the above results and summing the corresponding estimates, the theorem is then proved. The details of the proof may be found in Friedman [1], Showalter [1] or in Gilbarg-Trudinger [1].

We also need the following lemma, which is a generalization of the Riesz representation theorem and is known as the *Lax-Milgram lemma*. Recall that a *sesquilinear form* $B(u,v)$ on $V_1 \times V_2$, where V_1 and V_2 are vector spaces over C, is a functional which is linear in u and conjugate linear in v.

LEMMA 6.4. *Let H be a Hilbert space and let $B(\cdot,\cdot)$ be a sesquilinear form on $H \times H$. Suppose that there exist constants $c > 0$ and $C > 0$ such that*

$$(13) \qquad |B(u,v)| \leq C\|u\|\|v\|, \qquad u,v \text{ in } H,$$

$$(14) \qquad |B(u,u)| \geq c\|u\|^2, \qquad u \in H.$$

Then given a bounded linear functional F on H, there exist unique v and w in H with the property that

$$F(u) = B(u,v) = \overline{B(w,u)}, \qquad u \in H.$$

PROOF. For each $z \in H$, the map $u \to B(u,z)$ is a bounded linear functional on H. Hence, by the Riesz representation theorem, there exists a unique $s \in H$ such that

$$(15) \qquad B(u,z) = \langle u,s \rangle, \qquad u \in H.$$

Define $Az = s$. Then A is linear on H, $\langle u, Az \rangle = B(u,z)$ and it follows from (13) that A is bounded on H. Now by (14)

$$(16) \qquad \|Au\|\|u\| \geq |\langle Au, u \rangle| = |B(u,u)| \geq c\|u\|^2.$$

Hence A is injective and has a closed range. Since (16) also holds for A^* in place of A, the operator A^* is injective and

$$\operatorname{Im} A = (\operatorname{Ker} A^*)^\perp = H.$$

Therefore A is invertible. Given the bounded linear functional F on H, there exists a unique $y \in H$ such that

$$F(u) = \langle u, y \rangle, \qquad u \in H.$$

Since A is invertible, $Av = y$ for some $v \in H$. Therefore

$$F(u) = \langle u, y \rangle = \langle u, Av \rangle = B(u,v), \qquad u \in H.$$

The uniqueness of v is a consequence of (14). Finally, since $B_1(u,x) = \overline{B(x,u)}$ is a sesquilinear form on $H \times H$ satisfying (13) and (14), the result we just proved applied to F and B_1 ensures the existence of a unique $w \in H$ such that

$$F(u) = B_1(u,w) = \overline{B(w,u)}, \qquad u \in H. \quad \square$$

PROOF OF THEOREM 6.1. We write L in the divergence form (2). As before, L^* denotes the Lagrange adjoint of L. Define $B(\cdot,\cdot)$ on $H_1^0(\Omega) \times H_1^0(\Omega)$ by

$$(17) \qquad B(u,v) = -\sum_{i,j=1}^{n} \langle a_{ij} D_j u, D_i v \rangle_0 + \sum_{i=0}^{n} \langle a_i D_i u, v \rangle_0.$$

Clearly, B is sesquilinear. From (11) we obtain

$$(18) \qquad B(u,\varphi) = \langle u, L^*\varphi \rangle_0, \qquad u \in H_1^0(\Omega), \quad \varphi \in C_0^\infty(\Omega),$$

$$(19) \qquad B(\varphi,\varphi) = \langle L\varphi, \varphi \rangle_0, \qquad \varphi \in C_0^\infty(\Omega).$$

Hence, by Lemma 6.2, there exist constants $\alpha > 0$ and $\beta \geq 0$ such that

$$(20) \qquad \Re B(\varphi,\varphi) = \Re \langle L\varphi, \varphi \rangle_0 \geq \alpha \|\varphi\|_1^2 - \beta \|\varphi\|_0^2, \qquad \varphi \in C_0^\infty(\Omega).$$

It follows readily from the boundedness of each a_{ij} and a_i that for some constant γ,

$$(21) \qquad |B(u,v)| \leq \gamma \|u\|_1 \|v\|_1, \qquad u,v \in H_1^0(\Omega).$$

Since $C_0^\infty(\Omega)$ is dense in $H_1^0(\Omega)$, formulas (20) and (21) imply

$$(22) \qquad \Re B(u,u) \geq \alpha \|u\|_1^2 - \beta \|u\|_0^2, \qquad u \in H_1^0(\Omega).$$

Now take $\lambda_0 = \beta$, and let $\lambda \geq \lambda_0$. Define the sesquilinear form $B_\lambda(\cdot,\cdot)$ on $H_1^0(\Omega) \times H_1^0(\Omega)$ by

$$B_\lambda(u,v) = \lambda \langle u,v \rangle_0 + B(u,v).$$

The inequalities (21) and (22) imply that $B_\lambda(\cdot,\cdot)$ satisfies the hypotheses of Lemma 6.4 with $H = H_1^0(\Omega)$. Take $g \in L_2(\Omega)$. Then the linear functional

$$F_g(u) = \langle u, g \rangle_0, \qquad u \in H_1^0(\Omega),$$

is bounded on $H_1^0(\Omega)$. Thus, by Lemma 6.4, there exists $f \in H_1^0(\Omega)$ such that

$$F_g(u) = \overline{B_\lambda(f,u)} = \lambda \langle u, f \rangle_0 + \overline{B(f,u)}, \qquad u \in H_1^0(\Omega).$$

From (18) it follows that for each $\varphi \in C_0^\infty(\Omega)$,

$$\langle g, \varphi \rangle_0 = \overline{F_g(\varphi)} = \lambda \langle f, \varphi \rangle_0 + B(f, \varphi) = \langle f, (\lambda + L^*)\varphi \rangle_0.$$

In other words, $(\lambda + L)f \overset{w}{=} g$. Since $f \in H_1^0(\Omega)$ and L (and hence $\lambda + L$) is uniformly elliptic, we know from Theorem 6.3 that $f \in H_2(\Omega)$ and $(\lambda + L)f = g$. In particular, $f \in \mathcal{D}(A)$ and $(\lambda + A)f = g$. This proves that $\lambda + A$ is surjective.

Take $u \in \mathcal{D}(A) = H_1^0(\Omega) \cap H_2(\Omega)$. Formula (17) implies that

$$B(u, \varphi) = \langle Lu, \varphi \rangle_0, \qquad \varphi \in C_0^\infty(\Omega),$$

because $Lu \in L_2(\Omega)$. Now use (21) and the fact that $C_0^\infty(\Omega)$ is dense in $H_1^0(\Omega)$ to show that $B(u, u) = \langle Lu, u \rangle_0$. But then (22) yields

$$\begin{aligned}
(23) \qquad \|(\lambda + A)u\|_0 \|u\|_0 &\geq \Re \langle (\lambda + L)u, u \rangle_0 = \lambda \|u\|_0^2 + \Re B(u, u) \\
&\geq \alpha \|u\|_1^2 \geq \alpha \|u\|_0^2.
\end{aligned}$$

Here we used that $\lambda \geq \lambda_0 = \beta$. Since (23) holds for any $u \in \mathcal{D}(A)$, we have now proved that $\lambda + A$ has a bounded inverse. In particular (cf. Proposition 1.2) the operator A is closed.

It remains to prove that $(\lambda + A)^{-1}$ is a compact operator on $L_2(\Omega)$. Given a bounded sequence (g_n) in $L_2(\Omega)$, let $u_n = (\lambda + A)^{-1}g_n$. Then, by (23),

$$\|u_n\|_1^2 \leq \frac{1}{\alpha} \|g_n\|_0 \|u_n\|_0 \leq \frac{1}{\alpha} \|g_n\|_0 \|u_n\|_1.$$

Therefore (u_n) is a $\|\ \|_1$-bounded sequence in $H_1^0(\Omega)$. But then, by Theorem 5.4, the sequence (u_n) has a subsequence which converges in $L_2(\Omega)$, which proves the compactness of $(\lambda + A)^{-1}$ on $L_2(\Omega)$. $\square$

THEOREM 6.5. *Let $A\big(L_2(\Omega) \to L_2(\Omega)\big)$ be the Dirichlet operator associated with the differential expression L, and let L^* be the Lagrange adjoint of L. If L is uniformly elliptic, then A^* is the Dirichlet operator associated with L^*, i.e., $\mathcal{D}(A^*) = \mathcal{D}(A)$ and $A^*u = L^*u$.*

PROOF. Let $A_*\big(L_2(\Omega) \to L_2(\Omega)\big)$ be the Dirichlet operator associated with L^*. It follows from (12) that L^* is uniformly elliptic. Thus, by Theorem 6.1, there exists $\lambda \in \mathbf{R}$ such that $\lambda + A$ and $\lambda + A_*$ have bounded inverses on $L_2(\Omega)$. By (11) (applied to $\lambda + L^*$ in place of L) we have

$$(24) \qquad \langle (\lambda + A)\varphi, (\lambda + A_*)^{-1}f \rangle_0 = \langle \varphi, (\lambda + L^*)(\lambda + A_*)^{-1}f \rangle_0 = \langle \varphi, f \rangle_0$$

for any $f \in L_2(\Omega)$ and any $\varphi \in C_0^\infty(\Omega)$.

To prove the theorem it suffices to show that $(\lambda + A)C_0^\infty(\Omega)$ is dense in $L_2(\Omega)$. For if this is the case, then given $h \in L_2(\Omega)$ there exists a sequence (φ_n) in $C_0^\infty(\Omega)$ such that $(\lambda + A)\varphi_n \to h$ in $L_2(\Omega)$, and so by (24),

$$\begin{aligned}
\langle h, (\lambda + A_*)^{-1}f \rangle_0 &= \lim \langle (\lambda + A)\varphi_n, (\lambda + A_*)^{-1}f \rangle_0 \\
&= \lim \langle \varphi_n, f \rangle_0 = \langle (\lambda + A)^{-1}h, f \rangle_0,
\end{aligned}$$

for any $f \in L_2(\Omega)$. Hence

$$(\lambda + A^*)^{-1} = [(\lambda + A)^{-1}]^* = (\lambda + A_*)^{-1}.$$

Therefore $A^* = A_*$.

It remains to prove that $(\lambda + A)C_0^\infty(\Omega)$ is dense in $L_2(\Omega)$. Suppose not. Since $(\lambda + A)C_0^\infty(\Omega)$ is contained in $C_0^\infty(\Omega)$, Lemma 5.1 implies that there exists $\psi \neq 0$ in $C_0^\infty(\Omega)$ such that

$$(25) \qquad \langle (\lambda + A)\varphi, \psi \rangle_0 = 0, \qquad \varphi \in C_0^\infty(\Omega).$$

Let $B(\cdot, \cdot)$ be the sesquilinear form defined by (17). Then (11) and (18) imply that

$$(26) \qquad 0 = \langle (\lambda + A)\varphi, \psi \rangle_0 = \langle \varphi, (\lambda + L^*)\psi \rangle_0 = \lambda \langle \varphi, \psi \rangle_0 + B(\varphi, \psi)$$

for all $\varphi \in C_0^\infty(\Omega)$. In particular,

$$(27) \qquad \lambda \|\psi\|_0^2 + B(\psi, \psi) = 0.$$

By (23) (which holds for any $u \in \mathcal{D}(A)$) the left hand side of (27) dominates $\alpha\|\psi\|_1^2$ for $\lambda \geq \beta$. It follows that $\psi = 0$, which is a contradiction. $\square$

CHAPTER XV

FUNCTIONAL CALCULUS FOR UNBOUNDED OPERATORS

In the first two sections of this chapter the theory of Riesz projections and the functional calculus developed in Chapter I are extended to unbounded linear operators. This extension is quite straightforward. The next two sections concern a more difficult problem, namely, the case when the contour of integration goes through infinity. For the unbounded case (when infinity always belongs to the spectrum) the solution requires a spectral decomposition for the spectrum at infinity.

XV.1 INTRODUCTION OF THE FUNCTIONAL CALCULUS

Let X be a complex Banach space, and let $A(X \to X)$ be an unbounded linear operator with non-empty resolvent set $\rho(A)$. We assume that $\mathcal{D}(A) \neq X$. The purpose of this section is to develop an operational calculus for A based on the calculus for bounded operators presented in Section I.3. In fact, the bounded operator we choose is $(\alpha - A)^{-1}$, where α is fixed in $\rho(A)$. The spectrum $\sigma\big((\alpha - A)^{-1}\big)$ is compact, and, as we have seen, $\sigma(A)$ is closed but may be unbounded. Our first step is to "compactify" $\sigma(A)$ as follows. Let $\mathbf{C}_\infty$ be the extended complex plane, $\mathbf{C}_\infty = \mathbf{C} \cup \{\infty\}$, endowed with the usual topology (see [C], page 8). The set $\mathbf{C}_\infty$ is a compact topological space and the Möbius transformation

$$(1) \qquad \eta(\lambda) = (\alpha - \lambda)^{-1}$$

is a homeomorphism from $\mathbf{C}_\infty$ onto $\mathbf{C}_\infty$. We shall now show that

$$(2) \qquad \eta[\sigma(A) \cup \{\infty\}] = \sigma\big((\alpha - A)^{-1}\big).$$

First note that $\eta(\infty) = 0 \in \sigma\big((\alpha - A)^{-1}\big)$, because $\mathrm{Im}\big((\alpha - A)^{-1}\big)$ is not equal to X. Next we use the identity

$$(3) \qquad \lambda - A = (\alpha - \lambda)[(\alpha - \lambda)^{-1} - (\alpha - A)^{-1}](\alpha - A),$$

which holds for any $\lambda \neq \alpha$. Since $\alpha \notin \sigma(A)$, it follows from (3) that $\lambda \in \sigma(A)$ if and only if $\eta(\lambda) = (\alpha - \lambda)^{-1} \in \sigma\big((\alpha - A)^{-1}\big)$. From these remarks formula (2) is clear. From (2) and the fact that η is a homeomorphism we may conclude that $\sigma(A) \cup \{\infty\}$ is compact in $\mathbf{C}_\infty$.

Equality (2) also gives a clue to the definition of $f(A)$. Let $\mathcal{F}_\infty(A)$ denote the set of all complex functions that are analytic on an open set in $\mathbf{C}_\infty$ containing $\sigma(A)$ and ∞. Take $f \in \mathcal{F}_\infty(A)$. Then $f \circ \eta^{-1}$ is analytic on an open neighbourhood of $\sigma\big((\alpha - A)^{-1}\big)$. Hence the operator $(f \circ \eta^{-1})\big((\alpha - A)^{-1}\big) \in \mathcal{L}(X)$ can be defined as in Section I.3. This we take as our definition of $f(A)$. Thus

$$(4) \qquad f(A) := (f \circ \eta^{-1})\big((\alpha - A)^{-1}\big), \qquad f \in \mathcal{F}_\infty(A).$$

Given $f \in \mathcal{F}_\infty(A)$, there exists an open set $\Omega \subset \mathbf{C}_\infty$ such that $\sigma(A) \subset \Omega$, the complement of Ω is a compact subset of $\mathbf{C}$ and f is analytic on Ω. The argument given in Section I.1 shows that there exists an unbounded Cauchy domain Δ such that $\sigma(A) \subset \Delta \subset \overline{\Delta} \subset \Omega$. As before, $\partial\Delta$ is the oriented boundary of Δ. Thus $\partial\Delta$ is oriented in such a way that Δ is the inner domain of $\partial\Delta$. The following integral representation is the counter part to $f(A)$ when A is bounded. Since the formula does not involve α, it shows that $f(A)$ is independent of α.

PROPOSITION 1.1. *Suppose that f is analytic on an open set Ω in $\mathbf{C}_\infty$ containing $\sigma(A)$ and ∞. Let Δ be an unbounded Cauchy domain such that $\sigma(A) \subset \Delta \subset \overline{\Delta} \subset \Omega$. Then*

$$f(A) = f(\infty)I + \frac{1}{2\pi i}\int_{\partial\Delta} f(\lambda)(\lambda - A)^{-1}d\lambda,$$

where $\partial\Delta$ is the (oriented) boundary of Δ.

PROOF. Let $\alpha \in \rho(A)$ be fixed. Since $\lambda \to (\lambda - A)^{-1}$ is analytic on $\rho(A)$, we know from Cauchy's integral theorem that the above integral is unchanged if Δ is replaced by an unbounded Cauchy domain Δ', where $\sigma(A) \subset \Delta' \subset \overline{\Delta}' \subset \Delta$ and $\alpha \notin \overline{\Delta}'$. Hence we may assume $\alpha \notin \overline{\Delta}$. By (2), the set $\eta(\Delta \cup \{\infty\})$ is a Cauchy domain containing $\sigma((\alpha - A)^{-1})$ and $\Gamma = \eta(\partial\Delta)$ is its oriented boundary. Let $B = (\alpha - A)^{-1}$. It follows from (3) that if $\alpha \neq \lambda \in \rho(A)$ and $z = (\alpha - \lambda)^{-1}$, then

$$(\lambda - A)^{-1} = zB(z - B)^{-1} = z[-I + z(z - B)^{-1}].$$

By change of variable $\lambda = \eta^{-1}(z) = \alpha - z^{-1}$, Cauchy's integral formula and Section I.3 give

$$\frac{1}{2\pi i}\int_{\partial\Delta} f(\lambda)(\lambda - A)^{-1}d\lambda = \frac{1}{2\pi i}\int_{\Gamma} f(\eta^{-1}(z))\left(-z^{-1}I + (z - B)^{-1}\right)dz$$

$$= -f(\eta^{-1}(0))I + (f \circ \eta^{-1})(B)$$

$$= -f(\infty)I + f(A). \qquad \square$$

From the above results and Theorems I.3.1, I.3.3 we obtain the following result.

THEOREM 1.2. *If f and g are in $\mathcal{F}_\infty(A)$, then*

(a) $(f + g)(A) = f(A) + g(A)$; $(\alpha f)(A) = \alpha f(A)$, $\alpha \in \mathbf{C}$,

(b) $(fg)(A) = f(A)g(A)$,

(c) $\sigma(f(A)) = f[\sigma(A) \cup \{\infty\}]$.

To illustrate the operational calculus for unbounded operators, we consider an example. Let $X = L_1([0, \infty])$, and let A be the maximal operator corresponding to $\tau = \frac{d}{dt}$ and the interval $[0, \infty)$, i.e.,

$$\mathcal{D}(A) = \{y \in X \mid y \text{ is absolutely continuous on each}$$

$$\text{compact subinterval of } [0, \infty) \text{ and } y' \in X\},$$

$$Ay = y'.$$

We shall now show that

$$\sigma(A) = \{\lambda \mid \Re\lambda \le 0\}. \tag{5}$$

Given $g \in X$, the general solution to the differential equation

$$\lambda y - y' = g$$

is

$$y(t) = e^{\lambda t}\left[c - \int_0^t e^{-\lambda s} g(s)ds\right].$$

Suppose $\Re\lambda > 0$. If we take $c = \int_0^\infty e^{-\lambda s} g(s)ds$, then

$$y(t) = e^{\lambda t}\int_t^\infty e^{-\lambda s} g(s)ds = \int_0^\infty e^{-\lambda s} g(t + s)ds.$$

It is easy to see that y is in $\mathcal{D}(A)$ and, as noted above, $(\lambda - A)y = g$. Thus $\lambda - A$ is surjective. Moreover, $\lambda - A$ is injective, since $(\lambda - A)y = 0$ implies $y = ke^{\lambda t}$, which is not in X unless $k = 0$. Hence

$$((\lambda - A)^{-1}g)(t) = \int_0^\infty e^{-\lambda s} g(t + s)ds$$

and

$$\|(\lambda - A)^{-1}\| \le \frac{1}{\Re\lambda}.$$

If $\Re\lambda < 0$, then λ is an eigenvalue of A with eigenvector $e^{\lambda t}$. Hence $\lambda \in \sigma(A)$. Since $\sigma(A)$ is closed, we have $\{\lambda \mid \Re\lambda = 0\}$ is also in $\sigma(A)$, which establishes (5).

If $f \in \mathcal{F}_\infty(A)$, then there exist positive numbers r and ε with the properties that f is analytic on

$$\{\lambda \mid |\lambda| > r\} \cup \{\lambda \mid \Re\lambda < \varepsilon\}.$$

In particular, f is analytic on the closure of the Cauchy domain $\Delta = \{\lambda \mid |\lambda| > 2r\} \cup \{\lambda \mid \Re\lambda < \frac{\varepsilon}{2}\} \supset \sigma(A)$, and

$$(f(A)g)(t) = f(\infty)g(t) + \frac{1}{2\pi i}\int_{\partial\Delta} f(\lambda)\left[\int_0^\infty e^{-\lambda s} g(t + s)ds\right]d\lambda.$$

XV.2 RIESZ PROJECTIONS AND EIGENVALUES OF FINITE TYPE

The results in Sections I.2 and II.1 concerning spectral decompositions and eigenvalues of finite type can be extended to closed (linear) operators. We shall show how this can be accomplished.

Given an operator $A(X \to X)$, a linear manifold $M \subset X$ is called *A-invariant* if $A(M \cap \mathcal{D}(A)) \subset M$. In that case $A|M$ denotes the operator with domain $M \cap \mathcal{D}(A)$ and range in M.

THEOREM 2.1. *Suppose* $A(X \to X)$ *is a closed operator with spectrum* $\sigma(A) = \sigma \cup \tau$, *where* σ *is contained in a bounded Cauchy domain* Δ *such that* $\overline{\Delta} \cap \tau = \emptyset$. *Let* Γ *be the (oriented) boundary of* Δ. *Then*

(i) $P_\sigma = \frac{1}{2\pi i} \int_\Gamma (\lambda - A)^{-1} d\lambda$ *is a projection,*

(ii) *the subspaces* $M = \operatorname{Im} P_\sigma$ *and* $N = \operatorname{Ker} P_\sigma$ *are A-invariant,*

(iii) *the subspace* M *is contained in* $\mathcal{D}(A)$ *and* $A|M$ *is bounded,*

(iv) $\sigma(A|M) = \sigma$ *and* $\sigma(A|N) = \tau$.

PROOF. The proof of (i) is the same as the proof of Lemma I.2.1. Given $x \in X$,

$$(1) \qquad P_\sigma x = \frac{1}{2\pi i} \int_\Gamma (\lambda - A)^{-1} x \, d\lambda$$

and

$$(2) \qquad \int_\Gamma A(\lambda - A)^{-1} x \, d\lambda = \int_\Gamma [-x + \lambda(\lambda - A)^{-1} x] d\lambda.$$

An approximation of these integrals by Riemann-sums and use of the assumption that A is closed imply $P_\sigma x \in \mathcal{D}(A)$ and

$$(3) \qquad A P_\sigma x = \int_\Gamma [-x + \lambda(\lambda - A)^{-1} x] d\lambda.$$

Thus $M \subset \mathcal{D}(A)$. Now if $x \in M$, then $Ax = A P_\sigma x = P_\sigma Ax \in M$, which shows that M is A-invariant. Since $A P_\sigma x = P_\sigma Ax$ if $x \in \mathcal{D}(A)$, it also follows that N is A-invariant. Equality (3) implies that $A|M$ is bounded. The proof of (iv) is the same as the proof of Theorem I.2.2. $\square$

A point $\lambda_0 \in \sigma(A)$ is called an *eigenvalue of A of finite type* if λ_0 is an isolated point of $\sigma(A)$ and the associated projection $P_{\{\lambda_0\}}$ has finite rank. Here $P_{\{\lambda_0\}}$ is defined by (1) with $\sigma = \{\lambda_0\}$. Since $\{\lambda_0\} = \sigma(A| \operatorname{Im} P_{\{\lambda_0\}})$ and $\operatorname{Im} P_{\{\lambda_0\}}$ is finite dimensional (in $\mathcal{D}(A)$), it follows that λ_0 is an eigenvalue of A.

THEOREM 2.2. *Let* $A(X \to X)$ *be a closed operator, and let* λ_0 *be an eigenvalue of A of finite type. Then the resolvent* $(\lambda - A)^{-1}$ *admits an expansion of the form*

$$(\lambda - A)^{-1} = \frac{1}{\lambda - \lambda_0}P_{\{\lambda_0\}} + \sum_{k=1}^{p-1}(\lambda - \lambda_0)^{-k-1}B^k - \sum_{k=0}^{\infty}(\lambda - \lambda_0)^k T^{k+1}.$$

Here p is some positive integer, $B = (A - \lambda_0)P_{\{\lambda_0\}}$ and

$$T = \frac{1}{2\pi i}\int_\Gamma (\lambda - \lambda_0)^{-1}(A - \lambda)^{-1}d\lambda,$$

where Γ is the positively oriented boundary of a closed disc with centre λ_0 which is disjoint from $\sigma(A)\backslash\{\lambda_0\}$.

PROOF. Let $P_0 = P_{\{\lambda_0\}}$ and $A_0 = A|\operatorname{Im}P_0$. By Theorem 2.1, $\sigma(A_0) = \{\lambda_0\}$. Thus $\sigma(\lambda_0 - A_0) = \{0\}$, and therefore

$$(\lambda - A_0)^{-1} = \sum_{k=0}^{\infty}(\lambda - \lambda_0)^{-k-1}(A_0 - \lambda_0)^k$$

converges in $\mathcal{L}(M)$ for all $\lambda \neq \lambda_0$, where $M = \operatorname{Im}P_0$. Since

$$(\lambda - A)^{-1}P_0 x = (\lambda - A_0)^{-1}x = P_0(\lambda - A)^{-1}x, \qquad x \in M,$$

it follows that for $\lambda \notin \sigma(A)$,

$$(4) \qquad (\lambda - A)^{-1}P_0 = \frac{1}{\lambda - \lambda_0}P_0 + \sum_{k=1}^{\infty}(\lambda - \lambda_0)^{-k-1}B^k,$$

where

$$B := (A - \lambda_0)P_0 = \frac{1}{2\pi i}\int_\Gamma (\lambda - \lambda_0)(\lambda - A)^{-1}d\lambda \in \mathcal{L}(X).$$

Since B has finite rank, $B^p = 0$ for some positive integer p. Let T be as in the theorem. Since A is closed, the same is true for $A - \lambda_0$, and hence computations similar to those given in the proof of Lemma I.2.1 and Theorem I.2.2 show that

$$(A - \lambda)T = (A - \lambda_0)T - (\lambda - \lambda_0)T = I - P_0 - (\lambda - \lambda_0)T, \qquad P_0 T = 0.$$

Hence

$$(5) \qquad (A - \lambda)T = (I - P_0) - (\lambda - \lambda_0)(I - P_0)T = (I - P_0)[I - (\lambda - \lambda_0)T].$$

Now for λ sufficiently close to λ_0 and $\lambda \neq \lambda_0$, the operators $\lambda - A$ and $I - (\lambda - \lambda_0)T$ are invertible. Therefore from (5) we get

$$(6) \qquad (\lambda - A)^{-1}(I - P_0) = -T[I - (\lambda - \lambda_0)T]^{-1} = -\sum_{k=0}^{\infty}(\lambda - \lambda_0)^k T^{k+1}$$

for $\lambda \neq \lambda_0$ sufficiently close to λ_0. Since $(\lambda - A)^{-1} = (\lambda - A)^{-1}P_0 + (\lambda - A)^{-1}(I - P_0)$, the theorem is an immediate consequence of (4) and (6) and the observation made earlier that $B^p = 0$ for some positive integer p. $\square$

It can happen that the spectrum of an unbounded operator consists of eigenvalues of finite type only. For example, let $X = C[0,1]$. Define $A(X \to X)$ by

$$\mathcal{D}(A) = \{f \in X \mid f' \in X, \; f(0) = f(1)\},$$
$$Af = f'.$$

The spectrum of A consists of eigenvalues $2n\pi i$, $n = 0, \pm 1, \ldots$. If $\lambda \in \rho(A)$, then for all $g \in X$,

$$((\lambda - A)^{-1}g)(t) = e^{\lambda t}\left[c_\lambda - \int_0^t e^{-\lambda s}g(s)ds\right],$$

where $c_\lambda = e^\lambda(e^\lambda - 1)^{-1}\int_0^1 e^{-\lambda s}g(s)ds$. If we let Γ_n be a small circle about $\lambda_n = 2n\pi i$ which excludes the other eigenvalues of A, then the formula for the corresponding projection P_n is given by

$$(P_ng)(t) = \frac{1}{2\pi i}\left(\int_{\Gamma_n}(\lambda - A)^{-1}gd\lambda\right)(t) = e^{\lambda_n t}\int_0^1 e^{-\lambda_n s}g(s)ds.$$

The rank of P_n is 1. Thus $\sigma(A)$ consists only of eigenvalues of A of finite type.

The operator A above has the property that $(\lambda - A)^{-1}$ is compact for every $\lambda \in \rho(A)$. Operators with this property are called *operators with compact resolvent*. The partial differential operator appearing in Theorem XIV.6.1 is also an example of such an operator.

THEOREM 2.3. *Suppose $A(X \to X)$ has the property that $(\lambda_0 - A)^{-1}$ is compact for some $\lambda_0 \in \rho(A)$. Then the operator $(\lambda - A)^{-1}$ is compact for all $\lambda \in \rho(A)$. The spectrum of A does not have a limit point in $\mathbb{C}$ and every point in $\sigma(A)$ is an eigenvalue of A of finite type. For any $\lambda \in \mathbb{C}$, the operator $\lambda - A$ has closed range and*

$$(7) \qquad \dim \mathrm{Ker}(\lambda - A) = \mathrm{codim}\,\mathrm{Im}(\lambda - A) < \infty.$$

PROOF. By Proposition XIV.1.1,

$$(\mu - A)^{-1} = (\lambda_0 - A)^{-1} + (\lambda_0 - \mu)(\mu - A)^{-1}(\lambda_0 - A)^{-1}$$

which is compact for all $\mu \in \rho(A)$. Next we use the Möbius transformation $\tau(\lambda) = (\lambda_0 - \lambda)^{-1}$ and the following identity (see (2) in the previous section)

$$\tau[\sigma(A) \cup \{\infty\}] = \sigma((\lambda_0 - A)^{-1}).$$

Since τ is a homeomorphism and the operator $(\lambda_0 - A)^{-1}$ is compact, it follows that $\sigma(A)$ is a finite or countable set which does not have a limit point in $\mathbb{C}$. To see that each

$\eta \in \sigma(A)$ is an eigenvalue of A of finite type, it remains to prove that the corresponding Riesz projection $P_{\{\eta\}} = (2\pi i)^{-1} \int_\Gamma (\lambda - A)^{-1} d\lambda$ has finite rank. Now $(\lambda - A)^{-1}$ is compact for every $\lambda \in \rho(A)$. Therefore $P_{\{\eta\}}$ is the limit in $\mathcal{L}(X)$ of Riemann sums of compact operators. Hence $P_{\{\eta\}}$ is compact. Since this operator is also a projection, $\operatorname{Im} P_{\{\eta\}}$ is finite dimensional. To prove (7), let λ be arbitrary in $\mathbf{C}$. Note that

$$(8) \qquad (\lambda - A) = [I - (\lambda_0 - \lambda)(\lambda_0 - A)^{-1}](\lambda_0 - A).$$

Since $(\lambda_0 - A)^{-1}$ is compact, $I + (\lambda - \lambda_0)(\lambda_0 - A)^{-1}$ is a Fredholm operator with index zero by Corollary XI.4.3. It follows from (8) and the invertibility of $\lambda_0 - A$ that

$$\operatorname{Im}(\lambda - A) = \operatorname{Im}[I + (\lambda - \lambda_0)(\lambda_0 - A)^{-1}],$$
$$\dim \operatorname{Ker}(\lambda - A) = \dim \operatorname{Ker}[I + (\lambda - \lambda_0)(\lambda_0 - A)^{-1}]$$
$$= \operatorname{codim} \operatorname{Im}[I + (\lambda - \lambda_0)(\lambda_0 - A)^{-1}]$$
$$= \operatorname{codim} \operatorname{Im}(\lambda - A) < \infty. \qquad \square$$

XV.3 SPLITTING OF THE SPECTRUM AT INFINITY

Let X be a (complex) Banach space, and let $A(X \to X)$ be a closed linear operator such that the strip $|\Re\lambda| < h$ is in the resolvent set $\rho(A)$. Here h is a positive number which we shall keep fixed throughout the section. If A is everywhere defined (and hence bounded) our assumptions imply that there exists a decomposition, $X = X_- \oplus X_+$, such that X_- and X_+ are A-invariant subspaces of X and

$$(1.a) \qquad \sigma(A \mid X_-) \subset \{\lambda \in \mathbf{C} \mid \Re\lambda \leq -h\},$$

$$(1.b) \qquad \sigma(A \mid X_+) \subset \{\lambda \in \mathbf{C} \mid \Re\lambda \geq h\}.$$

This spectral decomposition of X is obtained by taking $X_- = \operatorname{Ker} P$ and $X_+ = \operatorname{Im} P$, where P is the Riesz projection corresponding to the part σ_+ of $\sigma(A)$ in the open right half plane, i.e.,

$$(2) \qquad P = \frac{1}{2\pi i} \int_\Gamma (\lambda - A)^{-1} d\lambda.$$

The contour Γ in (2) is in the open right half plane and contains σ_+ in its inner domain. In the case when A is not everywhere defined, it is a problem to make a decomposition of the space X with the properties (1.a) and (1.b). In general, the contour Γ appearing in (2) does not exist, and hence one cannot define a projection as in (2).

To make more transparent the difficulty which one encounters here, consider the operator A^{-1} (which is a well-defined bounded linear operator because of our assumption on $\rho(A)$). The spectrum of A^{-1} lies in the two closed discs which one obtains by applying the transformation $\lambda \mapsto \lambda^{-1}$ to the closed half planes $\Re\lambda \leq -h$ and $\Re\lambda \geq h$ (in the extended complex plane $\mathbf{C}_\infty$). In other words, $\sigma(A^{-1})$ lies in the set

$$(3) \qquad \left\{\lambda \,\Big|\, \left|\lambda - \frac{1}{2h}\right| \leq \frac{1}{h}\right\} \cup \left\{\lambda \,\Big|\, \left|\lambda + \frac{1}{2h}\right| \leq \frac{1}{h}\right\}.$$

Note that the point 0 is always in $\sigma(A^{-1})$ (if A is unbounded) and at the point 0 the two closed discs in (3) intersect. In general, 0 will not be an isolated point of $\sigma(A^{-1})$. In fact, it may happen that $\sigma(A^{-1})$ is precisely equal to the set (3). The problem is now to make a spectral decomposition of the space when the parts of the spectrum are not disjoint. For the original operator A this means that we have to split the spectrum at infinity. In this section we shall deal with a case for which the above mentioned problem can be solved.

For the operator A which we shall consider, the resolvent operator $(\lambda - A)^{-1}$ will be uniformly bounded on the strip $|\Re\lambda| < h$, that is,

$$(4) \qquad \sup_{|\Re\lambda|<h} \|(\lambda - A)^{-1}\| < \infty.$$

First, we shall introduce the candidates for the spaces X_- and X_+ for such an operator. We define N_- to be the set of all vectors $x \in X$ for which there exists an X-valued function φ_x^-, bounded and analytic on $\Re\lambda > -h$, which takes its values in $\mathcal{D}(A)$ and satisfies

$$(5) \qquad (\lambda - A)\varphi_x^-(\lambda) = x, \qquad \Re\lambda > -h.$$

Roughly speaking, N_- consists of all vectors $x \in X$ such that $(\lambda - A)^{-1}x$ has a bounded analytic continuation to the open half plane $\Re\lambda > -h$. The function φ_x^- (assuming it exists) is uniquely determined by x. Analogously, we let N_+ be the set of all vectors $x \in X$ for which there exists an X-valued function φ_x^+, bounded and analytic on $\Re\lambda < h$, which takes it values in $\mathcal{D}(A)$ and satisfies

$$(6) \qquad (\lambda - A)\varphi_x^+(\lambda) = x, \qquad \Re\lambda < h.$$

Also φ_x^+ is unique, provided it exists. Obviously, the sets N_- and N_+ are (possibly non-closed) linear manifolds of X. Their closures will be denoted by X_- and X_+, respectively. Thus, by definition,

$$(7) \qquad X_- = \overline{N_-}, \qquad X_+ = \overline{N_+}.$$

THEOREM 3.1. *Let $A(X \to X)$ be a densely defined closed linear operator for which condition (4) holds true for some $h > 0$, and let X_- and X_+ be the subspaces defined in (7). Then $X = \overline{X_- \oplus X_+}$, and furthermore,*

$$(8) \qquad X = X_- \oplus X_+$$

if and only if, for some (for each) $0 < \alpha < h$, the map

$$(9) \qquad x \mapsto \frac{-1}{2\pi i} \int_{\alpha-i\infty}^{\alpha+i\infty} \lambda^{-2}(\lambda - A)^{-1}A^2 x \, d\lambda, \qquad x \in \mathcal{D}(A^2),$$

extends to a bounded linear operator P on X, and in that case the spaces X_- and X_+ are invariant under A,

$$(10.a) \qquad \sigma(A|X_-) \subset \{\lambda \in \mathbf{C} \mid \Re\lambda \le -h\},$$

(10.b) $$\sigma(A|X_+) \subset \{\lambda \in \mathbf{C} \mid \Re\lambda \geq h\},$$

and P is the projection of X along X_- onto X_+.

For bounded linear operators Theorem 3.1 is trivial. First of all note that for a bounded linear operator A on X condition (4) is fulfilled for some $h > 0$ if and only if the imaginary axis $\Re\lambda = 0$ belongs to the resolvent set $\rho(A)$ of A. Furthermore, in that case, a straightforward argument based on Cauchy's theorem and the identity

$$\lambda^{-2}(\lambda - A)^{-1}A^2 = (\lambda - A)^{-1} - \lambda^{-1}I - \lambda^{-2}A$$

shows that the Riesz projection P corresponding to the part of $\sigma(A)$ in the open right half plane is given by the (improper) integral

$$P = \frac{-1}{2\pi i} \int\limits_{\alpha-i\infty}^{\alpha+i\infty} \lambda^{-2}(\lambda - A)^{-1}A^2 d\lambda,$$

for some $0 < \alpha < h$.

The proof of Theorem 3.1 will be based on two lemmas. In what follows $A(X \to X)$ is a (not necessarily densely defined) closed linear operator for which condition (4) holds true. In particular, the strip $|\Re\lambda| < h$ is in $\rho(A)$. We fix $0 < \alpha < h$ and introduce the following auxiliary operators:

(11) $$S_- = \frac{1}{2\pi i} \int\limits_{-\alpha-i\infty}^{-\alpha+i\infty} \lambda^{-2}(\lambda - A)^{-1} d\lambda,$$

(12) $$S_+ = \frac{-1}{2\pi i} \int\limits_{\alpha-i\infty}^{\alpha+i\infty} \lambda^{-2}(\lambda - A)^{-1} d\lambda.$$

The integrals in (11) and (12) have to be understood as improper integrals. Condition (4) guarantees that S_- and S_+ are well-defined bounded linear operators on X. We shall prove that S_- and S_+ commute with A. The latter means that $S_\pm \mathcal{D}(A) \subset \mathcal{D}(A)$ and $AS_\pm x = S_\pm Ax$ for each $x \in \mathcal{D}(A)$. To see this, let $T = A^{-1}$ be the bounded inverse of A. Since T commutes with the resolvent operator $(\lambda - A)^{-1}$, the operators S_- and S_+ commute with T. Take $x \in \mathcal{D}(A)$, and put $y = Ax$. Then $x = Ty$, and so $S_\pm x = S_\pm Ty = TS_\pm y \in \mathcal{D}(A)$. Also

$$AS_\pm x = ATS_\pm y = S_\pm y = S_\pm Ax.$$

LEMMA 3.2. *The operators S_- and S_+ commute, $S_-S_+ = S_+S_- = 0$ and* $\mathrm{Ker}\, S_- \cap \mathrm{Ker}\, S_+ = \{0\}$.

PROOF. To compute S_-S_+ we use the resolvent formula (Proposition

XIV.1.1) and use Fubini's theorem. We have

$$
\begin{aligned}
-S_- S_+ &= \left(\frac{1}{2\pi i}\right)^2 \int\limits_{-\alpha-i\infty}^{-\alpha+i\infty} \int\limits_{\alpha-i\infty}^{\alpha+i\infty} \lambda^{-2}\mu^{-2}(\lambda - A)^{-1}(\mu - A)^{-1}d\mu d\lambda \\
&= \left(\frac{1}{2\pi i}\right)^2 \int\limits_{-\alpha-i\infty}^{-\alpha+i\infty} \int\limits_{\alpha-i\infty}^{\alpha+i\infty} \lambda^{-2}\mu^{-2}(\mu - \lambda)^{-1}[(\lambda - A)^{-1} - (\mu - A)^{-1}]d\mu d\lambda \\
&= \left(\frac{1}{2\pi i}\right)^2 \int\limits_{-\alpha-i\infty}^{-\alpha+i\infty} \left(\int\limits_{\alpha-i\infty}^{\alpha+i\infty} \frac{d\mu}{\mu^2(\mu - \lambda)}\right)\lambda^{-2}(\lambda - A)^{-1}d\lambda \\
&\quad - \left(\frac{1}{2\pi i}\right)^2 \int\limits_{\alpha-i\infty}^{\alpha+i\infty} \left(\int\limits_{-\alpha-i\infty}^{-\alpha+i\infty} \frac{d\lambda}{\lambda^2(\mu - \lambda)}\right)\mu^{-2}(\mu - A)^{-1}d\mu.
\end{aligned}
$$

The use of Fubini's theorem is justified by the fact that

$$
\left\|\frac{(\mu - A)^{-1}}{\lambda^2\mu^2(\mu - \lambda)}\right\| \le \frac{\gamma}{2\alpha|\lambda|^2|\mu|^2}, \qquad \Re\lambda = -\alpha, \Re\mu = \alpha,
$$

where γ denotes the left hand side of (4). A standard argument of complex function theory shows that the integrals between parentheses are zero. Hence $S_- S_+ = 0$. Analogously, $S_+ S_- = 0$.

To prove that $\operatorname{Ker} S_- \cap \operatorname{Ker} S_+ = \{0\}$, let Γ be a circle around the origin with radius r, $0 < r < \alpha$. The orientation on Γ is counter clockwise. Recall that $T = A^{-1}$. By the operational calculus for closed linear operators,

$$
T^2 = A^{-2} = \frac{-1}{2\pi i}\int_\Gamma \lambda^{-2}(\lambda - A)^{-1}d\lambda.
$$

Using Cauchy's theorem and (4), one easily gets

$$
\int_\Gamma \lambda^{-2}(\lambda - A)^{-1}d\lambda = \int\limits_{\alpha-i\infty}^{\alpha+i\infty} \lambda^{-2}(\lambda - A)^{-1}d\lambda - \int\limits_{-\alpha-i\infty}^{-\alpha-i\infty} \lambda^{-2}(\lambda - A)^{-1}d\lambda,
$$

and thus $T^2 = S_- + S_+$. Hence $\operatorname{Ker} S_- \cap \operatorname{Ker} S_+ \subset \operatorname{Ker} T^2 = \{0\}$. $\square$

As before, let $T = A^{-1}$. Since T commutes with S_- and S_+, one has $T[\operatorname{Im} S_\pm] \subset \operatorname{Im} S_\pm$. Put

$$
(13) \qquad\qquad M_- = \overline{\operatorname{Im} S_-}, \qquad M_+ = \overline{\operatorname{Im} S_+}.
$$

Then M_- and M_+ are invariant subspaces for the bounded operator T. In fact, $TM_\pm \subset \mathcal{D}(A) \cap M_\pm$. Note that A maps TM_- into M_- and TM_+ into M_+. We define

$A_-(M_- \to M_-)$ and $A_+(M_+ \to M_+)$ to be the restrictions A to TM_- and TM_+, respectively. Thus

$$\mathcal{D}(A_\pm) = TM_\pm, \quad A_\pm x = Ax \qquad (x \in \mathcal{D}(A_\pm)).$$

From Lemma 3.2 we know that $S_+ S_- = 0$. This implies that $S_+[\overline{\operatorname{Im} S_-}] = \{0\}$, and hence, by definition, $S_+[M_-] = \{0\}$. In a similar way, $S_-[M_+] = \{0\}$. So, we have

$$M_- \subset \operatorname{Ker} S_+, \qquad M_+ \subset \operatorname{Ker} S_-.$$

LEMMA 3.3. *Suppose that $\mathcal{D}(A)$ is dense in X. Then the operators A_- and A_+ are closed and densely defined, their spectra are located as follows:*

$$(14) \qquad\qquad \sigma(A_-) \subset \{\lambda \in \mathbf{C} \mid \Re\lambda \le -h\},$$

$$(15) \qquad\qquad \sigma(A_+) \subset \{\lambda \in \mathbf{C} \mid \Re\lambda \ge h\},$$

and $\mathcal{D}(A_-^2) \subset N_-$ and $\mathcal{D}(A_+^2) \subset N_+$, where N_- and N_+ are as in (7).

PROOF. Let $T = A^{-1}$, and let M_- and M_+ be defined by (13). Obviously, A_- maps $\mathcal{D}(A_-)$ in a one-one way onto M_- and $A_-^{-1} = T|M_-$. Thus A_-^{-1} is a closed operator, and hence the same is true for A_-. Since $\operatorname{Im} T = \mathcal{D}(A)$ is dense in X, the space $\operatorname{Im} S_- \subset \overline{\operatorname{Im} S_- T}$. Now $S_- T = T S_-$, and so $\operatorname{Im} S_- T = T[\operatorname{Im} S_-] \subset TM_-$. We conclude that $M_- \subset \overline{TM_-}$, and thus $\mathcal{D}(A_-) = TM_-$ is dense in M_-. In a similar way one proves that A_+ is a closed densely defined linear operator.

Next we prove (14). First take $|\Re z| < h$. Then $z \in \rho(A)$ and $(z - A)^{-1}$ commutes with S_-. Hence $(z - A)^{-1}$ leaves M_- invariant. By the resolvent equation, we have $(z - A)^{-1} = -T + zT(z - A)^{-1}$, and so $(z - A)^{-1}$ maps M_- into $TM_- = \mathcal{D}(A_-)$. It follows that $z \in \rho(A_-)$ and $(z - A_-)^{-1} = (z - A)^{-1}|M_-$. As a second step we take $\Re z \ge h$, and put

$$(16) \qquad\qquad R(z) = \frac{-1}{2\pi i} \int\limits_{\alpha - i\infty}^{\alpha + i\infty} \frac{z^2}{\lambda^2(\lambda - z)}(\lambda - A)^{-1} d\lambda.$$

Since condition (4) is satisfied, $R(z)$ is a well-defined bounded linear operator on X. From

$$(z - A)\left[\frac{z^2}{\lambda^2(\lambda - z)}(\lambda - A)^{-1}\right] = -\frac{z^2}{\lambda^2}(\lambda - A)^{-1} + \frac{z^2}{\lambda^2(\lambda - z)}I,$$

it follows that

$$\frac{-1}{2\pi i} \int\limits_{\alpha - i\infty}^{\alpha + i\infty} (z - A)\left[\frac{z^2}{\lambda^2(\lambda - z)}(\lambda - A)^{-1}\right] d\lambda = -z^2 S_+ + I.$$

Now use that $z - A$ is a closed linear operator to conclude that $R(z)$ maps X into $\mathcal{D}(A)$ and $(z - A)R(z) = -z^2 S_+ + I$. Take $x \in M_-$. Recall that $M_- \subset \operatorname{Ker} S_+$, and so

$(z - A)R(z)x = x$. As we observed above, $(\lambda - A)^{-1}$ leaves M_- invariant. But then it is clear that $R(z)x \in M_-$. In turn, this gives that $AR(z)x = zR(z)x - x \in M_-$ too, and so $R(z)x \in TM_- = \mathcal{D}(A_-)$. Thus $(z - A_-)R(z)x = x$. In the case when $x \in \mathcal{D}(A_-) = TM_-$, then also $R(z)(z - A_-)x = x$. This follows from the fact that $R(z)$ commutes with T. Hence $z \in \rho(A_-)$ and $(z - A_-)^{-1} = R(z)|M_-$. We have now proved that (14) holds true, and we have shown that

$$(17) \qquad (z - A_-)^{-1} = \begin{cases} (z - A)^{-1}|M_-, & |\Re z| < h, \\ R(z)|M_-, & \Re z \geq h. \end{cases}$$

Analogous arguments yield (15).

To prove the inclusion $\mathcal{D}(A_-^2) \subset N_-$, take $x \in \mathcal{D}(A_-^2)$. For $\Re\lambda > -h$ define $\varphi_x^-(\lambda) = (\lambda - A_-)^{-1}x$. Obviously, φ_x^- is analytic on $\Re\lambda > -h$. Since $\mathcal{D}(A_-) \subset \mathcal{D}(A)$, the values of φ_x^- are in $\mathcal{D}(A)$ and

$$(\lambda - A)\varphi_x^-(\lambda) = (\lambda - A_-)\varphi_x^-(\lambda) = x, \qquad \Re\lambda > -h.$$

Condition (4) and formula (17) imply that φ_x^- is bounded on $|\Re\lambda| < h$. So, in order to prove that $x \in N_-$ it is sufficient to show that φ_x^- is bounded on $\Re\lambda \geq h$. Since $x \in \mathcal{D}(A_-^2)$, we have

$$\varphi_x^-(z) = z^{-1}x + z^{-2}A_-x + z^{-2}(z - A_-)^{-1}A_-^2x.$$

From formulas (16) and (17) we conclude that

$$z^{-2}(z - A_-)^{-1}A_-^2x = \frac{-1}{2\pi i} \int\limits_{\alpha-i\infty}^{\alpha+i\infty} \frac{1}{\lambda^2(\lambda - z)}(\lambda - A)^{-1}A_-^2x\,d\lambda$$

for $\Re z \geq h$. Together with (4) this gives that $z^{-2}(z - A_-)^{-1}A_-^2x$ is bounded on $\Re z \geq h$. Trivially, the same is true for the functions $z^{-1}x$ and $z^{-2}A_-x$. Hence φ_x^- is also bounded on $\Re z \geq h$. Thus $x \in N_-$. In a similar way, one proves that $\mathcal{D}(A_+^2) \subset N_+$. $\square$

PROOF OF THEOREM 3.1. The proof is divided into two parts. In the first part we show that Theorem 3.1 remains correct when X_- and X_+ are replaced by the spaces M_- and M_+, respectively. Here M_- and M_+ are defined by (13).

PART I. We already know that $M_- \subset \operatorname{Ker} S_+$ and $M_+ \subset \operatorname{Ker} S_-$. Now $\operatorname{Ker} S_- \cap \operatorname{Ker} S_+ = \{0\}$ (Lemma 3.2). Thus $M_- \cap M_+ = \{0\}$. So, in order to prove that

$$(18) \qquad X = \overline{M_- \oplus M_+},$$

it remains to show that $M_- + M_+$ is dense in X. For this we pass to the conjugate A' of A. The linear operator A' is closed and condition (4) remains valid when A is replaced by A' (cf., Proposition XIV.2.6). The conjugates S'_- and S'_+ of S_- and S_+ (see (11) and (12)) are given by

$$S'_- = \frac{1}{2\pi i} \int\limits_{-\alpha-i\infty}^{-\alpha+i\infty} \lambda^{-2}(\lambda - A')^{-1}d\lambda,$$

$$S'_+ = \frac{-1}{2\pi i} \int\limits_{\alpha-i\infty}^{\alpha+i\infty} \lambda^{-2}(\lambda - A')^{-1}d\lambda.$$

Maybe the domain $\mathcal{D}(A')$ of A' is not dense in the conjugate X' of X. However, in Lemma 3.2 we did not require $\mathcal{D}(A)$ to be dense in X. Therefore the lemma also applies with S'_- instead of S_- and S'_+ instead of S_+. Hence $\operatorname{Ker} S'_- \cap \operatorname{Ker} S'_+ = \{0\}$. But then $\operatorname{Im} S_- + \operatorname{Im} S_+$ is dense in X (use [GG], Theorem XI.5.1(i)). According to (13) this implies that $M_- + M_+$ is dense in X.

For $x \in \mathcal{D}(A^2)$, put

$$(19) \qquad Lx = \frac{-1}{2\pi i} \int\limits_{\alpha-i\infty}^{\alpha+i\infty} \lambda^{-2}(\lambda - A)^{-1} A^2 x \, d\lambda, \qquad x \in \mathcal{D}(A^2).$$

Condition (4) guarantees that Lx is well-defined. From the definition of S_+ (formula (12)) it follows that $Lx = S_+ A^2 x$. If $x \in \mathcal{D}(A^2_-)$, then $A^2 x \in M_- \subset \operatorname{Ker} S_+$ and so $Lx = 0$. Recall from the proof of Lemma 3.2 that $S_- + S_+ = T^2$, where $T = A^{-1}$. Thus $Lx = x - S_- A^2 x$, which implies that $Lx = x$ for $x \in \mathcal{D}(A^2_+)$.

Next, assume that there exists a bounded linear operator P on X such that $Px = Lx$ for $x \in \mathcal{D}(A^2)$. Then P vanishes on $\mathcal{D}(A^2_-)$ and coincides with the identity on $\mathcal{D}(A^2_+)$. From Lemma 3.3 we know that $T_- = A_-^{-1}$ is a well-defined bounded linear operator on M_- and $\operatorname{Im} T_- = \mathcal{D}(A_-)$ is dense in M_-. But then $\mathcal{D}(A^2_-) = \operatorname{Im} T_-^2$ is dense in M_-. Since P is bounded, we may conclude that P vanishes on M_-. Analogously, $\mathcal{D}(A^2_+)$ is dense in M_+, and so P coincides with the identity on M_+.

Take x in the closure of $M_- + M_+$, say

$$x = \lim_{k \to \infty} (x_k^- + x_k^+),$$

with $x_k^- \in M_-$ and $x_k^+ \in M_+$. Then $x_k^+ = P(x_k^- + x_k^+)$ converges to Px when k tends to infinity. Since M_+ is closed (see (13)), this implies that $Px \in M_+$. It also follows that x_k^- converges to $x - Px$ when k tends to infinity, and so $x - Px \in M_-$. But then $x \in M_- + M_+$, and we conclude that $M_- + M_+$ is closed. Together with (18) this yields that $X = M_- \oplus M_+$ and P is the projection of X along M_- onto M_+.

Conversely, suppose $X = M_- \oplus M_+$, and let P be the projection of X along M_- onto M_+. Then $Px = 0 = Lx$ for $x \in \mathcal{D}(A^2_-)$ and $Px = x = Lx$ for $x \in \mathcal{D}(A^2_+)$. So P and L coincide on $\mathcal{D}(A^2_-) \oplus \mathcal{D}(A^2_+)$. With $T = A^{-1}$ we have

$$\mathcal{D}(A^2) = \operatorname{Im} T^2 = T^2[M_- \oplus M_+]$$
$$= T^2 M_- \oplus T^2 M_+ = \mathcal{D}(A^2_-) \oplus \mathcal{D}(A^2_+).$$

Hence $Px = Lx$ for all $x \in \mathcal{D}(A^2)$, and thus the map (9) extends to a bounded linear operator on X.

From $X = M_- \oplus M_+$, we get

$$\mathcal{D}(A) = \operatorname{Im} T = TM_- \oplus TM_+ = \mathcal{D}(A_-) \oplus \mathcal{D}(A_+).$$

Here, as before, $T = A^{-1}$. It follows that $\mathcal{D}(A_-) = \mathcal{D}(A) \cap M_-$ and $\mathcal{D}(A_+) = \mathcal{D}(A) \cap M_+$. This implies that M_- and M_+ are invariant under A. Also, A_- and A_+ are the restrictions of A to M_- and M_+, respectively. Combining this with (14) and (15) we get

$$\sigma(A|M_-) = \sigma(A_-) \subset \{\lambda \in \mathbb{C} \mid \Re\lambda \le -h\},$$
$$\sigma(A|M_+) = \sigma(A_+) \subset \{\lambda \in \mathbb{C} \mid \Re\lambda \ge h\}.$$

This completes the first part of the proof.

PART II. Here we prove Theorem 3.1 the way it is stated. Take $x \in N_+$, and let φ_x^+ be a bounded, analytic X-valued function on $\Re\lambda < h$ such that

$$(\lambda - A)\varphi_x^+(\lambda) = x, \qquad \Re\lambda < h.$$

According to the definition of S_- (formula (11)) we have

$$S_- x = \frac{1}{2\pi i} \int\limits_{-\alpha-i\infty}^{-\alpha+i\infty} \lambda^{-2}\varphi_x^+(\lambda)d\lambda.$$

A straightforward argument, based on Cauchy's theorem and using the boundedness condition on φ_x^+, shows that the latter integral is zero. It follows that $N_+ \subset \operatorname{Ker} S_-$. Taking closures we get $X_+ \subset \operatorname{Ker} S_-$. Analogously one proves that $X_- \subset \operatorname{Ker} S_+$. But then Lemma 3.2 gives that $X_- \cap X_+ = \{0\}$. From Lemma 3.3 we know that $\mathcal{D}(A_-^2) \subset N_-$ and $\mathcal{D}(A_+^2) \subset N_+$. In Part I of the proof it was already noted that $\mathcal{D}(A_-^2)$ is dense in M_- and $\mathcal{D}(A_+^2)$ is dense in M_+. Hence $M_- \subset X_-$ and $M_+ \subset X_+$. Taking into account (18), one may now conclude that $X = \overline{X_- \oplus X_+}$.

Next, assume that there exists a bounded linear operator P on X such that $Px = Lx$ for all $x \in \mathcal{D}(A^2)$. Here Lx is defined by (19) and is equal to the right hand side of (9). As we have seen in Part I of the proof, this implies that $X = M_- \oplus M_+$ and that P is the projection of X along M_- onto M_+. Since $M_- \subset X_-$ and $M_+ \subset X_+$ and $X_- \cap X_+ = \{0\}$, it follows that $X_- = M_-$ and $X_+ = M_+$. So $X = X_- \oplus X_+$ and P is the projection of X along X_- onto X_+.

Finally, suppose that $X = X_- \oplus X_+$. As M_- is a closed subspace of X_- and M_+ is a closed subspace of X_+, the space $M_- \oplus M_+$ is closed. Together with (18) this implies $X = M_- \oplus M_+$. So again $X_- = M_-$ and $X_+ = M_+$. But then the desired conclusions are clear from what was established in Part I of the proof. $\square$

We shall refer to the spaces X_- and X_+ appearing in Theorem 3.1 as the *spectral subspaces of A corresponding to the left and right half plane, respectively.*

XV.4 A PERTURBATION THEOREM

In this section we show that the direct sum decomposition, $X = X_- \oplus X_+$, in Theorem 3.1 is preserved under certain perturbations of A.

THEOREM 4.1. *For $\nu = 1,2$ let $A_\nu(X \to X)$ be a densely defined closed linear operator such that for some $h > 0$ the strip $|\Re\lambda| < h$ is in the resolvent set $\rho(A_\nu)$ and*

$$(1) \qquad \sup_{|\Re\lambda| < h} \|(\lambda - A_\nu)^{-1}\| < \infty,$$

and let $X_{-}^{(\nu)}$, $X_{+}^{(\nu)}$ be the spectral subspaces of A_ν corresponding to the left and right half plane, respectively. Assume $\mathcal{D}(A_2^2) \subset \mathcal{D}(A_1^2)$ and

$$(2) \qquad \sup_{|\Re\lambda| < h} |\lambda|^2 \|(\lambda - A_2)^{-1} - (\lambda - A_1)^{-1}\| < \infty.$$

Then $X = X_{-}^{(1)} \oplus X_{+}^{(1)}$ implies $X = X_{-}^{(2)} \oplus X_{+}^{(2)}$.

PROOF. Choose $0 < \alpha < h$. For $\nu = 1,2$ define $L_\nu : \mathcal{D}(A_\nu^2) \to X$ by setting

$$L_\nu x = \frac{-1}{2\pi i} \int\limits_{\alpha - i\infty}^{\alpha + i\infty} \lambda^{-2}(\lambda - A_\nu)^{-1} A_\nu^2 x \, d\lambda, \qquad x \in \mathcal{D}(A_\nu^2).$$

Since $X = X_{-}^{(1)} \oplus X_{+}^{(1)}$, Theorem 3.1 implies that L_1 extends to a bounded linear operator on X. Hence, in order to prove the theorem, it suffices to show that there exists a constant γ_0 such that

$$\|L_2 x - L_1 x\| \le \gamma_0 \|x\|, \qquad x \in \mathcal{D}(A_2^2).$$

Take $x \in \mathcal{D}(A_2^2)$. Then $x \in \mathcal{D}(A_1^2)$, and for $\nu = 1,2$

$$(\lambda - A_\nu)^{-1} A_\nu^2 x = -A_\nu x - \lambda x + \lambda^2 (\lambda - A_\nu)^{-1} x.$$

Hence

$$L_2 x - L_1 x = \frac{1}{2\pi i} \int\limits_{\alpha - i\infty}^{\alpha + i\infty} \lambda^{-2}[(A_2 - A_1)x + \lambda^2(\lambda - A_1)^{-1}x - \lambda^2(\lambda - A_2)^{-1}x] \, d\lambda$$

$$= \frac{1}{2\pi i} \int\limits_{\alpha - i\infty}^{\alpha + i\infty} \frac{1}{\lambda^2}[\lambda^2(\lambda - A_1)^{-1}x - \lambda^2(\lambda - A_2)^{-1}x] \, d\lambda,$$

and it follows that

$$\|L_2 x - L_1 x\| \le \frac{\gamma}{2\pi} \left(\int\limits_{-\infty}^{\infty} \frac{ds}{\alpha^2 + s^2} \right) \|x\| = \frac{\gamma}{2\alpha} \|x\|,$$

where γ denotes the left hand side of (2). $\square$

When A_1 and A_2 are bounded operators, the conclusion of Theorem 4.1 is trivial and does not need condition (2). In fact, in the bounded case condition (2) is fulfilled automatically. This one sees by subtracting the Neumann series of A_1 and A_2, and noting that the resulting power series in λ^{-1} starts with a term involving λ^{-2}.

Let us illustrate Theorem 4.1 with an example. Take $X = L_1(\mathbf{R})$, and let $A(X \to X)$ be defined as follows. The domain $\mathcal{D}(A)$ of A consists of all functions $f \in X$

such that f is absolutely continuous on each compact subinterval of $[0, \infty)$ and on each compact subinterval of $(-\infty, 0]$ and, moreover, $f' \in X$. Note that functions in $\mathcal{D}(A)$ are allowed to have a jump discontinuity at 0. For $f \in \mathcal{D}(A)$ we set

$$(3) \qquad (Af)(t) = \begin{cases} f'(t) - qf(t), & t > 0, \\ f'(t) + qf(t), & t < 0. \end{cases}$$

Here q is a fixed positive real number. Applying the same reasoning as in the example after Theorem 1.2, one proves that $\rho(A)$ is precisely equal to the strip $|\Re\lambda| < q$ and for $\lambda \in \rho(A)$

$$((\lambda - A)^{-1}f)(s) = \begin{cases} \displaystyle\int_0^\infty e^{-(\lambda+q)t} f(t+s)dt, & s > 0, \\ \displaystyle -\int_{-\infty}^0 e^{-(\lambda-q)t} f(t+s)dt, & s < 0. \end{cases}$$

It follows that

$$\|(\lambda - A)^{-1}\| \le \max\{|\Re\lambda + q|^{-1}, |\Re\lambda - q|^{-1}\}, \qquad \lambda \in \rho(A).$$

Fix $0 < h < q$. Then condition (4) in Section XV.3 is satisfied. In this case the spectral subspaces X_- and X_+ are easily determined. In fact

$$X_- = \{f \in X \mid f = 0 \text{ a.e. on } [0, \infty)\},$$
$$X_+ = \{f \in X \mid f = 0 \text{ a.e. on } (-\infty, 0]\}.$$

It follows that for A defined by (3) the spectral decomposition

$$(4) \qquad X = X_- \oplus X_+$$

holds true.

Next, we consider a perturbation of the operator defined by (3), namely, the operator $A^\times = A + D$, where D is the rank one integral operator

$$(5) \qquad Df = \left(\int_{-\infty}^\infty se^{-q|s|} f(s)ds \right) g.$$

Here g is a function in $\mathcal{D}(A)$ which will remain fixed in what follows. Note that both g and Ag are in $L_1(\mathbf{R})$. Since D is bounded, $A^\times$ is a closed linear operator with domain $\mathcal{D}(A^\times) = \mathcal{D}(A)$. We shall prove that $A^\times$ has no spectrum on the imaginary axis if and only if

$$(6) \qquad \int_{-\infty}^\infty \lambda^{-2}(e^{-\lambda s} + \lambda s - 1)e^{-q|s|}g(s)ds \ne 1, \qquad \Re\lambda = 0.$$

Furthermore, we shall show that in that case the spectral decomposition (4) also holds true for $A^\times$ instead of A.

First note that

$$(7) \qquad \lambda - A^\times = [I - D(\lambda - A)^{-1}](\lambda - A), \qquad |\Re\lambda| < q.$$

It follows that for $|\Re\lambda| < q$ the operator $\lambda - A^\times$ is invertible if and only if $I - D(\lambda - A)^{-1}$ is invertible. Since D has rank one, the latter occurs if and only if $F\big((\lambda - A)^{-1}g\big) \neq 1$, where F is the continuous linear functional on $L_1(\mathbf{R})$ given by the integral in (5), that is,

$$F(f) = \int_{-\infty}^{\infty} s e^{-q|s|} f(s)\,ds.$$

A straightforward computation shows that for f in $L_1(\mathbf{R})$ and $|\Re\lambda| < q$

$$F\big((\lambda - A)^{-1}f\big) = \int_{-\infty}^{\infty} \lambda^{-2}(e^{-\lambda s} + \lambda s - 1)e^{-q|s|}f(s)\,ds.$$

In particular, $F\big((\lambda - A)^{-1}g\big)$ is equal to the left hand side of (6), and hence (6) is the necessary and sufficient condition in order that $\lambda - A^\times$ is invertible for $\Re\lambda = 0$.

Denote the left hand side of (6) by $d(\lambda)$. Note that the function $d(\cdot)$ is analytic in the strip $|\Re\lambda| < q$. Take $0 < \varepsilon < q$. The Riemann-Lebesgue lemma (see [R], Theorem 9.6) implies that

$$(8) \qquad \lim_{\lambda \to \infty, |\Re\lambda| \le \varepsilon} d(\lambda) = 0.$$

So there exists $\rho > 0$ such that $d(\lambda) \neq 1$ for $|\Re\lambda| \le \varepsilon$ and $|\lambda| \ge \rho$. It follows that $d(\cdot) - 1$ has only a finite number of zeros in the strip $|\Re\lambda| \le \varepsilon$.

Next, assume that $d(\lambda) \neq 1$ for $\Re\lambda = 0$. Thus $d(\cdot) - 1$ has no zeros on the imaginary axis. The result of the previous paragraph implies that we may choose $0 < h < q$ such that $d(\lambda) \neq 1$ for $|\Re\lambda| \le h$. But then it follows that $\lambda - A^\times$ is invertible for $|\Re\lambda| \le h$, and we can use (7) to show that for $f \in L_1(\mathbf{R})$

$$(9) \qquad \begin{aligned} (\lambda - A^\times)^{-1}f &= (\lambda - A)^{-1}f + (1 - d(\lambda))^{-1}\cdot \\ &\quad \cdot F\big((\lambda - A)^{-1}f\big)(\lambda - A)^{-1}g, \qquad |\Re\lambda| \le h. \end{aligned}$$

From formula (8) and $d(\lambda) \neq 1$ for $|\Re\lambda| \le h$ we conclude that $\big(1 - d(\cdot)\big)^{-1}$ is bounded on the strip $|\Re\lambda| \le h$. Also $\|(\cdot - A)^{-1}\|$ is a bounded function on $|\Re\lambda| \le h$. But then we can use (9) to show that

$$\sup_{|\Re\lambda| < h} \|(\lambda - A^\times)^{-1}\| < \infty.$$

In other words, condition (4) in Section XV.3 is also fulfilled for $A^\times$. Let $X_-^\times$ and $X_+^\times$ be the spectral subspaces of $A^\times$ corresponding to the left and right half plane, respectively. Theorem 3.1 tells us that $X = \overline{X_-^\times \oplus X_+^\times}$. We want to show

$$(10) \qquad X = X_-^\times \oplus X_+^\times.$$

To do this we apply Theorem 4.1. From (9) it follows that for $f \in L_1(\mathbf{R})$

$$\lambda^2[(\lambda - A^\times)^{-1}f - (\lambda - A)^{-1}f] =$$
$$= (1 - d(\lambda))^{-1}F(\lambda(\lambda - A)^{-1}f)\lambda(\lambda - A)^{-1}g, \qquad |\Re\lambda| \leq h.$$

Rewrite $\lambda(\lambda - A)^{-1}$ as $I + A(\lambda - A)^{-1}$. A simple computation shows that

$$F(Af) = -\int_{-\infty}^{\infty} e^{-q|t|}f(t)dt.$$

Thus $F \circ A$ extends to a bounded linear functional, G say, defined of $L_1(\mathbf{R})$. Next, recall that $g \in \mathcal{D}(A)$. Thus Ag is a well-defined element of $L_1(\mathbf{R})$ and $A(\lambda - A)^{-1}g = (\lambda - A)^{-1}(Ag)$. We conclude that

$$\lambda^2[(\lambda - A^\times)^{-1}f - (\lambda - A)^{-1}f] = (1 - d(\lambda))^{-1}\times$$
$$\times\,[F(f) + G((\lambda - A)^{-1}f)][g + (\lambda - A)^{-1}Ag], \qquad |\Re\lambda| \leq h.$$

Since $(1 - d(\cdot))^{-1}$ and $\|(\cdot - A)^{-1}\|$ are bounded on $|\Re\lambda| \leq h$, there exists a constant γ such that

$$\|\lambda^2[(\lambda - A^\times)^{-1}f - (\lambda - A)^{-1}f]\| \leq \gamma\|f\|, \qquad |\Re\lambda| \leq h.$$

In other words, condition (2) is fulfilled with $A_1 = A$ and $A_2 = A^\times$. It follows that (10) holds true.

CHAPTER XVI

UNBOUNDED SELFADJOINT OPERATORS

This chapter introduces the reader to unbounded selfadjoint operators on a complex Hilbert space. We give examples of such operators, describe their elementary properties, and prove the spectral theorem. The main method employed here is to transfer from the unbounded operator to its resolvent and to apply our results for bounded selfadjoint operators.

XVI.1 SELFADJOINT ORDINARY DIFFERENTIAL OPERATORS

Let H be a complex Hilbert space. An operator $T(H \to H)$ is called *selfadjoint* if T is densely defined and $T^* = T$. There are many important examples of unbounded selfadjoint operators. This section concerns selfadjointness for ordinary differential operators of the type considered in Theorem XIV.4.4.

THEOREM 1.1. *Let τ be the differential expression $\tau = \sum_{j=0}^{n} a_j(t)D^j$, where $a_j \in C^j([a,b])$, $0 \leq j \leq n$, and $a_n(t) \neq 0$ for all $a \leq t \leq b$. Let $T\big(L_2([a,b]) \to L_2([a,b])\big)$ be the restriction of $T_{\max,\tau}$ to those $g \in \mathcal{D}(T_{\max,\tau})$ which satisfy the following boundary conditions*

$$(1) \qquad B_i(g) = \sum_{j=0}^{n-1} \alpha_{ij} g^{(j)}(a) + \sum_{j=0}^{n-1} \beta_{ij} g^{(j)}(b) = 0, \qquad 1 \leq i \leq k.$$

Put

$$(2) \qquad N_1 = [\alpha_{ij}]_{i=1,\ j=0}^{k,\ n-1}, \qquad N_2 = [\beta_{ij}]_{i=1,\ j=0}^{k,\ n-1}.$$

Assume that the rank of the $k \times 2n$ matrix $N = [\ N_1 \quad N_2\ ]$ is k. Then T is selfadjoint if and only if the following three conditions are satisfied:

(i) $\tau = \tau^*$,

(ii) $k = n$,

(iii) $N_1\big(F(a)^{-1}\big)^* N_1^* = N_2\big(F(b)^{-1}\big)^* N_2^*$.

Here $F(t)$ is the $n \times n$ matrix $[F_{mj}(t)]_{m,j=0}^{n-1}$ with

$$(3) \qquad F_{mj}(t) = \begin{cases} \sum_{k=j}^{n-m-1} (-1)^k \binom{k}{j} \overline{a}_{m+k+1}^{(k-j)}(t), & m+j \leq n-1, \\ 0, & m+j > n-1. \end{cases}$$

For the proof of Theorem 1.1 we need the following lemma.

LEMMA 1.2. *Let $f_1, \ldots, f_k, g$ be linear functionals on a vector space V over* C. *Suppose $\bigcap_{j=1}^{k} \operatorname{Ker} f_j \subset \operatorname{Ker} g$. Then $g \in \operatorname{sp}\{f_1, \ldots, f_k\}$.*

PROOF. Let $M = \{(f_1(x), \ldots, f_k(x)) \mid x \in V\}$. Clearly, M is a subspace of $\mathbf{C}^k$. Define a linear functional F on M by

$$F((f_1(x), \ldots, f_k(x))) = g(x), \qquad x \in V.$$

The functional F is well-defined since $(f_1(x), \ldots, f_k(x)) = (f_1(y), \ldots, f_k(y))$ implies that $x - y \in \bigcap_{j=1}^{k} \operatorname{Ker} f_j \subset \operatorname{Ker} g$, whence $g(x) = g(y)$. Let $\widetilde{F}$ be any linear extension of F to all of $\mathbf{C}^k$, and let $e_1, \ldots, e_k$ be the standard basis for $\mathbf{C}^k$. Then for all $x \in V$,

$$g(x) = \widetilde{F}(f_1(x), \ldots, f_k(x)) = \widetilde{F}\left(\sum_{j=1}^{k} f_j(x)e_j\right) = \left(\sum_{j=1}^{k} \widetilde{F}(e_j)f_j\right)(x). \quad \square$$

PROOF OF THEOREM 1.1. From Theorem XIV.4.4 (and the remark in the one but last paragraph of Section XIV.4) we know that the adjoint T^* is the restriction of $T_{\max,\tau^*}$ to those $f \in \mathcal{D}(T_{\max,\tau^*})$ which satisfy the adjoint boundary conditions

$$B_i^{\#}(f) = \sum_{j=0}^{n-1} \alpha_{ij}^{\#} f^{(j)}(a) + \sum_{j=0}^{n-1} \beta_{ij}^{\#} f^{(j)}(b) = 0, \qquad 1 \leq i \leq 2n - \ell,$$

where $\ell = \operatorname{rank}[\, N_1 \quad N_2 \,]$. Note that $\ell = k$ by our assumption on $\operatorname{rank}[\, N_1 \quad N_2 \,]$.

Suppose $T = T^*$. Since $C_0^{\infty}([a, b])$ is contained in the domain of $T = T^*$, we have $\tau\varphi = \tau^*\varphi$ for each $\varphi \in C_0^{\infty}([a, b])$. By Green's formula ((9) in Section XIV.4)

$$\langle \tau f, \varphi \rangle = \langle f, \tau^*\varphi \rangle = \langle f, \tau\varphi \rangle = \langle \tau^* f, \varphi \rangle$$

for all $f \in C^n([a, b])$ and $\varphi \in C_0^{\infty}([a, b])$. Since $C_0^{\infty}([a, b])$ is dense in $L_2([a, b])$, we conclude that $\tau f = \tau^* f$ for $f \in C^n([a, b])$. Recall that τ^* is of the form $\tau^* = \sum_{j=0}^{n} b_j(t)D^n$ with $b_j \in C^j([a, b])$ for $0 \leq j \leq n$. By choosing f to be the polynomials t^j, $0 \leq j \leq n$, we obtain $a_j = b_j$, $0 \leq j \leq n$. Thus (i) holds.

To prove (ii) note that the assumption $\operatorname{rank}[\, N_1 \quad N_2 \,] = k$ implies (see the paragraph after the proof of Theorem XIV.4.4) that the linear functionals $\{B_1, \ldots, B_k\}$ are linearly independent on $\mathcal{D} = \mathcal{D}(T_{\max,\tau})$. Also (see the second paragraph after the proof of Theorem XIV.4.4) the functionals $\{B_1^{\#}, \ldots, B_{2n-k}^{\#}\}$ are linearly independent on $\mathcal{D}$. (Recall that $\mathcal{D}$ is also equal to $\mathcal{D}(T_{\max,\tau^*})$.) Since $T = T^*$, Theorem XIV.4.4 shows that $\bigcap_{i=1}^{k} \operatorname{Ker} B_i = \bigcap_{j=1}^{2n-\ell} \operatorname{Ker} B_j^{\#}$. Hence

$$(4) \qquad\qquad \operatorname{span}\{B_1, \ldots, B_k\} = \operatorname{span}\{B_1^{\#}, \ldots, B_{2n-k}^{\#}\}$$

by Lemma 1.2. Therefore $k = 2n - k$ or $k = n$.

Next, we prove (iii). By (4) and $k = n$ there exist constants c_{ij}, $1 \leq i, j \leq n$, such that $B_i = \sum_{j=1}^{n} c_{ij} B_j^{\#}$ for $i = 1, \ldots, n$. Since both $\{B_i\}_{i=1}^{n}$ and $\{B_i^{\#}\}_{i=1}^{n}$ are linearly independent the matrix $C = [c_{ij}]_{i,j=1}^n$ is invertible. Let p_{aj} and p_{bj}, $0 \leq j \leq n-1$, be the polynomials defined by formulas (20) and (21) in Section XIV.4. Then

$$\alpha_{ir} = B_i(p_{ar}) = \sum_{j=1}^n c_{ij} B_j^{\#}(p_{ar}) = \sum_{j=1}^n c_{ij} \alpha_{jr}^{\#},$$

$$\beta_{ir} = B_i(p_{br}) = \sum_{j=1}^n c_{ij} B_j^*(p_{br}) = \sum_{j=1}^n c_{ij} \beta_{jr}^{\#},$$

for $1 \leq i \leq n$ and $0 \leq r \leq n - 1$. It follows that

$$(5) \qquad\qquad [\, N_1 \quad N_2 \,] = C[\, N_1^{\#} \quad N_2^{\#} \,],$$

where $N_1^{\#} = [\alpha_{ij}^{\#}]_{i=1}^n,\ _{j=0}^{n-1}$ and $N_2^{\#} = [\beta_{ij}^{\#}]_{i=1}^n,\ _{j=0}^{n-1}$. Now recall (see formula (26) in Section XIV.4) that

$$(6) \qquad\qquad [\, N_1^{\#} \quad N_2^{\#} \,] = [\, G_1^* \quad G_2^* \,] \begin{bmatrix} -F(a) & 0 \\ 0 & F(b) \end{bmatrix},$$

where $F(t)$ is as in the theorem and G_1 and G_2 are $n \times n$ matrices such that $N_1 G_1 + N_2 G_2 = 0$. We know (formula (13) in Section XIV.4) that $F(a)$ and $F(b)$ are invertible. Thus (use (5) and (6))

$$0 = N_1 G_1 + N_2 G_2 = [\, N_1 \quad N_2 \,] \begin{bmatrix} G_1 \\ G_2 \end{bmatrix}$$

$$= [\, N_1 \quad N_2 \,] \begin{bmatrix} -F(a)^{-1} & 0 \\ 0 & F(b)^{-1} \end{bmatrix}^* \begin{bmatrix} N_1^* \\ N_2^* \end{bmatrix} (C^{-1})^*,$$

which yields (iii).

To prove the reverse implication, assume that (i), (ii) and (iii) hold. Define $n \times n$ matrices D_1 and D_2 by setting

$$\begin{bmatrix} D_1 \\ D_2 \end{bmatrix} = \begin{bmatrix} -F(a)^{-1} & 0 \\ 0 & F(b)^{-1} \end{bmatrix}^* \begin{bmatrix} N_1^* \\ N_2^* \end{bmatrix}.$$

Then $N_1 D_1 + N_2 D_2 = 0$ because of (iii). By (ii)

$$\text{rank} \begin{bmatrix} D_1 \\ D_2 \end{bmatrix} = \text{rank}[\, N_1 \quad N_2 \,] = k = n.$$

It follows that there exists an invertible $n \times n$ matrix E such that

$$\begin{bmatrix} D_1 \\ D_2 \end{bmatrix} = \begin{bmatrix} G_1 \\ G_2 \end{bmatrix} E,$$

where G_1 and G_2 are as in formula (15) of Section XIV.4. Now use (6) to show that

$$[\, N_1 \quad N_2 \,] = [\, D_1^* \quad D_2^* \,] \begin{bmatrix} -F(a) & 0 \\ 0 & F(b) \end{bmatrix}$$

$$= E^*[\, G_1^* \quad G_2^* \,] \begin{bmatrix} -F(a) & 0 \\ 0 & F(b) \end{bmatrix} = E^*[\, N_1^{\#} \quad N_2^{\#} \,].$$

It follows that $\mathrm{span}\{B_i\}_{i=1}^n = \mathrm{span}\{B_i^\#\}_{i=1}^n$ because E is invertible. Since $\tau = \tau^*$ (condition (i)), Theorem XIV.4.4 implies $T = T^*$. $\quad\square$

COROLLARY 1.3. *Let τ be the differential expression given by*

$$\tau f = D(pf') + qf,$$

where $p \in C^2([a,b])$ and $q \in C([a,b])$ are real-valued functions with $p(t) \neq 0$ for all $t \in [a,b]$. Let $T\big(L_2([a,b]) \to L_2([a,b])\big)$ be the restriction of $T_{\max,\tau}$ to those $g \in \mathcal{D}(T_{\max,\tau})$ which satisfy the boundary conditions:

$$B_1 g = \alpha_{10} g(a) + \alpha_{11} g'(a) + \beta_{10} g(b) + \beta_{11} g'(b) = 0,$$
$$B_2 g = \alpha_{20} g(a) + \alpha_{21} g'(a) + \beta_{20} g(b) + \beta_{21} g'(b) = 0,$$

where each α_{ij} and β_{ij} is a real number. Then T is selfadjoint if and only if the rank of $[[\alpha_{ij}][\beta_{ij}]]$ is 2 and

$$(7) \qquad \frac{1}{p(a)} \det \begin{bmatrix} \alpha_{10} & \alpha_{11} \\ \alpha_{20} & \alpha_{21} \end{bmatrix} = \frac{1}{p(b)} \det \begin{bmatrix} \beta_{10} & \beta_{11} \\ \beta_{20} & \beta_{21} \end{bmatrix}.$$

PROOF. It was shown in the example at the end of Section XIV.4 that $\tau = \tau^*$ and

$$F(t) = \begin{bmatrix} 0 & p(t) \\ -p(t) & 0 \end{bmatrix}.$$

But then condition (iii) in the previous theorem reduces to

$$\frac{1}{p(a)} d_1 \begin{bmatrix} 0 & 1 \\ -1 & 0 \end{bmatrix} = \frac{1}{p(b)} d_2 \begin{bmatrix} 0 & 1 \\ -1 & 0 \end{bmatrix},$$

where $d_1 = \det[\alpha_{ij}]$ and $d_2 = \det[\beta_{ij}]$. Hence (iii) is equivalent to (7), and the corollary follows. $\quad\square$

XVI.2   AN EXAMPLE FROM PARTIAL DIFFERENTIAL EQUATIONS

THEOREM 2.1. *Let Ω be a bounded open set in $\mathbf{R}^n$ with boundary of class C^2. Let*

$$L = \sum_{i,j=1}^n D_i\big(a_{ij}(x) D_j\big) + a_0(x),$$

where a_0 is a real-valued function in $C^\infty(\overline{\Omega})$ and $\overline{a_{ij}} = a_{ji} \in C^\infty(\overline{\Omega})$, $1 \leq i,j \leq n$. Assume that L is uniformly elliptic. Then the operator $A\big(L_2(\Omega) \to L_2(\Omega)\big)$ defined by

$$\mathcal{D}(A) = H_0^1(\Omega) \cap H_2(\Omega), \qquad Au = Lu$$

is selfadjoint.

PROOF. Note that A is the Dirichlet operator on $L_2(\Omega)$ associated with L. Since the boundary of Ω is of class C^2, we can apply the results of Section XIV.6. In particular, Theorem XIV.6.5 tells us that A^* is the Dirichlet operator on $L_2(\Omega)$ associated with the Lagrange adjoint L^* of L because L is uniformly elliptic. Recall that

$$L^* = \sum_{i,j=1}^{n} D_j\big(\overline{a}_{ij}(x)D_i\big) + \overline{a}_0(x).$$

Our condition on the coefficients of L imply that $L = L^*$, and hence $A = A^*$. $\square$

The above theorem also holds true if $-L$ is uniformly elliptic. In particular, the theorem applies to the Laplacian $\Delta = \sum_{i=1}^{n} D_i^2$, because $-\Delta$ is uniformly elliptic and its coefficients have the desired symmetry properties. Thus, if $L = \Delta$, then the operator A defined in (1) is selfadjoint.

XVI.3 SPECTRUM AND CAYLEY TRANSFORM

Throughout the remaining part of this chapter H denotes a complex Hilbert space.

THEOREM 3.1. *If $A(H \to H)$ is selfadjoint, then A is a closed operator, its spectrum $\sigma(A) \subset \mathbf{R}$ and*

$$(1) \qquad \|(\lambda - A)^{-1}\| \le |\operatorname{Im}\lambda|^{-1}, \qquad \operatorname{Im}\lambda \neq 0.$$

PROOF. The operator A is closed by Proposition XIV.2.2. Let $\lambda = a + ib$, $b \neq 0$ with a and b real. Since $a - A$ is selfadjoint,

$$(2) \qquad \|(\lambda - A)x\|^2 = \|(a - A)x\|^2 + |b|^2\|x\|^2 \ge |b|^2\|x\|^2.$$

Therefore $\lambda - A$ is injective. From (2) and the fact that $\lambda - A$ is closed, it follows that $\operatorname{Im}(\lambda - A)$ is closed. Hence

$$\operatorname{Im} A = (\operatorname{Ker} A)^{\perp} H,$$

by Proposition XIV.2.5. Therefore $\lambda \in \rho(A)$, the resolvent set of A, and (1) is a consequence of (2). $\square$

Let $A(H \to H)$ be densely defined, and assume that for some real λ the operator $\lambda - A$ is invertible. Then $\lambda - A^*$ is invertible and

$$(3) \qquad (\lambda - A^*)^{-1} = [(\lambda - A)^{-1}]^*,$$

because of Proposition XIV.2.6. It follows that A is selfadjoint if and only if the bounded operator $(\lambda - A)^{-1}$ is selfadjoint.

The transformation of an unbounded operator into a bounded one is often useful. There is a specific transformation of this type which is widely used. We have in mind the so-called Cayley transform. This transformation may be viewed as the operator-valued version of the Möbius transformation

$$(4) \qquad w = \frac{z + i}{z - i},$$

which maps the real line onto the unit circle. If we replace in (4) the variable z by a selfadjoint operator A, then the resulting operator

$$(5) \qquad\qquad U = (A + i)(A - i)^{-1}$$

is called the *Cayley transform* of A. Note that $(A - i)^{-1}$ in (5) is well-defined because of Theorem 3.1.

THEOREM 3.2. *If $A(H \to H)$ is a selfadjoint operator, then its Cayley transform is unitary.*

PROOF. By Theorem 3.1 the operator $A + i$ maps the domain of A in a one-one way onto H. It follows that the Cayley transform U is invertible on H. Also, given $y \in H$, there exist $x \in \mathcal{D}(A)$ such that $(A - i)x = y$. So, by (2),

$$\|Uy\|^2 = \|(A + i)x\|^2 = \|Ax\|^2 + \|x\|^2 = \|(A - i)x\|^2 = \|y\|^2.$$

Hence U is an isometry and invertible. Therefore U is unitary. $\square$

If U is the Cayley transform of the selfadjoint operator $A(H \to H)$, then A may be recovered from U via the following formula:

$$(6) \qquad\qquad Ax = i(U - I)^{-1}(U + I)x, \qquad x \in \mathcal{D}(A).$$

To see this, note that

$$U - I = \{(A + i) - (A - i)\}(A - i)^{-1} = 2i(A - i)^{-1},$$

and hence $U - I$ maps H in a one-one way onto the domain of A. Also, for $x \in \mathcal{D}(A)$, we have $(U - I)Ax = i(U + I)x$, which implies (6).

THEOREM 3.3. *Let $A(H \to H)$ be closed and densely defined. Then*

(a) $I + A^*A$ *is invertible with* $\|(I + A^*A)^{-1}\| \le 1$,

(b) A^*A *is selfadjoint.*

PROOF. Given $z \in H$, Proposition XIV.2.1 ensures the existence of vectors $u \in \mathcal{D}(A)$ and $v \in \mathcal{D}(A^*)$ such that

$$(z, 0) = (u, Au) + (A^*v, -v)$$

or

$$z = u + A^*v, \qquad 0 = Au - v.$$

Thus $z = (I + A^*A)u$, and hence $I + A^*A$ is surjective. Suppose $x \in \mathcal{D}(A^*A)$. Then

$$\|(I + A^*A)x\|\|x\| \ge \langle (I + A^*A)x, x \rangle = \|x\|^2 + \|Ax\|^2 \ge \|x\|^2,$$

and therefore (a) holds.

Put $T = (I + A^*A)^{-1}$. For x and y in H we have

$$\langle Tx, y \rangle = \langle Tx, (I + A^*A)Ty \rangle = \langle Tx, Ty \rangle + \langle ATx, ATy \rangle$$

and

$$\langle x, Ty \rangle = \langle (I + A^*A)Tx, Ty \rangle = \langle Tx, Ty \rangle + \langle ATx, ATy \rangle.$$

Thus T is selfadjoint and

$$\overline{\mathcal{D}(A^*A)} = \overline{\operatorname{Im} T} = H.$$

Proposition XIV.2.6 now implies that A^*A is selfadjoint. $\square$

XVI.4 SYMMETRIC OPERATORS

In a number of important cases a densely defined operator on a Hilbert space H possesses the main feature of selfadjointness, namely

$$(1) \qquad \langle Ax, y \rangle = \langle x, Ay \rangle, \qquad x, y \in \mathcal{D}(A).$$

Such an operator is called *symmetric*. Note that A is symmetric if and only if $A \subset A^*$ (i.e., A is a restriction of A^*). In this case A is closable and its minimal closed linear extension $\overline{A}$ is again symmetric since $\overline{A} \subset A^* = (\overline{A})^*$, by Proposition XIV.2.4.

Let $\tau = \sum_{k=0}^{n} a_k(t) D^k$ with $a_k \in C^k([a, b])$ for $0 \leq k \leq n$ and $a_n(t) \neq 0$ for $a \leq t \leq b$. Assume $\tau = \tau^*$, where τ^* is the Lagrange adjoint of τ (see Section XIV.4). Then the minimal operator $T_{\min,\tau}$ is symmetric but not selfadjoint on $L_2([a, b])$. Indeed,

$$(2) \qquad T_{\min,\tau} \subset T_{\max,\tau} = T^*_{\min,\tau} \, ,$$

by Theorem XIV.4.1 (and $\tau = \tau^*$). However, $T_{\min,\tau} \neq T_{\max,\tau}$ because the constant function $1 \in \mathcal{D}(T_{\max,\tau})$ and $1 \notin \mathcal{D}(T_{\min,\tau})$, by the second part of Proposition XIV.3.5. The operator $T_{\max,\tau}$ is not symmetric as $T^*_{\max,\tau} = T_{\min,\tau}$ is not an extension of $T_{\max,\tau}$.

THEOREM 4.1. *Let $A(H \to H)$ be symmetric. Then the following three statements are equivalent:*

(a) *A is selfadjoint,*

(b) $\mathrm{Im}(A \pm i) = H$,

(c) *A is closed and* $\mathrm{Ker}(A^* \pm i) = \{0\}$.

PROOF. (a) $\Rightarrow$ (b). This is a trivial consequence of Theorem 3.1.

(b) $\Rightarrow$ (c). By Proposition XIV.2.5,

$$(3) \qquad \mathrm{Ker}(A^* \pm i) = \mathrm{Im}(A \mp i)^{\perp} = \{0\}.$$

Since A is symmetric, the operator $A^* + i$ is an extension of $A + i$. But $A + i$ is surjective. So $\mathrm{Ker}(A^* + i) = \{0\}$ implies $A^* = A$. In particular, A is closed.

(c) $\Rightarrow$ (a). Since A is symmetric, equality (2) in Section XVI.3 holds. In particular, $\|(A + i)x\| \geq \|x\|$ for all $x \in \mathcal{D}(A)$. This, together with the fact that A is closed, implies that $\mathrm{Im}(A + i)$ is closed. Hence

$$\mathrm{Im}(A + i) = \mathrm{Ker}(A^* - i)^{\perp} = H.$$

Since $A^* + i$ is injective and an extension of the surjective operator $A + i$, it follows that $A = A^*$. $\square$

We note that Theorem 4.1 remains valid if i and $-i$ are replaced by λ and $\overline{\lambda}$, resp., for some nonreal λ.

PROPOSITION 4.2. *A symmetric operator $A(H \to H)$ which is surjective is selfadjoint. In particular, if the resolvent set $\rho(A)$ of a symmetric operator $A(H \to H)$ contains a real number, then A is selfadjoint.*

PROOF. Take $y \in \mathcal{D}(A^*)$. Since A is surjective, there exists $z \in \mathcal{D}(A)$ such that $Az = A^*y$. Hence for any $x \in \mathcal{D}(A)$,

$$\langle Ax, y \rangle = \langle x, A^*y \rangle = \langle x, Az \rangle = \langle Ax, z \rangle.$$

It follows that $y = z$ because $\operatorname{Im} A = H$. Thus $\mathcal{D}(A^*) \subset \mathcal{D}(A)$. But $A \subset A^*$. Therefore $A = A^*$.

For the second part, it suffices to note that given $r \in \mathbf{R}$ the operator $r - A$ is symmetric (selfadjoint) if and only if A is symmetric (selfadjoint). $\square$

In the theory of symmetric operators one of the important problems is to extend a symmetric operator to a selfadjoint one. We shall not deal with this problem, but we shall make a few remarks to illustrate some of the results; for the complete theory the reader is referred to the books Riesz-Sz.-Nagy [1] and Dunford-Schwartz [1]. One of the main tools to treat the above extension problem is a generalization of the Cayley transform, which is used to reduce the problem to the problem of extending a partial isometry to a unitary operator. This reduction is effective and leads to a full solution (see Riesz-Nagy [1]).

It turns out that the selfadjoint extension problem for symmetric operators is not always solvable. In fact, the general theory shows (see the books referred to above) that a symmetric operator $T(H \to H)$ has a selfadjoint extension if and only if

$$(4) \qquad \dim \operatorname{Ker}(T^* - i) = \dim \operatorname{Ker}(T^* + i).$$

Let us apply this result to the minimal operator $T_{\min,\tau,J}$ with $\tau = iD$. Since $\tau = \tau^*$, Theorem XIV.4.1 implies that $T_{\min,\tau,J}$ is a symmetric operator on $L_2(J)$. Next, we specify the interval J. First, take $J = [0, \infty)$. The operator $T_{\min,\tau,[0,\infty)}$ does not have a selfadjoint extension. Indeed

$$\operatorname{Ker}(T^*_{\min,\tau,[0,\infty)} - i) = \{0\},$$
$$\operatorname{Ker}(T^*_{\min,\tau,[0,\infty)} + i) = \{\alpha e^{-t} \mid \alpha \in \mathbf{C}\},$$

and hence (4) is violated. On the other hand, $T_{\min,\tau,J}$ has a selfadjoint extension for any compact interval J. To see this, take $J = [a, b]$, and let $T(L_2([a, b]) \to L_2([a, b]))$ be the restriction of $T_{\max,\tau}$ to those $g \in \mathcal{D}(T_{\max,\tau})$ such that $g(a) + g(b) = 0$. Then Theorem 1.1 implies that T is a selfadjoint extension of $T_{\min,\tau,[a,b]}$.

XVI.5 UNBOUNDED SELFADJOINT OPERATORS WITH A COMPACT INVERSE

The Dirichlet operator A in Theorem 2.1 has the property that for all sufficiently large λ in $\mathbf{R}$ the operator $(\lambda + A)^{-1}$ is a compact selfadjoint operator. This is an immediate consequence of Theorem 2.1, Theorem XIV.6.1 and formula (3) in Section XVI.3. Also, if T is the differential operator defined in Theorem 1.1, then T has a compact selfadjoint inverse provided conditions (i)–(iii) in Theorem 1.1 and condition (2) in Theorem XIV.3.1 are fulfilled.

In this section we employ the spectral theory of compact selfadjoint operators (see, e.g., [GG], Section III.5) to derive the spectral properties of a selfadjoint operator with a compact inverse. For an unbounded operator $A(H \to H)$ with domain $\mathcal{D}(A)$ a vector x is called an *eigenvector* of A with *eigenvalue* λ if $0 \neq x \in \mathcal{D}(A)$ and $Ax = \lambda x$.

THEOREM 5.1. *Suppose $A(H \to H)$ is selfadjoint and has a compact inverse. Then*

(a) *there exists an orthonormal basis $\{\varphi_1, \varphi_2, \ldots\}$ for H consisting of eigenvectors of A. If $\mu_1, \mu_2, \cdots$ are the corresponding eigenvalues, then each μ_j is real and $|\mu_j| \to \infty$, provided $\dim H = \infty$. The number of repetitions of μ_j in the sequence $\mu_1, \mu_2, \cdots$ is precisely $\dim \operatorname{Ker}(\mu_j - A)$,*

(b) $\mathcal{D}(A) = \{v \in H \mid \sum_j |\mu_j|^2 |\langle v, \varphi_j \rangle|^2 < \infty\}$,

(c) $Av = \sum_j \mu_j \langle v, \varphi_j \rangle \varphi_j, \ v \in \mathcal{D}(A)$.

PROOF. Put $T = A^{-1}$. Formula (3) in Section XVI.3 implies that T is selfadjoint. Thus T is a compact selfadjoint operator with $\operatorname{Ker} T = \{0\}$. Hence there exists an orthonormal basis $\{\varphi_1, \varphi_2, \ldots\}$ of H consisting of eigenvectors of T with corresponding real (nonzero) eigenvalues $\lambda_1, \lambda_2, \cdots$ such that $\lambda_j \to 0$ if $\dim H = \infty$. Also the number of repetitions of each λ_j in the sequence $\lambda_1, \lambda_2, \ldots$ is precisely equal to $\dim \operatorname{Ker}(\lambda_j - T)$ (see [GG], Section III.6). The above results, together with the observation that $T\varphi_j = \lambda_j \varphi_j$ if and only if $A\varphi_j = \lambda_j^{-1} \varphi_j$, proves (a) with $\mu_j = \lambda_j^{-1}$.

Next, assume that $v \in H$ is such that

$$(1) \qquad \sum_j |\mu_j|^2 |\langle v, \varphi_j \rangle|^2 < \infty.$$

Then

$$v = \sum_j \langle v, \varphi_j \rangle \varphi_j = A^{-1} \left(\sum_j \mu_j \langle v, \varphi_j \rangle \varphi_j \right) \in \mathcal{D}(A).$$

Conversely, if $v \in \mathcal{D}(A)$, then

$$Av = \sum_j \langle Av, \varphi_j \rangle \varphi_j = \sum_j \mu_j \langle Av, T\varphi_j \rangle \varphi_j$$

$$= \sum_j \mu_j \langle TAv, \varphi_j \rangle \varphi_j = \sum_j \mu_j \langle v, \varphi_j \rangle \varphi_j.$$

The latter identity proves (c). It also shows that v satisfies (1), and hence (b) is proved. $\square$

The operator A in Theorem 5.1 has a compact resolvent, and so Theorem XV.2.3 applies to A. Hence, if A is selfadjoint and has a compact inverse, then the spectrum $\sigma(A)$ of A is a finite or countable set consisting of eigenvalues of finite type only. From Theorem 5.1(a) and (c) it follows that

$$\sigma(A) = \{\mu_j \mid j = 1, 2, \ldots\}.$$

XVI.6 THE SPECTRAL THEOREM FOR UNBOUNDED SELF-ADJOINT OPERATORS

In this section the spectral theorem for bounded selfadjoint operators is extended to unbounded ones. We begin with the definition of a resolution of the identity. A family $\{E(t)\}_{t\in\mathbf{R}}$ of orthogonal projections on the Hilbert space H is called a *resolution of the identity* on H if

(P_1) $\operatorname{Im} E(s) \subset \operatorname{Im} E(t)$ whenever $s \leq t$,

(P_2) $\operatorname{Im} E(s) = \cap\{\operatorname{Im} E(t) \mid t > s\}$,

(P_3) $\cap\{\operatorname{Im} E(t) \mid t \in \mathbf{R}\} = \{0\}$,

(P_4) $\operatorname{span}\{\operatorname{Im} E(t) \mid t \in \mathbf{R}\}$ is dense in H.

Note that this definition is an extension of the definition of a bounded resolution of the identity given in Section V.3. The conditions (P_1) and (P_2) are the same as the conditions (C_1) and (C_2) in Section V.3, but (P_3) and (P_4) are weaker than the corresponding conditions (C_3) and (C_4). In fact, in this section the resolutions of the identity are not required to be supported by a bounded interval (as is the case in Section V.3).

THEOREM 6.1. *Let $A(H \to H)$ be a selfadjoint operator. Then there exists a unique resolution of the identity $\{E(t)\}_{t\in\mathbf{R}}$ such that*

(a) $\mathcal{D}(A) = \{x \in H \mid \int_{\mathbf{R}} \lambda^2 d\langle E(\lambda)x, x\rangle < \infty\}$,

(b) $Ax = \lim_{N\to\infty}\big(\int_{-N}^{N} \lambda dE(\lambda)\big)x, \qquad x \in \mathcal{D}(A)$.

Furthermore, if S is a bounded linear operator on H commuting with A, i.e.,

$$(1) \qquad\qquad S\mathcal{D}(A) \subset \mathcal{D}(A), \quad SAx = ASx \qquad (x \in \mathcal{D}(A)),$$

then S commutes with $E(t)$ for each $t \in \mathbf{R}$.

Let us explain the meaning of the integrals appearing in Theorem 6.1. Let $\{E(t)\}_{t\in\mathbf{R}}$ be a resolution of the identity. Take a bounded interval $[\alpha, \beta]$, and let $f\colon[\alpha, \beta] \to \mathbf{C}$ be a continuous function. Then

$$(2) \qquad \int_{\alpha}^{\beta} f(\lambda)dE(\lambda) := \lim_{\nu(P)\to 0} S_\tau(f, P).$$

Here P is a partition, $\alpha = \lambda_0 < \lambda_1 < \cdots < \lambda_n = \beta$, of the interval $[\alpha, \beta]$, the number $\nu(P)$ is the maximal length of the subintervals $[\lambda_{j-1}, \lambda_j]$, the symbol τ stands for a set, $\tau = \{t_1, \ldots, t_n\}$, of points $t_j \in [\lambda_{j-1}, \lambda_j]$, and

$$S_\tau(f, P) = \sum_{j=1}^{n} f(t_j)\big(E(\lambda_j) - E(\lambda_{j-1})\big).$$

Note that $S_\tau(f, P)$ is a bounded linear operator on H. The usual argument from the theory of Riemann-Stieltjes integration shows that the limit in the right hand side of (2) exists in the norm of $\mathcal{L}(H)$ (cf., the text after formula (4) in Section V.3), and hence the left hand side of (2) is a well-defined bounded linear operator on H. Now take $\alpha = N$, $\beta = -N$ and $f(\lambda) = \lambda$, and one sees that the second integral in Theorem 6.1 is a well-defined operator on H.

The first integral in Theorem 6.1 has to be understood as an improper Stieltjes integral. Note that the integrator $\langle E(\cdot)x, x \rangle$ is a non-negative monotonically increasing function and the integrand λ^2 is non-negative on **R**. Thus

$$(3) \qquad \int_{-\infty}^{\infty} \lambda^2 d\langle E(\lambda)x, x \rangle$$

is a non-negative number if the integral converges and is $+\infty$ otherwise. Condition (a) in Theorem 6.1 states that $x \in \mathcal{D}(A)$ if and only if the integral (3) is convergent.

There are various ways to prove Theorem 6.1. A direct way is to give an explicit formula for the projection $E(t)$ in terms of contour integrals of the type appearing in Section V.2. This means to prove the analogues of Theorems V.2.2 and V.3.2 for unbounded selfadjoint operators. Here we shall follow a less direct approach and prove Theorem 6.1 by reduction to the bounded case. For this purpose we need two lemmas.

The first lemma shows that (a) and (b) in Theorem 6.1 define a selfadjoint operator.

LEMMA 6.2. *Let* $\{E(t)\}_{t \in \mathbf{R}}$ *be a resolution of the identity on the Hilbert space* H, *and let* $T(H \to H)$ *be defined by*

$$\mathcal{D}(T) = \left\{ x \in H \mid \int_{-\infty}^{\infty} \lambda^2 d\langle E(\lambda)x, x \rangle < \infty \right\},$$

$$Tx = \lim_{N \to \infty} \left(\int_{-N}^{N} \lambda\, dE(\lambda) \right) x.$$

Then T is a selfadjoint operator.

PROOF. The proof is divided into four parts.

Part (i). First we show that for each $x \in H$

$$(4) \qquad \lim_{t \downarrow -\infty} E(t)x = 0, \qquad \lim_{t \uparrow \infty} E(t)x = x.$$

From the property (P_1) of a resolution of the identity it follows that $\|E(\cdot)x\|^2$ is a monotonically increasing function. This function is also bounded, because $\|E(t)\| \leq 1$ for each t. Hence we may conclude that the limits

$$(5) \qquad \lim_{t \downarrow -\infty} \|E(t)x\|^2, \qquad \lim_{t \uparrow \infty} \|E(t)x\|^2$$

exist. Property (P_1) also shows that the vectors $E(t)x - E(s)x$ and $E(s)x$ are orthogonal whenever $t \geq s$, and thus, by the Pythagorean theorem,

$$\|E(t)x - E(s)x\|^2 = \|E(t)x\|^2 - \|E(s)x\|^2, \qquad t \geq s.$$

Since H is a Hilbert space, the last identity and (5) imply that the limits

$$x_0 = \lim_{t\downarrow-\infty} E(t)x, \qquad x_1 = \lim_{t\uparrow\infty} E(t)x$$

exist in the norm of H. Obviously, $x_0 \in \operatorname{Im} E(t)$ for each t, and hence $x_0 = 0$ by property (P_3). Furthermore, the vector $x - x_1$ is orthogonal to $\operatorname{Im} E(t)$ for all t, and hence $x = x_1$ by property (P_4).

Part (ii). In this part we show that

$$(6) \qquad \mathcal{D}(T) = \left\{ x \in H \mid \lim_{N\to\infty} \left(\int_{-N}^{N} \lambda dE(\lambda) \right) x \text{ exists in } H \right\}.$$

Given $x \in H$, put $y_N = (\int_{-N}^{N} \lambda dE(\lambda))x$. From the definition of the integral it follows that

$$(7) \qquad y_N = \lim_{\nu(P)\to 0} \sum_{j=1}^{n} t_j \big(E(\lambda_j)x - E(\lambda_{j-1})x \big),$$

where P is the partition $-N = \lambda_0 < \lambda_1 < \cdots < \lambda_n = N$ with width $\nu(P)$ and $\lambda_{j-1} \leq t_j \leq \lambda_j$ for $j = 1,\ldots,n$. Since the vectors are mutually orthogonal, we see that

$$\begin{aligned}
\|y_N\|^2 &= \lim_{\nu(P)\to 0} \sum_{j=1}^{n} t_j^2 \|E(\lambda_j)x - E(\lambda_{j-1})x\|^2 \\
&= \lim_{\nu(P)\to 0} \sum_{j=1}^{n} t_j^2 \{\langle E(t_j)x, x\rangle - \langle E(t_{j-1})x, x\rangle\} \\
&= \int_{-N}^{N} \lambda^2 d\langle E(\lambda)x, x\rangle.
\end{aligned}$$

Now, assume that x belongs to the right hand side of (6). Then the sequence $(\|y_N\|)_{N=1}^{\infty}$ converges and hence

$$\int_{-\infty}^{\infty} \lambda^2 d\langle E(\lambda)x, x\rangle = \lim_{N\to\infty} \|y_N\|^2 < \infty.$$

Thus $x \in \mathcal{D}(T)$. On the other hand if $x \in \mathcal{D}(T)$, then for $N > M \geq 1$

$$y_N - y_M = \left(\int_{-N}^{N} \lambda dE(\lambda) \right) x - \left(\int_{-M}^{M} \lambda dE(\lambda) \right) x \to 0$$

as $N, M \to \infty$, because

$$\left\| \left(\int_c^d \lambda dE(\lambda) \right) x \right\|^2 = \int_c^d \lambda^2 d\langle E(\lambda)x, x \rangle \to 0$$

as c, d both go to ∞ or both to $-\infty$. Hence the sequence (y_N) converges in H which shows that x belongs to the right hand side of (6).

Part (iii). From (6) we conclude that T is a well-defined linear operator. Let us show that its domain is dense in H. For $R > 0$ we write Q_R for the orthogonal projection $E(R) - E(-R)$. Take $x \in X$. The identities in (4) imply that $\|x - Q_R x\| \to 0$ if $R \to \infty$. Thus to prove that T is densely defined, it suffices to show that $Q_R x \in \mathcal{D}(T)$ for $R > 0$. Fix $R > 0$. Then

$$(8) \qquad \left(\int_{-N}^N \lambda dE(\lambda) \right) Q_R x = \left(\int_{-R}^R \lambda dE(\lambda) \right) x, \qquad N > R.$$

Thus for N sufficiently large the left hand side of (8) is independent of N. According to (6), this implies that $Q_R x \in \mathcal{D}(T)$, and hence $\mathcal{D}(T)$ is dense in H.

Part (iv). We prove that T is selfadjoint. Take x, y in X. A consideration of Riemann-Stieltjes sums leads to

$$(9) \qquad \int_{-N}^N \lambda d\langle E(\lambda)x, y \rangle = \langle \left(\int_{-N}^N \lambda dE(\lambda) \right) x, y \rangle$$

for each $N > 0$. In particular, the integral in the left hand side of (9) exists. Now, take $x, y \in \mathcal{D}(T)$. Then

$$\langle Tx, y \rangle = \lim_{N \to \infty} \langle \left(\int_{-N}^N \lambda dE(\lambda) \right) x, y \rangle$$

$$= \lim_{N \to \infty} \int_{-N}^N \lambda d\langle E(\lambda)x, y \rangle$$

$$= \lim_{N \to \infty} \int_{-N}^N \lambda d\langle x, E(\lambda)y \rangle$$

$$= \lim_{N \to \infty} \langle x, \left(\int_{-N}^N \lambda dE(\lambda) \right) y \rangle = \langle x, Ty \rangle.$$

Thus T is symmetric. Finally, take $z \in \mathcal{D}(T^*)$. To finish the proof we have to show that $z \in \mathcal{D}(T)$. As in Part (iii) of the proof, let Q_R denote the projection $E(R) - E(-R)$.

Take an arbitrary $x \in X$. Since $Q_R x \in \mathcal{D}(T)$, we have

$$\langle x, Q_R T^* z \rangle = \langle T Q_R x, z \rangle$$
$$= \langle \left(\int_{-R}^{R} \lambda dE(\lambda) \right) x, z \rangle$$
$$= \langle x, \left(\int_{-R}^{R} \lambda dE(\lambda) \right) z \rangle.$$

Here we used (8) to get the second equality; the third equality follows by applying twice (9). We know that $Q_R T^* z \to T^* z \ (R \to \infty)$ in the norm of H. Thus

$$\lim_{R \to \infty} \langle x, \left(\int_{-R}^{R} \lambda dE(\lambda) \right) z \rangle = \langle x, T^* z \rangle.$$

This equality holds for each $x \in X$. Now, take for x the vector

$$x_N = \left(\int_{-N}^{N} \lambda dE(\lambda) \right) z.$$

Note that

$$\langle x_N, \left(\int_{-R}^{R} \lambda dE(\lambda) \right) z \rangle = \langle x_N, x_N \rangle, \qquad R > N.$$

It follows that

$$\|x_N\|^2 = \lim_{R \to \infty} \langle x_N, \left(\int_{-R}^{R} \lambda dE(\lambda) \right) z \rangle$$
$$= \langle x_N, T^* z \rangle.$$

But then we may conclude that $\|x_N\| \leq \|T^* z\|$, and we see that

$$(10) \qquad \int_{-N}^{N} \lambda^2 d\langle E(\lambda) z, z \rangle = \|x_N\|^2 \leq \|T^* z\|.$$

This holds for each $N > 0$. Since the third term in (10) is a fixed finite number, independent of N, we have proved that $z \in \mathcal{D}(T)$. $\square$

For the second lemma we need some additional notation. Let $H_1, H_2, \ldots$ be a sequence of Hilbert spaces. By $\bigoplus_{j=1}^{\infty} H_j$ we denote the space consisting of all sequences

$x = (x_1, x_2, \dots)$ such that $x_j \in H_j$ $(j = 1, 2, \dots)$ and $\sum_{j=1}^{\infty} \|x_j\|^2 < \infty$. The space $\bigoplus_{j=1}^{\infty} H_j$ is a Hilbert space with the inner product defined by

$$\langle x, y \rangle = \sum_{j=1}^{\infty} \langle x_j, y_j \rangle.$$

Here $\langle x_j, y_j \rangle$ denotes the inner product of the vectors x_j and y_j in the space H_j.

Now, for $j = 1, 2, \dots$ let A_j be a bounded linear operator on H_j. With the sequence $A_1, A_2, \dots$ we associate an operator A $\left(\bigoplus_{j=1}^{\infty} H_j \to \bigoplus_{j=1}^{\infty} H_j \right)$ by setting

$$\mathcal{D}(A) = \left\{ x = (x_1, x_2, \dots) \in \bigoplus_{j=1}^{\infty} H_j \mid \sum_{j=1}^{\infty} \|A_j x_j\|^2 < \infty \right\},$$
$$A x = (A_1 x_1, A_2 x_2, \dots), \qquad x \in \mathcal{D}(A).$$

It is a simple matter to check that A is a well-defined linear operator. Since all sequences $(x_1, x_2, \dots)$ with a finite number of non-zero entries are in $\mathcal{D}(A)$, the operator A is densely defined. The operator A is also closed. Indeed, if $x(1), x(2), \dots$ is a sequence in $\mathcal{D}(A)$, $x(n) = (x_1(n), x_2(n), \dots)$, such that

$$x(n) \to x, \quad A x(n) \to y \qquad (n \to \infty),$$

then for $j = 1, 2, \dots$

$$A_j x_j = \lim_{n \to \infty} A_j x_j(n) = \lim_{n \to \infty} \big(A x(n) \big)_j = y_j,$$

which implies that $x \in \mathcal{D}(A)$ and $A x = y$. We shall denote the operator A by $\operatorname{diag}(A_j)_{j=1}^{\infty}$. If the sequence $A_1, A_2, \dots$ is uniformly bounded, i.e.,

$$(11) \qquad\qquad\qquad \sup\{\|A_j\| \mid j = 1, 2, \dots\} < \infty,$$

then $\operatorname{diag}(A_j)_{j=1}^{\infty}$ is a bounded operator defined on the whole space. (The converse of the latter statement is also true.)

LEMMA 6.3. *For $j = 1, 2, \dots$ let $A_j \colon H_j \to H_j$ be a bounded selfadjoint operator acting on the Hilbert space H_j. Let $\{E_j(t)\}_{t \in \mathbf{R}}$ be the resolution of the identity for A_j $(j = 1, 2, \dots)$. Put*

$$A = \operatorname{diag}(A_j)_{j=1}^{\infty}, \qquad E(t) = \operatorname{diag}\big(E_j(t) \big)_{j=1}^{\infty}.$$

Then A is a selfadjoint operator, $\{E(t)\}_{t \in \mathbf{R}}$ is a resolution of the identity and

$$(12) \qquad\qquad \mathcal{D}(A) = \left\{ x \in \bigoplus_{j=1}^{\infty} H_j \mid \int_{-\infty}^{\infty} \lambda^2 d\langle E(\lambda) x, x \rangle < \infty \right\},$$

$$(13) \qquad Ax = \lim_{N \to \infty} \left(\int_{-N}^{N} \lambda \, dE(\lambda) \right) x, \qquad x \in \mathcal{D}(A).$$

PROOF. The proof is divided into three parts. First we show that the operator A is selfadjoint.

Part (i). Take x, y in $\mathcal{D}(A)$. Then

$$\langle Ax, y \rangle = \sum_{j=1}^{\infty} \langle A_j x_j, y_j \rangle$$
$$= \sum_{j=1}^{\infty} \langle x_j, A_j y_j \rangle = \langle x, Ay \rangle.$$

The second equality holds true because A_j is selfadjoint for each j. The above calculation shows that $A \subset A^*$. To prove the reverse inclusion, take $z \in \mathcal{D}(A^*)$. Fix a positive integer k, and let x be an arbitrary element of H_k. Consider the sequence $\widehat{x} = (\delta_{jk} x)_{j=1}^{\infty}$. Obviously, $\widehat{x} \in \mathcal{D}(A)$. Write $[A^* z]_k$ for the k-th entry of the sequence $A^* z$. Then

$$\langle x, A_k z_k \rangle = \langle A_k x, z_k \rangle = \langle A\widehat{x}, z \rangle = \langle \widehat{x}, A^* z \rangle = \langle x, [A^* z]_k \rangle.$$

This holds for each $x \in H_k$, and thus $A_k z_k = [A^* z]_k$. But then

$$\sum_{k=1}^{\infty} \|A_k z_k\|^2 = \sum_{k=1}^{\infty} \|[A^* z]_k\|^2 = \|A^* z\|^2 < \infty.$$

Hence $z \in \mathcal{D}(A)$, and it follows that $A = A^*$.

Part (ii). In this part we prove that $\{E(t)\}_{t \in \mathbf{R}}$ is a resolution of the identity. We know that $E_j(t)$ is an orthogonal projection for each j. It follows that

$$\sup\{\|E_j(t)\| \mid j = 1, 2, \ldots\} \leq 1.$$

Hence (cf., the remark made at the end of the paragraph preceding the present lemma) the operator $E(t)$ is a bounded operator on the whole space $\bigoplus_{j=1}^{\infty} H_j$. Since each $E_j(t)$ is selfadjoint, also $E(t)$ is selfadjoint (by the result proved under Part (i)). From

$$E(t)^2 = \mathrm{diag}\big(E_j(t)^2\big)_{j=1}^{\infty} = \mathrm{diag}\big(E_j(t)\big)_{j=1}^{\infty} = E(t),$$

we see that $E(t)$ is a projection. So $\{E(t)\}_{t \in \mathbf{R}}$ is a family of orthogonal projections. Note that

$$(14) \qquad \mathrm{Im}\, E(t) = \{x = (x_1, x_2, \ldots) \mid x_j \in \mathrm{Im}\, E_j(t), \ j = 1, 2, \ldots\}.$$

As conditions (C_1) and (C_2) in Section V.3 are fulfilled for each of the resolutions $\{E_j(t)\}_{t \in \mathbf{R}}$, the identity (14) implies that (P_1) and (P_2) hold for the family $\{E(t)\}_{t \in \mathbf{R}}$. Next, assume $x \in \mathrm{Im}\, E(t)$ for all $t \in \mathbf{R}$. To prove that condition (P_3) holds we have to show that $x = 0$. Consider the k-th entry x_k of x. According to (14), the vector $x_k \in \operatorname{Im} E_k(t)$ for all $t \in \mathbf{R}$. But the resolution $\{E_k(t)\}_{t \in \mathbf{R}}$ is supported by a compact interval. Hence $E_k(t) = 0$ for t sufficiently small, which shows that $x_k = 0$. This holds for each k. Hence $x = 0$ and (P_3) is established. Finally, let $y \perp \operatorname{Im} E(t)$ for all $t \in \mathbf{R}$. To prove (P_4), it suffices to show that $y = 0$. Fix a positive integer k, and let t be an arbitrary real number. Take $x \in \operatorname{Im} E_k(t)$. Then $\widehat{x} := (\delta_{jk} x)_{j=1}^{\infty}$ belongs to $\operatorname{Im} E(t)$ by (14), and thus

$$\langle y_k, x \rangle = \langle y, \widehat{x} \rangle = 0.$$

It follows that $y_k \perp \operatorname{Im} E_k(t)$ for all $t \in \mathbf{R}$. Now use again that the resolution $\{E_k(t)\}_{t \in \mathbf{R}}$ is supported by a compact interval. Thus $E_k(t) = I$ for t sufficiently large, which implies that $y_k = 0$. But k is arbitrary. So $y = 0$, and condition (P_4) holds true.

Part (iii). Next, we prove the identities (12) and (13). Let $T(\bigoplus_{j=1}^{\infty} H_j \to \bigoplus_{j=1}^{\infty} H_j)$ be the operator defined by the right hand sides of (12) and (13). Thus the domain of T is given by the right hand side of (12) and its action by the right hand side of (13). We want to prove that $T = A$. First, we show that $T \subset A$. Take $x \in \mathcal{D}(T)$. For each k the vector x_k belongs to H_k. Since $\{E_k(t)\}_{t \in \mathbf{R}}$ is the resolution of the identity for the selfadjoint operator A_k, we have

$$(15) \qquad \|A_k x_k\|^2 = \int_{\alpha}^{\beta} \lambda^2 d\langle E_k(\lambda) x_k, x_k \rangle,$$

provided

$$\alpha < m(A_k) = \inf\{\langle A_k z, z \rangle \mid z \in H_k, \ \|z\| = 1\},$$
$$\beta > M(A_k) = \sup\{\langle A_k z, z \rangle \mid z \in H_k, \ \|z\| = 1\}.$$

Note that the left hand side of (15) is independent of α and β, and hence in (15) we may replace α by $-\infty$ and β by ∞. It follows that

$$\sum_{k=1}^{p} \|A_k x_k\|^2 = \sum_{k=1}^{p} \int_{-\infty}^{\infty} \lambda^2 d\langle E_k(\lambda) x_k, x_k \rangle$$
$$= \int_{-\infty}^{\infty} \lambda^2 d\left(\sum_{k=1}^{p} \langle E_k(\lambda) x_k, x_k \rangle \right)$$
$$\leq \int_{-\infty}^{\infty} \lambda^2 d\langle E(\lambda) x, x \rangle < \infty,$$

which holds for $p = 1, 2, \ldots$. Hence $\sum_{k=1}^{\infty} \|A_k x_k\|^2 < \infty$, and thus $x \in \mathcal{D}(A)$. We proceed by showing $Tx = Ax$. Let $[\cdot]_k$ denote the map which assigns to a vector in $\bigoplus_{j=1}^{\infty} H_j$ its k-th coordinate. Then

$$[Tx]_k = \lim_{N \to \infty} \left[\left(\int_{-N}^{N} \lambda dE(\lambda) \right) x \right]_k$$

(16)

$$= \lim_{N \to \infty} \left(\int_{-N}^{N} \lambda dE_k(\lambda) \right) x_k.$$

From the spectral theorem for bounded selfadjoint operators (Section V.4) we know that

$$\int_{-N}^{N} \lambda dE_k(\lambda) = A_k$$

for $N > M(A_k)$ and $-N < m(A_k)$. Thus the second limit in (16) is equal to $A_k x_k$. This holds for each k. Thus $Tx = Ax$. We have now proved that $T \subset A$. By Lemma 6.2 the operator T is selfadjoint. Also, A is selfadjoint. Thus $A = A^* \subset T^* = T$. Therefore, $T = A$. $\square$

PROOF OF THEOREM 6.1. Let $A(H \to H)$ be selfadjoint. Introduce the operator $B = T^*T$, where $T := (A - i)^{-1}$. Theorem 3.1 implies that T is a well-defined bounded linear operator on H and $\|T\| \leq 1$. Hence B is a well-defined, bounded non-negative selfadjoint operator and $\|B\| \leq 1$. In what follows $\{F(t)\}_{t \in \mathbf{R}}$ is the bounded resolution of the identity for B. Since $\|B\| \leq 1$, we have

$$m(B) = \inf\{\langle Bx, x \rangle \mid \|x\| = 1\} \geq 0,$$
$$M(B) = \sup\{\langle Bx, x \rangle \mid \|x\| = 1\} \leq 1,$$

and hence the resolution for B is supported by the interval $[0, 1]$. Note that the operator T is injective, and therefore B is also injective, which implies (cf. Corollary V.5.2) that $F(0)$ is the zero operator.

For $n = 1, 2, \ldots$ let H_n be the range of the orthogonal projection $F(1/n) - F(1/(n + 1))$. Obviously, each H_n is a subspace (closed linear submanifold) of H, and hence each H_n is a Hilbert space in its own right. We claim that the Hilbert spaces H and $\bigoplus_{j=1}^{\infty} H_j$ are isometrically isomorphic. Indeed, define

(17)
$$J : \bigoplus_{j=1}^{\infty} H_j \to H, \quad Jx = J(x_1, x_2, \ldots) = \sum_{j=1}^{\infty} x_j.$$

To see that the map J is well-defined, take $x = (x_1, x_2, \ldots) \in \bigoplus_{j=1}^{\infty} H_j$, and put $s_n = \sum_{j=1}^{n} x_j$ for $n \geq 1$. First, note that the spaces $H_1, H_2, \ldots$ are mutually orthogonal, because

$$\left\{ F\left(\frac{1}{j}\right) - F\left(\frac{1}{j+1}\right) \right\} \left\{ F\left(\frac{1}{k}\right) - F\left(\frac{1}{k+1}\right) \right\} = 0, \qquad j \neq k.$$

In particular, $x_j \perp x_k$ for $j \neq k$. Thus for $m > n$ we have

$$\|s_m - s_n\|^2 = \left\| \sum_{j=n+1}^{m} x_j \right\|^2 = \sum_{j=n+1}^{m} \|x_j\|^2,$$

which goes to zero as $n, m \to \infty$. Hence the sequence (s_n) is a Cauchy sequence, and therefore the series $\sum_{j=1}^{\infty} x_j$ converges in H. Furthermore,

$$\|Jx\|^2 = \left\| \sum_{j=1}^{\infty} x_j \right\|^2 = \sum_{j=1}^{\infty} \|x_j\|^2 = \|x\|,$$

and so J is a well-defined isometry. The latter fact implies that $\operatorname{Im} J$ is closed. Thus, in order to prove that $\operatorname{Im} J = H$, it suffices to show that $x = 0$ whenever $x \perp \operatorname{Im} J$. To do this, take $x \perp \operatorname{Im} J$. Then $x \perp H_n$ for each n, and hence $x \perp \operatorname{Im}\{F(1) - F(1/n)\}$ for each n. As $F(1) = I$, we conclude that $x \in \operatorname{Im} F(1/n)$ for $n = 1, 2, \ldots$. According to conditions (C_1) and (C_2) in the definition of a bounded resolution (see Section V.3), this yields $x \in \operatorname{Im} F(0)$. But $F(0)$ is the zero operator. So $x = 0$, and we have shown that the map J in (17) has the desired properties.

Next, we prove that

$$(18) \qquad\qquad H_n \subset \mathcal{D}(A), \qquad A H_n \subset H_n.$$

Note that $T^* = (A + i)^{-1}$ by Proposition XIV.2.6 and the selfadjointness of A. Thus $B = (A+i)^{-1}(A-i)^{-1}$, and hence $\operatorname{Im} B \subset \mathcal{D}(A)$. Since $F(t)$ is the orthogonal projection of H onto the spectral subspace of B associated with $(-\infty, t]$ (by Theorem V.3.2), the space $\operatorname{Im} F(t)$ is invariant under B. This implies that for each n the space H_n is invariant under B and, according to Theorem V.5.1,

$$\sigma(B|H_n) \subset \sigma\left(B\Big|\operatorname{Im}\left\{F\left(\frac{1}{n}\right) - F\left(\frac{1}{n+1} - 0\right)\right\}\right) \subset \left[\frac{1}{n+1}, \frac{1}{n}\right].$$

In particular, $B|H_n$ is invertible, and hence $H_n \subset \operatorname{Im} B \subset \mathcal{D}(A)$ for each n.

To prove the second inclusion in (18), observe that T commutes with B. Hence T commutes with each $F(t)$ by the second part of Theorem V.3.2, and therefore $T H_n \subset H_n$ for each n. Fix n, and take $x \in H_n$. Then $x = Bz$ for some $z \in H_n$ by the result of the previous paragraph, and thus

$$Ax = ABz = A(A + i)^{-1}(A - i)^{-1}z = Tz - iBz \in H_n,$$

which proves $A H_n \subset H_n$.

For each $j \geq 1$ let $A_j \colon H_j \to H_j$ be the restriction of A to H_j. Since A is closed, the same is true for A_j, and hence by the closed graph theorem A_j is a bounded linear operator on H_j for each j. Also, A_j is selfadjoint for each j. The next step of the proof is to apply Lemma 6.3. First, note that

$$(19) \qquad\qquad A = J\{\operatorname{diag}(A_j)_{j=1}^{\infty}\}J^{-1},$$

where J is the invertible isometry defined by (17). To prove (19), put $C = J^{-1}AJ$, and write D for $\operatorname{diag}(A_j)_{j=1}^{\infty}$. Both C and D are selfadjoint operators acting in $\bigoplus_{j=1}^{\infty} H_j$. We want to show that $C = D$. Take $x = (x_1, x_2, \ldots)$ in $\mathcal{D}(D)$, and put $x(n) = (x_1, \ldots, x_n, 0, 0, \ldots)$ for $n = 1, 2, \ldots$. Since $H_j \subset \mathcal{D}(A)$ for $j \geq 1$, we have $x_1 + \cdots + x_n \in \mathcal{D}(A)$, and hence $x(n) \in \mathcal{D}(C)$. Moreover,

$$C x(n) = (A_1 x_1, \ldots, A_n x_n, 0, 0, \ldots) = D x(n).$$

This holds for each $n \geq 1$. Also

$$x(n) \to x, \quad C x(n) = D x(n) \to Dx \qquad (n \to \infty).$$

It follows that $x \in \mathcal{D}(C)$ and $Cx = Dx$. Thus C is an extension of D, but then $C = D$, because both C and D are selfadjoint.

Formula (19) suggests how to construct the desired resolution of the identity for A. Indeed, put

(20) $$E(t) = J\bigl\{\operatorname{diag}\bigl(E_j(t)\bigr)_{j=1}^{\infty}\bigr\} J^{-1}, \qquad t \in \mathbf{R},$$

where $\{E_j(t)\}_{t\in\mathbf{R}}$ is the bounded resolution of the identity for the bounded selfadjoint operator A_j $(j \geq 1)$. Since J is an invertible isometry, Lemma 6.3 and formula (19) imply that $\{E(t)\}_{t\in\mathbf{R}}$ is a resolution of the identity such that (a) and (b) in Theorem 6.1 hold true.

We have now completed the proof of the first part of Theorem 6.1. It remains to prove the uniqueness and the statement about commutativity. We proceed with the latter. Let $S: H \to H$ be a bounded linear operator commuting with A (i.e., formula (1) holds). We want to show that S commutes with each $E(t)$. To do this note that (1) implies that S commutes with the bounded operators $(A - i)^{-1}$ and $(A + i)^{-1}$. It follows that S commutes with B. But then we can use the last part of Theorem V.3.2 to show that S commutes with the projections $F(1/n) - F(1/(n+1))$, $n = 1, 2, \ldots$. In particular, the spaces $H_1, H_2, \ldots$ are invariant under S. Put

$$S_j = S|H_j : H_j \to H_j, \qquad j = 1, 2, \ldots .$$

Since S commutes with A, it follows that S_j commutes with A_j for each $j \geq 1$. Now, apply Theorem V.3.2 again. We obtain that for each $t \in \mathbf{R}$ the operator S_j commutes with $E_j(t)$. Take $x \in H$. We may write $x = \sum_{j=1}^{\infty} x_j$ with $x_j \in H_j$ for $j \geq 1$. Then

$$\begin{aligned}
SE(t)x &= \sum_{j=1}^{\infty} S_j E_j(t) x_j \\
&= \sum_{j=1}^{\infty} E_j(t) S_j x_j = E(t) Sx, \qquad t \in \mathbf{R},
\end{aligned}$$

and so S commutes with $E(t)$ for each $t \in \mathbf{R}$.

Finally, we prove uniqueness. Let $\{G(t)\}_{t\in\mathbf{R}}$ be a second resolution of the identity such that (a) and (b) in Theorem 6.1 hold with G in place of E. We want to show that $G(\cdot) = E(\cdot)$. Fix $t \in \mathbf{R}$. First we shall show that $G(t)$ commutes with A. Take $x \in \mathcal{D}(A)$. Then

$$\langle G(\lambda)G(t)x, G(t)x \rangle = \left\{ \begin{array}{ll} \langle G(\lambda)x, x \rangle, & \lambda < t, \\ \langle G(t)x, x \rangle, & \lambda \geq t, \end{array} \right.$$

and hence

$$\langle G(\lambda)G(t)x, G(t)x \rangle \leq \langle G(\lambda)x, x \rangle, \qquad \lambda \in \mathbf{R}.$$

It follows that

$$\int_{-\infty}^{\infty} \lambda^2 d\langle G(\lambda)G(t)x, G(t)x \rangle \leq \int_{-\infty}^{\infty} \lambda^2 d\langle G(\lambda)x, x \rangle < \infty.$$

Thus by condition (a) in Theorem 6.1 (with E replaced by G) we obtain $G(t)x \in \mathcal{D}(A)$. Furthermore, since (b) in Theorem 6.1 holds for G in place of E,

$$\begin{aligned} AG(t)x &= \lim_{N\to\infty} \left(\int_{-N}^{N} \lambda dG(\lambda) \right) G(t)x \\ &= \lim_{N\to\infty} G(t)\left(\int_{-N}^{N} \lambda dG(\lambda) \right) x \\ &= G(t)Ax. \end{aligned}$$

Here we used that $G(t)$ commutes with $G(\lambda)$ for each $\lambda \in \mathbf{R}$.

Since $G(t)$ commutes with A, it follows that $G(t)$ commutes with the bounded selfadjoint operator B. As before, by the last part of Theorem V.3.2, this implies that the spaces $H_1, H_2, \ldots$ are invariant under $G(t)$. Put

$$G_j(t) = G(t)|H_j \colon H_j \to H_j, \qquad j = 1, 2, \ldots .$$

By condition (a) in Theorem 6.1 (with G in place of E)

$$\begin{aligned} \|Ax\|^2 &= \lim_{N\to\infty} \left\| \left(\int_{-N}^{n} \lambda dG(\lambda) \right) x \right\|^2 \\ &= \lim_{N\to\infty} \int_{-N}^{N} \lambda^2 d\langle G(\lambda)x, x \rangle, \qquad x \in \mathcal{D}(A). \end{aligned}$$

Thus

$$(21) \qquad \int_{-\infty}^{\infty} \lambda^2 d\langle G_j(\lambda)x, x \rangle \leq \|A_j\|^2 \|x\|^2, \qquad x \in H_j,$$

which implies that $G_j(\lambda) = 0$ on the interval $-\infty < \lambda < -\|A_j\|$ and $G_j(\lambda)$ is the identity operator on H_j for $\|A_j\| < \lambda < \infty$. Indeed, if $x_0 \in \operatorname{Im} G_j(\lambda_0)$, where $\lambda_0 < -\|A_j\|$, then $\langle G_j(\lambda)x_0, x_0 \rangle = \|x_0\|^2$ for $\lambda \geq \lambda_0$, and hence

$$
\begin{aligned}
\|A_j\|^2\|x_0\|^2 &\geq \int_{-\infty}^{\infty} \lambda^2 d\langle G_j(\lambda)x_0, x_0 \rangle \\
&= \int_{-\infty}^{\lambda_0} \lambda^2 d\langle G_j(\lambda)x_0, x_0 \rangle \\
&\geq \lambda_0^2 \int_{-\infty}^{\lambda_0} d\langle G_j(\lambda)x_0, x_0 \rangle = \lambda_0^2\|x_0\|^2.
\end{aligned}
$$

Since $\lambda_0^2 > \|A_j\|^2$, this yields $x_0 = 0$. Similarly, if $\lambda_0 > \|A_j\|$ and $x_0 \perp \operatorname{Im} G_j(\lambda_0)$, then $\langle G(\lambda)x_0, x_0 \rangle = 0$ for $\lambda \leq \lambda_0$,

$$
\begin{aligned}
\|A_j\|^2\|x_0\|^2 &\geq \int_{-\infty}^{\infty} \lambda^2 d\langle G_j(\lambda)x_0, x_0 \rangle \\
&= \int_{\lambda_0}^{\infty} \lambda^2 d\langle G_j(\lambda)x_0, x_0 \rangle \geq \lambda_0^2\|x_0\|^2,
\end{aligned}
$$

and thus $x_0 = 0$. It follows that $\{G_j(t)\}_{t \in \mathbf{R}}$ is a bounded resolution of the identity supported by the compact interval $[-\|A_j\|, \|A_j\|]$. Condition (b) in Theorem 6.1 (with G in place of E) implies that

$$
A_j x = \lim_{N \to \infty} \left(\int_{-N}^{N} \lambda dG_j(\lambda) \right) x, \, x \in H_j,
$$

and hence we may conclude (by Theorem V.4.2) that $\{G_j(t)\}_{t \in \mathbf{R}}$ is the resolution of the identity of the bounded selfadjoint operator A_j. In other words, $G_j(t) = E_j(t)$ for each $t \in \mathbf{R}$. This holds for each $j \geq 1$. Take $x \in H$, and write $x = \sum_{j=1}^{\infty} x_j$ with $x_j \in H_j$ for $j \geq 1$. Then

$$
G(t)x = \sum_{j=1}^{\infty} G_j(t)x_j = \sum_{j=1}^{\infty} E_j(t)x_j = E(t)x
$$

for each $t \in \mathbf{R}$, which completes the proof. $\square$

The resolution $\{E(t)\}_{t \in \mathbf{R}}$ appearing in Theorem 6.1 will be referred to as the *resolution of the identity for A*. The operator-valued function which assigns to each left open interval $(a, b]$ the orthogonal projection $E(b) - E(a)$ will be called the *spectral measure* of A and will be denoted by E_A. In other words,

$$
E_A((a, b]) = E(b) - E(a), \qquad a < b.
$$

The extension of the spectral measure to other sets will be discussed later (in Volume II) in the section about the spectral theorem for normal operators.

XVI.7 AN ILLUSTRATIVE EXAMPLE

In this section we compute the spectral measure for a particular unbounded selfadjoint operator. We have in mind the operator $T(L_2(\mathbf{R}) \to L_2(\mathbf{R}))$ defined by

$$(1) \qquad Tf = if', \qquad f \in \mathcal{D}(T),$$

where $\mathcal{D}(T)$ consists of all $f \in L_2(\mathbf{R})$ such that f is absolutely continuous on every compact interval of $\mathbf{R}$ and $f' \in L_2(\mathbf{R})$. In other words, $T = T_{\max,\tau}$ on $L_2(\mathbf{R})$ with $\tau = iD$. We shall prove the following two theorems.

THEOREM 7.1. *The operator T is selfadjoint, $\sigma(T) = \mathbf{R}$ and the spectral measure E_T of T is given by*

$$(2) \qquad (E_T((a,b])g)(t) = \frac{1}{2\pi} \int_{-\infty}^{\infty} \frac{e^{i(s-t)b} - e^{i(s-t)a}}{i(s-t)} g(s)\,ds, \ \text{a.e.},$$

where $g \in L_2(\mathbf{R})$ and $-\infty < a < b < \infty$.

THEOREM 7.2. *The operator T is unitarily equivalent to the multiplication operator $M(L_2(\mathbf{R}) \to L_2(\mathbf{R}))$ defined by*

$$(3a) \qquad \mathcal{D}(M) = \{f \in L_2(\mathbf{R}) \mid tf(t) \in L_2(\mathbf{R})\},$$

$$(3b) \qquad (Mf)(t) = tf(t).$$

More precisely

$$(4) \qquad T = UMU^{-1},$$

where U is the Fourier transformation on $L_2(\mathbf{R})$, i.e.,

$$(Uf)(t) = \frac{1}{\sqrt{2\pi}} \int_{-\infty}^{\infty} e^{-its} f(s)\,ds.$$

First let us prove that T is selfadjoint. Note that

$$(5) \qquad \lim_{t \to \pm\infty} f(t) = 0, \qquad f \in \mathcal{D}(T).$$

Indeed, integration by parts yields

$$\int_0^t f'(s)\overline{f(s)}\,ds = |f(t)|^2 - |f(0)|^2 - \int_0^t f(s)\overline{f'(s)}\,ds,$$

which implies that $\lim_{t \to \pm\infty} |f(t)|$ exists. Since $|f|^2$ is integrable, this limit must be zero. Thus we can use integration by parts to show that for every $f, g \in \mathcal{D}(T)$

$$\langle Tf, g \rangle = \langle if', g \rangle = \langle f, ig' \rangle = \langle f, Tg \rangle.$$

It follows that $T \subset T^*$. Next, let τ be the differential expression $\tau = iD$. Since $T = T_{\max,\tau}$ on $L_2(\mathbf{R})$ and $\tau = \tau^*$, Theorem XIV.4.1 yields

$$T^* = (T_{\max,\tau})^* = T_{\min,\tau^*} = T_{\min,\tau}$$

$$\subset T_{\max,\tau} = T,$$

and we have $T^* = T$. (This also proves that $T_{\min,iD} = T_{\max,iD}$ which is never the case for compact intervals.)

PROOF OF THEOREM 7.2. First we establish the following formulas. For every $f \in L_2(\mathbf{R})$,

$$(6) \qquad (Uf)(t) = \frac{1}{\sqrt{2\pi}} \frac{d}{dt} \int_{-\infty}^{\infty} \frac{e^{-its} - 1}{is} f(s)ds, \quad \text{a.e.,}$$

$$(7) \qquad (U^{-1}f)(t) = \frac{1}{\sqrt{2\pi}} \frac{d}{dt} \int_{-\infty}^{\infty} \frac{e^{its} - 1}{is} f(s)ds, \quad \text{a.e. .}$$

Let (f_n) be a sequence of functions in $L_2(\mathbf{R})$ with compact support which converges in $L_2(\mathbf{R})$ to f. For each $t > 0$, define ψ_t to be the characteristic function on $[0, t]$. Then by Fubini's theorem,

$$\int_0^t (Uf)(s)ds = \lim_{n \to \infty} \langle Uf_n, \psi_t \rangle$$

$$(8) \qquad
\begin{aligned}
&= \lim_{n \to \infty} \frac{1}{\sqrt{2\pi}} \int_0^t \int_{-\infty}^{\infty} e^{-is\alpha} f_n(\alpha) d\alpha\, ds \\
&= \lim_{n \to \infty} \frac{1}{\sqrt{2\pi}} \int_{-\infty}^{\infty} \left[\int_0^t e^{-is\alpha} ds \right] f_n(\alpha) d\alpha \\
&= \lim_{n \to \infty} \frac{1}{\sqrt{2\pi}} \int_{-\infty}^{\infty} \frac{e^{-it\alpha} - 1}{-i\alpha} f_n(\alpha) d\alpha.
\end{aligned}$$

Now for every $t \in \mathbf{R}$ the function

$$(9) \qquad h_t(s) = \begin{cases} \frac{e^{its} - 1}{is}, & s \neq 0, \\ t, & s = 0, \end{cases}$$

is in $L_2(\mathbf{R})$. This follows from the continuity of h_t on $\mathbf{R}$ and the inequality

$$|h_t(s)| \leq \frac{2}{|s|}, \qquad |s| \geq 1.$$

From (8) we get

$$(10) \qquad \int_0^t (Uf)(s)ds = \lim_{n\to\infty} \frac{1}{\sqrt{2\pi}}\langle f_n, h_t\rangle = \frac{1}{\sqrt{2\pi}}\langle f, h_t\rangle$$

$$= \frac{1}{\sqrt{2\pi}} \int_{-\infty}^{\infty} \frac{e^{-it\alpha} - 1}{-i\alpha} f(\alpha)d\alpha.$$

Since Uf is in $L_2([0,t]) \subset L_1([0,t])$, the left hand side of (10) is differentiable for almost every $t \in \mathbf{R}$ and (6) follows. The proof for $t < 0$ is similar. Formula (7) is obtained by applying the above argument to U^{-1} with $-i$ replaced by i.

We are now prepared to prove formula (4). Given $f \in \mathcal{D}(T)$, we have from (7) that

$$(U^{-1}Tf)(t) = \frac{1}{\sqrt{2\pi}} \frac{d}{dt} \int_{-\infty}^{\infty} \frac{e^{its} - 1}{s} f'(s)ds.$$

Since the function h_t, defined by (9), has a continuous derivative on every interval, integration by parts and (5) give

$$(U^{-1}Tf)(t) = -\frac{1}{\sqrt{2\pi}} \frac{d}{dt} \int_{-\infty}^{\infty} \frac{itse^{its} - (e^{its} - 1)}{s^2} f(s)ds$$

$$= \frac{1}{\sqrt{2\pi}} \frac{d}{dt} t \int_{-\infty}^{\infty} \frac{e^{its} - 1}{is} f(s)ds + \frac{1}{\sqrt{2\pi}} \frac{d}{dt} \int_{-\infty}^{\infty} \frac{e^{its} - 1 - its}{s^2} f(s)ds$$

$$(11) \qquad = \frac{1}{\sqrt{2\pi}} t \frac{d}{dt} \int_{-\infty}^{\infty} \frac{e^{its} - 1}{is} f(s)ds + \frac{1}{\sqrt{2\pi}} \int_{-\infty}^{\infty} \frac{e^{its} - 1}{is} f(s)ds$$

$$- \frac{1}{\sqrt{2\pi}} \int_{-\infty}^{\infty} \frac{e^{its} - 1}{is} f(s)ds = t(U^{-1}f)(t).$$

Thus $U^{-1}f \in \mathcal{D}(M)$ and $MU^{-1}f = U^{-1}Tf$. We have now proved that $T \subset UMU^{-1}$. Note that M is symmetric. Since U is unitary, it follows that UMU^{-1} is also symmetric. Hence

$$(12) \qquad UMU^{-1} \subset (UMU^{-1})^* \subset T^* = T,$$

because T is selfadjoint, and (4) is proved. $\square$

Since U is unitary and T is selfadjoint, the identity (4) implies that M is selfadjoint. Next, we prove that the resolution of the identity for the selfadjoint operator M is given by

$$(13) \qquad G(t)f = \psi_{(-\infty,t]}f, \qquad f \in L_2(\mathbf{R}),$$

where $\psi_{(-\infty,t]}$ is the characteristic function of the interval $(-\infty, t]$. It is clear that $\{G(t)\}_{t\in\mathbf{R}}$ is a resolution of the identity on $L_2(\mathbf{R})$. Let us prove that (a) and (b) in Theorem 6.1 hold. We start by showing that

$$(14) \qquad \mathcal{D}(M) = \mathcal{D} := \left\{ f \in L_2(\mathbf{R}) \mid \int_{-\infty}^{\infty} \lambda^2 d\langle G(\lambda)f, f\rangle < \infty \right\}.$$

Given $\varepsilon > 0$, let $-N = \lambda_0 < \lambda_1 < \cdots < \lambda_p = N$ be a partition of $[-N, N]$ with $|\lambda_j^2 - \lambda_{j-1}^2| < \varepsilon$ for $j = 1, \ldots, p$. Take $f \in L_2(\mathbf{R})$. Then

$$\left| \int_{-N}^{N} \lambda^2 |f(\lambda)|^2 d\lambda - \sum_{j=1}^{p} \lambda_j^2 \{\langle G(\lambda_j)f, f\rangle - \langle G(\lambda_{j-1})f, f\rangle\} \right|$$

$$= \left| \sum_{j=1}^{p} \int_{\lambda_{j-1}}^{\lambda_j} (\lambda^2 - \lambda_j^2)|f(\lambda)|^2 d\lambda \right| \le \varepsilon \|f\|^2.$$

It follows that

$$\int_{-N}^{N} \lambda^2 |f(\lambda)|^2 d\lambda = \int_{-N}^{N} \lambda^2 d\langle G(\lambda)f, f\rangle$$

for each N. Thus $f \in \mathcal{D}(M)$ if and only if $f \in \mathcal{D}$. Now, take $f \in \mathcal{D}(M)$, and let $N = \lambda_0 < \lambda_1 < \cdots < \lambda_p = N$ be a partition such that $\lambda_j - \lambda_{j-1} < \varepsilon$ for $j = 1, \ldots, p$. Then

$$\left\| Mf - \sum_{j=1}^{p} \lambda_j \{G(\lambda_j) - G(\lambda_{j-1})\}f \right\|^2$$

$$= \sum_{j=1}^{p} \int_{\lambda_{j-1}}^{\lambda_j} |\lambda - \lambda_j|^2 |f(\lambda)|^2 d\lambda + \int_{-\infty}^{-N} \lambda^2 |f(\lambda)|^2 d\lambda + \int_{N}^{\infty} \lambda^2 |f(\lambda)|^2 d\lambda$$

$$\le \varepsilon^2 \|f\|^2 + \varepsilon,$$

for N sufficiently large. This shows that

$$Mf = \lim_{N\to\infty} \left(\int_{-N}^{N} \lambda dG(\lambda) \right) f, \qquad f \in \mathcal{D}(M).$$

This result combined with (14) proves that (a) and (b) in Theorem 6.1 hold for M and the projections (13). Therefore $\{G(t)\}_{t\in\mathbf{R}}$ is the resolution of the identity for M.

PROOF OF THEOREM 7.1. We already showed that $T = T^*$. Let us determine $\sigma(T)$. Since $T = T^*$, we know that $\sigma(T) \subset \mathbf{R}$, by Theorem 3.1. Take $\lambda \in \mathbf{R}$.

We shall prove that $\lambda - T$ is not surjective (and hence $\sigma(T) = \mathbf{R}$). Choose $\alpha < 0$ such that

$$(15) \qquad \int_{-\infty}^{\alpha} s^{-1} e^{i\lambda s} ds \neq 0,$$

and let

$$(16) \qquad g(t) = \begin{cases} t^{-1}, & -\infty < t \leq \alpha, \\ 0, & \text{otherwise.} \end{cases}$$

Then $g \in L_2(\mathbf{R})$. Suppose $g \in \text{Im}(\lambda - T)$. Then $(\lambda - T)f = g$ for some $f \in \mathcal{D}(T)$. In particular, $f' = -i\lambda f + ig$. It follows that for any $a \in \mathbf{R}$

$$(17) \qquad f(t) = e^{-i\lambda(t-a)} f(a) + i \int_{a}^{t} e^{-i\lambda(t-s)} g(s) ds.$$

Now use (15) and take in (17) the limit for $a \to -\infty$. Then (5) implies that

$$f(t) = i e^{-i\lambda t} \left(\int_{-\infty}^{\alpha} s^{-1} e^{i\lambda s} ds \right), \qquad t > \alpha.$$

Then $|f(t)| = c$ for $t > \alpha$, where c is a constant which is different from zero (because of (15)). But this contradicts (5). Thus $g \notin \text{Im}(\lambda - T)$, and $\lambda - T$ is not surjective.

Finally, we prove formula (2). Since T and M are unitarily equivalent via (4), it is rather straightforward to determine the resolution of the identity for T. For $t \in \mathbf{R}$ define

$$(18) \qquad E(t)f = UG(t)U^{-1}f, \qquad f \in L_2(\mathbf{R}),$$

where $\{G(t)\}_{t \in \mathbf{R}}$ is the resolution of the identity for M, which is defined by (13). Since U is unitary, $\{E(t)\}_{t \in \mathbf{R}}$ is a resolution of the identity, and by considering Riemann-Stieltjes sums it follows from (4) and (18) that $\{E(t)\}_{t \in \mathbf{R}}$ satisfies (a) and (b) in Theorem 6.1 (for T in place of A). Hence $\{E(t)\}_{t \in \mathbf{R}}$ is the resolution of the identity for T.

Now, let us compute $E_T((a, b]) = E(b) - E(a)$. Take $f \in C_0^{\infty}(\mathbf{R})$. Fubini's theorem gives

$$\left(E_T((a, b]) f \right)(t) = \left(U\{G(b) - G(a)\} U^{-1} f \right)(t)$$

$$= \frac{1}{\sqrt{2\pi}} \int_{-\infty}^{\infty} e^{-its} \psi_{[a,b]}(s) (U^{-1} f)(s) ds$$

$$= \frac{1}{2\pi} \int\limits_{-\infty}^{\infty} e^{-its} \left(\int\limits_{-\infty}^{\infty} e^{is\alpha} \psi_{[a,b]}(s) f(\alpha) d\alpha \right) ds$$

$$\text{(19)} \qquad = \frac{1}{2\pi} \int\limits_{-\infty}^{\infty} \left[\int\limits_{a}^{b} e^{i(\alpha-t)s} ds \right] f(\alpha) d\alpha$$

$$= \frac{1}{2\pi} \int\limits_{-\infty}^{\infty} h_t(\alpha) f(\alpha) d\alpha,$$

where

$$h_t(\alpha) = \frac{e^{i(\alpha-t)b} - e^{i(\alpha-t)a}}{\alpha - t}, \qquad \alpha \neq t.$$

Note that h_t is bounded on $0 < |\alpha - t| \leq 1$ and $|h_t(\alpha)| \leq 2|\alpha - t|^{-1}$ for $|\alpha - t| \geq 1$. We conclude that $h_t \in L_2(\mathbf{R})$. Now take $g \in L_2(\mathbf{R})$. Since $C_0^\infty(\mathbf{R})$ is dense in $L_2(\mathbf{R})$ and $E_T\big((a,b]\big)$ is a bounded operator, there exists a sequence (f_n) in $C_0^\infty(\mathbf{R})$ such that $f_n \to g$ and $E_T\big((a,b]\big) f_n \to E_T\big((a,b]\big) g$ in the norm of $L_2(\mathbf{R})$. But then

$$E_T\big((a,b]\big) g = \lim_{n\to\infty} E_T\big((a,b]\big) f_n = \lim_{n\to\infty} \frac{1}{2\pi} \langle f_n, \overline{h}_t \rangle = \langle g, \overline{h}_t \rangle,$$

which proves (2). $\square$

Let T be the operator appearing in Theorem 7.1. We know that any non-real λ belongs to the resolvent set of T. A simple computation, using (5) and (17) shows that for $\lambda \notin \mathbf{R}$

$$\text{(20)} \qquad ((\lambda - T)^{-1} g)(t) = \begin{cases} i \int\limits_{-\infty}^{t} e^{-i\lambda(t-s)} g(s) ds, & \operatorname{Im} \lambda < 0, \\[2mm] -i \int\limits_{t}^{\infty} e^{-i\lambda(t-s)} g(s) ds, & \operatorname{Im} \lambda > 0, \end{cases}$$

where g is an arbitrary element of $L_2(\mathbf{R})$. For later purposes (see Section XVII.2) we note that $\operatorname{Im}(\lambda - T)$ is not closed for real λ. To see this, take $\lambda \in \mathbf{R}$. If $g = 0$ in (17), then f in (17) must also be zero, because $e^{-i\lambda(t-a)} f(a)$ is in $L_2(\mathbf{R})$ only if $f(a) = 0$. It follows that $\lambda - T$ is injective, and hence, using $T = T^*$,

$$\overline{\operatorname{Im}(\lambda - T)} = \operatorname{Ker}(\lambda - T)^{\perp} = L_2(\mathbf{R}).$$

Since $\lambda - T$ is not surjective, this implies that the range of $\lambda - T$ is not closed.

CHAPTER XVII

UNBOUNDED FREDHOLM OPERATORS AND
PERTURBATION THEOREMS

In this chapter the perturbation theory developed for bounded Fredholm operators in Sections XI.3 and XI.4 is extended to closed unbounded Fredholm operators. For the additive perturbation theorems this is accomplished via the simple device of renorming the domain of the operator T. In this way T becomes a bounded Fredholm operator to which the theorems of Chapter XI are applicable. We start with some properties and examples of the graph norm. The third and fourth section contain the main perturbation theorems. Section 5 presents a completeness theorem for certain compact perturbations of unbounded selfadjoint operators.

XVII.1 THE GRAPH NORM

Throughout this chapter X and Y are complex Banach spaces. Given an operator $T(X \to Y)$ the *graph norm* $\| \ \|_T$ on $\mathcal{D}(T)$ is defined by

$$\|x\|_T = \|x\| + \|Tx\|.$$

This norm is also referred to as the *T-norm* on $\mathcal{D}(T)$. In what follows, X_T denotes $\mathcal{D}(T)$ endowed with the graph norm. Note that $T \colon X_T \to Y$ is bounded. Since X_T is linearly isometric to the graph of T, it is clear that X_T is a Banach space if and only if T is closed.

An operator $B(X \to Y)$ is called *T-bounded* if

(i) $\mathcal{D}(T) \subset \mathcal{D}(B)$ and

(ii) B is bounded on X_T.

The latter condition means that there exists a constant $\gamma \geq 0$ such that

$$\|Bx\| \leq \gamma(\|x\| + \|Tx\|), \qquad x \in \mathcal{D}(T).$$

Obviously, a bounded operator on X is T-bounded.

PROPOSITION 1.1. *Suppose that $T(X \to Y)$ is closed and $B(X \to Y)$ is closable with $\mathcal{D}(T) \subset \mathcal{D}(B)$. Then B is T-bounded.*

PROOF. Let B_0 be the restriction of B to $\mathcal{D}(T)$, and consider B_0 as an operator of X_T into Y. We claim that B_0 is closable. Indeed, if $x_n \to 0$ in X_T and $B_0 x_n \to y$ in Y, then $\| \cdot \| \leq \| \cdot \|_T$ implies that $x_n \to 0$ in $\mathcal{D}(T)$ and hence $y = 0$, because B is closable. Since $B_0 \colon X_T \to Y$, its closed linear extension coincides with B_0. So B_0 is closed. But X_T is a Banach space, and thus B_0 is bounded by the closed graph theorem.
$\square$

As a consequence of Proposition 1.1, we show that lower order differential operators are bounded relative to those of higher order. To be precise, we have the following result.

COROLLARY 1.2. *Let T and B be the maximal operator corresponding to a compact interval J and the differential expressions*

$$\tau = D^n + \sum_{k=0}^{n-1} a_k(t)D^k, \qquad \nu = D^m + \sum_{j=0}^{m-1} b_j(t)D^j,$$

respectively, where $m < n$, and each $a_k \in L_1(J)$, each $b_j \in L_2(J)$. Then B is T-bounded.

PROOF. By Proposition XIV.3.3, the operators T and B are closed. Moreover $\mathcal{D}(T) \subset \mathcal{D}(B)$. Indeed, if $g \in \mathcal{D}(T)$, then $g^{(n-1)}$ is absolutely continuous on J and therefore so is $g^{(m-1)}$, since $m < n$. Because of the condition on the coefficients of ν, the function $\nu(g)$ is in $L_2(J)$. Thus $g \in \mathcal{D}(B)$, and B is T-bounded by Proposition 1.1. $\square$

Let T and B be linear operators from X to Y. We say that B is T-*compact* if

(i) $\mathcal{D}(T) \subset \mathcal{D}(B)$ and

(ii) B is compact on X_T.

If we require that the coefficients a_k of τ in the above corollary be bounded, then we have the following result.

THEOREM 1.3. *Let T and B be the maximal operators corresponding to a compact interval J and the differential expressions*

$$\tau = D^n + \sum_{k=0}^{n-1} a_k(t)D^k, \qquad \nu = \sum_{k=0}^{m} b_j(t)D^j,$$

respectively, where $m < n$ and each $a_k \in L_\infty(J)$, each $b_j \in L_2(J)$. Then B is T-compact.

First we prove the following two lemmas which are of interest in their own right.

LEMMA 1.4. *Let $\tau = D^n + \sum_{k=0}^{n-1} a_k(t)D^k$, where $a_k \in L_\infty([a,b])$. Let T be the maximal operator corresponding to τ and $[a,b]$. Then there exists a constant $c \geq 0$ so that for all $f \in \mathcal{D}(T)$*

$$(1) \qquad \|f^{(k)}\| \leq c(\|f\| + \|Tf\|), \qquad 0 \leq k \leq n.$$

PROOF. For $0 \leq k \leq n-1$ let B_k be the maximal operator corresponding to D^k and $[a,b]$. Apply Corollary 1.2 to T and B_k. It follows that there exists a constant $c_1 \geq 0$ such that for all $f \in \mathcal{D}(T)$

$$(2) \qquad \|f^{(k)}\| \leq c_1(\|f\| + \|Tf\|), \qquad 0 \leq k \leq n-1.$$

But then

$$\|f^{(n)}\| = \left\| \tau f - \sum_{k=0}^{n-1} a_k(t) f^k \right\|$$

$$(3) \qquad \leq \|Tf\| + \sum_{k=0}^{n-1} \|a_k\|_\infty \|f^{(k)}\|$$

$$\leq \left(1 + \sum_{k=0}^{n-1} \|a_k\|_\infty c_1\right)(\|f\| + \|Tf\|).$$

Inequality (1) now follows from (2) and (3). $\square$

LEMMA 1.5. *Given $\varepsilon > 0$ there exists a constant K such that*

$$(4) \qquad |f(t)| \leq \varepsilon \|f'\| + K\|f\|, \qquad a \leq t \leq b,$$

for all absolutely continuous functions f on $[a, b]$ such that $f' \in L_2([a, b])$.

PROOF. Let $J_1 = [a, c]$, $J_2 = [c, b]$, where $a < c < b$. Choose $\eta > 0$ so that $a \leq c - \eta$ and $c + \eta \leq b$. Let φ be in $C^1([0, \eta])$ with the following properties: $0 \leq \varphi \leq 1$, $\varphi(0) = 1$ and $\varphi(\eta) = 0$. Then for $t \in J_1$

$$f(t) = -\int_0^\eta \frac{d}{ds}[\varphi(s)f(t+s)]ds$$

$$= -\int_0^\eta \varphi(s)f'(t+s)ds - \int_0^\eta \varphi'(s)f(t+s)ds.$$

By the Cauchy-Schwartz inequality,

$$(5) \qquad |f(t)| \leq \eta^{1/2}\|f'\| + \alpha\eta^{1/2}\|f\|, \qquad t \in J_1,$$

where $\alpha = \max\{|\varphi'(s)| \mid 0 \leq s \leq \eta\}$. Similarly,

$$f(t) = -\int_0^\eta \frac{d}{ds}[\varphi(s)f(t-s)]ds, \qquad t \in J_2,$$

and hence also

$$(6) \qquad |f(t)| \leq \eta^{1/2}\|f'\| + \alpha\eta^{1/2}\|f\|, \qquad t \in J_2.$$

Without loss of generality we assume that $\eta < \varepsilon^2$. Then (5) and (6) give the desired inequality (4). $\square$

PROOF OF THEOREM 1.3. Let (g_j) be a bounded sequence in X_T. By Lemma 1.4 applied to $f = g_j$ there exists $\gamma \geq 0$ such that

$$(7) \qquad \|g_j^{(k)}\| \leq \gamma(\|g_j\| + \|Tg_j\|), \qquad 0 \leq k \leq n,$$

and all $j \geq 1$. Next apply Lemma 1.5 with $\varepsilon = 1$ to $f = g_j^{(k)}$, where $0 \leq k \leq n - 1$, and use (7). It follows that there exists a constant M so that for all $t \in J$

$$(8) \qquad |g_j^{(k)}(t)| \leq M(\|g_j\| + \|Tg_j\|), \qquad 0 \leq k \leq n - 1, \; j \geq 1.$$

From (7) it also follows that for $t, s \in J$, $s < t$,

$$(9) \qquad
\begin{aligned}
|g_j^{(k)}(t) - g_j^{(k)}(s)| &= \left| \int_s^t g_j^{(k+1)}(\alpha)\, d\alpha \right| \\
&\leq |t - s|^{1/2} \|g_j^{(k+1)}\| \\
&\leq \gamma |t - s|^{1/2}(\|g_j\| + \|Tg_j\|), \qquad 0 \leq k \leq n - 1, \; j \geq 1.
\end{aligned}$$

Since $(g_j)_{j=1}^{\infty}$ is bounded in X_T, the inequalities in (8) and (9) imply that for $0 \leq k \leq n-1$ the sequence $(g_j^{(k)})_{j=1}^{\infty}$ is uniformly bounded and equicontinuous. Hence by the Ascoli-Arzela theorem ([W], Section 43), the sequence (g_j) has a subsequence $(g_{j,0})$ which converges uniformly on J. Again, by the Ascoli-Arzela theorem, the sequence $(g'_{j,0})$ has a subsequence $(g'_{j,1})$ which converges uniformly on J. So, $(g_{j,1})$ and $(g'_{j,1})$ converge uniformly on J. Continuing in this manner, a subsequence (f_j) of (g_j) is obtained with the property that $(f_j^{(k)})_{j=1}^{\infty}$ converges uniformly for $0 \leq k \leq n - 1$. Therefore

$$
\begin{aligned}
\|Bf_k - Bf_j\| &\leq \sum_{k=0}^{m} \|b_k(f_i^{(k)} - f_j^{(k)})\| \\
&\leq \left(\max_{t \in J} |f_i^{(k)}(t) - f_j^{(k)}(t)| \right) \left(\sum_{k=0}^{m} \|b_k\| \right) \\
&\to 0 \qquad (i, j \to \infty),
\end{aligned}
$$

which shows that the subsequence (Bf_j) of (Bg_j) converges. Hence $B: X_T \to Y$ is compact, and therefore B is T-compact. $\square$

XVII.2   FREDHOLM OPERATORS AND ESSENTIAL SPECTRUM

As before, X and Y are complex Banach spaces. An operator $T(X \to Y)$ is called a *Fredholm operator* if T is closed and the integers

$$n(T) := \dim \operatorname{Ker} T, \qquad d(T) := \dim(Y/\operatorname{Im} T)$$

are finite. In that case $\operatorname{ind} T = n(T) - d(T)$ is called the *index* of T. Evidently, $\operatorname{Im} T$, $n(T)$ and $d(T)$ are independent of any norm on $\mathcal{D}(T)$. Thus if we let T_1 be the operator T acting on $\mathcal{D}(T)$ endowed with the graph norm, then $\operatorname{Im} T = \operatorname{Im} T_1$ is closed by Corollary XI.2.3.

As for bounded operators (see Section XI.5) the *essential spectrum* of an operator $T(X \to X)$ is defined to be the set of all $\lambda \in \mathbf{C}$ such that $\lambda - T$ is not a Fredholm operator. This set is denoted by $\sigma_{\text{ess}}(T)$. As the following examples show it may happen that $\sigma_{\text{ess}}(T)$ is empty.

Let $T_{\max}$ be the maximal operator corresponding to a compact interval $J = [a, b]$ and the differential expression

$$(1) \qquad \tau = D^n + \sum_{k=0}^{n-1} a_k(t) D^k, \quad a_k \in C^k([a, b]), \qquad 0 \leq k \leq n-1.$$

Then $T_{\max}\big(L_2([a, b]) \to L_2([a, b])\big)$ is a Fredholm operator with $n(T_{\max}) = n$ and $d(T_{\max}) = 0$, by Corollary XIV.3.2 and Proposition XIV.3.3. Next we apply this result to $\lambda - T_{\max}$ in place of $T_{\max}$. It follows that $\sigma_{\text{ess}}(T_{\max}) = \emptyset$. Also the minimal operator $T_{\min}$ corresponding to the differential expression τ in (1) is a Fredholm operator. In fact, $n(T_{\min}) = 0$ and $d(T_{\min}) = n$, by Proposition XIV.3.5 and Corollary XIV.4.3. Note that $T_{\min}$ is closed by definition and $\operatorname{Im} T_{\min}$ is closed (according to Proposition XIV.3.5). By replacing $T_{\min}$ by $\lambda - T_{\min}$, we conclude that also $\sigma_{\text{ess}}(T_{\min}) = \emptyset$.

For differential operators defined by differential expressions on an infinite interval the situation is different. In general such operators have a nonempty essential spectrum. For example, let $T\big(L_2(\mathbf{R}) \to L_2(\mathbf{R})\big)$ be the maximal operator corresponding to $\tau = iD$. In Section XVI.7 we showed that $\operatorname{Im}(\lambda - T)$ is not closed for $\lambda \in \mathbf{R}$ and

$$n(\lambda - T) = d(\lambda - T) = 0, \qquad \operatorname{Im} \lambda \neq 0.$$

Since T is a maximal operator, $\lambda - T$ is closed for each λ (cf. Proposition XIV.3.3). Hence $\sigma_{\text{ess}}(T) = \mathbf{R}$. Other examples of this type will be given in the next chapter.

THEOREM 2.1. *Let $T(X \to X)$ be an operator with a nonempty resolvent set, and let Ω be an open connected subset of $\mathbf{C} \backslash \sigma_{\text{ess}}(T)$. If $\Omega \cap \rho(T) \neq \emptyset$, then $\sigma(T) \cap \Omega$ is a finite or countable set, with no accumulation point in Ω, consisting of eigenvalues of T of finite type.*

PROOF. Let X_T be the space $\mathcal{D}(T)$ endowed with the T-norm, and for $\lambda \in \Omega$ define $W(\lambda): X_T \to X$ by $W(\lambda)x = \lambda x - Tx$ for each $x \in X_T$. From

$$\|W(\lambda)x\| \leqq |\lambda| \|x\| + \|Tx\| \leq (1 + |\lambda|)\|x\|_T,$$

we conclude that $W(\lambda)$ is a bounded operator from the Banach space X_T into the Banach space X which depends analytically on the parameter λ. Moreover, by our hypothesis, $W(\lambda)$ is a Fredholm operator for each $\lambda \in \Omega$. Since

$$\sigma(T) \cap \Omega = \{\lambda \in \Omega \mid W(\lambda) \text{ not invertible}\},$$

it follows from Corollary XI.8.4 that $\sigma(T) \cap \Omega$ is a finite or countable set with no accumulation point in Ω. Take $\lambda_0 \in \sigma(T) \cap \Omega$, and let $F: X_T \to X$ be defined by $Fx = x$. For $\lambda \neq \lambda_0$ and λ sufficiently close to λ_0 we have

$$(2) \qquad (\lambda - T)^{-1} = FW(\lambda)^{-1} = \sum_{n=-q}^{\infty} (\lambda - \lambda_0)^n FA_n,$$

where $A_{-1}, \ldots, A_{-q}$ are operators of finite rank. Here we used the second part of Corollary XI.8.4 and the fact that $F\colon X_T \to X$ is bounded. Thus λ_0 is an isolated point of $\sigma(T)$ and by (2) the associated Riesz projection $P_{\{\lambda_0\}}$ is equal to FA_{-1}. In particular, rank $P_{\{\lambda_0\}}$ is finite, and hence λ_0 is an eigenvalue of finite type. $\square$

XVII.3 THE PRODUCT THEOREM

The aim of this section is to extend Theorem XI.3.2 to unbounded Fredholm operators. Given two operators $T(X \to Y)$ and $S(Z \to X)$, where Z is a Banach space, the *product* $TS(Z \to Y)$ is the operator defined by

$$\mathcal{D}(TS) = \{z \in \mathcal{D}(S) \mid Sz \in \mathcal{D}(T)\},$$

$$(TS)z = T(Sz), \qquad z \in \mathcal{D}(TS).$$

We shall prove the following theorem.

THEOREM 3.1. *Let $T(X \to Y)$ be a densely defined Fredholm operator, and let $S(Z \to X)$ be a Fredholm operator, where Z is a Banach space. Then TS is a Fredholm operator and*

(1) $$\mathrm{ind}(TS) = \mathrm{ind}\, T + \mathrm{ind}\, S.$$

For the proof of Theorem 3.1 we need the following proposition.

PROPOSITION 3.2. *Let $T(X \to Y)$ be a closed operator with closed range and $\dim \mathrm{Ker}\, T < \infty$. Let C be a closed opertor with domain in a complex Banach space Z and range in X. Then TC is a closed operator.*

PROOF. Let (z_j) be a sequence in $\mathcal{D}(TC)$, and assume that $z_n \to z$ in Z and $TCz_n \to y$ in Y. Put $x_j = Cz_j$, $j = 1, 2, \ldots$, and consider the sequence $([x_j])$ in the quotient space $\mathcal{D}(T)/\mathrm{Ker}\, T$. Let X_T be the domain $\mathcal{D}(T)$ endowed with the graph norm, and let $T_1\colon X_T \to Y$ be the restriction of T to $\mathcal{D}(T)$. Then T_1 has a closed range, and we can apply Theorem XI.2.1 to show that the sequence $([x_j])$ is a Cauchy sequence in the quotient space $X_T/\mathrm{Ker}\, T$. The identity map from $\mathcal{D}(T)/\mathrm{Ker}\, T$ into $X_T/\mathrm{Ker}\, T$ is continuous, because

$$\begin{aligned}
\|[x]\|_T &= \inf\{\|x - z\|_T \mid z \in \mathrm{Ker}\, T\} \\
&= \inf\{\|x - z\| + \|Tx\| \mid z \in \mathrm{Ker}\, T\} \\
&= \|[x]\| + \|Tx\|, \qquad x \in \mathcal{D}(T).
\end{aligned}$$

Hence $([x_j])$ is also a Cauchy sequence in $\mathcal{D}(T)/\mathrm{Ker}\, T$, which therefore converges to some $[x] \in X/\mathrm{Ker}\, T$. Consequently there is a sequence (u_n) in $\mathrm{Ker}\, T$ such that $x_n + u_n \to x$ if $n \to \infty$.

Next we show that (u_n) is bounded. Assume this is not the case. Then there exists a subsequence $(u_{n'})$ of (u_n) such that $0 < \|u_{n'}\| \to \infty$ if $n' \to \infty$. Since $(\|u_{n'}\|^{-1} u_{n'})$ is a bounded sequence in the finite dimensional space $\mathrm{Ker}\, T$, it has a subsequence which converges to some $v \in \mathrm{Ker}\, T$. By passing to this subsequence we may assume that $\|u_{n'}\|^{-1} u_{n'} \to v$. It follows that $\|v\| = 1$. Note that

$$\|u_{n'}\|^{-1} z_{n'} \to 0,$$

$$C(\|u_{n'}\|^{-1}z_{n'}) = \|u_{n'}\|^{-1}(x_{n'} + u_{n'}) - \|u_{n'}\|^{-1}u_{n'} \to -v,$$

because $\|u_{n'}\| \to \infty$. Since C is closed, we conclude that $v = 0$, which contradicts $\|v\| = 1$.

Hence (u_n) is bounded. Since $\operatorname{Ker} T$ is finite dimensional, there exists a subsequence (u_{n_j}) of (u_n) which converges to some $u \in \operatorname{Ker} T$. Therefore $x_{n_j} \to x - u$. We now have

$$z_{n_j} \to z, \qquad Cz_{n_j} = x_{n_j} \to x - u, \qquad TCz_{n_j} \to y$$

for $j \to \infty$. But the operators C and T are closed. Hence $z \in \mathcal{D}(C)$, the vector $Cz = x - u \in \mathcal{D}(T)$ and $y = T(x - u) = TCz$. Hence TC is closed. $\square$

PROOF OF THEOREM 3.1. By Proposition 3.2 the operator TS is closed. The quotient space $\operatorname{Ker}(TS)/\operatorname{Ker} S$ is isomorphic to $\operatorname{Im} S \cap \operatorname{Ker} T$ under the map $[x] \mapsto Sx$. Hence

$$(2) \qquad n(TS) = n(S) + \dim(\operatorname{Im} S \cap \operatorname{Ker} T) < \infty.$$

Put $N_1 = \operatorname{Im} S \cap \operatorname{Ker} T$. Since $\operatorname{Ker} T$ is finite dimensional,

$$(3) \qquad \operatorname{Ker} T = N_1 \oplus N_2,$$

for some finite dimensional subspace N_2. Obviously, $\operatorname{Im} S \cap N_2 = \{0\}$. Furthermore $\operatorname{Im} S \oplus N_2$ is closed, because $\operatorname{Im} S$ is closed and $\dim N_2 < \infty$ (see [GG], Theorem IX.2.5).

Next, we prove that there exists a finite dimensional subspace N_3 such that

$$(4) \qquad (\operatorname{Im} S \oplus N_2) \oplus N_3 = X, \qquad N_3 \subset \mathcal{D}(T).$$

Put $X_0 = \operatorname{Im} S \oplus N_2$, and let $k = \dim X/X_0$. Note that $k \le \operatorname{codim} \operatorname{Im} S < \infty$. If $k = 0$, then we take $N_3 = \{0\}$ in (4). Assume $k > 0$. Since $\overline{\mathcal{D}(T)} = X$ and X_0 is closed, $\mathcal{D}(T)$ is not entirely contained in X_0. So there exists a vector $x_1 \in \mathcal{D}(T)$ such that $x_1 \notin X_0$. Put $X_1 = X_0 \oplus \operatorname{span}\{x_1\}$. Then X_1 is closed and $\dim X/X_1 = k - 1$. Thus we can repeat the above reasoning for X_1 in place of X_0. Proceeding in this way we find in k steps vectors $x_1, \dots, x_k$ in $\mathcal{D}(T)$ such that $X = X_0 \oplus \operatorname{span}\{x_1, \dots, x_k\}$. Put $N_3 = \operatorname{span}\{x_1, \dots, x_k\}$ and (4) is fulfilled.

The space N_3 is isomorphic to the quotient space $\operatorname{Im} T/\operatorname{Im}(TS)$ under the map $u \mapsto [Tu]$, because of (4). Indeed, if $x \in \mathcal{D}(T)$, then (4) implies that $x = Sz + v + u$, where $z \in \mathcal{D}(S)$, $v \in N_2 \subset \operatorname{Ker} T$ and $u \in N_3$. It follows that $Sz = x - v - u \in \mathcal{D}(T)$ and $T(Sz) = Tx - Tu$, which shows that $[Tx] = [Tu]$. Furthermore, if $[Tu] = [0]$ for $u \in N_3$, then

$$u \in \operatorname{Im} S + \operatorname{Ker} T = \operatorname{Im} S \oplus N_2,$$

and hence $u = 0$. So $u \mapsto [Tu]$ has the desired properties, and thus

$$(5) \qquad d(TS) = d(T) + \dim N_3 < \infty.$$

We have now proved that TS is a Fredholm operator. To prove the index formula (1) note that $\dim N_1 = n(T) - \dim N_2$, by (3), and $\dim N_3 = d(S) - \dim N_2$, by

(4). So, using formulas (2) and (5), we have

$$\begin{aligned}
\operatorname{ind}(TS) &= n(TS) - d(TS) \\
&= n(S) + \dim N_1 - d(T) - \dim N_3 \\
&= n(T) + n(S) - d(T) - d(S) \\
&= \operatorname{ind} T + \operatorname{ind} S. \quad \square
\end{aligned}$$

Simple finite dimensional examples already show that the index formula (1) for TS may fail to hold true if the density condition on the domain of T is not fulfilled. Indeed, let $S : \mathbf{C}^2 \to \mathbf{C}^2$ be defined by $S(x_1, x_2) = (x_1, 0)$, and let $T(\mathbf{C}^2 \to \mathbf{C}^2)$ be the operator with

$$\mathcal{D}(T) = \operatorname{span}\{(1, 0)\}, \qquad T(x_1, 0) = (x_1, 0).$$

Then T and S are both Fredholm operators, $\operatorname{ind} T = -1$ and $\operatorname{ind} S = 0$. Note that $TS = S$. Thus $\operatorname{ind}(TS) = 0 \neq -1 = \operatorname{ind} T + \operatorname{ind} S$.

XVII.4 ADDITIVE PERTURBATIONS

In this section we employ the graph norm to extend the perturbation theorems of Section XI.4 to unbounded Fredholm operators. For two operators $T(X \to Y)$ and $B(X \to Y)$ such that $\mathcal{D}(T) \subset \mathcal{D}(B)$, the operator $T + B$ is defined by

$$(1) \qquad\qquad \mathcal{D}(T + B) = \mathcal{D}(T), \qquad (T + B)x = Tx + Bx.$$

LEMMA 4.1. *Let $T(X \to Y)$ be a closed operator, and let $B(X \to Y)$ have the following properties:*

(i) $\mathcal{D}(T) \subset \mathcal{D}(B)$,

(ii) *there exist numbers a and b, $b < 1$, such that*

$$\|Bx\| \le a\|x\| + b\|Tx\|, \qquad x \in \mathcal{D}(T).$$

Then $T + B$ is closed.

PROOF. For all $x \in \mathcal{D}(T)$,

$$(2) \qquad\qquad \|(T + B)x\| \le a\|x\| + (1 + b)\|Tx\|,$$

$$(3) \qquad \|(T + B)x\| \ge \|Tx\| - \|Bx\| \ge -a\|x\| + (1 - b)\|Tx\|.$$

It follows that

$$(4) \qquad\qquad \|Tx\| \le (1 - b)^{-1}\big(a\|x\| + \|(T + B)x\|\big).$$

Suppose $x_n \to x$ and $(T+B)x_n \to y$. It is clear from (4) that (Tx_n) is a Cauchy sequence which therefore converges to some $z \in Y$. Since T is closed, $x \in \mathcal{D}(T) = \mathcal{D}(T + B)$ and $Tx = z$. By (2),

$$\|(T + B)(x_n - x)\| \le a\|x_n - x\| + (1 + b)\|Tx_n - Tx\| \to 0.$$

Hence $(T+B)x_n \to (T+B)x$ which shows that $y = (T+B)x$, and thus $T+B$ is closed.
$\square$

Let $T(X \to Y)$ be a Fredholm operator, and let T_1 be the operator T restricted to $\mathcal{D}(T)$ with the T-norm. Then T_1 is a bounded Fredholm operator. Thus T_1 has a bijection $\widetilde{T_1}$ associated with it which is defined at the beginning of Section XI.3. We shall refer to $\widetilde{T_1}$ as a *bijection associated* with T and denote it by $\widetilde{T}$. We now have the following generalization of Theorem XI.4.1 and Corollary XI.8.3.

THEOREM 4.2. *Let $T(X \to Y)$ be a Fredholm operator, and let $\widetilde{T}$ be a bijection associated with T. Suppose $B(X \to Y)$ is an operator with $\mathcal{D}(B) \supset \mathcal{D}(T)$ such that for some $0 \le \gamma \le 1$*

$$\|Bx\| \le \gamma \min(1, \|\widetilde{T}^{-1}\|^{-1})(\|x\| + \|Tx\|), \qquad x \in \mathcal{D}(T).$$

Then $T + B$ is a Fredholm operator with

(i) $n(T + B) \le n(T)$,

(ii) $d(T + B) \le d(T)$,

(iii) $\operatorname{ind}(T + B) = \operatorname{ind} T$,

and there exist $\varepsilon > 0$ and integers n_0 and d_0 such that

$$(5a) \qquad\qquad n(T) \ge n_0 = n(T + \lambda B), \qquad 0 < |\lambda| < \varepsilon,$$

$$(5b) \qquad\qquad d(T) \ge d_0 = d(T + \lambda B), \qquad 0 < |\lambda| < \varepsilon.$$

PROOF. Since $\|Bx\| \le \gamma(\|x\| + \|Tx\|)$ for all $x \in \mathcal{D}(T)$ and $\gamma < 1$, Lemma 4.1 implies that $T + B$ is closed. Let X_T be the space $\mathcal{D}(T)$ endowed with the T-norm. Define T_1 and B_1 to be the restrictions of T and B, respectively, to X_T. Then X_T is a Banach space, $T_1 : X_T \to Y$ is a Fredholm operator and $B_1 : X_T \to Y$ is a bounded linear operator such that $\|B_1\| < \|\widetilde{T}^{-1}\|^{-1}$. Statements (i), (ii) and (iii) follow now immediately from Theorem XI.4.1, and formulas (5a) and (5b) are a consequence of Corollary XI.8.3. $\square$

THEOREM 4.3. *Suppose that $T(X \to Y)$ is a Fredholm operator and $B(X \to Y)$ is T-compact. Then*

(i) *$T + B$ is a Fredholm operator,*

(ii) *$\operatorname{ind}(T + B) = \operatorname{ind} T$.*

Furthermore, there exists a finite or countable subset Σ of $\mathbf{C}$ which has no accumulation point inside $\mathbf{C}$ and there exist constants n_0 and d_0 such that

$$(6a) \qquad \dim \operatorname{Ker}(T + \lambda B) = n_0, \quad \operatorname{codim} \operatorname{Im}(T + \lambda B) = d_0, \qquad \lambda \in \mathbf{C} \backslash \Sigma,$$

$$(6b) \qquad \dim \operatorname{Ker}(T + \lambda B) > n_0, \quad \operatorname{codim} \operatorname{Im}(T + \lambda B) > d_0, \qquad \lambda \in \Sigma.$$

PROOF. Let X_T be the space $\mathcal{D}(T)$ endowed with the T-norm. Define T_1 and B_1 to be the restrictions of T and B, respectively, to X_T. Then T_1 is a Fredholm operator and B_1 is compact. Thus we can apply Theorem XI.4.2 to show that $T_1 + B_1$ is a Fredholm operator and $\operatorname{ind}(T_1 + B_1) = \operatorname{ind} T_1$. It follows that

$$n(T + B) = n(T_1 + B_1) < \infty, \qquad d(T + B) = d(T_1 + B_1) < \infty,$$

$$\operatorname{ind}(T + B) = \operatorname{ind}(T_1 + B_1) = \operatorname{ind} T_1 = \operatorname{ind} T.$$

So to prove (i) and (ii) it remains to show that $T+B$ is a closed operator. Let $E(X \to X_T)$ be defined by

$$\mathcal{D}(E) = \mathcal{D}(T), \qquad Ex = x.$$

Since $E^{-1}: X_T \to X$ is a bounded operator, the graph of E^{-1} is closed in $X_T \times X$, and hence the graph of E is closed in $X \times X_T$. Thus E is a closed operator. Note that $T + B = (T_1 + B_1)E$. Since $T_1 + B_1$ has closed range and $\dim \operatorname{Ker}(T_1 + B_1) < \infty$, Proposition 3.2 yields that $T + B$ is closed.

To prove the final statement of the theorem, note that $W(\lambda) = T_1 + \lambda B_1$ is an analytic function on $\mathbf{C}$ whose values are bounded Fredholm operators from X_T into Y (because of Theorem XI.4.2). This allows us to apply Theorem XI.8.2 to get the formulas (6a) and (6b). $\square$

Theorem 4.3 yields the following corollary for the essential spectrum.

COROLLARY 4.4. *Let $T(X \to X)$ and $B(X \to X)$ be operators, and assume that B is T-compact. Then*

$$(7) \qquad\qquad \sigma_{\mathrm{ess}}(T + B) \subset \sigma_{\mathrm{ess}}(T).$$

PROOF. Take $\lambda \notin \sigma_{\mathrm{ess}}(T)$. Thus $\lambda - T$ is a Fredholm operator. From

$$\|x\| + \|Tx\| \le (1 + |\lambda|)(\|x\| + \|(\lambda - T)x\|)$$

it follows that B is also $(\lambda - T)$-compact. But then we can apply Theorem 4.3 to show that $\lambda \notin \sigma_{\mathrm{ess}}(T + B)$. $\square$

XVII.5 A COMPLETENESS THEOREM

Throughout this section H is a complex Hilbert space. Let $A(H \to H)$ be an operator. As for bounded operators, a vector $x \in H$ is called a *generalized eigenvector* of A corresponding to λ_0 if for some $k \ge 1$

$$(1) \qquad\qquad (\lambda_0 - A)^{k+1} x = 0, \qquad (\lambda_0 - A)^k x \ne 0.$$

In that case $x_0 = (\lambda_0 - A)^k x$ is an eigenvector and λ_0 is the corresponding eigenvalue. Note that (1) requires that the vectors $x, (\lambda_0 - A)x, \ldots, (\lambda_0 - A)^k x$ all belong to $\mathcal{D}(A)$. The set of eigenvectors and generalized eigenvectors of A is said to be *complete* if the linear span of these vectors is dense in H. (To avoid confusion let us remark that by definition every eigenvector is also a generalized eigenvector, and hence the part in the preceding sentence referring to the eigenvectors could be omitted.) In this section we shall prove completeness for certain differential operators. We begin with a general theorem.

THEOREM 5.1. *Let $T(H \to H)$ be a selfadjoint operator with a compact inverse, and let $B(H \to H)$ be T-compact. Assume that the sequence of eigenvalues $\mu_1, \mu_2, \ldots$ of T (multiplicities taken into account) satisfies the condition*

$$(2) \qquad \sum_{j=1}^{\infty} \left(\frac{1}{\mu_j}\right)^p < \infty$$

for some $p \geq 1$, and assume that $I + BT^{-1}$ is invertible. Then the entire spectrum of $T + B$ consists of eigenvalues of finite type and the set of eigenvectors and generalized eigenvectors of $T + B$ is complete.

PROOF. We may represent $T + B$ in the form

$$(3) \qquad T + B = (I + BT^{-1})T.$$

Since B is T compact, the operator $BT^{-1} \colon H \to H$ is compact. To see this, let H_T be the space $\mathcal{D}(T)$ endowed with the T-norm. Let T_1 and B_1 be the operators T and B, respectively, considered as operators from the Banach space H_T into H. By our hypotheses, B_1 is compact and T_1 has a bounded inverse. Thus $B_1 T_1^{-1}$ is compact. But $BT^{-1} = B_1 T_1^{-1}$, and therefore BT^{-1} is a compact operator on H.

Since $I + BT^{-1}$ and T are invertible, (3) implies that

$$(4) \qquad (T + B)^{-1} = T^{-1}(I + BT^{-1})^{-1}.$$

The first factor in the right hand side of (4) is a compact operator on H and the second factor is a bounded linear operator on H. We conclude that $T + B$ has a compact inverse, and hence $T + B$ has a compact resolvent. So, by Theorem XV.2.3, the entire spectrum of $T + B$ consists of eigenvalues of finite type.

Next, we apply the Keldysh theorem for completeness (Theorem X.4.1) to $A = (T + B)^{-1}$. Put $K = T^{-1}$ and $S = BT^{-1}$. Then the operator K is a compact selfadjoint operator on H, the space $\operatorname{Ker} K$ consists of the zero vector only, and (2) implies that for the sequence $\lambda_1, \lambda_2, \ldots$ of eigenvalues of K (multiplicities taken into account)

$$\sum_{j=1}^{\infty} |\lambda_j|^p < \infty,$$

where p is a real number ≥ 1. The operator S is compact and $I + S$ is invertible in $\mathcal{L}(H)$. Since $A = K(I + S)^{-1}$ by (4), the Keldysh theorem implies that the set of eigenvectors and generalized eigenvectors of A is complete.

Now, let x be an eigenvector or generalized eigenvector of A corresponding to the eigenvalue λ_0. Thus (1) holds for some $k \geq 0$. Then $\lambda_0 \neq 0$ (because A is injective) and $x = \lambda_0^{-1} A q(A) x$, where $q(A)$ is some polynomial in A. The latter identity implies that $x \in \operatorname{Im} A^n = \mathcal{D}((T+B)^n)$ for each n, and hence

$$(5) \qquad (T+B)^j (\lambda_0 - A)^j x = \lambda_0^j [(T+B) - \lambda_0^{-1}]^j x$$

for $j = 0, 1, 2, \dots$. Recall that $T + B$ is injective. Thus (1) and (5) imply that

$$[(T+B) - \lambda_0^{-1}]^{k+1} x = 0, \qquad [(T+B) - \lambda_0^{-1}]^k x \neq 0.$$

Hence x is also an eigenvector or generalized eigenvector of $T + B$, and thus we have completeness for $T + B$. $\square$

In the next corollary we prove completeness for a class of differential operators acting in $L_2^n([a, b])$, the space of all $\mathbf{C}^n$-valued functions with components in $L_2([a, b])$. The space $L_2^n([a, b])$ is a Hilbert space with inner product

$$\langle f, g \rangle = \sum_{j=1}^{n} \int_a^b f_j(t) \overline{g_j(t)} dt,$$

where f_j and g_j are the j-th components of f and g, respectively. A function $f \in L_2^n([a, b])$ is said to be absolutely continuous if each of its components is absolutely continuous. The derivative of a $\mathbf{C}^n$-valued function is also defined component wise.

COROLLARY 5.2. *Let* $A\big(L_2^n([a, b]) \to L_2^n([a, b])\big)$ *be defined by*

$$\mathcal{D}(A) = \{ f \mid f \text{ absolutely continuous, } f' \in L_2^n([a, b]),\ N_1 f(a) + N_2 f(b) = 0 \},$$

$$Af = if' + B(t)f.$$

Here $B(\cdot)$ *is a continuous* $n \times n$ *matrix function on* $[a, b]$ *and* N_1, N_2 *are* $n \times n$ *matrices such that*

$$(6) \qquad \det(N_1 + N_2) \neq 0, \qquad N_1 N_1^* = N_2 N_2^*.$$

Let $U(t)$ *be the unique continuous* $n \times n$ *matrix function such that*

$$U(t) = I_n + i \int_a^t B(s) U(s) ds, \qquad a \leq t \leq b.$$

If, in addition, $\det\big(N_1 + N_2 U(b)\big) \neq 0$, *then the spectrum of* A *consists of eigenvalues of finite type only and the set of eigenvectors and generalized eigenvectors of* A *is complete.*

PROOF. Introduce the following auxiliary operators T and B:

$$\mathcal{D}(T) = \mathcal{D}(A), \qquad Tf = if',$$

$$B : L_2^n([a, b]) \to L_2^n([a, b]), \qquad (Bf)(t) = B(t)f(t).$$

Then $A = T + B$. To prove the corollary it suffices to show that T and B satisfy the hypotheses of Theorem 5.1.

The first condition in (6) implies that T is invertible and one computes that its inverse is given by

$$(T^{-1}g)(t) = \int_a^b k(t,s)g(s)ds, \qquad a \le t \le b,$$

where

$$k(t,s) = \begin{cases} -i(N_1 + N_2)^{-1}N_1, & a \le s < t \le b, \\ +i(N_1 + N_2)^{-1}N_2, & a \le t < s \le b. \end{cases}$$

In particular, T^{-1} is a Hilbert-Schmidt operator, and hence formula (2) holds with $p = 2$. The second condition in (6) implies that $N_2^*(N_1^* + N_2^*)^{-1} = (N_1 + N_2)^{-1}N_1$, and therefore $k(t,s)^* = k(s,t)$, which implies that T^{-1} is selfadjoint. Hence T is an unbounded selfadjoint operator.

The operator B is bounded and BT^{-1} is compact, because T^{-1} is compact. The condition $\det(N_1 + N_2U(b)) \ne 0$ implies that $T + B$ is invertible. Indeed, take $g \in L_2^n([a,b])$, and let us solve the equation

$$(7) \qquad\qquad if' + B(t)f = g$$

with f in the domain of $A = T + B$. The general solution of (7) is given by

$$(8) \qquad f(t) = U(t)x - iU(t)\int_a^t U(s)^{-1}g(s)ds, \qquad a \le t \le b.$$

Since f has to satisfy the boundary conditions, one sees that x in (8) is uniquely determined and given by

$$x = i\big(N_1 + N_2U(b)\big)^{-1}N_2U(b)\int_a^b U(s)^{-1}g(s)ds.$$

Thus (7) has a unique solution in $\mathcal{D}(T + B)$ for each $g \in L_2^n([a,b])$, and therefore $T + B$ is invertible. Thus $I + BT^{-1}$ is invertible, and we have proved that T and B have the desired properties. $\square$

For other completeness theorems for unbounded operators (including partial differential operators) we refer to the book of A.S. Markus [1].

CHAPTER XVIII

A CLASS OF ORDINARY DIFFERENTIAL OPERATORS
ON A HALF LINE

Ordinary differential operators on a half line differ considerably from their counterparts on a finite interval. In this chapter these differences are illustrated for a specific class of differential operators on $[0, \infty)$. The operators involved do not have a compact resolvent. Their spectra and essential spectra are described. Also, the Green's function and the Fredholm characteristics are computed explicitly. The first four sections concern first order constant coefficient differential operators. Applications to Wiener-Hopf integral equations appear in the fifth section. In the last section the results are extended to higher order differential operators on $[0, \infty)$.

XVIII.1 DEFINITION AND ADJOINT

The differential operators considered in this chapter act in the space $L_2^n([0, \infty))$, the Hilbert space of all $\mathbf{C}^n$-valued functions whose components are square (Lebesgue) integrable on $[0, \infty)$. Here n is a positive integer which we shall keep fixed throughout the chapter. Recall (see Sections XII.1 and XII.2) that for f and g in $L_2^n([0, \infty))$

$$\langle f, g \rangle = \int_0^\infty \langle f(t), g(t) \rangle dt,$$

where the inner product under the integral sign is the usual inner product in $\mathbf{C}^n$.

Consider the following initial value problem for $\mathbf{C}^n$-valued functions:

$$(1) \qquad \begin{cases} f'(t) = -Af(t), & t \geq 0, \\ f(0) \in L. \end{cases}$$

Here A is an $n \times n$ matrix with complex entries and L is a subspace of $\mathbf{C}^n$. With (1) we associate an operator T with domain and range in $L_2^n([0, \infty))$ as follows. The domain $\mathcal{D}(T)$ of T consists of all $f \in L_2^n([0, \infty))$ such that (each component of) f is absolutely continuous on compact intervals of $[0, \infty)$, the initial value $f(0) \in L$ and the derivative f' (which is defined component wise) is in $L_2^n([0, \infty))$. The action of T is given by:

$$(2) \qquad (Tf)(t) = f'(t) + Af(t), \qquad t \geq 0, \text{ a.e. }.$$

Obviously, $T\big(L_2^n([0, \infty)) \to L_2^n([0, \infty))\big)$ and T is linear. We shall refer to T as the *differential operator in $L_2^n([0, \infty))$ associated with* (1). If $L = \mathbf{C}^n$, then T may be viewed as the maximal operator corresponding to the differential expression $\tau f = f' + Af$ and the interval $[0, \infty)$.

THEOREM 1.1. *The differential operator T in $L_2^n([0,\infty))$ associated with (1) is a densely defined closed operator and $T^* = -S$, where S is the differential operator in $L_2^n([0,\infty))$ associated with*

$$(3) \qquad \begin{cases} f'(t) = A^*f(t), & t \geq 0, \\ f(0) \in L^\perp. \end{cases}$$

Here A^* is the adjoint of the matrix A in (1) and $L^\perp$ is the orthogonal complement in $\mathbf{C}^n$ of the subspace L in (1).

PROOF. Let $\mathcal{D}_0$ be the space of all $\mathbf{C}^n$-valued functions f on $[0,\infty)$ with the property that each component of f is a C^∞-function with compact support in the open interval $(0,\infty)$. Obviously, $f(0) = 0$ for each $f \in \mathcal{D}_0$. It follows that $\mathcal{D}_0 \subset \mathcal{D}(T)$. Since $\mathcal{D}_0$ is dense in $L_2^n([0,\infty))$ (cf., Lemma XIV.5.1) we conclude that T is densely defined.

Let S be the differential operator in $L_2^n([0,\infty))$ associated with (3). We want to show that $T^* = -S$. First we prove that T^* is an extension of $-S$. Take $f \in \mathcal{D}(T)$ and $g \in \mathcal{D}(S)$. Let us show that

$$(4) \qquad \langle f', g \rangle = -\langle f, g' \rangle.$$

The function $\langle f(\cdot), g(\cdot) \rangle$ is a finite sum of products of functions which are absolutely continuous on each compact subinterval of $[0,\infty)$. Thus $\langle f(\cdot), g(\cdot) \rangle$ is differentiable a.e. on $[0,\infty)$ and

$$(5) \qquad \frac{d}{dt}\langle f(t), g(t) \rangle = \langle f'(t), g(t) \rangle + \langle f(t), g'(t) \rangle, \qquad 0 \leq t < \infty, \text{ a.e. .}$$

Note that $f(0)\perp g(0)$. Thus $\langle f(0), g(0) \rangle = 0$ and integrating (5) over $0 \leq t \leq c$ yields

$$\langle f(c), g(c) \rangle = \int_0^c \langle f'(t), g(t) \rangle dt + \int_0^c \langle f(t), g'(t) \rangle dt.$$

This holds for each $c > 0$. According to our hypotheses, f' and g' are in $L_2^n([0,\infty))$, and so

$$(6) \qquad \lim_{c \to \infty} \langle f(c), g(c) \rangle = \langle f', g \rangle + \langle f, g' \rangle.$$

Since f and g are in $L_2^n([0,\infty))$, the function $\langle f(\cdot), g(\cdot) \rangle$ is integrable on $[0,\infty)$, which implies that the limit in the left hand side of (6) is zero, and (4) is proved. By using (4), we get

$$\begin{aligned}
\langle Tf, g \rangle &= \int_0^\infty \langle f'(t) + Af(t), g(t) \rangle dt \\
&= \int_0^\infty \langle f'(t), g(t) \rangle dt + \int_0^\infty \langle Af(t), g(t) \rangle dt \\
&= -\int_0^\infty \langle f(t), g'(t) \rangle dt + \int_0^\infty \langle f(t), A^*g(t) \rangle dt \\
&= \langle f, -Sg \rangle,
\end{aligned}$$

which yields that $g \in \mathcal{D}(T^*)$ and $T^*g = -Sg$. Thus $-S \subset T^*$.

Next, we prove that $T^* = -S$. Take $h \in \mathcal{D}(T^*)$. It suffices to show that $h \in \mathcal{D}(-S)$. So we have to prove that h is absolutely continuous on each compact subinterval of $[0, \infty)$, the derivative $h' \in L_2^n([0, \infty))$ and $h(0) \perp L$. Fix an arbitrary positive number c. For $j = 1, \ldots, n$ let $\mathcal{X}_j$ be the $\mathbf{C}^n$-valued function defined by

$$\mathcal{X}_j(t) = \begin{bmatrix} \delta_{1j} \\ \vdots \\ \delta_{nj} \end{bmatrix}, \qquad 0 \leq t \leq c,$$

where δ_{ij} is the Kronecker delta. The subspace of $L_2^n([0, c])$ spanned by $\mathcal{X}_1, \ldots, \mathcal{X}_n$ will be denoted by $\mathcal{M}$. Take $u \in L_2^n([0, c])$ such that $u \perp \mathcal{M}$, and put

$$g(t) = \begin{cases} \int\limits_0^t u(s)ds, & 0 \leq t \leq c, \\ 0, & t > c. \end{cases}$$

For the j-th component g_j of g we have

$$g_j(t) = \int\limits_0^t \langle u(t), \mathcal{X}_j(t) \rangle dt, \qquad 0 \leq t \leq c.$$

It follows that $g_j(c) = 0$ for $j = 1, \ldots, n$. Hence $g(c) = 0$, and we may conclude that g is absolutely continuous on each compact subinterval of $[0, \infty)$. Obviously, $g(0) = 0$ and $g' \in L_2^n([0, \infty))$. Thus $g \in \mathcal{D}(T)$ and

$$\int\limits_a^c \langle g(t), (T^*h)(t) \rangle dt = \int\limits_0^\infty \langle g(t), (T^*h)(t) \rangle dt$$

$$= \int\limits_0^\infty \langle g'(t) + Ag(t), h(t) \rangle dt$$

$$= \int\limits_0^c \langle g'(t), h(t) \rangle dt + \int\limits_0^c \langle g(t), A^*h(t) \rangle dt.$$

Here we used that g and g' are zero on (c, ∞). Put

$$v(t) = \int\limits_0^t \{(T^*h)(s) - A^*h(s)\}ds, \qquad 0 \leq t \leq c.$$

Then the result of the previous calculation can be summarized as:

$$(7) \qquad \int\limits_0^c \langle g'(t), h(t) \rangle dt = \int\limits_0^c \langle g(t), v'(t) \rangle dt.$$

Since $g(0) = g(c) = 0$ and $g'(t) = u(t)$ a.e. on $0 \leq t \leq c$, partial integration in (7) yields

$$\int_0^c \langle u(t), h(t) + v(t) \rangle dt = 0,$$

which shows that on $[0, c]$ the function $h + v$ is orthogonal to u. But u is an arbitrary element of the orthogonal complement of $\mathcal{M}$ in $L_2^n([0, c])$. We conclude that on $[0, c]$ the function $h + v$ is a linear combination of the functions $\mathcal{X}_1, \ldots, \mathcal{X}_n$. Note that the functions $\mathcal{X}_1, \ldots, \mathcal{X}_n$ and v are absolutely continuous on $[0, c]$. Moreover, $\mathcal{X}_1', \ldots, \mathcal{X}_n'$ are zero. It follows that h is absolutely continuous on $[0, c]$ and

$$h'(t) = A^* h(t) - (T^* h)(t), \qquad 0 \leq t \leq c, \text{ a.e. } .$$

This holds for each $c > 0$. Thus h is absolutely continuous on each compact interval and

$$(9) \qquad h'(t) = A^* h(t) - (T^* h)(t), \qquad 0 \leq t < \infty, \text{ a.e. } .$$

Since the right hand side of (9) is in $L_2^n([0, \infty))$, we also have $h' \in L_2^n([0, \infty))$. To prove that $h(0) \in L^\perp$, take $x \in L$, and let

$$f(t) = \begin{cases} \int_t^1 x dt, & 0 \leq t \leq 1, \\ \\ 0, & t > 1. \end{cases}$$

Then $f \in \mathcal{D}(T)$, and thus

$$\int_0^1 \langle f(t), h'(t) \rangle dt = \int_0^\infty \langle f(t), h'(t) \rangle dt$$

$$= \int_0^\infty \langle f(t), A^* h(t) - (T^* h)(t) \rangle dt$$

$$= \int_0^\infty \langle Af(t), h(t) \rangle dt - \int_0^\infty \langle (Tf)(t), h(t) \rangle dt$$

$$= -\int_0^\infty \langle f'(t), h(t) \rangle dt$$

$$= -\int_0^1 \langle f'(t), h(t) \rangle dt.$$

Since $f(1) = 0$, the above calculation and the formula for partial integration shows that

$$\langle x, h(0) \rangle = \langle f(0), h(0) \rangle = 0.$$

This holds for each $x \in L$, and so $h(0) \in L^\perp$.

We have now proved that $T^* = -S$. Note that S is an operator of the same type as T. Thus we may apply to S the results proved so far. It follows that $T^{**} = -S^* = T$, and therefore T is closed (by Proposition XIV.2.2). $\square$

XVIII.2 INVERTIBILITY AND GREEN'S FUNCTION

Let T be the differential operator in $L_2^n([0,\infty))$ associated with

$$
(1) \qquad \begin{cases} f'(t) = -Af(t), & t \geq 0, \\ f(0) \in L. \end{cases}
$$

As in the previous section, A is assumed to be an $n \times n$ matrix and L is a subspace of $\mathbf{C}^n$. In this section we analyze the invertibility of T and compute its resolvent kernel (Green's function).

THEOREM 2.1. *Let T be the differential operator in $L_2^n([0,\infty))$ associated with* (1). *Then T is invertible if and only if A has no eigenvalues on the imaginary axis and*

$$
(2) \qquad \mathbf{C}^n = L \oplus \operatorname{Ker} Q,
$$

where Q is the Riesz projection of A corresponding to the eigenvalues of A in the open left half plane. Furthermore, in that case the inverse of T is the integral operator on $L_2^n([0,\infty))$ given by

$$
(T^{-1}g)(t) = \int_0^\infty \gamma(t,s)g(s)ds, \qquad 0 \leq t < \infty,
$$

with

$$
(3) \qquad \gamma(t,s) = \begin{cases} e^{-tA}(I - \Pi)e^{sA}, & 0 \leq s < t < \infty, \\ -e^{-tA}\Pi e^{sA}, & 0 \leq t < s < \infty, \end{cases}
$$

where Π is the projection of $\mathbf{C}^n$ onto L along $\operatorname{Ker} Q$.

PROOF. We split the proof into four parts. The first part concerns a general statement about T.

Part (i). First we show that for $f \in \mathcal{D}(T)$ and $Tf = g$ the following equality holds true:

$$
(4) \qquad Qf(0) = - \int_0^\infty Qe^{sA}g(s)ds.
$$

Since $Tf = g$, we have $f'(t) = -Af(t) + g(t)$ a.e. on $[0,\infty)$, and hence

$$
(5) \qquad f(t) = e^{-tA}f(0) + e^{-tA}\int_0^t e^{sA}g(s)ds, \qquad t \geq 0.
$$

Multiplying (5) from the left by Qe^{tA} yields

$$
(6) \qquad Qe^{tA}f(t) = Qf(0) + \int_0^t Qe^{sA}g(s)ds, \qquad t \geq 0.
$$

Recall that Q is the Riesz projection of A corresponding to the eigenvalues in the open left half plane. This implies (cf. Lemma I.5.3) that $\|Qe^{tA}\| \leq \gamma e^{-\delta t}$ for some constants $\gamma \geq 0$ and $\delta > 0$. It follows that $Qe^{tA}h(t)$ is integrable on $0 \leq t < \infty$ for any $h \in L_2^n([0,\infty))$, in particular, for $h = f$ and $h = g$. The latter implies that the right hand side of (6) has a limit for $t \to \infty$. Since the left hand side of (6) is integrable on $[0,\infty)$, this limit must be zero, which proves (4).

Part (ii). In this part we assume that A has no eigenvalues on the imaginary axis and that T is invertible. We shall show that (2) holds. Take $x \in L \cap \operatorname{Ker} Q$, and consider the function $f(t) = e^{-tA}x$. Since A has no eigenvalues on the imaginary axis, $x \in \operatorname{Ker} Q$ implies that $f \in L_2^n([0,\infty))$. But then $f \in \mathcal{D}(T)$ and $Tf = 0$. Therefore, by the injectivity of T, we have $f = 0$. It follows that $x = f(0) = 0$, and thus $L \cap \operatorname{Ker} Q = \{0\}$.

Next, take $y \in \mathbf{C}^n$, and consider the function

$$(7) \qquad g_k(t) = \begin{cases} Ay, & 0 \leq t \leq k, \\ 0, & t > k, \end{cases}$$

where k is an arbitrary positive integer. Since T is surjective, there exists $f_k \in \mathcal{D}(T)$ such that $Tf_k = g_k$. It follows (from (4)) that

$$Qf_k(0) = -\int_0^\infty Qe^{sA}g_k(s)ds = Qy - Qe^{kA}y.$$

We know that $f_k(0) \in L$. Thus

$$Qy - Qe^{kA}y = f_k(0) - (I - Q)f_k(0) \in L + \operatorname{Ker} Q.$$

The function Qe^{tA} is exponentially decaying on $0 \leq t < \infty$. Hence $Qe^{kA}y \to 0$ if $k \to \infty$. Now use that $L + \operatorname{Ker} Q$ is closed (because of finite dimensionality), and it follows that $Qy \in L + \operatorname{Ker} Q$. But then $y = Qy + (I - Q)y \in L + \operatorname{Ker} Q$. Recall that y is an arbitrary vector in $\mathbf{C}^n$. So we may conclude that $L + \operatorname{Ker} Q = \mathbf{C}^n$.

Part (iii). In this part we assume that T is invertible. We shall show that A has no eigenvalues on the imaginary axis. For $\alpha \in \mathbf{R}$ and α sufficiently small the operator $T - \alpha I$ is again invertible (cf. Proposition XIV.1.1). Note that $\mathcal{D}(T - \alpha I) = \mathcal{D}(T)$ and for $f \in \mathcal{D}(T - \alpha I)$

$$([T - \alpha I]f)(t) = f'(t) + (A - \alpha I_n)f(t), \qquad t \geq 0, \text{ a.e.} .$$

Here I_n is the $n \times n$ identity matrix. It follows that $T - \alpha I$ is the differential operator on $L_2^n([0,\infty))$ associated with the initial value problem (1) with $A - \alpha I_n$ in place of A. Now take $\alpha \neq 0$ and α sufficiently small. Then $A - \alpha I_n$ has no eigenvalues on the imaginary axis. So, by the result of Part (ii) of the proof,

$$(8) \qquad \mathbf{C}^n = L \oplus \operatorname{Ker} Q_\alpha,$$

where Q_α is the Riesz projection of A corresponding to the eigenvalues in the open half plane $\Re\lambda < \alpha$. We conclude that for $0 < \alpha \in \mathbf{R}$ and α sufficiently small, $\operatorname{Ker} Q_\alpha = \operatorname{Ker} Q_{-\alpha}$. But for $\alpha > 0$ the space $\operatorname{Ker} Q_{-\alpha}$ contains the eigenvectors of A corresponding to eigenvalues on the imaginary axis, while these vectors are not in $\operatorname{Ker} Q_\alpha$. Therefore A has no eigenvalues on the imaginary axis.

The results proved in Parts (ii) and (iii) show that for the invertibility of T it is necessary that A has no eigenvalues on the imaginary axis and the decomposition (2) holds. The final part of the proof concerns the sufficiency of these conditions and the formula for the inverse.

Part (iv). Assume that A has no eigenvalue on the imaginary axis and that (2) holds. First, we show that T is injective. Assume $Tf = 0$. By (4) and (5) we have $f(t) = e^{-tA}f(0)$ with $f(0) \in \operatorname{Ker} Q$. Since $f \in \mathcal{D}(T)$, also $f(0) \in L$, and hence $f(0) = 0$ by (2). We conclude that $f = 0$, which shows that T is injective. Next, take $g \in L_2^n([0,\infty))$, and put

$$x = -\Pi \int\limits_0^\infty Q e^{sA} g(s)\, ds,$$

where Π is the projection of $\mathbf{C}^n$ onto L along $\operatorname{Ker} Q$. Note that $Q e^{sA} g(s)$ is integrable on $[0,\infty)$, and thus x is a well-defined vector of L. Define

$$(9) \qquad f(t) = e^{-tA}x + e^{-tA}\int\limits_0^t e^{sA} g(s)\, ds, \qquad 0 \le t < \infty.$$

Then f is absolutely continuous on compact subintervals of $[0,\infty)$, the initial vector $f(0) = x \in L$ and

$$(10) \qquad f'(t) = -Af(t) + g(t), \qquad 0 \le t < \infty.$$

To prove that $f \in \mathcal{D}(T)$, we have to show that $f \in L_2^n([0,\infty))$. To do this, note that $Q\Pi = Q$. This allows us to rewrite f in the form

$$(11) \qquad f(t) = -e^{-tA}(I - Q)\Pi \int\limits_0^\infty Q e^{sA} g(s)\, ds + \int\limits_0^\infty h(t-s)g(s)\, ds, \qquad 0 \le t < \infty,$$

where

$$(12) \qquad h(t) = \begin{cases} e^{-tA}(I - Q), & t \ge 0, \\ -e^{-tA}Q, & t < 0. \end{cases}$$

Here we used that A commutes with Q. Note that h is a matrix-valued kernel function, which is continuous on $\mathbf{R}\backslash\{0\}$, has a jump discontinuity at 0 and

$$\|h(t)\| \le ce^{-d|t|}, \qquad t \in \mathbf{R},$$

for some constants $c \geq 0$ and $d > 0$. It follows that the integral in the right hand side of (11) defines a bounded linear operator on $L_2^n([0, \infty))$ (cf., Sections XII.1 and XII.2). The same is true for the first term in the right hand side of (11), because $e^{-tA}(I - Q)$ and Qe^{tA} are matrices with square integrable entries. In particular, $f \in \mathcal{D}(T)$.

We have now proved that T is bijective and $T^{-1}g$ is the function defined by the right hand side of (11). Thus T^{-1} is a bounded linear operator on $L_2^n([0, \infty))$, and so T is invertible. It remains to establish formula (3) for the resolvent kernel. To obtain (3) one uses the fact that $\Pi(I - Q) = 0$ and $Q(I - \Pi) = 0$. These two equalities imply that $(I - Q)\Pi Q + Q = \Pi$. Since $T^{-1}g$ is given by the right hand side of (11), the latter identity, $Qe^{tA} = e^{tA}Q$ for all $t \in \mathbf{R}$ and formulas (11) and (12) yield the desired representation (3). $\square$

XVIII.3 THE SPECTRUM

Let T be the differential operator in $L_2^n([0, \infty))$ associated with

$$(1) \qquad \begin{cases} f'(t) = -Af(t), & t \geq 0, \\ f(0) \in L. \end{cases}$$

As in the previous section, A is assumed to be an $n \times n$ matrix and L is a subspace of $\mathbf{C}^n$. The next theorem describes the spectrum of T.

THEOREM 3.1. *Let T be the differential operator in $L_2^n([0, \infty))$ associated with (1), and let $a_1, \ldots, a_s$ be the real parts of the eigenvalues of A ordered increasingly. Put $a_0 = -\infty$ and $a_{s+1} = \infty$. Then the spectrum $\sigma(T)$ is the entire complex plane or consists of the closed half planes $\Re\lambda \leq a_{j-1}$ and $\Re\lambda \geq a_j$ for some $1 \leq j \leq s + 1$.*

PROOF. Assume $\sigma(T) \neq \mathbf{C}$. Take $\lambda_0 \in \rho(T)$, and write $\lambda_0 = a + ib$ with a and b real. The operator $T - \lambda_0 I$ is the differential operator associated with the initial value problem (1) with $A - \lambda_0 I$ in place of A (cf. Part (iii) of the proof of Theorem 2.1). Since $T - \lambda_0 I$ is invertible, we can apply Theorem 2.1 to show that A has no eigenvalues on the line $\Re\lambda = a$ and

$$(2) \qquad \mathbf{C}^n = L \oplus \operatorname{Ker} Q_a,$$

where Q_a is the Riesz projection of A corresponding to the eigenvalues in the half plane $\Re\lambda < a$. In particular, $a_{j-1} < a < a_j$ for some $j \in \{1, 2, \ldots, s+1\}$.

In what follows, Q_α denotes the Riesz projection of A corresponding to the eigenvalues in $\Re\lambda < \alpha$. We have

$$(3a) \qquad \operatorname{Ker} Q_\alpha \subset \operatorname{Ker} Q_a, \quad \operatorname{Ker} Q_\alpha \neq \operatorname{Ker} Q_a, \qquad \alpha > a_j,$$

$$(3b) \qquad \operatorname{Ker} Q_\alpha = \operatorname{Ker} Q_a, \qquad a_{j-1} < \alpha \leq a_j,$$

$$(3c) \qquad \operatorname{Ker} Q_\alpha \supset \operatorname{Ker} Q_a, \quad \operatorname{Ker} Q_\alpha \neq \operatorname{Ker} Q_a, \qquad \alpha \leq a_{j-1}.$$

But then we see from (2) that

$$(4) \qquad L + \operatorname{Ker} Q_\alpha \neq \mathbf{C}^n \ (\alpha > a_j), \qquad L \cap \operatorname{Ker} Q_\alpha \neq \{0\} \ (\alpha \leq a_{j-1}).$$

Now, take $\lambda \in \mathbf{C}$, and put $\alpha = \Re\lambda$. Then Q_α is the Riesz projection of $A - \lambda$ corresponding to the eigenvalues in the open left half plane. Thus, by Theorem 2.1 and (4), the operator $T - \lambda I$ is not invertible if $\Re\lambda > a_j$ or $\Re\lambda \leq a_{j-1}$. Since $\sigma(T)$ is closed, we conclude the the closed half planes $\Re\lambda \geq a_j$ and $\Re\lambda \leq a_{j-1}$ are in $\sigma(T)$.

Finally, take $a_{j-1} < \Re\lambda < a_j$, and put $\alpha = \Re\lambda$. Then $A - \lambda I_n$ has no eigenvalues on the imaginary axis and (2) holds with α in place of a. It follows from Theorem 2.1 that $T - \lambda I$ is invertible, and hence $\lambda \notin \sigma(T)$. $\square$

The results of this and the previous section are of particular interest for a number of different choices of L. Here we illustrate this with two corollaries; the first concerns the case $L = \{0\}$ and in the second we take $L = \mathbf{C}^n$.

COROLLARY 3.2. *Let T be the differential operator in $L_2^n([0,\infty))$ associated with* (1). *Assume $L = \{0\}$. Then the spectrum of T consists of the closed half plane $\Re\lambda \geq a$, where $a = \min\{\Re\lambda \mid \lambda \in \sigma(A)\}$. Furthermore, for $\Re\lambda < a$*

$$([\lambda - T]^{-1}f)(t) = -\int_0^t e^{(t-s)(\lambda - A)} f(s)ds, \qquad 0 \leq t < \infty.$$

PROOF. By Theorem 2.1, the operator T is invertible if and only if all the eigenvalues of A are in the open right half plane and in that case

$$(T^{-1}f)(t) = \int_0^t e^{-(t-s)A} f(s)ds, \qquad 0 \leq t < \infty.$$

Now, apply this result with T and A replaced by $T - \lambda I$ and $A - \lambda I_n$, respectively, and the corollary follows. $\square$

The second corollary is proved in a similar way; we omit the details.

COROLLARY 3.3. *Let T be the differential operator in $L_2^n([0,\infty))$ associated with* (1). *Assume $L = \mathbf{C}^n$. Then the spectrum of T consists of the closed half plane $\Re\lambda \leq b$, where $b = \max\{\Re\lambda \mid \lambda \in \sigma(A)\}$. Furthermore, for $\Re\lambda > b$*

$$([\lambda - T]^{-1}f)(t) = \int_t^\infty e^{(t-s)(\lambda - A)} g(s)ds, \qquad 0 \leq t < \infty.$$

XVIII.4 FREDHOLM CHARACTERISTICS

This section concerns the Fredholm properties of the differential operator associated with

$$(1) \qquad \begin{cases} f'(t) = -Af(t), & t \geq 0, \\ f(0) \in L. \end{cases}$$

As in the previous section, A is an $n \times n$ matrix and L is a subspace of $\mathbf{C}^n$. We shall prove the following two theorems.

THEOREM 4.1. *Let T be the differential operator in $L_2^n([0,\infty))$ associated with (1). Then T is a Fredholm operator if and only if A has no eigenvalues on the imaginary axis, and in that case*

$$\text{(2)} \qquad \operatorname{ind} T = \dim L - \dim M,$$

where M is the space spanned by the eigenvectors and generalized eigenvectors of A corresponding to eigenvalues of A in the open left half plane. Furthermore, the essential spectrum of T consists of the union of the lines parallel to the imaginary axis through the eigenvalues of A.

THEOREM 4.2. *Let T be the differential operator in $L_2^n([0,\infty))$ associated with (1), and assume that A has no eigenvalues on the imaginary axis. Let Q be the Riesz projection corresponding to the eigenvalues of A in the open left half plane, and put $M = \operatorname{Im} Q$, $N = \operatorname{Ker} Q$. Then T is a Fredholm operator and*

$$\text{(3)} \qquad \operatorname{Ker} T = \{ f \mid f(t) = e^{-tA} x, \ x \in L \cap N \},$$

$$\text{(4)} \qquad \operatorname{Im} T = \left\{ g \in L_2^n([0,\infty)) \mid \int_0^\infty Q e^{sA} g(s) ds \in L + N \right\},$$

$$\text{(5)} \qquad n(T) = \dim(L \cap N), \qquad d(T) = \dim \mathbf{C}^n/(L + N),$$

$$\text{(6)} \qquad \operatorname{ind} T = \dim L - \dim M.$$

Furthermore, let Γ be the bounded linear operator on $L_2^n([0,\infty))$ defined by

$$(\Gamma f)(t) = \int_0^\infty \gamma(t,s) f(s) ds, \qquad t \geq 0,$$

with

$$\gamma(t,s) = \begin{cases} e^{-tA}[I - Q - (I - Q)S^+ Q]e^{sA}, & 0 \leq s < t < \infty, \\ -e^{-tA}[Q + (I - Q)S^+ Q]e^{sA}, & 0 \leq t < s < \infty, \end{cases}$$

where $S^+ : M \to L$ is a generalized inverse of the operator $S = Q|L : L \to M$ in the weak sense (i.e., $SS^+ S = S$). Then

$$\text{(7)} \qquad T = T\Gamma T, \quad \Gamma f = \Gamma T \Gamma f \qquad ((\Gamma f)(0) \in L).$$

It is convenient to prove the second theorem first.

PROOF OF THEOREM 4.2. The proof is split into four parts. In the first part we recall some facts proved in Section 2.

Part (i). For $f \in \mathcal{D}(T)$ and $Tf = g$ the following identities hold true:

$$(8) \qquad Qf(0) = -\int_0^\infty Qe^{sA}g(s)ds,$$

$$(9) \qquad f(t) = e^{-tA}(I - Q)f(0) + \int_0^\infty h(t - s)g(s)ds, \qquad 0 \le t < \infty,$$

where h is the $n \times n$ matrix function defined by formula (12) in Section 2. Since A has no eigenvalues on the imaginary axis, we know that

$$\|h(t)\| \le ce^{-d|t|}, \qquad t \in \mathbf{R},$$

for some positive constants c and d. Hence the integral in (9) is well-defined. Note that formula (8) is proved in Part (i) of the proof of Theorem 2.1. Formula (9) follows from (8) and the identity (5) in Section 2.

Part (ii). In this part we prove (3) and (4). Assume that $f \in \operatorname{Ker} T$. Then $g = Tf = 0$, and (9) implies that $f(t) = e^{-tA}z$ for some $z \in \operatorname{Ker} Q = N$. Also, $z = f(0) \in L$. So $z \in L \cap N$. Conversely, if $f(t) = e^{-tA}z$ with $z \in L \cap N$, then $f \in \mathcal{D}(T)$ and $Tf = 0$. Thus $f \in \operatorname{Ker} T$ and (3) is proved.

Assume $g \in \operatorname{Im} T$. So $g = Tf$ for some $f \in \mathcal{D}(T)$. Formula (8) tells us that

$$(10) \qquad \int_0^\infty Qe^{sA}g(s)ds \in L + N.$$

Conversely, if $g \in L_2^m([0, \infty))$ and (10) holds, then there exists $x \in L$ such that Qx is equal to the left hand side of (10). Using this x, put

$$(11) \qquad f(t) = -e^{-tA}(I - Q)x + \int_0^\infty h(t - s)g(s)ds, \qquad 0 \le t < \infty.$$

Then $f \in L_2^n([0, \infty))$ and

$$f(t) = -e^{-tA}x + e^{-tA}\int_0^\infty Qe^{sA}g(s)ds + \int_0^\infty h(t - s)g(s)ds$$

$$= -e^{-tA}x + e^{-tA}\int_0^t e^{sA}g(s)ds, \qquad 0 \le t < \infty.$$

Here we used that Q commutes with A. From the above calculation it is clear that f is absolutely continuous on each compact subinterval of $[0, \infty)$, the vector $f(0) = -x \in L$ and $f' = -Af + g$. Thus $f \in \mathcal{D}(T)$ and $Tf = g$. Hence $g \in \operatorname{Im} T$, and we have proved (4).

Part (iii). Next, we prove (5) and (6). Note that the map $x \mapsto e^{-tA}x$ is an isomorphism from $L \cap N$ onto $\operatorname{Ker} T$, because of (3). It follows that the first identity in (5) holds. To prove the second, define

$$J \colon L_2^n([0, \infty)) \to \frac{\mathbf{C}^n}{L + N}, \qquad Jg = [\int_0^\infty Q e^{sA} g(s)\,ds],$$

where $[y]$ denotes the coset $y + (L + N)$ for any $y \in \mathbf{C}^n$. According to (4) we have $\operatorname{Ker} J = \operatorname{Im} T$. So to prove the second identity in (5) it suffices to show that J is surjective. Take $x \in \mathbf{C}^n$. Then $(I - Q)x \in N$, and so $[x] = [Qx]$. Put

$$g_k(t) = \begin{cases} -Ax & \text{for} \quad 0 \le t \le k, \\ 0 & \text{for} \quad t > k. \end{cases}$$

Then $g_k \in L_2^n([0, \infty))$ for $k = 1, 2, \dots$ and

$$\tag{12} Jg_k = -[\int_0^k Q e^{sA} Ax\,ds]$$

$$= [Qx - Q e^{kA} x] \to [Qx], \qquad k \to \infty.$$

Here we used that $Q e^{tA} x$ is exponentially decaying. Since $\mathbf{C}^n/(L + N)$ is finite dimensional, the image of J is closed, and thus (12) implies that $[x] = [Qx] \in \operatorname{Im} J$. Thus J is surjective, which finishes the proof of (5).

From Theorem 1.1 we know that T is closed. Thus the identities in (5) imply that T is Fredholm and

$$\begin{aligned} \operatorname{ind} T &= \dim(L \cap N) - \{n - \dim(L + N)\} \\ &= \dim L + \dim N - n \\ &= \dim L - \dim M, \end{aligned}$$

because M and N are complementary subspaces of $\mathbf{C}^n$.

Part (iv). It remains to establish (7). Take $g \in \operatorname{Im} T$. We know (see (4)) that

$$\tag{13} z := \int_0^\infty Q e^{sA} g(s)\,ds \in L + N.$$

Thus $z = Qz \in QL = \operatorname{Im} S$. Pur $x = S^+ z$. Then $x \in L$ and

$$Qx = QS^+ z = SS^+ z = z,$$

because $SS^+S = S$. Thus $x \in L$ and Qx is equal to the integral in (13). Now define f by the right hand side of (11). Then (as we have seen before), the function $f \in \mathcal{D}(T)$ and $Tf = g$. Because of the special form of x, the function f may be rewritten as

$$f(t) = -e^{-tA}(I - Q)S^+ \int_0^\infty Qe^{sA}g(s)ds + \int_0^\infty h(t - s)g(s)ds$$

(14)

$$= \int_0^\infty \gamma(t,s)g(s)ds = (\Gamma g)(t), \qquad 0 \le t < \infty.$$

We conclude that $T\Gamma g = g$ for each $g \in \operatorname{Im} T$, and hence the first identity in (7) holds.

For any $f \in L_2^n([0, \infty))$ the function Γf is absolutely continuous on each compact subinterval of $[0, \infty)$ and $(\Gamma f)' = -A(\Gamma f) + f$. Thus, if, in addition, $(\Gamma f)(0) \in L$, then $\Gamma f = \mathcal{D}(T)$ and $T\Gamma f = f$, which proves the second identity in (7). Note that the first part of (14) implies that Γ is a bounded linear operator on $L_2^n([0, \infty))$. $\square$

PROOF OF THEOREM 4.1. Assume that A has no eigenvalues on the imaginary axis. Note that the space M is precisely the image of the projection Q appearing in Theorem 4.2. So we know from Theorem 4.2 that T is a Fredholm operator with index given by (2).

To prove the converse statement, let us assume that A has eigenvalues on the imaginary axis. We have to show that T is not a Fredholm operator. By contradiction, assume that T is a Fredholm operator. Then, by Theorem XVII.4.2, for $\alpha \in \mathbf{R}$, α sufficiently small, the operator $T + \alpha I$ is again a Fredholm operator and

(15) $$\operatorname{ind}(T + \alpha I) = \operatorname{ind} T.$$

Note that $\mathcal{D}(T + \alpha I) = \mathcal{D}(T)$ and for $f \in \mathcal{D}(T + \alpha I)$

$$((T + \alpha I)f)(t) = f'(t) + (A + \alpha I_n)f(t), \qquad t \ge 0, \text{ a.e. .}$$

Here I_n is the $n \times n$ identity matrix. It follows that $T + \alpha I$ is the differential operator on $L_2^n([0, \infty))$ associated with the initial value problem (1) with $A + \alpha I_n$ in place of A. Now take $\alpha \ne 0$ and α sufficiently small. Then $A + \alpha I_n$ has no eigenvalues on the imaginary axis. So, by what we have proved so far,

(16) $$\operatorname{ind}(T + \alpha I) = \dim L - \dim M_\alpha,$$

where M_α is the spectral subspace spanned by the eigenvectors and generalized eigenvectors of A corresponding to eigenvalues in the half plane $\Re\lambda < \alpha$. By comparing (15) and (16) we see that $\dim M_\alpha = \dim M_{-\alpha}$. But this contradicts the fact that A has eigenvalues on the imaginary axis. Thus T is not a Fredholm operator.

It remains to prove the statement about the essential spectrum of T. Take $\mu \in \mathbf{C}$. We have already seen that $T - \mu I$ is the differential operator associated with the initial value problem (1) with $A - \mu I_n$ in place of A. So according to the first part of the theorem, the operator $T - \mu I$ is Fredholm if and only if $A - \mu I_n$ has no eigenvalue on the imaginary axis. Thus $\mu \in \sigma_{\mathrm{ess}}(T)$ if and only if there exists $a \in \mathbf{R}$ such that ia is an eigenvalue of $A - \mu I_n$, or, equivalently, $ia + \mu$ is an eigenvalue of A. It follows that $\mu \in \sigma_{\mathrm{ess}}(T)$ if and only if the line through μ parallel to the imaginary axis contains an eigenvalue of A. In other words, $\sigma_{\mathrm{ess}}(T)$ is precisely the union of the lines parallel to the imaginary axis through the eigenvalues of A. $\square$

XVIII.5 APPLICATIONS TO WIENER-HOPF OPERATORS

In this section the results of the previous sections are used to analyse Wiener-Hopf operators. This analysis provides an alternative way to derive the inversion and Fredholm properties of Wiener-Hopf operators on $L_2^m([0,\infty))$ with a rational symbol (Theorems XIII.7.1 and 8.1). Also an application to so-called first kind Wiener-Hopf operators is included.

Let $S: L_2^m([0,\infty)) \to L_2^m([0,\infty))$ be a Wiener-Hopf operator with a rational $m \times m$ matrix symbol W. Thus

$$(1) \qquad (S\varphi)(t) = \varphi(t) - \int_0^\infty k(t-s)\varphi(s)ds, \qquad 0 \le t < \infty,$$

and

$$(2) \qquad W(\lambda) = I_m - \int_{-\infty}^\infty e^{i\lambda t}k(t)ds, \qquad \lambda \in \mathbf{R}$$

(cf., Section XIII.1). Since the symbol W is rational, we know from Section XIII.4 that W can be written in *realized form*, that is,

$$(3) \qquad W(\lambda) = I + C(\lambda - A)^{-1}B, \qquad \lambda \in \mathbf{R},$$

where A is a square matrix of order n, say, which has no eigenvalues on the real line, and B and C are matrices of sizes $n \times m$ and $m \times n$, respectively. First, we employ (3) to derive a special representation of S.

LEMMA 5.1. *Let S be the Wiener-Hopf operator on $L_2^m([0,\infty))$ with the symbol W given in the realized form (3). Then*

$$(4) \qquad S = I - iM_C T^{-1} M_B,$$

where M_B and M_C are the multiplication operators defined by

$$(5) \qquad M_B: L_2^m([0,\infty)) \to L_2^n([0,\infty)), \quad (M_B f)(t) = Bf(t) \qquad 0 \le t < \infty,$$

$$(6) \qquad M_C: L_2^n([0,\infty)) \to L_2^m([0,\infty)), \quad (M_C f)(t) = Cf(t), \qquad 0 \le t < \infty,$$

and T is the differential operator in $L_2^n([0,\infty))$ associated with

(7)
$$\begin{cases} f'(t) = -iAf(t), & t \geq 0, \\ f(0) \in \operatorname{Im} P. \end{cases}$$

Here P is the Riesz projection of A corresponding to the eigenvalues in the open upper half plane.

PROOF. First, we prove that T is invertible. By our assumption on A the matrix iA has no eigenvalues on the imaginary axis and the Riesz projection of iA corresponding to the eigenvalues in the open left half plane is precisely P. Now, apply Theorem 2.1 with iA in place of A and with $L = \operatorname{Im} P$. In this case the matching condition in formula (2) of Section 2 is automatically fulfilled. It follows that the differential operator T associated with (7) is invertible. Furthermore, for the case considered here, the projection Π appearing in Theorem 2.1 is equal to P. But then we may conclude from Theorem 2.1 that

(8)
$$(T^{-1}g)(t) = \int_0^\infty h(t-s)g(s)\,ds, \qquad 0 \leq t < \infty,$$

where

(9)
$$h(t) = \begin{cases} e^{-itA}(I - P), & t > 0, \\ -e^{-itA}P, & t < 0. \end{cases}$$

Here we used that P and e^{-itA} commute.

Next, recall from Theorem XIII.4.2 that the realization (3) implies that the kernel function k of S in (1) is given by

$$k(t) = \begin{cases} iCe^{-itA}(I - P)B, & t > 0, \\ -iCe^{-itA}PB, & t < 0. \end{cases}$$

But then we see from (9) that $k(\cdot) = Ch(\cdot)B$, which, according to (8) and (1), implies that (4) holds. $\square$

The representation (4) will allow us to describe the inversion and Fredholm properties of the Wiener-Hopf operator S in terms of the differential operator $T^\times$ associated with the following initial value problem:

(10)
$$\begin{cases} f'(t) = -i(A - BC)f(t), & t \geq 0, \\ f(0) \in \operatorname{Im} P, \end{cases}$$

where P is as in (7).

THEOREM 5.2. *Let S be the Wiener-Hopf operator on $L_2^m([0,\infty))$ with the symbol W given in the realized form (3), and let $T^\times$ be the differential operator in*

$L_2^n([0, \infty))$ *associated with* (10). *Then* S *is invertible if and only if* $T^\times$ *is invertible, and in that case*

$$(11) \qquad S^{-1} = I + i M_C (T^\times)^{-1} M_B,$$

where M_B *and* M_C *are as in* (5) *and* (6), *respectively.*

PROOF. We shall apply Lemma 5.1. Let T be the differential operator associated with the problem (7). By Lemma 5.1 the operator T is invertible. Note that T and $T^\times$ have the same domain and

$$(12) \qquad T^\times f = Tf - i M_B M_C f, \qquad f \in \mathcal{D}(T) = \mathcal{D}(T^\times).$$

It follows that the operator $S^\times = T^\times T^{-1}$ is a well-defined bounded linear operator on $L_2^n([0, \infty))$, which is invertible if and only if $T^\times$ is invertible. According to (12), we have

$$(13) \qquad S^\times = I - M_B (i M_C T^{-1}).$$

The formulas (4) and (13) imply that the operators S and $S^\times$ are matricially coupled (see Section III.4). Hence, we can apply Corollary III.4.3 to show that S is invertible if and only if $S^\times$ is invertible, and in that case

$$S^{-1} = I + i M_C T^{-1} (S^\times)^{-1} M_B.$$

Since $S^\times = T^\times T^{-1}$, this proves the theorem. $\square$

Let $T^\times$ be the differential operator associated with (10). Apply Theorem 2.1 to $T^\times$. This yields necessary and sufficient conditions for the invertibility of $T^\times$ and (when these conditions are fulfilled) an explicit formula for $(T^\times)^{-1}$. Using these results in Theorem 5.2 yields an inversion theorem for Wiener-Hopf operators which is precisely, Theorem XIII.7.1.

The next theorem concerns the Fredholm properties.

THEOREM 5.3. *Let* S *be the Wiener-Hopf operator on* $L_2^m([0, \infty))$ *with the symbol* W *given in the realized form* (3), *and let* $T^\times$ *be the differential operator in* $L_2^n([0, \infty))$ *associated with* (10). *Then*

$$(14) \qquad \operatorname{Ker} S = \{ Cf \mid f \in \operatorname{Ker} T^\times \},$$

$$(15) \qquad \operatorname{Im} S = \{ g \in L_2^m([0, \infty)) \mid Bg \in \operatorname{Im} T^\times \},$$

$$(16) \qquad \dim \operatorname{Ker} S = \dim \operatorname{Ker} T^\times, \qquad \operatorname{codim} \operatorname{Im} S = \operatorname{codim} \operatorname{Im} T^\times,$$

and the operator S *is Fredholm if and only if* $T^\times$ *is a Fredholm operator.*

PROOF. We use the same terminology as in the proof of Theorem 5.2. In particular,

$$(17) \qquad S^\times := T^\times T^{-1} = I - M_B (i M_C T^{-1}).$$

From (4) and (17) we deduce that S and $S^\times$ are matricially coupled, and hence, by Corollary III.4.3,

$$\operatorname{Ker} S = \{CT^{-1}h \mid h \in \operatorname{Ker} S^\times\},$$

$$\operatorname{Im} S = \{g \in L_2^m([0,\infty)) \mid Bg \in \operatorname{Im} S^\times\},$$

$$\dim \operatorname{Ker} S = \dim \operatorname{Ker} S^\times, \qquad \operatorname{codim} \operatorname{Im} S = \operatorname{codim} \operatorname{Im} S^\times.$$

Since $\operatorname{Ker} S^\times = T \operatorname{Ker} T^\times$ and $\operatorname{Im} S^\times = \operatorname{Im} T^\times$, the previous identities prove (14), (15) and (16). As $T^\times$ is closed (by Theorem 1.1), formula (16) implies that S is Fredholm if and only if the same is true for $T^\times$. $\square$

Note that the Fredholm properties of $T^\times$ may be derived from Theorems 4.1 and 4.2. Hence, the latter theorems, together with Theorem 5.3, yield the Fredholm properties of the Wiener-Hopf operator S. The final results are the same as the ones in Theorem XIII.8.1.

As a further application of the results proved in the previous sections, we compute the spectrum of the Wiener-Hopf operator

$$(Hf)(t) = \int_0^\infty h(t - s)f(s)ds, \qquad 0 \le t < \infty,$$

with kernel function

$$h(t) = \begin{cases} e^{-itA}(I - P), & t > 0, \\ -e^{-itA}P, & t < 0, \end{cases}$$

where A is an $n \times n$ matrix without eigenvalues on the real line and P is the Riesz projection of A corresponding to the eigenvalues in the open upper half plane. The operator H is a so-called *first kind Wiener-Hopf operator* acting on $L_2^n([0,\infty))$; its symbol is given by

$$\int_{-\infty}^\infty e^{i\lambda t} h(t)dt = i(\lambda - A)^{-1}.$$

THEOREM 5.4. *Assume that the $n \times n$ matrix A has no eigenvalues on the real line, and let H be the first kind Wiener-Hopf operator on $L_2^n([0,\infty))$ with symbol $i(\lambda - A)^{-1}$. Then the spectrum of H is the union of the closed discs*

$$\left| \lambda + \frac{1}{2a_+} \right| \le \frac{1}{2a_+}, \qquad \left| \lambda + \frac{1}{2a_-} \right| \le \frac{1}{2|a_-|},$$

where

$$(18a) \qquad a_+ = \min\{\Im\lambda \mid \lambda \in \sigma(A),\ \Im\lambda > 0\},$$

$$(18b) \qquad a_- = \max\{\Im\lambda \mid \lambda \in \sigma(A),\ \Im\lambda < 0\}.$$

PROOF. Let P be the Riesz projection of A corresponding to the eigenvalues in the open upper half plane, and consider the differential operator T on $L_2^n([0, \infty))$ associated with

$$(19) \qquad \begin{cases} f'(t) = -iAf(t), & t \geq 0, \\ f(0) \in \operatorname{Im} P. \end{cases}$$

From the first part of the proof of Lemma 5.1 we know that T is invertible and $T^{-1} = H$. Thus

$$(20) \qquad \sigma(H) = \left\{ \frac{1}{\lambda} \;\middle|\; \lambda \in \sigma(T) \right\} \cup \{0\}.$$

Since iA has no eigenvalues on the imaginary axis, Theorem 3.1 (applied to (19)) shows that

$$\sigma(T) = \{\lambda \mid \Re\lambda \leq -a_+\} \cup \{\lambda \mid \Re\lambda \geq -a_-\},$$

where a_+ and a_- are defined by (18a) and (18b). Now use (20) and the theorem is proved. $\square$

XVIII.6 HIGHER ORDER DIFFERENTIAL OPERATORS ON A HALF LINE

Throughout this section τ is the differential expression

$$(1) \qquad \tau = D^n + a_{n-1}D^{n-1} + \cdots + a_1 D + a_0,$$

where $D = \frac{d}{dt}$ and $a_0, \ldots, a_{n-1}$ are given complex numbers. Furthermore, L will be a given subspace in $\mathbf{C}^n$. With τ and L we associated a differential operator $T\big(L_2([0, \infty)) \to L_2([0, \infty))\big)$ as follows:

$$\mathcal{D}(T) = \left\{ u \in \mathcal{D}(T_{\max}) \;\middle|\; \begin{bmatrix} u(0) \\ \vdots \\ u^{(n-1)}(0) \end{bmatrix} \in L \right\},$$

$$Tu = \tau u, \qquad u \in \mathcal{D}(T).$$

Here $T_{\max}$ is the maximal operator in $L_2([0, \infty))$ corresponding to τ and the interval $[0, \infty)$. By definition, the function $u^{(n-1)}$ exists for each $u \in \mathcal{D}(T_{\max})$ and $u^{(n-1)}$ is absolutely continuous on the compact subintervals of $[0, \infty)$. Thus the vector

$$\big(u(0), u'(0), \ldots, u^{(n-1)}(0)\big)^T,$$

appearing in the definition of $\mathcal{D}(T)$, is well-defined. We shall refer to T as the *differential operator in $L_1([0, \infty))$ defined by τ and the subspace L.*

The polynomial $p(\lambda) = \lambda^n + \sum_{\nu=0}^{n-1} a_\nu \lambda^\nu$ is called the *symbol* of the differential expression τ. It determines τ uniquely via $\tau = p(D)$.

We shall prove the following three theorems.

THEOREM 6.1. *Let T be the differential operator in $L_2([0,\infty))$ defined by the differential expression τ and the subspace L. Put*

$$
(2) \qquad H = \begin{bmatrix} 0 & 1 & & \\ & & \ddots & \\ & & & 1 \\ -a_0 & -a_1 & \cdots & -a_{n-1} \end{bmatrix}.
$$

The operator T is a closed operator, which is invertible if and only if the following two conditions are satisfied:

(i) the symbol $p(\cdot)$ of τ has no zeros on the imaginary axis,

(ii) $\mathbf{C}^n = L \oplus N$, where N is the spectral subspace of H spanned by the eigenvectors and generalized eigenvectors of H corresponding to eigenvalues in $\Re\lambda < 0$.

Furthermore, in that case

$$
(T^{-1}v)(s) = \int_0^\infty \gamma(t,s)v(s)ds, \qquad 0 \le t < \infty,
$$

with

$$
\gamma(t,s) = \begin{cases} Ce^{tH}(I-\Pi)e^{-sH}B, & 0 \le s < t < \infty, \\ -Ce^{tH}\Pi e^{-sH}B, & 0 \le t < s < \infty, \end{cases}
$$

where

$$
(3) \qquad B = \begin{bmatrix} 0 \\ \vdots \\ 0 \\ 1 \end{bmatrix}, \qquad C = [\, 1 \quad 0 \quad \cdots \quad 0\,],
$$

and Π is the projection of $\mathbf{C}^n$ onto L along N.

THEOREM 6.2. *Let T be the differential operator in $L_2([0,\infty))$ defined by the differential expression τ and the subspace L. Then T is Fredholm if and only if the symbol $p(\cdot)$ of τ has no zeros on the imaginary axis, and in that case*

$$
(4) \qquad \operatorname{ind} T = \dim L - m,
$$

where m is the number of zeros (multiplicities taken into account) of $p(\cdot)$ in the open right half plane.

THEOREM 6.3. *Assume that the differential operator T in $L_2([0,\infty))$ defined by the differential expression τ and the subspace L is Fredholm. Put*

$$
H = \begin{bmatrix} 0 & 1 & & \\ & & \ddots & \\ & & & 1 \\ -a_0 & -a_1 & \cdots & -a_{n-1} \end{bmatrix}, \qquad B = \begin{bmatrix} 0 \\ \vdots \\ 0 \\ 1 \end{bmatrix}
$$

$$C = [\, 1 \quad 0 \quad \cdots \quad 0 \,],$$

and let Q be the Riesz projection of H corresponding to the eigenvalues in $\Re\lambda > 0$. Then

$$(5) \qquad \operatorname{Ker} T = \{u \mid u(t) = Ce^{tH}x, \ x \in L \cap \operatorname{Ker} Q\},$$

$$(6) \qquad \operatorname{Im} T = \left\{ v \in L_2([0,\infty)) \ \middle| \ \int_0^\infty Q e^{-tH} B v(s)\, ds \in L + \operatorname{Ker} Q \right\},$$

$$(7) \qquad n(T) = \dim(L \cap \operatorname{Ker} Q), \qquad d(T) = \dim \frac{\mathbf{C}^n}{L + \operatorname{Ker} Q}$$

$$(8) \qquad \operatorname{ind} T = \dim L - \dim \operatorname{Im} Q.$$

Furthermore, let Γ be the bounded linear operator on $L_2^n([0,\infty))$ defined by

$$(\Gamma v)(t) = \int_0^\infty \gamma(t,s) v(s)\, ds, \qquad t \geq 0,$$

with

$$(9) \qquad \gamma(t,s) = \begin{cases} C e^{tH}[I - Q - (I - Q)R^+ Q] e^{-sH} B, & 0 \leq s < t < \infty, \\ -C e^{tH}[Q + (I - Q)R^+ Q] e^{-sH} B, & 0 \leq t < s < \infty, \end{cases}$$

where $R^+ \colon \operatorname{Im} Q \to L$ is a generalized inverse of the operator $R = Q|L \colon L \to \operatorname{Im} Q$ in the weak sense (i.e., $RR^+R = R$). Then

$$(10) \qquad T = T\Gamma T.$$

We shall need the following basic fact about the domain of $T_{\max}$.

PROPOSITION 6.4. *Let $T_{\max}$ be the maximal operator in $L_2([0,\infty))$ corresponding to τ and the interval $[0,\infty)$. Then*

$$(11) \qquad u \in \mathcal{D}(T_{\max}) \Rightarrow \widehat{u} = \begin{bmatrix} u \\ u' \\ \vdots \\ u^{(n-1)} \end{bmatrix} \in L_2^n([0,\infty)).$$

The implication (11) also holds for differential expressions τ of which the coefficients are L_∞-functions on $[0,\infty)$ (see, e.g., S. Goldberg [1], Theorem VI.6.2).

To prove Theorems 6.1–6.3 and Proposition 6.4 we shall employ the differential operator S in $L_2^n([0,\infty))$ associated with

$$(12) \qquad \begin{cases} f'(t) = H f(t), & t \geq 0, \\ f(0) \in L. \end{cases}$$

Here H is the $n \times n$ matrix defined by (2). The operator S may be viewed as a linearization of the differential operator T defined by τ and L. In what follows the main idea is to apply first the theorems of Sections 2 and 4 to S and next to use the results to derive the corresponding theorems for T. We begin with a lemma.

LEMMA 6.5. *Let T be the differential operator in $L_2([0,\infty))$ defined by τ and L, and let S be the differential operator in $L_2^n([0,\infty))$ associated with (12). Let B and C be as in (3). Assume $f \in \mathcal{D}(S)$ and $Sf = Bv$. Then $Cf \in \mathcal{D}(T)$ and $TCf = v$.*

PROOF. Let $f_1, \ldots, f_n$ be the components of f. Then

$$(13) \qquad Sf = f' - Hf = \begin{bmatrix} f_1' - f_2 \\ \vdots \\ f_{n-1}' - f_n \\ f_n' + a_0 f_1 + a_1 f_2 + \cdots + a_{n-1} f_n \end{bmatrix}.$$

Since the first $n - 1$ components of Bv are zero, the identity $Sf = Bv$ implies that

$$(14) \qquad f_j = f_1^{(j-1)}, \quad j = 1, \ldots, n-1, \qquad \tau(f_1) = v.$$

Hence, also

$$(15) \qquad \begin{bmatrix} f_1(0) \\ f_1'(0) \\ \vdots \\ f_1^{(n-1)}(0) \end{bmatrix} = \begin{bmatrix} f_1(0) \\ f_2(0) \\ \vdots \\ f_n(0) \end{bmatrix} \in L.$$

Put $u = Cf$. In other words, $u = f_1$. The first $n - 1$ identities in (14) and formula (15) imply that $u \in \mathcal{D}(T)$. The last identity in (14) shows that $Tu = v$. $\square$

PROOF OF PROPOSITION 6.4. Let p be the symbol of τ. Since p is a polynomial, the set $\{p(\lambda) \mid \lambda \in i\mathbf{R}\}$ is not the entire complex plane, and hence we can find a complex number c such that $p(\cdot) - c$ has no zeros on the imaginary axis. Obviously, the domain of $T_{\max}$ does not change if τ is replaced by $\tau - c$. Thus for the proof of (11) we may assume without loss of generality that the symbol p has no zeros on the imaginary axis.

Put $g = BT_{\max}u$, where B is as in (3). Let $S_{\max}$ be the differential operator in $L_2^n([0,\infty))$ associated with (12) and with $L = \mathbf{C}^n$. Since $\det(\lambda - H) = p(\lambda)$, the matrix H has no eigenvalues on the imaginary axis, and hence we can apply Theorem 4.2 (with $A = -H$ and $L = \mathbf{C}^n$) to show that $S_{\max}$ is surjective. Thus there exists $f \in \mathcal{D}(S_{\max})$ such that $S_{\max}f = BT_{\max}u$. But then Lemma 6.5 shows that $Cf \in \mathcal{D}(T)$ and $T_{\max}(Cf) = T_{\max}u$. Thus $Cf - u \in \operatorname{Ker} T_{\max}$. From the proof of Lemma 6.5 (see formula (14)) we know that the first $n - 1$ derivatives of Cf are in $L_2([0,\infty))$. Thus it remains to prove (11) for $u \in \operatorname{Ker} T_{\max}$.

Take $u \in \operatorname{Ker} T_{\max}$. Thus $\tau(u) = 0$. Let b be an arbitrary positive number, and consider the differential expression τ on $[0, b]$. According to formula (4) in Section XIV.3 there exists a vector x_b such that

$$u(t) = Ce^{tH}x_b, \qquad 0 \le t \le b,$$

where as before, C is given by the second identity in (3). Note that

$$
(16) \qquad \begin{bmatrix} C \\ CH \\ \vdots \\ CH^{n-1} \end{bmatrix}
$$

is the $n \times n$ identity matrix. It follows that

$$
\begin{bmatrix} u(0) \\ u'(0) \\ \vdots \\ u^{(n-1)}(0) \end{bmatrix} = \begin{bmatrix} C \\ CH \\ \vdots \\ CH^{n-1} \end{bmatrix} x_b = x_b.
$$

Thus x_b does not depend on b, and we may conclude that $u(t) = Ce^{tH}x$ for some $x \in \mathbb{C}^n$ and all $t \geq 0$. Write $x = x_0 + x_1$ with $x_0 \in \operatorname{Im} Q$ and $x_1 \in \operatorname{Ker} Q$, where Q is the Riesz projection of H corresponding to the eigenvalues in the open right half plane. Note that $u_1(t) = Ce^{tH}x_1$, $0 \leq t < \infty$, and all its derivatives are in $L_2([0, \infty))$. In fact, since $H^j x_1$ belongs to $\operatorname{Ker} Q$, we conclude (see Lemma I.5.4) that

$$
u_1^{(j)}(t) = Ce^{tH}(H^j x_1), \qquad 0 \leq t < \infty,
$$

is exponentially decaying. To finish the proof we show that $x_0 = 0$.

In order to do this, consider the function

$$
(17) \qquad r(\lambda) = \int_0^\infty e^{-\lambda t} Ce^{tH}x_0 \, dt, \qquad \Re\lambda > 0.
$$

The function $e^{-\lambda t}$ is in $L_2([0, \infty))$ for each λ in open right half plane. Also,

$$
Ce^{tH}x_0 = u(t) - u_1(t), \qquad 0 \leq t < \infty,
$$

is in $L_2([0, \infty))$. It follows that the right hand side of (17) is well-defined. From complex function theory (see, e.g., [R], Section 19.1) we know that $r(\cdot)$ is analytic in $\Re\lambda > 0$. Choose $\gamma > 0$ such that $H - \gamma$ has all its eigenvalues in the open left half plane. Then $e^{(H-\lambda)t}$ is exponentially decaying for each $\Re\lambda \geq \gamma$ by Lemma I.5.4, and hence

$$
\int_0^\infty e^{-\lambda t} Ce^{tH}x_0 \, dt = C(H - \lambda)^{-1} e^{(H-\lambda)t} x_0 \Big|_{t=0}^{t=\infty}
$$

$$
= C(\lambda - H)^{-1} x_0, \qquad \Re\lambda \geq \gamma > 0.
$$

It follows that

$$
(18) \qquad r(\lambda) = C(\lambda - H)^{-1} x_0, \qquad \Re\lambda \geq \gamma > 0.
$$

Since $x_0 \in \operatorname{Im} Q$, the function $C(\lambda - H)^{-1} x_0$ is rational and has its poles in $\Re \lambda > 0$. On the other hand, according to (18), the function $C(\lambda - H)^{-1} x_0$ has an analytic continuation on $\Re \lambda > 0$. Therefore $C(\lambda - H)^{-1} x_0$ is analytic on the entire complex plane. From the Neumann series expansion we see that

$$C(\lambda - H)^{-1} x_0 \to 0, \qquad |\lambda| \to \infty.$$

But then Liouville's theorem implies that $C(\lambda - H)^{-1} x_0$ is identically zero, and thus (again use the Neumann series expansion) $C H^j x_0 = 0$ for $j = 0, 1, 2, \dots$. Now recall that the matrix in (16) is the $n \times n$ identity matrix. Thus

$$x_0 = \begin{bmatrix} C \\ CH \\ \vdots \\ CH^{n-1} \end{bmatrix} x_0 = 0,$$

and the proof of (11) is completed. $\quad\square$

The next lemma will allow us to describe the Fredholm properties of T in terms of those of S.

LEMMA 6.6. *Let T be the differential operator in $L_2([0,\infty))$ defined by τ and L, and let S be the differential operator in $L_2^n([0,\infty))$ associated with (12). Let B and C be as in (3). Then*

$$\text{(19)} \qquad\qquad \operatorname{Ker} T = \{C f \mid f \in \operatorname{Ker} S\},$$

$$\text{(20)} \qquad\qquad \operatorname{Im} T = \{v \in L_2([0,\infty)) \mid B v \in \operatorname{Im} S\},$$

$$\text{(21)} \qquad\qquad \dim \operatorname{Ker} T = \dim \operatorname{Ker} S, \qquad \operatorname{codim} \operatorname{Im} T = \operatorname{codim} \operatorname{Im} S.$$

PROOF. Take $u \in \operatorname{Ker} T$. Then $u \in \mathcal{D}(T) \subset \mathcal{D}(T_{\max})$, and hence (by Proposition 6.4)

$$\widehat{u} := \begin{bmatrix} u \\ u' \\ \vdots \\ u^{(n-1)} \end{bmatrix} \in L_2^n([0,\infty)).$$

Since $u \in \mathcal{D}(T)$, we have $\widehat{u}(0) \in L$. Hence $\widehat{u} \in \mathcal{D}(S)$, and (13) (with f replaced by $\widehat{u}$) yields $S\widehat{u} = 0$. Thus $u = C\widehat{u}$ with $\widehat{u} \in \operatorname{Ker} S$. Conversely, assume $S f = 0$. Then Lemma 6.5 implies that $C f \in \operatorname{Ker} T$. Furthermore, from the proof of Lemma 6.5 (see formula (14)) we see that the j-th component of f is equal to $(Cf)^{(j-1)}$ for $j = 1, \dots, n$. Thus $C f = 0$ implies $f = 0$. We conclude that C maps $\operatorname{Ker} S$ in a one-one way onto $\operatorname{Ker} T$, which proves (19) and the first identity in (21).

To prove (20), let $v = T u$ for some $u \in \mathcal{D}(T)$. As in the first paragraph of the present proof one shows that

$$\widehat{u} := \begin{bmatrix} u \\ u' \\ \vdots \\ u^{(n-1)} \end{bmatrix} \in \mathcal{D}(S).$$

Thus we can apply (13) with $f = \widehat{u}$, which yields

$$S\widehat{u} = \begin{bmatrix} 0 \\ \vdots \\ 0 \\ Tu \end{bmatrix} = BTu = Bv.$$

Thus $Bv \in \operatorname{Im} S$. Conversely, if $Bv = Sf$ for some $f \in \mathcal{D}(S)$, then $v = T(Cf)$, by Lemma 6.5. Hence $v \in \operatorname{Im} T$ and (20) is proved.

The proof of the second identity in (21) is based on the following equality:

$$(22) \qquad L_2^n([0,\infty)) = \operatorname{Im} S + \left\{ \begin{bmatrix} 0 \\ \vdots \\ 0 \\ v \end{bmatrix} \,\middle|\, v \in L_2([0,\infty)) \right\}.$$

Formula (13) implies that the right hand side of (22) does not depend on the coefficients $a_0, \ldots, a_{n-1}$ of τ. Thus to prove (22) we may choose the coefficients $a_0, \ldots, a_{n-1}$ as we like. It follows that for the proof of (22) we may assume without loss of generality that the symbol p of τ has all its zeros in the open left half plane. Then all the eigenvalues of H are in the open left half plane, and we can apply Theorem 4.2 (with $A = -H$) to show that $\operatorname{Im} S = L_2^n([0,\infty))$, which proves (22). Now, let

$$J : \frac{L_2([0,\infty))}{\operatorname{Im} T} \to \frac{L_2^n([0,\infty))}{\operatorname{Im} S}$$

be defined by

$$J(v + \operatorname{Im} T) = \begin{bmatrix} 0 \\ \vdots \\ 0 \\ v \end{bmatrix} + \operatorname{Im} S.$$

Then J is injective because of (20), and the equality (22) shows that J is surjective. Thus J is a linear bijective map, and therefore the second identity in (21) holds true. $\square$

PROOF OF THEOREM 6.1. First we show that T is closed. Let $u_1, u_2, \ldots$ be a sequence in $\mathcal{D}(T)$ such that $u_j \to u$ and $Tu_j \to v$ in $L_2([0,\infty))$. Since $T_{\max}$ is an extension of T which is closed (by Proposition XIV.3.3), we conclude that $u \in \mathcal{D}(T_{\max})$ and $\tau(u) = v$. It remains to prove that

$$(23) \qquad \begin{bmatrix} u(0) \\ u'(0) \\ \vdots \\ u^{(n-1)}(0) \end{bmatrix} \in L.$$

To do this we restrict our attention to the compact interval $[0,1]$. Let $\widehat{T}$ and $\widehat{T_0}$ be the operators in $L_2([0,1])$ defined by

$$
\begin{cases}
\mathcal{D}(\widehat{T}) = \left\{ g \in \mathcal{D}(T_{\max,\tau,[0,1]}) \,\middle|\, \begin{bmatrix} g(0) \\ g'(0) \\ \vdots \\ g^{(n-1)}(0) \end{bmatrix} \in L \right\}, \\[4pt]
\widehat{T}g = \tau(g);
\end{cases}
$$

$$
\begin{cases}
\mathcal{D}(\widehat{T}_0) = \left\{ g \in \mathcal{D}(T_{\max,\tau,[0,1]}) \,\middle|\, \begin{bmatrix} g(0) \\ g'(0) \\ \vdots \\ g^{(n-1)}(0) \end{bmatrix} = 0 \right\}, \\[4pt]
\widehat{T}_0 g = \tau(g).
\end{cases}
$$

Obviously,

$$
G(T_{\max,\tau,[0,1]}) \supset G(\widehat{T}) \supset G(\widehat{T}_0).
$$

The argument in the first paragraph of the proof of Proposition XIV.3.3 shows that the graphs $G(\widehat{T}_0)$ and $G(T_{\max,\tau,[0,1]})$ are closed. Moreover $G(\widehat{T}_0)$ has finite codimension in $G(T_{\max,\tau,[0,1]})$. Thus $G(\widehat{T})$ is a finite dimensional extension of the closed linear manifold $G(\widehat{T}_0)$, which implies that $\widehat{T}$ is closed. Let $\widehat{v}$, $\widehat{u}$, $\widehat{u}_1$, $\widehat{u}_2$, ... be the restrictions of the functions v, u, u_1, u_2, ..., respectively, to the interval $[0,1]$. Then $\widehat{u}_1, \widehat{u}_2, \ldots$ is a sequence in $\mathcal{D}(\widehat{T})$, $\widehat{u}_j \to \widehat{u}$ and $\widehat{T}\widehat{u}_j \to \widehat{v}$ in $L_2([0,1])$. Since $\widehat{T}$ is closed, we conclude that $\widehat{u} \in \mathcal{D}(\widehat{T})$, and hence (23) is fulfilled.

According to Lemma 6.6 the operator T is invertible if and only if S is invertible. By Theorem 2.1 (applied to $-H$ instead of A) the latter happens if and only if H has no eigenvalues on the imaginary axis and

$$
\mathbf{C}^n = L \oplus \operatorname{Ker} Q,
$$

where Q is the Riesz projection of H corresponding to the eigenvalues of H in the open right half plane. Now note that the eigenvalues of H are precisely the zeros of the symbol p of τ and $\operatorname{Ker} Q$ is equal to the space N appearing in condition (ii) of Theorem 6.1. Thus the above arguments show that the conditions (i) and (ii) of Theorem 6.1 are necessary and sufficient for the invertibility of T.

Finally, assume that T is invertible, and let us derive the formula for its inverse. Take $v \in L_2([0,\infty))$, and put $u = Tv$. Then, as was shown in the second paragraph of the proof of Lemma 6.6,

$$
\widehat{u} := \begin{bmatrix} u \\ u' \\ \vdots \\ u^{(n-1)} \end{bmatrix} \in \mathcal{D}(S), \qquad S\widehat{u} = BTu = Bv.
$$

Now S is invertible, and so

$$
(24) \qquad\qquad u = C\widehat{u} = CS^{-1}Bv.
$$

Theorem 2.1 (applied to $A = -H$) yields the formula for S^{-1}. By using this formula in the third term of (24), one obtains the expression of T^{-1} in Theorem 6.1. $\square$

PROOF OF THEOREM 6.2. First we apply Lemma 6.6. Note that both T and S are closed operators. Thus Lemma 6.6 shows that T is Fredholm if and only if S is Fredholm, and in that case $\operatorname{ind} T = \operatorname{ind} S$. Next, we apply Theorem 4.1 with A replaced by $-H$. It follows that S is Fredholm if and only if H has no eigenvalues on the imaginary axis, and in that case

$$\operatorname{ind} S = \dim L - \dim M,$$

where M is the space spanned by the eigenvectors and generalized eigenvectors of H corresponding to the eigenvalues in $\Re\lambda > 0$. Since $\det(\lambda - H) = p(\lambda)$, it follows that H has no eigenvalues on the imaginary axis if and only if p has no zeros on the imaginary axis. Furthermore, $\dim M$ is precisely equal to the number m appearing in (4). With these remarks the theorem is proved. $\square$

PROOF OF THEOREM 6.3. By Lemma 6.6 and Theorem 4.3 (applied with A replaced by $-H$) formulas (5), (6), (7) and (8) are evident.

To prove (10), let $\widetilde{\gamma}(t,s)$ be the function defined by the right hand side of (9) with B and C omitted. Then

$$(\widetilde{\Gamma}f)(t) = \int_0^\infty \widetilde{\gamma}(t,s)f(s)ds, \qquad t \geq 0,$$

defines a bounded linear operator on $L_2^n([0,\infty))$, and from Theorem 4.3 (applied to $A = -H$) we know that $S = S\widetilde{\Gamma}S$. Obviously, $\Gamma v = C\widetilde{\Gamma}Bv$ for any $v \in L_2([0,\infty))$. Now, take $u \in \mathcal{D}(T)$, and put $v = Tu$. By Proposition 6.4

$$\widehat{u} := \begin{bmatrix} u \\ u' \\ \vdots \\ u^{(n-1)} \end{bmatrix} \in \mathcal{D}(S),$$

and according to formula (13) (with $\widehat{u}$ in place of f) we have $S\widehat{u} = Bv$. Thus

$$Bv = S\widehat{u} = (S\widetilde{\Gamma}S)\widehat{u} = S(\widetilde{\Gamma}Bv),$$

which, by Lemma 6.5, implies that $\Gamma v = C\widetilde{\Gamma}Bv \in \mathcal{D}(T)$ and $T\Gamma v = v$. Since $v = Tu$, the latter identity proves (10). $\square$

We conclude this section with a corollary about the essential spectrum.

COROLLARY 6.7. *Let T be the differential operator in $L_2([0,\infty))$ defined by τ and L, and let $p(\cdot)$ be the symbol of τ. Then*

$$(25) \qquad\qquad \sigma_e(T) = \{p(\lambda) \mid \Re\lambda = 0\}.$$

PROOF. Let c be an arbitrary complex number. Then $T - cI$ is the differential operator defined by $\tau - c$ and L. The symbol of $\tau - c$ is the polynomial $p(\cdot) - c$. Thus, by Theorem 6.2, we have $c \in \sigma_e(T)$ if and only if $p(\cdot) - c$ has a zero on the imaginary axis or, equivalently, if and only if c belongs to the set in the right hand side of (25). $\square$

CHAPTER XIX
STRONGLY CONTINUOUS SEMIGROUPS

This chapter contains a short introduction to the theory of strongly continuous semigroups. Such semigroups stem from the study of certain partial differential equations, which are recast in the form

(1)
$$\begin{cases} u'(t) = Au(t), & 0 \le t < \infty, \\ u(0) = x, \end{cases}$$

where u maps $[0, \infty)$ into a certain Banach space. When A is unbounded, strongly continuous semigroups can be used sometimes to give a meaning to the well-known formula $e^{tA}x$ for the solution of (1).

This chapter consists of seven sections. The first three contain the main results about the connections between strongly continuous semigroups and the equation (1), which is referred to as the *abstract Cauchy problem*. The next three sections treat different classes of strongly continuous semigroups, namely contraction, unitary and compact semigroups. The final section concerns an application to linear transport theory.

XIX.1 THE ABSTRACT CAUCHY PROBLEM

Throughout this chapter X is a complex Banach space. Let $A(X \to X)$ be a linear operator with domain $\mathcal{D}(A)$ in X, and let x be a vector in X. The initial value problem

(1)
$$\begin{cases} u'(t) = Au(t), & 0 \le t < \infty, \\ u(0) = x, \end{cases}$$

is called *the abstract Cauchy-problem associated with A.* An X-valued function u on $0 \le t < \infty$ is said to be a *solution* of (1) if $u(0) = x$ and for each $t \ge 0$ the vector $u(t) \in \mathcal{D}(A)$,

$$u'(t) = \lim_{h \to 0} \frac{1}{h}\big(u(t+h) - u(t)\big)$$

exists and $u'(t) = Au(t)$. As usual, $u'(0)$ is the right hand side derivative at 0. A solution u of (1) is said to be *continuously differentiable* (or shortly: a *C^1-solution*) if, in addition, $u' \colon [0, \infty) \to X$ is a continuous function. For the existence of a solution it is necessary that $x \in \mathcal{D}(A)$. In what follows we shall deal mainly with C^1-solutions.

Abstract Cauchy problems arise in a natural way from certain initial value problems involving partial differential equations. An illustration of this is the following. Consider the (so-called wave) equation:

(2)
$$\begin{cases} \dfrac{\partial v}{\partial t}(t,s) = \dfrac{\partial v}{\partial s}(t,s), & 0 \le t < \infty,\ s \in \mathbf{R}, \\ v(0,s) = f(s), & s \in \mathbf{R}. \end{cases}$$

Let us assume that f is a given differentiable function on $\mathbf{R}$. Then $u(t,s) = f(t+s)$, $t \in [0,\infty)$, $s \in \mathbf{R}$, is a solution to (2). Moreover, it is the only solution. Indeed, suppose $w(t,s)$ is also a solution of (2). Let $h(t,s) = u(t,s) - w(t,s)$. Then $\frac{\partial h}{\partial t} = \frac{\partial h}{\partial s}$ on $[0,\infty) \times \mathbf{R}$ and $h(0,s) = 0$, $s \in \mathbf{R}$. By the chain rule

$$\frac{\partial}{\partial \alpha} h(\alpha, \beta - \alpha) = \frac{\partial h}{\partial t}(\alpha, \beta - \alpha) - \frac{\partial h}{\partial s}(\alpha, \beta - \alpha) = 0,$$

and hence $h(\alpha, \beta - \alpha) = g(\beta)$ for some differentiable function g. But $0 = h(0,\beta) = g(\beta)$, $\beta \in \mathbf{R}$, and thus $h = 0$.

To put (2) in the abstract form (1), let us assume that f and f' are bounded and uniformly continuous on $\mathbf{R}$. Take X to be the Banach space $BUC(\mathbf{R})$ of all bounded, uniformly continuous complex-valued functions on $\mathbf{R}$ endowed with the supremum norm, and define $A(X \to X)$ by setting

$$\mathcal{D}(A) = \{g \in X \mid g' \text{ exists}, \ g' \in X\}, \qquad Ag = g'.$$

Then we may rewrite (2) as

$$(3) \qquad \begin{cases} u'(t) = Au(t), & 0 \le t < \infty, \\ u(0) = f. \end{cases}$$

Consider the function $u \colon [0,\infty) \to X$ defined by

$$u(t)(s) = u(t,s) = f(t+s), \qquad s \in \mathbf{R}.$$

The function u is a C^1-solution of the abstract Cauchy problem (3). To see this, note that $u(0) = f$. Furthermore, because of our conditions on f, for each $t \in [0,\infty)$ the function $u(t) \in \mathcal{D}(A)$ and

$$(4) \qquad \big(Au(t)\big)(s) = \frac{d}{ds} u(t)(s) = f'(t+s), \qquad s \in \mathbf{R}.$$

On the other hand,

$$\left\| \frac{1}{h}\big(u(t+h) - u(t)\big) - f'(t + \cdot) \right\| = \sup_{s \in \mathbf{R}} \left| \frac{1}{h}\big(f(t+h+s) - f(t+s)\big) - f'(t+s) \right|$$

$$= \sup_{s \in \mathbf{R}} \left| \frac{1}{h} \int_{t+s}^{t+s+h} [f'(\alpha) - f'(t+s)] d\alpha \right|.$$

Since f' is uniformly continuous on $\mathbf{R}$, it follows that the last term converges to zero as $h \to \infty$. Hence u is differentiable on $0 \le t < \infty$ (as an X-valued function) and $u'(t)(s) = f'(t+s)$, $s \in \mathbf{R}$. Thus $u'(t) = Au(t)$, by (4). Also, given $\varepsilon > 0$, the uniform continuity of f' implies

$$\|u'(t_1) - u'(t_2)\| = \sup_{s \in \mathbf{R}} \left| \big(Au(t_1)\big)(s) - \big(Au(t_2)\big)(s) \right|$$

$$= \sup_{s \in \mathbf{R}} |f'(t_1 + s) - f'(t_2 + s)| < \varepsilon,$$

provided that $|t_1 - t_2|$ is sufficiently small. Hence u' is continuous (in fact, uniformly continuous) on $\mathbf{R}$. Thus u solves (3).

Now suppose that $v(\cdot)$ is a second solution of (3). Define $v(t, s) = v(t)(s)$. Then, $v(0, s) = v(0)(s) = f(s)$ and

$$\left| \frac{1}{h} \left(v(t + h, s) - v(t, s) \right) - (Av(t))(s) \right| \le \left\| \frac{1}{h} \left(v(t + h) - v(t) \right) - Av(t) \right\| \to 0, \qquad h \to 0,$$

which implies that

$$\frac{\partial v}{\partial t}(t, s) = \left(Av(t) \right)(s) = \frac{d}{ds} v(t)(s) = \frac{\partial}{\partial s} v(t, s).$$

We have shown that $v(t, s)$ satisfies (2). But (2) has a unique solution. So also (3) is uniquely solvable.

Theorem I.5.2 shows that the Cauchy problem (1) has a unique solution $u(t; x) = e^{tA}x$ for every $x \in X$ whenever A is a bounded linear operator on X. This result has the following generalization.

THEOREM 1.1. *Let $A(X \to X)$ be densely defined. Suppose that there exist numbers ω and M such that $\lambda \in \rho(A)$ and*

$$(5) \qquad \|(\lambda - A)^{-k}\| \le M(\lambda - \omega)^{-k}, \qquad k = 1, 2, \ldots,$$

whenever ($\lambda \in \mathbf{R}$ and) $\lambda > \omega$. Then the abstract Cauchy problem (1) has one and only one solution for each $x \in \mathcal{D}(A)$. This solution is continuously differentiable and is given by

$$(6) \qquad u(t; x) = \lim_{\lambda \to \infty} e^{tA_\lambda}x, \qquad A_\lambda := \lambda A(\lambda - A)^{-1}.$$

PROOF. The proof is divided into six parts. The first four parts show that the function defined in (6) is a C^1-solution of (1); the last two parts establish the uniqueness of the solution. In what follows we assume that the condition of the theorems are fulfilled.

Part (i). First we show that for each $y \in X$

$$(7) \qquad \lambda(\lambda - A)^{-1}y \to y \qquad (\lambda \in \mathbf{R},\ \lambda \to \infty).$$

Take $y \in X$. For any $z \in \mathcal{D}(A)$

$$\|\lambda(\lambda - A)^{-1}z - z\| = \|(\lambda - A)^{-1}Az\|$$
$$\le M(\lambda - \omega)^{-1}\|Az\|, \qquad \lambda > \omega.$$

It follows that for $\lambda > \omega$

$$\|\lambda(\lambda - A)^{-1}y - y\| \le \|\lambda(\lambda - A)^{-1}z - z\| + \|\lambda(\lambda - A)^{-1}(y - z)\| + \|y - z\|$$
$$\le \frac{M}{\lambda - \omega}\|Az\| + \left(\frac{\lambda M}{\lambda - \omega} + 1 \right)\|y - z\|.$$

Note that $\lambda(\lambda - \omega)^{-1}$ is a decreasing function on $\omega < \lambda < \infty$. Now, let $\varepsilon > 0$ be given. Since $\overline{\mathcal{D}(A)} = X$, we may choose a vector $z_0 \in \mathcal{D}(A)$ such that

$$(2M + 1)\|y - z\| < \frac{1}{2}\varepsilon.$$

Next, choose $N > 2\omega$ such that $M(\lambda - \omega)^{-1}\|Az_0\| < \frac{1}{2}\varepsilon$ for $\lambda > N$. Then the above calculations show that

$$\|\lambda(\lambda - A)^{-1}y - y\| < \varepsilon \qquad (\lambda > N),$$

and (7) is proved.

Let A_λ be as in (6). By applying (7) to $y = Ax$ we see that

$$(8) \qquad A_\lambda x = \lambda A(\lambda - A)^{-1}x = \lambda(\lambda - A)^{-1}Ax \to Ax \qquad (\lambda \to \infty)$$

for each $x \in \mathcal{D}(A)$.

Part (ii). Note that

$$A_\lambda = \lambda^2(\lambda - A)^{-1} - \lambda I \in \mathcal{L}(X),$$

and hence e^{tA_λ} is a well-defined bounded linear operator on X. In this part we show that for each $y \in Y$

$$(9) \qquad \lim_{\lambda \to \infty} e^{tA_\lambda}y$$

exists and the convergence in (9) is uniform on bounded subintervals of $0 \le t < \infty$.

For $y \in Y$, we have

$$
\begin{aligned}
\|e^{tA_\lambda}y\| &= e^{-\lambda t}\|e^{t\lambda^2(\lambda-A)^{-1}}y\| \\
&\le e^{-\lambda t}\sum_{\nu=0}^{\infty} \frac{(t\lambda^2)^\nu}{\nu!}\|(\lambda - A)^{-\nu}\|\|y\| \\
&\le e^{-\lambda t}\sum_{\nu=0}^{\infty} \frac{(t\lambda^2)^\nu}{\nu!}(\lambda - \omega)^{-\nu}M\|y\| \\
&= M\|y\|\exp\left(\frac{\lambda\omega}{\lambda - \omega}t\right), \qquad \lambda > \omega.
\end{aligned}
$$

(10)

Take $\lambda > \mu > \omega$, then $\mu\omega(\mu - \omega)^{-1} \ge \lambda\omega(\lambda - \omega)^{-1}$, and hence (10) implies that

$$\|e^{(t-s)A_\lambda}e^{sA_\mu}\| \le M^2\exp\left(\frac{\mu\omega}{\mu - \omega}t\right), \qquad 0 \le s \le t.$$

Since the operators A_λ, A_μ, e^{tA_λ}, e^{tA_μ} commute, it follows that

$$
\begin{aligned}
\|e^{tA_\mu}y - e^{tA_\lambda}y\| &= \left\|\int_0^t \frac{d}{ds}(e^{(t-s)A_\lambda}e^{sA_\mu})y\,ds\right\| \\
&= \left\|\int_0^t e^{(t-s)A_\lambda}e^{sA_\mu}(A_\mu y - A_\lambda y)\,ds\right\| \\
&\le M^2\exp\left(\frac{\mu\omega t}{\mu - \omega}\right)\|A_\mu y - A_\lambda y\|t, \qquad \lambda > \mu > \omega.
\end{aligned}
$$

(11)

Now take $x \in \mathcal{D}(A)$ and $0 \le t \le \tau$. Then (10) and (11) imply that for $\lambda > \mu > a > \omega$

$$\|e^{tA_\mu}y - e^{tA_\lambda}y\| \le \|e^{tA_\mu}x - e^{tA_\lambda}x\| + \|e^{tA_\mu}(y - x)\| + \|e^{tA_\lambda}(y - x)\|$$
$$\le M \exp\left(\frac{a\omega}{a - \omega}\tau\right)(\tau M\|A_\mu x - A_\lambda x\| + 2\|y - x\|).$$

Now use (8) and the fact that $\mathcal{D}(A)$ is dense in X. It follows that the limit (9) exists and the convergence is uniform on every bounded t-interval of $[0, \infty)$.

Part (iii). For each $y \in X$ put

$$(12) \qquad\qquad T(t)y := \lim_{\lambda \to \infty} e^{tA_\lambda}y, \qquad t \ge 0.$$

By the preceding part of the proof, $T(t)$ is a well-defined linear transformation on X. From (10) it follows that

$$\|T(t)y\| = \lim_{\lambda \to \infty} \|e^{tA_\lambda}y\| \le \lim_{\lambda \to \infty} M\|y\| \exp\left(\frac{\lambda\omega}{\lambda - \omega}t\right)$$
$$= M\|y\|e^{\omega t}.$$

Hence $T(t) \in \mathcal{L}(X)$ and

$$(13) \qquad\qquad \|T(t)\| \le Me^{\omega t}, \qquad t \ge 0.$$

Since the convergence in (12) is uniform on bounded subintervals of $0 \le t < \infty$, the map $t \mapsto T(t)y$ is continuous from $[0, \infty)$ into X. In particular,

$$(14) \qquad\qquad \lim_{t \downarrow 0} T(t)y = T(0)y = y, \qquad y \in X.$$

Also, for all $y \in X$,

$$T(t + s)y = \lim_{\lambda \to \infty} e^{(t+s)A_\lambda} = \lim_{\lambda \to \infty} e^{tA_\lambda}e^{sA_\lambda}y,$$

and by (10)

$$\|e^{tA_\lambda}e^{sA_\lambda}y - T(t)T(s)y\|$$
$$\le \|e^{tA_\lambda}e^{sA_\lambda}y - e^{tA_\lambda}T(s)y\| + \|e^{tA_\lambda}T(s)y - T(t)T(s)y\|$$
$$\le M\|e^{sA_\lambda}y - T(s)y\| \exp\left(\frac{\lambda\omega}{\lambda - \omega}t\right) + \|e^{tA_\lambda}T(s)y - T(t)T(s)y\| \to 0, \qquad \lambda \to \infty.$$

Therefore

$$(15) \qquad\qquad T(t + s) = T(t)T(s), \qquad t \ge 0, \; s \ge 0.$$

Part (iv). Next we prove that $u(t; x) = T(t)x$ is a C^1-solution of (1) for every $x \in \mathcal{D}(A)$. Let $x \in \mathcal{D}(A)$ be fixed. First we show that $T(t)Ax = AT(t)x$ for $t \ge 0$. If $\mu > \omega$, then

$$(\mu - A)^{-1}e^{tA_\lambda} = e^{tA_\lambda}(\mu - A)^{-1}, \qquad \lambda > \omega.$$

Put $y = (\mu - A)x$. Taking limits as $\lambda \to \infty$ gives

$$(\mu - A)^{-1}T(t)y = T(t)(\mu - A)^{-1}y = T(t)x.$$

Hence $T(t)x \in \mathcal{D}(A)$ and

$$(\mu - A)T(t)x = T(t)y = T(t)(\mu - A)x,$$

which shows that

$$(16) \qquad AT(t)x = T(t)Ax, \qquad t \geq 0, \ x \in \mathcal{D}(A).$$

From (10) it follows that

$$\|e^{tA_\lambda}A_\lambda x - T(t)Ax\| \leq M\|A_\lambda x - Ax\| \exp\left(\frac{\lambda\omega}{\lambda - \omega}t\right) + \|e^{tA_\lambda}Ax - T(t)Ax\|.$$

Hence $e^{tA_\lambda}A_\lambda x \to T(t)Ax$ if $\lambda \to \infty$ and the convergence is uniform on each bounded subinterval of $0 \leq t < \infty$. It follows that

$$(17) \qquad \begin{aligned} T(t)x - T(t_0)x &= \lim_{\lambda \to \infty} e^{tA_\lambda}x - e^{t_0 A_\lambda}x \\ &= \lim_{\lambda \to \infty} \int_{t_0}^{t} \frac{d}{ds}e^{sA_\lambda}x\,ds \\ &= \lim_{\lambda \to \infty} \int_{t_0}^{t} e^{sA_\lambda}A_\lambda x\,ds \\ &= \int_{t_0}^{t} T(s)Ax\,ds. \end{aligned}$$

Since the function $s \mapsto T(s)Ax$ is continuous on $[0, \infty)$, formula (16) implies that

$$\begin{aligned} \frac{1}{t - t_0}(T(t)x - T(t_0)x) &= \frac{1}{t - t_0}\int_{t_0}^{t} T(s)Ax\,ds \\ &\to T(t_0)Ax = AT(t_0)x, \qquad t \to t_0. \end{aligned}$$

We conclude that

$$(18) \qquad \frac{d}{dt}T(t)x = AT(t)x = T(t)Ax, \qquad t \geq 0,$$

which shows that $u(t) = u(t; x) = T(t)x$ is a C^1-solution of the abstract Cauchy problem (1). Indeed, $u(0) = T(0)x = x$ by (14), the vector $u(t) = T(t)x \in \mathcal{D}(A)$ for every

$t \in [0, \infty)$ because of (16), and the function u is continuously differentiable on $[0, \infty)$ and satisfies the first equation in (1) because of (18).

Part (v). To establish the uniqueness of the solution we need the following general remark. Let J be a subinterval of $[0, \infty)$, and let $g: J \to X$ be a differentiable function such that $g(s) \in \mathcal{D}(A)$ for every $s \in J$. Then $s \mapsto T(s)g(s)$ is differentiable on J and

$$(19) \qquad \frac{d}{ds}T(s)g(s) = T(s)Ag(s) + T(s)g'(s), \qquad s \in J.$$

To prove (19) take $s \in J$ and $0 \neq h \in \mathbf{R}$. Then

$$\frac{1}{h}\{T(s+h)g(s+h) - T(s)g(s)\}$$

$$(20) \qquad = \frac{1}{h}\{T(s+h) - T(s)\}g(s) + T(s)\left\{\frac{g(s+h) - g(s)}{h}\right\}$$

$$+ \frac{1}{h}\{T(s+h) - T(s)\}\{g(s+h) - g(s)\}.$$

So to establish (19) it suffices to show that the last term in (20) goes to zero if $h \to 0$. But

$$\frac{1}{h}\{T(s+h) - T(s)\}\{g(s+h) - g(s)\}$$

$$= \{T(s+h) - T(s)\}\left\{\frac{g(s+h) - g(s)}{h} - g'(s)\right\}$$

$$+ \{T(s+h) - T(s)\}g'(s) \to 0, \qquad h \to 0,$$

because of (13) and the continuity of the function $s \mapsto T(s)y$ for any $y \in X$.

Part (vi). Suppose that v is another solution of (1). We have to show that $v(t) = T(t)x$. To do this we fix $t > 0$ and apply the remark made in the preceding part to the function $g_0(s) = v(t - s)$ with $s \in J = [0, t]$. Note that g_0 is differentiable and $g_0(s) \in \mathcal{D}(A)$ for $0 \leq s \leq t$. Thus we may conclude that

$$\frac{d}{ds}T(s)v(t - s) = T(s)Av(t - s) + T(s)\bigl(-v'(t - s)\bigr)$$

$$= T(s)Av(t - s) - T(s)Av(t - s) = 0$$

for $0 \leq s \leq t$. It follows that

$$T(t)x - v(t) = T(s)v(t - s)\Big|_{s=0}^{s=t}$$

$$= \int_0^t \frac{d}{ds}T(s)v(t - s)ds = 0,$$

and hence $v(t) = T(t)x$, which proves that (1) is uniquely solvable. $\square$

The operator A_λ appearing in (6) is called the *Yoshida approximant* of A. This terminology is justified by formula (8).

XIX.2　GENERATORS OF STRONGLY CONTINUOUS SEMI-GROUPS

In the previous section we have seen that the solution $u(t; x)$ of the abstract Cauchy problem gives rise to a family $T(t)$, $0 \le t < \infty$, in $\mathcal{L}(X)$, the set of all bounded linear operators on the Banach space X, such that

(i) $T(t + s) = T(t)T(s)$, $s, t \in [0, \infty)$,

(ii) $T(0) = I$,

(iii) $\lim_{t \downarrow 0} T(t)x = x$, $x \in X$.

A family $T(t) \in \mathcal{L}(X)$, $0 \le t < \infty$, with the above properties is called a *strongly continuous semigroup* on X (shortly: a C_0 *semigroup*). We shall now show that, conversely, a C_0 semigroup $T(\cdot)$ gives rise to an operator A which satisfies the hypotheses of Theorem 1.1 and the unique solution to the associated abstract Cauchy problem is $u(t; x) = T(t)x$ for each $x \in \mathcal{D}(A)$.

LEMMA 2.1. *If $T(\cdot)$ is a strongly continuous semigroup on X, then there exist real numbers M and ω such that*

$$(1) \qquad \|T(t)\| \le Me^{\omega t}, \qquad 0 \le t < \infty.$$

PROOF. There exists $\delta > 0$ such that

$$(2) \qquad M := \sup\{\|T(t)\| \mid 0 \le t \le \delta\} < \infty.$$

If this is not the case, then there exists a sequence (t_n) such that $t_n \downarrow 0$ and $\|T(t_n)\| \ge n$ for $n = 1, 2, \ldots$. Hence we may infer from the uniform boundedness principle that for some $x \in X$, the sequence $(\|T(t_n)x\|)$ is unbounded. But this contradicts property (iii) in the definition of a strongly continuous semigroup. Thus there exists $\delta > 0$ such that (2) holds. Now, take $t = n\delta + r$, where $0 \le r < \delta$ and n is a nonnegative integer. Since $T(t)$ is a semigroup

$$\|T(t)\| \le \|T(\delta)^n\| \|T(r)\| \le M^{n+1}$$
$$\le MM^{t/\delta} = Me^{\omega t},$$

with $\omega := \delta^{-1} \log M$. $\square$

For the strongly continuous semigroup $T(\cdot)$ constructed in the proof of Theorem 1.1 we have seen that

$$(3) \qquad \lim_{h \downarrow 0} \frac{1}{h}(T(h)x - x) = Ax, \qquad x \in \mathcal{D}(A).$$

In the following theorem this identity is the starting point for our definition of A.

THEOREM 2.2. *Suppose that $T(\cdot)$ is a strongly continuous semigroup on X. Define $A(X \to X)$ by*

$$\mathcal{D}(A) = \left\{ x \in X \mid \lim_{h \downarrow 0} \frac{1}{h}(T(h)x - x) \text{ exists} \right\},$$

$$Ax = \lim_{h \downarrow 0} \frac{1}{h}(T(h)x - x).$$

Then A is a closed densely defined operator and the abstract Cauchy problem associated with A has the unique solution $u(t; x) = T(t)x$ for each $x \in \mathcal{D}(A)$. Furthermore, if M and ω are real numbers such that (1) holds, then for $\Re\lambda > \omega$

    (a) $\lambda \in \rho(A)$,

    (b) $(\lambda - A)^{-1}x = \int\limits_0^\infty e^{-\lambda t}T(t)x\,dt$, $x \in X$,

    (c) $\|(\lambda - A)^{-n}\| \leq M(\Re\lambda - \omega)^{-n}$, $n = 1, 2, \ldots$.

PROOF. The proof is split into six parts. The first four parts concern the abstract Cauchy problem associated with A and in the two remaining parts we prove the statements (a)–(c). In what follows we assume that the conditions of the theorem are fulfilled.

    Part (i). The first step is to prove that the map $t \mapsto T(t)$ is strongly continuous on $[0, \infty)$, that is, given $x \in X$ we have to show that $t \mapsto T(t)x$ is a continuous map from $[0, \infty)$ into X. By property (iii) of a strongly continuous semigroup, $t \mapsto T(t)x$ is continuous from the right at $t_0 = 0$. Take $t_0 > 0$. Then, by property (iii) applied to $y = T(t_0)x$,

$$\lim_{t \downarrow t_0} T(t)x - T(t_0)x = \lim_{t \downarrow t_0} T(t - t_0)y - y = 0.$$

For $t < t_0$ we have

(4) $$T(t)x - T(t_0)x = T(t)\big(x - T(t_0 - t)x\big).$$

By Lemma 2.1 the function $t \mapsto T(t)$ is uniformly bounded in the operator norm on compact subintervals of $[0, \infty)$. It follows (again use property (iii)) that for $t \uparrow t_0$ the right hand side of (4) goes to zero. Thus $T(t)x - T(t_0)x \to 0$ whenever $t \to t_0$.

    Part (ii). Next we show that $\overline{\mathcal{D}(A)} = X$. It follows from the strong continuity of $T(\cdot)$ that for each $x \in X$

(5) $$x_t := \frac{1}{t}\int\limits_0^t T(s)x\,ds \to T(0)x = x, \qquad t \downarrow 0.$$

Moreover, x_t is in $\mathcal{D}(A)$, because for $t > 0$

$$\frac{1}{h}\big(T(h)x_t - x_t\big) = \frac{1}{th}\int\limits_0^t T(h+s)x\,ds - \frac{1}{th}\int\limits_0^t T(s)x\,ds$$

$$= \frac{1}{th}\int\limits_h^{t+h} T(s)x\,ds - \frac{1}{th}\int\limits_0^t T(s)x\,ds$$

$$= \frac{1}{th}\int\limits_t^{t+h} T(s)x\,ds - \frac{1}{th}\int\limits_0^h T(s)x\,ds$$

$$\to \frac{1}{t}\big(T(t)x - x\big), \qquad h \downarrow 0.$$

Thus $x_t \in \mathcal{D}(A)$, and hence $x \in \overline{\mathcal{D}(A)}$ by (5). We conclude that $\overline{\mathcal{D}(A)} = X$. It is clear that A is linear.

Part (iii). In this part we show that for each $x \in \mathcal{D}(A)$ the vector $T(t)x \in \mathcal{D}(A)$ and

$$(6) \qquad \frac{d}{dt}T(t)x = T(t)Ax = AT(t)x, \qquad 0 \le t < \infty.$$

To do this, note that

$$(7) \quad \frac{1}{h}\big(T(h)T(t)x - T(t)x\big) - T(t)Ax = T(t)\left[\frac{1}{h}(T(h)x - x) - Ax\right] \to 0, \qquad h \downarrow 0.$$

This shows that $T(t)x \in \mathcal{D}(A)$ and $AT(t)x = T(t)Ax$. From (7) it also follows that $T(t)x$ has a right-sided derivative equal to $T(t)Ax$. On the other hand

$$\frac{1}{h}\big(T(t+h)x - T(t)x\big) - T(t)Ax$$
$$= T(t+h)\left\{\frac{T(-h)x - x}{-h} - Ax\right\} + \big(T(t+h) - T(t)\big)Ax \to 0, \qquad h \uparrow 0,$$

because $x \in \mathcal{D}(A)$, the map $t \mapsto T(t)$ is uniformly bounded in the operator norm on compact subintervals of $[0, \infty)$ and $t \mapsto T(t)y$ is continuous. Thus (6) is established.

From (6) it follows that for $x \in \mathcal{D}(A)$ the function $u(t; x) := T(t)x$ is a solution of the abstract Cauchy problem associated with A. To see that there is no other solution with the same initial value x one applies the same reasoning as used in Part (vi) of Theorem 1.1. Note that the general remark made in Part (v) of the proof of Theorem 1.1 is valid whenever $T(\cdot)$ is a strongly continuous semigroup.

Part (iv). We show that A is closed. To do this, suppose that $x_1, x_2, \ldots$ is a sequence in $\mathcal{D}(A)$ such that $x_n \to x$ and $Ax_n \to y$. From (6) we get

$$T(t)x_n - x_n = \int_0^t \frac{d}{ds}T(s)x_n ds = \int_0^t T(s)Ax_n ds$$
$$\to \int_0^t T(s)y ds, \qquad n \to \infty.$$

Here we used that for $0 \le s \le t$

$$\|T(s)Ax_n - T(s)y\| \le Me^{\omega t}\|Ax_n - y\|.$$

On the other hand, $T(t)x_n - x_n \to T(t)x - x$. Thus

$$\frac{1}{t}\big(T(t)x - x\big) = \frac{1}{t}\int_0^t T(s)y ds \to T(0)y = y, \qquad t \downarrow 0.$$

Hence $x \in \mathcal{D}(A)$ and $Ax = y$, which shows that A is closed.

Part (v). It remains to establish (a)–(c). First we prove (a) and (b). For each $x \in X$ and $\Re\lambda > \omega$, the *Laplace transform*

$$(8) \qquad R(\lambda)x := \int_0^\infty e^{-\lambda t} T(t) x \, dt$$

(considered as an indefinite Riemann-Stieltjes integral) exists. Indeed, since $T(t)x$ is continuous on $0 \le t < \infty$, the integrand in (8) is a continuous function of t, and from $\|T(t)x\| \le M e^{\omega t}\|x\|$ and $\Re\lambda > \omega$ it follows that the norm of the integrand is majorized by an exponential decaying function. Let us show that $R(\lambda)x \in \mathcal{D}(A)$. We have

$$
\begin{aligned}
\frac{1}{h}\{T(h)R(\lambda)x - R(\lambda)x\} &= \frac{1}{h}\int_0^\infty e^{-\lambda t}T(t+h)x\,dt - \frac{1}{h}\int_0^\infty e^{-\lambda t}T(t)x\,dt \\
&= \frac{1}{h}e^{\lambda h}\int_h^\infty e^{-\lambda t}T(t)x\,dt - \frac{1}{h}\int_0^\infty e^{-\lambda t}T(t)x\,dt \\
&= \frac{1}{h}(e^{\lambda h}-1)\int_0^\infty e^{-\lambda t}T(t)x\,dt - e^{\lambda h}\left(\frac{1}{h}\int_0^h e^{-\lambda t}T(t)x\,dt\right) \\
&\to \lambda R(\lambda)x - x, \qquad h \downarrow 0.
\end{aligned}
$$

Hence $R(\lambda)x$ is in $\mathcal{D}(A)$ and $AR(\lambda)x = \lambda R(\lambda)x - x$, or $(\lambda - A)R(\lambda)x = x$ for all $x \in X$. On the other hand, given $x \in \mathcal{D}(A)$, we have from (6) that

$$R(\lambda)Ax = \int_0^\infty e^{-\lambda t}T(t)Ax\,dt = \int_0^\infty e^{-\lambda t}AT(t)x\,dt = AR(\lambda)x.$$

The last equality may be seen by considering Riemann sums corresponding to the integrals

$$\int_0^N e^{-\lambda t}T(t)x\,dt, \qquad \int_0^N e^{-\lambda t}AT(t)x\,dt,$$

and using the fact that A is closed. Hence (a) and (b) are proved.

Part (vi). We prove the statement (c). Define a new norm $\|\cdot\|_1$ on X by setting

$$\|x\|_1 = \sup\{e^{-\omega s}\|T(s)x\| \mid 0 \le s < \infty\}.$$

Since $T(0) = I$ and (1) holds, it is clear that

$$(9) \qquad \|x\| \le \|x\|_1 \le M\|x\|, \qquad x \in X.$$

Take $\Re\lambda > \omega$. It follows from (b) that

$$\|(\lambda - A)^{-1}x\|_1 = \sup_{s\geq 0}\left\|\int_0^\infty e^{-\omega s}T(s)e^{-\lambda t}T(t)x\,dt\right\|$$

$$= \sup_{s\geq 0}\left\|\int_0^\infty e^{(\omega-\lambda)t}e^{-\omega(s+t)}T(t+s)x\,dt\right\|$$

$$\leq \int_0^\infty e^{(\omega-\Re\lambda)t}\|x\|_1\,dt = \frac{1}{\Re\lambda - \omega}\|x\|_1.$$

But then $\|(\lambda - A)^{-n}\|_1 \leq (\Re\lambda - \omega)^{-n}$, and by (9)

$$\|(\lambda - A)^{-n}x\| \leq \|(\lambda - A)^{-n}x\|_1 \leq (\Re\lambda - \omega)^{-n}\|x\|_1$$
$$\leq M(\Re\lambda - \omega)^{-n}\|x\|,$$

which proves (c). $\quad\square$

The operator A defined in Theorem 2.2 is called the (*infinitesimal*) *generator* of the C_0 semigroup $T(\cdot)$.

Let us reconsider Theorem 1.1 in terms of Theorem 2.2. If we start with an operator A which satisfies the hypotheses of Theorem 1.1, then A determines a C_0 semigroup $T(\cdot)$ via the formula

$$T(t)x = \lim_{\lambda\to\infty} e^{tA_\lambda}x, \qquad A_\lambda := \lambda A(\lambda - A)^{-1}.$$

Now $T(\cdot)$ has a generator, call it A_1. How does A compare with A_1? From formula (3) and the definition of a generator it follows that A is a restriction of A_1. On the other hand, by Theorem 2.2 (and Lemma 2.1) there exists λ such that both $\lambda - A$ and $\lambda - A_1$ are invertible. But then

$$X = (\lambda - A)\mathcal{D}(A) = (\lambda - A_1)\mathcal{D}(A) \subset (\lambda - A_1)\mathcal{D}(A_1) = X,$$

which implies that $\mathcal{D}(A) = \mathcal{D}(A_1)$, and thus $A = A_1$. This observation, Theorem 2.2 and the proof of Theorem 1.1 yield the following fundamental result.

THEOREM 2.3 (Hille-Yoshida-Phillips). *An operator* $A(X \to X)$ *is the infinitesimal generator of a strongly continuous semigroup* $T(\cdot)$ *satisfying*

$$\|T(t)\| \leq Me^{\omega t}, \qquad t \geq 0,$$

if and only if A *is a closed densely defined operator and for every real* $\lambda > \omega$ *one has* $\lambda \in \rho(A)$ *with*

$$(10) \qquad \|(\lambda - A)^{-n}\| \leq M(\lambda - \omega)^{-n}, \qquad n = 1, 2, \ldots .$$

Let $X = BUC(\mathbf{R})$. The *translation semigroup* $T(\cdot)$ on X, which is defined by

$$(11) \qquad (T(t)f)(s) = f(t + s), \qquad s \in \mathbf{R}, \ t \geq 0,$$

is strongly continuous. Let B be its generator. From the example in Section 1 it follows that B is an extension of the operator A given by

$$(12) \qquad \mathcal{D}(A) = \{g \in X \mid g' \in X\}, \qquad Ag = g'.$$

A simple computation shows that each $\lambda > 0$ is in $\rho(A)$. Thus for $\lambda \in \mathbf{R}$ sufficiently large both $\lambda - A$ and $\lambda - B$ are invertible, and hence $\mathcal{D}(A) = \mathcal{D}(B)$. We conclude that the operator A in (12) is the generator of the translation semigroup on $X = BUC(\mathbf{R})$.

The result of the previous paragraph can be used to prove the classical *Weierstrass approximation theorem* as follows. Let f be a continuous real-valued function on $[0, 1]$. Extend f to $\mathbf{R}$ by letting $f(t) = f(0)$ for $t \leq 0$ and $f(t) = f(1)$ for $t \geq 1$. This extension we still denote by f. Now apply Theorem 1.1 to the operator A in (12). Then

$$(13) \qquad \lim_{\lambda \to \infty} e^{tA_\lambda} f = T(t)f,$$

where $T(t)$ is given by (11). From the proof of Theorem 1.1 we know that in (13) the convergence is uniform in t on $[0, 1]$. Also the series

$$\sum_{k=0}^{\infty} \frac{t^k}{k!} A_\lambda^k f = e^{tA_\lambda} f$$

converges uniformly on $0 \leq t \leq 1$. Thus, given $\varepsilon > 0$, there exists an integer N and a real number $\lambda > 0$ such that

$$\sup_{s \in \mathbf{R}} \left| (T(t)f)(s) - \sum_{k=0}^{N} \frac{t^k}{k!} (A_\lambda^k f)(s) \right| < \varepsilon, \qquad 0 \leq t \leq 1.$$

Setting $s = 0$ yields

$$\left| f(t) - \sum_{k=0}^{N} c_k t^k \right| < \varepsilon, \qquad 0 \leq t \leq 1, \quad c_k = \frac{1}{k!} (A_\lambda^k f)(0),$$

which is the desired approximation.

Let $X = L_p([0, \infty))$ with $1 \leq p < \infty$. Next we shall show that the maximal operator $A(X \to X)$ corresponding to $\tau = \frac{d}{dt}$ is the generator of the semigroup

$$(14) \qquad (T(t)f)(s) = f(t + s), \qquad t \geq 0, \ s \geq 0.$$

Recall that $\mathcal{D}(A)$ consists of those $f \in X$ which are absolutely continuous on every compact subinterval of $[0, \infty)$ and $Af = f'$. First, we prove that the semigroup (14) is strongly continuous. For each $t \geq 0$ the operator $T(t)$ is bounded on $X = L_p([0, \infty))$

and $\|T(t)\| \leq 1$. Take $f \in X$, and let $\varepsilon > 0$. Choose a continuous function φ on $[0, \infty)$ with support in a bounded interval $[0, a]$ such that $\|f - \varphi\| < \varepsilon/3$. Since φ is uniformly continuous on $[0, \infty)$, we can find $\delta > 0$ such that $0 \leq t < \delta$ implies that

$$\sup_{s \geq 0} \left| (T(t)\varphi)(s) - \varphi(s) \right| < \frac{1}{3}\varepsilon a^{-1/p}.$$

But then for $0 \leq t < \delta$

$$\|T(t)f - f\| \leq \|T(t)\| \|f - \varphi\| + \|T(t)\varphi - \varphi\| + \|\varphi - f\| < \varepsilon,$$

which proves the strong continuity of the semigroup.

Let A_0 be the generator of $T(t)$. We have to show that A_0 is equal to the maximal operator A. Take $f \in \mathcal{D}(A)$. Then

$$\left\| \frac{1}{h}(T(h)f - f) - Af \right\|^p = \int_0^\infty \left| \frac{f(t+h) - f(t)}{h} - f'(t) \right|^p dt$$

$$= \int_0^\infty \left| \frac{1}{h} \int_0^h [f'(t+s) - f'(t)]ds \right|^p dt$$

$$= \int_0^\infty \left| \int_0^1 [f'(t+xh) - f'(t)]dx \right|^p dt$$

$$\leq \int_0^\infty \int_0^1 |f'(t+xh) - f'(t)|^p \, dx \, dt$$

$$= \int_0^1 \int_0^\infty |f'(t+xh) - f'(t)|^p \, dt \, dx$$

$$= \int_0^1 \|T(xh)f' - f'\|^p dx \to 0 \qquad (h \downarrow 0).$$

This shows that $f \in \mathcal{D}(A_0)$ and $A_0 f = Af$. It remains to show that $\mathcal{D}(A_0) \subset \mathcal{D}(A)$. Take $g \in \mathcal{D}(A_0)$. Let

$$u = A_0 g = \lim_{h \downarrow 0} \frac{1}{h}(T(h)g - g).$$

For any $0 \leq a < t < \infty$, Hölder's inequality implies that convergence in $L_p([a,t])$ implies convergence in $L_1([a,t])$. Hence

$$\int_a^t u(s)ds = \lim_{h \downarrow 0} \int_0^t \frac{1}{h}(g(s+h) - g(s))ds$$

$$= \lim_{h \downarrow 0} \left(\frac{1}{h} \int_a^t g(s+h)ds - \frac{1}{h} \int_a^t g(s)ds \right)$$

$$= \lim_{h \downarrow 0} \left(\frac{1}{h} \int_{a+h}^{t+h} g(s)ds - \frac{1}{h} \int_a^t g(s)ds \right)$$

$$= \lim_{h \downarrow 0} \left(\frac{1}{h} \int_t^{t+h} g(s)ds - \frac{1}{h} \int_a^{a+h} g(s)ds \right).$$

Since g is integrable on bounded subintervals of $[0, \infty)$, there exists a set $E \subset [0, \infty)$ of measure zero such that

$$\lim_{h \downarrow 0} \frac{1}{h} \int_b^{b+h} g(s)ds = g(b), \qquad b \notin E.$$

But then we may conclude that

$$\int_a^t u(s)ds = g(t) - g(a), \qquad t, a \notin E.$$

If we define $\widetilde{g}(t) = \int_a^t u(s)ds - g(a)$, then $\widetilde{g} = g$ as elements in $X = L_p([0, \infty))$. Furthermore, $\widetilde{g} \in \mathcal{D}(A)$, which proves that $\mathcal{D}(A_0) \subset \mathcal{D}(A)$. Thus $A = A_0$.

Additional examples of generators of C_0 semigroups appear in the next sections.

XIX.3 THE CAUCHY PROBLEM REVISITED

The following theorem makes precise the relationship between generators of strongly continuous semigroups and the Cauchy problem.

THEOREM 3.1. *Let $A(X \to X)$ be a densely defined operator with a non-empty resolvent set. A necessary and sufficient condition that the Cauchy problem*

$$\begin{cases} u'(t) = Au(t), & 0 \le t < \infty, \\ u(0) = x, \end{cases}$$

has a unique continuously differentiable solution for every $x \in \mathcal{D}(A)$ is that A is the generator of a strongly continuous semigroup.

PROOF. The sufficiency is evident from Theorems 2.2 and 1.1. The proof of the necessity is split into five parts. In what follows, the function $u(\cdot; x)$ is the unique C^1-solution to the Cauchy problem above.

Part (i). We have to associate with A a C_0 semigroup. First we restrict our attention to the domain of A. Define $T(t)\colon \mathcal{D}(A) \to \mathcal{D}(A)$ by $T(t)x = u(t;x)$, $0 \le t < \infty$. It follows from the uniqueness of the solution that $T(t)$ is linear. Furthermore, $T(t)$ is a semigroup in t. Indeed, for each $x \in \mathcal{D}(A)$ and $s \ge 0$ fixed, $w(t) = T(t)T(s)x$ and $v(t) = T(t+s)x$ are C^1-solutions to the Cauchy problem with initial condition $T(s)x$. Hence $w(t) = v(t)$, and thus $T(t+s) = T(t)T(s)$. Clearly, $T(0)x = x$ for each $x \in \mathcal{D}(A)$.

Part (ii). Let X_A denote the space $\mathcal{D}(A)$ endowed with the graph norm $\|x\|_A = \|x\| + \|Ax\|$. Since $\rho(A) \ne \emptyset$, the operator A is closed, and therefore X_A is a Banach space. We shall show that $T(\cdot)$ is a strongly continuous semigroup on X_A. Since $u(t;x)$ is continuously differentiable on $[0,\infty)$, we have

$$(1) \qquad \lim_{t\downarrow 0} T(t)x = \lim_{t\downarrow 0} u(t;x) = u(0;x) = x,$$

$$(2) \qquad \lim_{t\downarrow 0} AT(t)x = \lim_{t\downarrow 0} u'(t;x) = u'(0;x).$$

The fact that A is closed implies that $Ax = u'(0;x)$, and hence (1) and (2) show that

$$\|T(t)x - x\|_A \to 0 \qquad (t \downarrow 0).$$

It remains to prove that $T(t) \in \mathcal{L}(X_A)$. Let $\mathcal{F} = C([0,1], X_A)$ be the Banach space of all continuous functions from $[0,1]$ into X_A endowed with the supremum norm

$$\|\|f\|\|_A = \max\{\|f(t)\|_A \mid 0 \le t \le 1\}.$$

Define the linear operator $S\colon X_A \to \mathcal{F}$ by $(Sx)(t) = u(t;x)$. The function Sx is in $\mathcal{F}$, because $u(t;x)$ is continuously differentiable on $0 \le t < \infty$ and

$$\|(Sx)(t_1) - (Sx)(t_2)\|_A = \|u(t_1;x) - u(t_2;x)\| + \|Au(t_1;x) - Au(t_2;x)\|$$
$$= \|u(t_1;x) - u(t_2;x)\| + \|u'(t_1;x) - u'(t_2;x)\|.$$

Let us prove that S is a closed operator. Suppose $\|x_n - x\|_A \to 0$ and $\|\|Sx_n - g\|\|_A \to 0$ for $n \to \infty$. Put $u_n(t) = u(t;x_n)$. It follows from the definitions of the norms $\|\cdot\|_A$ and $\|\|\cdot\|\|_A$ that

$$(3) \qquad u_n'(t) = Au_n(t) \to Ag(t) \qquad (n \to \infty)$$

uniformly on $0 \le t \le 1$ and

$$g(t) = \lim_{n\to\infty} u_n(t) = \lim_{n\to\infty} \left(u_n(0) + \int_0^t u_n'(s)\,ds \right)$$

$$= \lim_{n\to\infty} \left(x_n + \int_0^t Au_n(s)\,ds \right)$$

$$= x + \int_0^t Ag(s)\,ds, \qquad 0 \le t \le 1.$$

Since the convergence in (3) is uniform on $0 \leq t \leq 1$, the continuity of u'_n implies the continuity of Ag, and therefore $g'(t) = Ag(t)$, $0 \leq t \leq 1$, and $g(0) = x$. Put

$$w(t) = \begin{cases} g(t), & 0 \leq t \leq 1, \\ u(t-1; g(1)), & t \geq 1. \end{cases}$$

Then $w(\cdot)$ is a C^1-solution of the Cauchy problem with initial value $w(0) = g(0) = x$. Hence, by the uniqueness of the C^1-solutions, $w(t) = u(t; x)$ for $t \geq 0$. Therefore,

$$(Sx)(t) = u(t; x) = w(t) = g(t), \qquad 0 \leq t \leq 1,$$

which shows that S is closed on the Banach space X_A. Consequently, S is bounded and for $0 \leq t \leq 1$,

$$(4) \qquad \|T(t)x\|_A = \|(Sx)(t)\|_A \leq \|\|Sx\|\|_A \leq \|\|S\|\|\|x\|_A, \qquad x \in X_A.$$

For $t \geq 0$, we have $t = n + \delta$, where n is a nonnegative integer and $0 \leq \delta < 1$. Since $T(\cdot)$ is a semigroup on $\mathcal{D}(A)$, it follows from (4) that

$$(5) \qquad \begin{aligned} \|T(t)x\|_A &= \|T(1)^n T(\delta)x\|_A \leq \|S\|^{n+1}\|x\|_A \\ &\leq \|S\|\|S\|^t\|x\|_A = \|S\|e^{\omega t}\|x\|_A, \qquad 0 \leq t < \infty, \; x \in X_A, \end{aligned}$$

where $\omega = \log\|S\|$.

Part (iii). Next, we show that $T(t)Ax = AT(t)x$ for every $x \in \mathcal{D}(A^2)$. For such a vector x we have

$$(6) \qquad \begin{aligned} T(t)Ax = u(t; Ax) &= Ax + \int_0^t u'(s; Ax)ds \\ &= Ax + \int_0^t Au(s; Ax)ds, \qquad 0 \leq t < \infty. \end{aligned}$$

Put

$$(7) \qquad w(t) = x + \int_0^t u(s; Ax)ds, \qquad 0 \leq t < \infty.$$

Since A is closed, (6) and (7) imply that

$$\begin{aligned} Aw(t) &= A\left(x + \int_0^t u(s; Ax)ds \right) \\ &= Ax + \int_0^t Au(s; Ax)ds = u(t; Ax). \end{aligned}$$

Thus $Aw(\cdot)$ is continuous and

$$w'(t) = u(t; Ax) = Aw(t), \qquad w(0) = x.$$

By uniqueness, $w(t) = u(t; x)$, and

$$(8) \qquad T(t)Ax = Aw(t) = Au(t; x) = AT(t)x, \qquad x \in \mathcal{D}(A^2).$$

Part (iv). In this part we extend the semigroup $T(\cdot)$ to all of X. By hypothesis, there exists $\lambda_0 \in \rho(A)$. Given $x \in \mathcal{D}(A)$, the vector $y := (\lambda_0 - A)^{-1}x$ is in $\mathcal{D}(A^2)$. Thus (5), (8) and the boundedness of the operators $(\lambda_0 - A)^{-1}$ and $A(\lambda_0 - A)^{-1}$ imply that for some positive constants α_0, α_1, α_2,

$$(9) \qquad \begin{aligned} \|T(t)x\| &= \|(\lambda_0 - A)T(t)y\| \le \alpha_0\|T(t)y\|_A \\ &\le \alpha_1 e^{\omega t}\|y\|_A \le \alpha_2 e^{\omega t}\|x\|, \qquad 0 \le t < \infty. \end{aligned}$$

Since $\mathcal{D}(A)$ is dense in X, formula (9) shows that $T(t)$ can be extended to a bounded linear operator on X which we still denote by $T(t)$. Since $T(\cdot)$ is a semigroup on $\mathcal{D}(A)$, a continuity argument shows that the same holds true for $T(\cdot)$ on X. Also, by continuity, $T(0)$ is the identity operator on X. From

$$\|T(t)z - z\| \le \|T(t)z - T(t)x\| + \|T(t)x - x\| + \|x - z\|$$
$$\le (1 + \alpha_2 e^{\omega t})\|x - z\| + \|T(t)x - x\|$$

and the fact that $T(\cdot)$ is strongly continuous on X_A, it follows that $T(\cdot)$ is also a strongly continuous semigroup on X.

Part (v). The proof of the theorem is complete once it is shown that the infinitesimal generator A_0 of the semigroup $T(\cdot)$ on X is equal to A. For each $x \in \mathcal{D}(A)$, we have

$$\frac{1}{t}(T(t)x - x) = \frac{1}{t}(u(t, x) - x)$$
$$\to u'(0, x) = Ax, \qquad t \downarrow 0.$$

Thus $x \in \mathcal{D}(A_0)$ and $A_0 x = Ax$. Hence $A \subset A_0$.

It remains to prove that $\mathcal{D}(A_0) \subset \mathcal{D}(A)$. Let $x \in X$ be given. Since

$$(\lambda_0 - A)^{-1}\mathcal{D}(A) \subset \mathcal{D}(A^2)$$

and $\mathcal{D}(A)$ is dense in X, it follows that $\mathcal{D}(A^2)$ is also dense in X. Therefore there exists a sequence (x_n) in $\mathcal{D}(A^2)$ which converges to x. By Theorem 2.2 and formula (8), there exists $\lambda_0 \in \rho(A_0)$ such that for $n = 1, 2, \ldots$

$$(\lambda - A_0)^{-1}x_n = \int_0^\infty e^{-\lambda t}T(t)x_n\,dt,$$

$$(\lambda - A_0)^{-1}A_0 x_n = (\lambda - A_0)^{-1}Ax_n = \int_0^\infty e^{-\lambda t}T(t)Ax_n\,dt$$

$$= \int_0^\infty A\big(e^{-\lambda t}T(t)x_n\big)\,dt.$$

Since A is closed, we conclude that

$$A(\lambda - A_0)^{-1}x_n = (\lambda - A_0)^{-1}A_0 x_n$$
$$= A_0(\lambda - A_0)^{-1}x_n \to A_0(\lambda - A_0)^{-1}x.$$

Also, $(\lambda - A_0)^{-1}x_n \to (\lambda - A_0)^{-1}x$. Again we use that A is closed. It follows that $(\lambda - A_0)^{-1}x$ is in $\mathcal{D}(A)$. Thus $\mathcal{D}(A_0) = \mathrm{Im}(\lambda - A_0)^{-1} \subset \mathcal{D}(A)$, and we have proved that $A = A_0$. $\square$

The next theorem gives the solution of the inhomogeneous Cauchy problem.

THEOREM 3.2. *Let $A(X \to X)$ be the infinitesimal generator of the strongly continuous semigroup $T(\cdot)$. Suppose that $f:[0,\infty) \to X$ is continuously differentiable on $[0,\infty)$. Then for each $x \in \mathcal{D}(A)$, there exists a unique solution to*

$$(10) \qquad \begin{cases} u'(t) = Au(t) + f(t), & 0 \le t < \infty, \\ u(0) = x. \end{cases}$$

This solution is continuously differentiable and is given by

$$(11) \qquad u(t) = T(t)x + \int_0^t T(t - s)f(s)ds, \qquad t \ge 0.$$

PROOF. Put

$$v(t) = \int_0^t T(t - s)f(s)ds = \int_0^t T(s)f(t - s)ds, \qquad t \ge 0.$$

Take $h > 0$, fix $t \ge 0$, and let $0 \le \alpha \le t$. Note that

$$\frac{1}{h}\{f(\alpha + h) - f(\alpha)\} - f'(\alpha) = \frac{1}{h}\int_\alpha^{\alpha+h} \{f'(s) - f'(\alpha)\}ds \to 0 \qquad (h \downarrow 0).$$

In fact, since f' is uniformly continuous on compact subintervals of $[0,\infty)$, the above convergence is uniform on $0 \le \alpha \le t$. It follows that

$$\frac{1}{h}\{v(t + h) - v(t)\} = \int_0^t T(s)\left[\frac{1}{h}\{f(t - s + h) - f(t - s)\}\right]ds$$

$$+ \frac{1}{h}\int_t^{t+h} T(s)f(t + h - s)ds$$

$$\to \int_0^t T(s)f'(t - s)ds + T(t)f(0), \qquad h \downarrow 0.$$

On the other hand,

$$\frac{1}{h}\{v(t+h)-v(t)\} = \frac{1}{h}\int_0^{t+h} T(t-s+h)f(s)ds - \frac{1}{h}\int_0^t T(t-s)f(s)ds$$

$$= \frac{1}{h}\{T(h)v(t)-v(t)\} + \frac{1}{h}\int_t^{t+h} T(t-s+h)f(s)ds.$$

Since

$$\frac{1}{h}\int_t^{t+h} T(t-s+h)f(s)ds \to T(0)f(t), \qquad h\downarrow 0,$$

we conclude that

$$\lim_{h\downarrow 0}\frac{1}{h}\{T(h)v(t)-v(t)\} = \lim_{h\downarrow 0}\frac{1}{h}\{v(t+h)-v(t)\} - f(t).$$

A similar calculation holds true for $t > 0$ and $h\uparrow 0$. It follows that

$$v(t)\in \mathcal{D}(A), \quad Av(t) = v'(t) - f(t), \qquad t\geq 0.$$

Now let u be the function defined by (11), where $x\in\mathcal{D}(A)$. Then $u(0) = x$ and

$$u'(t) = \frac{d}{dt}T(t)x + v'(t) = AT(t)x + Av(t) + f(t)$$

$$= Au(t) + f(t), \qquad t\geq 0.$$

Thus u is a solution of (10). Since $v'(\cdot)$ and $AT(\cdot)x = T(\cdot)Ax$ are continuous functions, the same holds true for u'.

Suppose that $w(\cdot)$ is another solution to (10). Then $u(\cdot) - w(\cdot)$ is a solution to the Cauchy problem

(12)
$$\begin{cases} y'(t) = Ay(t), & 0\leq t < \infty, \\ y(0) = 0. \end{cases}$$

Hence $u(t) - w(t) = 0$ for all $0\leq t < \infty$ by the uniqueness of the solution to (12). $\square$

XIX.4   DISSIPATIVE OPERATORS AND CONTRACTION SEMIGROUPS

A strongly continuous semigroup $T(\cdot)$ is called a *contraction semigroup* if $\|T(t)\| \leq 1$ for all $t\geq 0$. The semigroup $T(\cdot)$ defined on $X = BUC(\mathbf{R})$ or on $X = L_p([0,\infty))$, $1\leq p < \infty$, by

$$(T(t)f)(s) = f(t+s), \qquad t\geq 0,$$

is an example of a contraction semigroup.

THEOREM 4.1. *An operator $A(X \to X)$ is the infinitesimal generator of a contraction semigroup if and only if A is densely defined, $(0, \infty)$ belongs to the resolvent set of A and*

$$(1) \qquad \|(\lambda - A)^{-1}\| \leq \lambda^{-1}, \qquad \lambda > 0.$$

PROOF. The result is an immediate consequence of Theorem 2.3. One only has to remark that (1) implies that for $n = 1, 2, \ldots$

$$\|(\lambda - A)^{-n}\| \leq \lambda^{-n}, \qquad \lambda > 0. \quad \square$$

Dissipative operators arise naturally in the study of contraction semigroups. We first consider the Hilbert space case. Let H be a Hilbert space. An operator $A(H \to H)$ is called *dissipative* if

$$\Re\langle Ax, x \rangle \leq 0, \qquad x \in \mathcal{D}(A).$$

THEOREM 4.2. *Let H be a Hilbert space, and let $A(H \to H)$ be densely defined. Then A is the generator of a contraction semigroup if and only if A is dissipative and there exists a $\lambda_0 > 0$ in $\rho(A)$.*

PROOF. Suppose A is dissipative and $\lambda_0 \in \rho(A)$. Take $\lambda > 0$ and $x \in \mathcal{D}(A)$. Then

$$(2) \qquad \|(\lambda - A)x\|^2 = \|Ax\|^2 - 2\lambda\Re\langle Ax, x \rangle + \lambda^2\|x\|^2 \geq \lambda^2\|x\|^2.$$

Hence

$$(3) \qquad \|(\lambda_0 - A)^{-1}\| \leq \lambda_0^{-1}.$$

Now for any λ,

$$(\lambda - A) = [I + (\lambda - \lambda_0)(\lambda_0 - A)^{-1}](\lambda_0 - A).$$

Therefore, it follows from (3) that $\lambda_1 \in \rho(A)$ if $|\lambda_1 - \lambda_0| < \lambda_0$. If we replace λ_0 by $\lambda_1 = \lambda_0 \pm \frac{1}{2}\lambda_0$ in the above argument, then we have that $\lambda_2 \in \rho(A)$ if $|\lambda_2 - \lambda_1| < \lambda_1$. Continuing in this manner, it follows that $(0, \infty) \subset \rho(A)$ and (1) holds. Hence A is the generator of a contraction semigroup by Theorem 4.1.

Conversely, let A be the generator of a contraction semigroup $T(\cdot)$. Then for $x \in \mathcal{D}(A)$

$$\Re\langle T(h)x - x, x \rangle = \Re\langle T(h)x, x \rangle - \|x\|^2 \leq 0.$$

Hence

$$\Re\langle Ax, x \rangle = \lim_{h \downarrow 0} \Re\langle \frac{T(h)x - x}{h}, x \rangle \leq 0, \qquad x \in \mathcal{D}(A).$$

Also, $(0, \infty) \subset \rho(A)$ by Theorem 4.1. $\quad \square$

The above theorem can be extended to an arbitrary Banach space X in the following manner. For each $x \in X$, the Hahn-Banach theorem implies the existence of a functional $F_x \in X'$ such that

$$(4) \qquad \|F_x\|^2 = \|x\|^2 = F_x(x).$$

An operator $A(X \to X)$ is called *dissipative* if for each $x \in \mathcal{D}(A)$ there exists an $F_x \in X'$ such that (4) holds and

$$(5) \qquad \Re F_x(Ax) \le 0, \qquad x \in \mathcal{D}(A).$$

If X is a Hilbert space, then the two notions of dissipativeness coincide. To see this one needs only to note that for X a Hilbert space, formula (4) implies that $F_x(v) = \langle v, x \rangle$.

THEOREM 4.3. *A densely defined operator $A(X \to X)$ in a Banach space X is the generator of a contraction semigroup if and only if A is dissipative and there exists a $\lambda_0 > 0$ in $\rho(A)$.*

PROOF. Suppose A is dissipative, and let $0 < \lambda_0 \in \rho(A)$. Given $x \in X$, let $F_x \in X'$ satisfy (4) and (5). Hence for $\lambda > 0$,

$$\|x\| \|\lambda x - Ax\| \ge \Re F_x(\lambda x - Ax)$$
$$= \lambda \|x\|^2 - \Re F_x(Ax) \ge \lambda \|x\|^2.$$

Thus

$$(6) \qquad \|(\lambda I - A)x\| \ge \lambda \|x\|, \qquad \lambda > 0,$$

and the arguments used in the first paragraph of the proof of Theorem 4.2 show that $(0, \infty) \subset \rho(A)$ and

$$\|(\lambda - A)^{-1}\| \le \lambda^{-1}, \qquad \lambda > 0.$$

Hence A is the generator of a contraction semigroup, by Theorem 4.1.

Conversely, let A be the generator of a contraction semigroup $T(\cdot)$. For each $x \in X$ there exists an $F_x \in X'$ which satisfies (4) (by the Hahn-Banach theorem). Since $\|T(t)\| \le 1$, we have

$$\Re F_x(T(h)x - x) = \Re F_x(T(h)x) - \|x\|^2 \le 0.$$

Therefore for $x \in \mathcal{D}(A)$

$$\Re F_x(Ax) = \lim_{h \downarrow 0} \Re F_x \left(\frac{1}{h} \{T(h)x - x\} \right) \le 0.$$

Also $(0, \infty) \subset \rho(A)$ by Theorem 4.1. $\square$

Applications of Theorem 4.2 yield the following results.

PROPOSITION 4.4. *Let $\tau = a_2(t)D^2 + a_1(t)D + a_0(t)$, where a_j is a real-valued function in $C^j([a,b])$, $0 \le j \le 2$, and let $A(L_2([a,b]) \to L_2([a,b]))$ be defined by*

$$\mathcal{D}(A) := \{ f \in L_2([a,b]) \mid f' \text{ is absolutely continuous,}$$
$$f'' \in L_2([a,b]), \qquad f(a) = f(b) = 0 \},$$
$$Af = \tau f.$$

Then A is the generator of a contraction semigroup whenever $a_2(t) > 0$ and $a_2''(t) - a_1'(t) + 2a_2(t) \le 0$ for $a \le t \le b$.

PROOF. Given $f \in \mathcal{D}(A)$,

$$\tau f = \big(a_2(t)f'\big)' + \big(a_1(t) - a_2'(t)\big)f' + a_0(t)f. \tag{7}$$

Integration by parts yields

$$\langle (a_2 f')', f \rangle = \int_a^b \big(a_2(t)f'(t)\big)'\overline{f(t)}\,dt$$

$$= -\int_a^b a_2(t)|f'(t)|^2\,dt \le 0. \tag{8}$$

Here we used the boundary conditions $f(a) = f(b) = 0$. Now

$$\frac{d}{dt}|f(t)|^2 = \frac{d}{dt}f(t)\overline{f(t)} = 2\Re f'(t)\overline{f(t)}.$$

Hence

$$\Re\langle (a_1 - a_2')f', f \rangle = \Re \int_a^b \big(a_1(t) - a_2'(t)\big)f'(t)\overline{f(t)}\,dt$$

$$= \frac{1}{2}\int_a^b \big(a_1(t) - a_2'(t)\big)\frac{d}{dt}|f(t)|^2\,dt \tag{9}$$

$$= \frac{1}{2}\int_a^b \big(a_2''(t) - a_1'(t)\big)|f(t)|^2\,dt,$$

by partial integration. By (7), (8) and (9),

$$\Re\langle Af, f \rangle \le \frac{1}{2}\int_a^b \big(a_2''(t) - a_1'(t) + 2a_0(t)\big)|f(t)|^2\,dt \le 0.$$

Thus A is dissipative. If we can show that $1 \in \rho(A)$, then the proof of the proposition is complete by Theorem 4.2.

We know (see formula (2)) that $\|(I - A)x\| \ge \|x\|$. In particular, $I - A$ is injective. Let us prove that $T = I - A$ is a closed operator. Let ρ be the differential expression $1 - \tau$. The operator T is an extension of $T_{\min,\rho}$. Since $\operatorname{codim}\operatorname{Im} T_{\min,\rho} = 2$ and T is injective, $\dim \mathcal{D}(T)/\mathcal{D}(T_{\min,\rho}) \le 2$. It follows that for the graphs a similar inequality holds true. But $G(T_{\min,\rho})$ is closed, because $T_{\min,\rho}$ is closed by definition. It follows that $G(T)$ is closed, and thus T is a closed operator.

It remains to prove that $T = I - A$ is surjective. The operator $T_{\max,\rho}$ is an extension of T. Take $f \in \mathcal{D}(T_{\max,\rho})$. Then there exist complex numbers c_0, c_1 such that $g = f - c_0 - c_1 t \in \mathcal{D}(T)$. In fact, one may take

$$c_0 = \frac{af(b) - bf(a)}{a - b}, \qquad c_1 = \frac{f(a) - f(b)}{a - b}.$$

It follows that $\dim \mathcal{D}(T_{\max,\rho})/\mathcal{D}(T) \leq 2$. Since $\operatorname{Ker} T_{\max,\rho}$ is two dimensional and T is injective, we conclude that

$$\mathcal{D}(T_{\max,\rho}) = \operatorname{Ker} T_{\max,\rho} \oplus \mathcal{D}(T).$$

The latter identity and the fact that $T_{\max,\rho}$ is surjective imply that $\operatorname{Im} T = L_2([a,b])$.
$\square$

THEOREM 4.5. *Let Ω be a bounded open set in $\mathbf{R}^n$ with boundary of class C^2. Let*

$$L = \sum_{i,j=1}^{n} a_{ij}(x)D_i D_j + \sum_{i=0}^{n} a_i(x)D_i, \qquad x \in \Omega,$$

where each a_{ij} and a_i are in $C^\infty(\overline{\Omega})$. Assume that L is uniformly elliptic. Define the operator $A\big(L_2(\Omega) \to L_2(\Omega)\big)$ by

$$\mathcal{D}(A) = H_1^0(\Omega) \cap H_2(\Omega), \qquad Au = Lu.$$

There exists $\lambda_0 \in \mathbf{R}$ such that if $\lambda \geq \lambda_0$, then $-\lambda - A$ is a generator of a contraction semigroup. Moreover, the operator $-A$ is the generator of a strongly continuous semigroup.

PROOF. By formula (20) in Section XIV.6, there exists $\lambda_0 \in \mathbf{R}$ such that $-\lambda - A$ is dissipative for all $\lambda \geq \lambda_0$. By Theorem XIV.6.1 we may assume that $-\lambda - A$ is invertible for $\lambda \geq \lambda_0$. But then $-\lambda - A$ is the generator of a contraction semigroup $T_\lambda(\cdot)$ by Theorem 4.2. Now $e^{\lambda t}T_\lambda(\cdot)$ is a strongly continuous semigroup. Its generator is $-A$. Indeed, since

$$\frac{1}{t}(e^{\lambda t}T_\lambda(t)x - x) = e^{\lambda t}\left\{\frac{1}{t}\big(T_\lambda(t)x - x\big)\right\} + \frac{1}{t}(e^{\lambda t} - 1)x,$$

the domain of the generator of $e^{\lambda t}T_\lambda(t)$ is the same as $\mathcal{D}(-\lambda - A) = \mathcal{D}(A)$, and for $x \in \mathcal{D}(A)$

$$\lim_{t\downarrow 0}\frac{1}{t}\big(e^{\lambda t}T_\lambda(t)x - x\big) = (-\lambda - A)x + \lambda x = -Ax. \quad \square$$

PROPOSITION 4.6. *If A is the generator of a contraction semigroup $T(\cdot)$ on a Hilbert space H, then $T(\cdot)^*$ is a contraction semigroup with generator A^*.*

PROOF. For each $x \in H$,

$$\|T(t)^*x - x\|^2 = \|T(t)^*x\|^2 - 2\Re\langle x, T(t)x\rangle + \|x\|^2$$
$$\leq 2\|x\|^2 - 2\Re\langle x, T(t)x\rangle \to 0, \qquad t \downarrow 0.$$

Also $T(0)^* = I$,

$$T(t)^* T(s)^* = \left(T(s)T(t)\right)^* = T(t+s)^*,$$

and $\|T(t)^*\| = \|T(t)\| \leq 1$. Hence $T(t)^*$ is a contraction semigroup. Let B be its generator.

If $x \in \mathcal{D}(A)$ and $y \in \mathcal{D}(B)$, then

$$\langle Ax, y \rangle = \lim_{h\downarrow 0} \frac{1}{h} \langle T(h)x - x, y \rangle$$

$$= \lim_{h\downarrow 0} \frac{1}{h} \langle x, T(h)^* y - y \rangle = \langle x, By \rangle.$$

Therefore $y \in \mathcal{D}(A^*)$ and $A^* y = By$. In other words, $B \subset A^*$. It remains to show that $\mathcal{D}(A^*) \subset \mathcal{D}(B)$. Since A and B are generators of C_0 semigroups, there exists $\lambda \in \rho(A) \cap \rho(B)$. This implies that

$$X = (\lambda - B)\mathcal{D}(B) \subset (\lambda - A^*)\mathcal{D}(B),$$

whence $\mathcal{D}(B) \supset (\lambda - A^*)^{-1} X = \mathcal{D}(A^*)$. $\square$

XIX.5 UNITARY SEMIGROUPS

Throughout this section H is a complex Hilbert space. A strongly continuous semigroup $T(\cdot)$ on H is called *unitary (isometric)* if $T(t)$ is a unitary operator (an isometry) for every $t \geq 0$. Our aim is to characterize the generators of the isometric and unitary semigroups. The next theorem will be one of the main results.

THEOREM 5.1.   *An operator $A(H \rightarrow H)$ is the generator of a unitary semigroup if and only if $A = iB$ for some selfadjoint operator B.*

For the proof of Theorem 5.1 we need the following lemma and some other preliminary results.

LEMMA 5.2.   *Let $A(H \rightarrow H)$ be the generator of a strongly continuous semigroup $T(\cdot)$. Then for each $x \in \mathcal{D}(A)$*

$$\frac{d}{dt}\|T(t)x\|^2 = 2\Re\langle AT(t)x, x \rangle, \qquad t \geq 0.$$

PROOF. Since

$$\frac{1}{h}\left(\|T(t+h)x\|^2 - \|T(t)x\|^2\right)$$

$$= \langle \frac{1}{h}(T(t+h)x - T(t)x), T(t+h)x \rangle + \langle T(t)x, \frac{1}{h}(T(t+h)x - T(t)x) \rangle,$$

it follows that for each $x \in \mathcal{D}(A)$

$$\lim_{h\rightarrow 0} \frac{1}{h}\left(\|(T(t+h)x\|^2 - \|T(t)x\|^2\right) = \langle AT(t)x, T(t)x \rangle + \langle T(t)x, AT(t)x \rangle$$

$$= 2\Re\langle AT(t)x, T(t)x \rangle.$$

Here we used that $T(\cdot)x$ is continuous and

$$\frac{1}{h}(T(t+h)x - T(t)x) \to AT(t)x, \qquad h \to 0,$$

which follows from Theorem 2.2. $\square$

THEOREM 5.3. *Let $A(H \to H)$ be the generator of a contraction semigroup $T(\cdot)$. Then the following statements are equivalent:*

(a) *$T(t)$ is an isometry for $t \geq 0$,*

(b) *$\|(\lambda - A)x\| \geq -\lambda\|x\|$ for all $\lambda < 0$ and $x \in \mathcal{D}(A)$,*

(c) *$A \subset -A^*$, that is, iA is symmetric.*

PROOF. (a) $\Rightarrow$ (c). Let $x \in \mathcal{D}(A)$. Since $\|T(t)x\| = \|x\|$ for all $t \geq 0$, Lemma 5.2 implies that $\Re\langle AT(t)x, T(t)x\rangle = 0$. In particular, $\Re\langle Ax, x\rangle = 0$. It follows that $\langle Ax, x\rangle = -\langle x, Ax\rangle$ for each $x \in \mathcal{D}(A)$, which is equivalent to (c).

(c) $\Rightarrow$ (b). For $x \in \mathcal{D}(A)$ and $\lambda \in \mathbf{R}$, we have

(1)
$$\|(\lambda - A)x\|^2 = \lambda^2\|x\|^2 - 2\lambda\Re\langle Ax, x\rangle + \|Ax\|^2.$$

According to our hypothesis, $\Re\langle Ax, x\rangle = 0$. But then we can use (1) to derive (b).

(b) $\Rightarrow$ (a). Again we use (1). Take $x \in \mathcal{D}(A)$. Our hypothesis and (1) imply that $\|Ax\|^2 \geq 2\lambda\Re\langle Ax, x\rangle$ for each $\lambda < 0$. It follows that $\Re\langle Ax, x\rangle \geq 0$. Now, recall that $T(t)$ maps $\mathcal{D}(A)$ into $\mathcal{D}(A)$. So we replace x by $T(t)x$ in the preceding argument, and hence

$$\Re\langle AT(t)x, T(t)x\rangle \geq 0, \qquad t \geq 0.$$

By Lemma 5.2 this yields $\|T(t)x\| \geq \|T(0)x\| = \|x\|$. On the other hand $\|T(t)x\| \leq \|x\|$, because $T(\cdot)$ is a contraction semigroup. Therefore, $T(t)$ is an isometry on $\mathcal{D}(A)$. By continuity, $\|T(t)x\| = \|x\|$ for each $x \in X$ since $\overline{\mathcal{D}(A)} = X$. $\square$

THEOREM 5.4. *Let $A(H \to H)$ be the generator of a contraction semigroup $T(\cdot)$. Then the following statements are equivalent:*

(a) *$T(t)$ is unitary for $t \geq 0$,*

(b) *if $0 \neq \lambda \in \mathbf{R}$, then $\lambda \in \rho(A)$ and*

$$\|(\lambda - A)^{-1}\| \leq |\lambda|^{-1},$$

(c) *$A = -A^*$, that is, iA is selfadjoint.*

PROOF. (a) $\Rightarrow$ (c). Since $T(t)$ is unitary, both $T(t)$ and $T(t)^*$ are isometries. By Proposition 4.6 the family $T(t)^*$, $0 \leq t < \infty$, is a contraction semigroup with generator A^*. Hence $A \subset -A^*$ and $A^* \subset -A^{**}$ by Theorem 5.3. This proves (c), because $A^{**} = A$.

(c) $\Rightarrow$ (b). Since A and $A^* = -A$ are generators of contraction semigroups, Theorem 4.1 applied to A and $-A$ implies (b).

(b) $\Rightarrow$ (a). From our hypothesis it follows that $\lambda < 0$ implies $\lambda \in \rho(A^*)$ and

$$\|(\lambda - A^*)^{-1}\| = \|[(\lambda - A)^{-1}]^*\| = \|(\lambda - A)^{-1}\| < -\lambda^{-1}.$$

Hence condition (b) in Theorem 5.3 is satisfied for both A and A^*. Therefore $T(t)$ and $T(t)^*$ are isometries, and thus $T(t)$ is unitary. $\square$

PROOF OF THEOREM 5.1. Suppose $A = iB$, where $B = B^*$. Then $1 \in \rho(A)$, by Theorem XVI.3.1. Since $\Re\langle Ax, x\rangle = 0$, Theorem 4.2 implies that A is the generator of a contraction semigroup. But then we can use Theorem 5.4 to show that $T(\cdot)$ is unitary, because $A = -A^*$.

Conversely, if A generates a unitary semigroup, then $A = -A^*$ by Theorem 5.4. Therefore $A = iB$, where $B = -iA$ is selfadjoint. $\square$

COROLLARY 5.5. *Let τ be the differential expression given by*

$$\tau f = D(pf') + qf,$$

where $p \in C^2([a,b])$ and $q \in C([a,b])$ are real-valued functions with $p(t) \neq 0$, $t \in [a,b]$. Let $T\big(L_2([a,b]) \to L_2([a,b])\big)$ be the restriction of $T_{\max,\tau}$ to those $g \in \mathcal{D}(T_{\max,\tau})$ which satisfy the boundary conditions

$$\alpha_{10}g(a) + \alpha_{11}g'(a) + \beta_{10}g(b) + \beta_{11}g'(b) = 0,$$
$$\alpha_{20}g(a) + \alpha_{21}g'(a) + \beta_{20}g(b) + \beta_{22}g'(b) = 0,$$

where each α_{ij} and β_{ij} is a real number. Suppose that

$$\mathrm{rank} \begin{bmatrix} \alpha_{10} & \alpha_{11} & \beta_{10} & \beta_{11} \\ \alpha_{20} & \alpha_{21} & \beta_{20} & \beta_{21} \end{bmatrix} = 2$$

and

$$\frac{1}{p(a)} \det \begin{bmatrix} \alpha_{10} & \alpha_{11} \\ \alpha_{20} & \alpha_{21} \end{bmatrix} = \frac{1}{p(b)} \det \begin{bmatrix} \beta_{10} & \beta_{11} \\ \beta_{20} & \beta_{21} \end{bmatrix}.$$

Then iT is the generator of a unitary semigroup.

PROOF. By Corollary XVI.1.3 the operator T is selfadjoint, and hence Theorem 5.1 gives the desired result. $\square$

COROLLARY 5.6. *Let Ω be a bounded open set in $\mathbf{R}^n$ with boundary of class C^2. Let*

$$L = \sum_{i,j=1}^{n} D_i\big(a_{ij}(x)D_1\big) + a_0(x),$$

where a_0 is a real-valued function in $C^\infty(\overline{\Omega})$ and $a_{ij} = a_{ji} \in C^\infty(\overline{\Omega})$, $i, j = 1, \ldots, n$. Suppose that L is uniformly elliptic. Then the operator $A\big(L_2(\Omega) \to L_2(\Omega)\big)$ defined by

$$\mathcal{D}(A) = H_0^1(\Omega) \cap H_2(\Omega), \qquad Au = iLu$$

is the generator of a unitary semigroup.

PROOF. Apply Theorem 5.1 and Theorem XVI.2.1. $\square$

XIX.6 COMPACT SEMIGROUPS

A strongly continuous semigroup $T(\cdot)$ on X is called *compact* if $T(t)$ is a compact operator on X for every $t > 0$. Here and in what follows, X is a complex Banach space. In this section we characterize compact semigroups and present some examples.

THEOREM 6.1. *Let $T(\cdot)$ be a strongly continuous semigroup on X with generator A. Then $T(\cdot)$ is a compact semigroup if and only if the map $t \mapsto T(t)$ is continuous from $0 < t < \infty$ into $\mathcal{L}(X)$ and $(\lambda - A)^{-1}$ is compact for some $\lambda \in \rho(A)$.*

PROOF. By Lemma 2.1, there exist constants M and $\omega > 0$ such that $\|T(t)\| \leq Me^{\omega t}$ for $0 \leq t < \infty$. Suppose that $T(\cdot)$ is a compact semigroup. Given $\varepsilon > 0$ and K a compact subset of X, there exists $\delta = \delta(K, \varepsilon)$, $0 < \delta \leq 1$, such that

$$(1) \qquad \|T(h)y - y\| < \varepsilon, \qquad y \in K, \ 0 \leq h \leq \delta.$$

Indeed, since K is compact, the set K is totally bounded (see [W], 24B), and hence there exist $y_1, \ldots, y_n$ in K with the property that the open balls $\|y - y_j\| < \varepsilon(2Me^\omega + 2)^{-1}$, $j = 1, \ldots, n$, cover K. Let $y \in K$. Then

$$(2) \qquad \begin{aligned} \|T(h)y - y\| &\leq \|T(h)y - T(h)y_i\| + \|T(h)y_i - y_i\| + \|y_i - y\| \\ &\leq (Me^\omega + 1)\|y_i - y\| + \|T(h)y_i - y_i\|, \qquad 0 \leq h \leq 1. \end{aligned}$$

Fix i such that $\|y_i - y\| < \varepsilon(2Me^\omega + 2)^{-1}$. Next, choose $0 < \delta \leq 1$ such that

$$\|T(h)y_i - y_i\| < \varepsilon/3.$$

Then, with this choice of δ, the inequality (2) implies (1).

Now let $a > 0$ be arbitrary, and apply (1) to the set $K = \overline{T(a)S}$, where S is the closed unit ball of X. Our hypothesis on $T(\cdot)$ implies that K is compact. Hence we have from (1) that there exists $\delta(a) = \delta(a, \varepsilon)$, $0 < \delta(a) \leq 1$, so that

$$(3) \qquad \|T(h)T(a)x - T(a)x\| < \varepsilon, \qquad \|x\| \leq 1, \ 0 \leq h \leq \delta(a).$$

It follows that $\|T(h + a) - T(a)\| \leq \varepsilon$ if $0 \leq h \leq \delta(a)$. This shows that the map $t \mapsto T(t)$ is a $\mathcal{L}(X)$-valued function which is continuous from the right on $0 < t < \infty$.

To prove that $t \mapsto T(t)$ is also continuous from the left, let $t_0 > 0$ be given. Take $a > 0$ such that $0 < a < t_0 < a + 1$. Choose $0 < \eta < \delta(a)$ with $t_0 - \eta \geq a$. Then for $t_0 - \eta \leq t \leq t_0$ we have $0 \leq t_0 - t \leq \delta(a)$, and hence by (3)

$$\begin{aligned} \|T(t)x - T(t_0)x\| &= \|T(t - a)\{T(t_0 - t)T(a)x - T(a)x\}\| \\ &\leq \varepsilon Me^{\omega(t - a)}, \qquad \|x\| \leq 1. \end{aligned}$$

But $0 \leq t - a \leq 1$. Thus

$$\|T(t) - T(t_0)\| \leq \varepsilon \sup_{0 \leq s \leq 1} Me^{\omega s}, \qquad t_0 - \eta \leq t \leq t_0,$$

which proves the left continuity.

Next we show that $R(\lambda) = (\lambda - A)^{-1}$ is compact if $\lambda > \omega$. Since $t \to T(t)$ is continuous in the norm topology and $\|e^{-\lambda t}T(t)\| \leq Me^{(\omega-\lambda)t}$, it follows from Theorem 2.2(b) that for $\lambda > \omega$

$$R(\lambda) = \int_0^\infty e^{-\lambda t}T(t)dt$$

converges in the operator norm topology. Now

$$R_\varepsilon(\lambda) = \int_\varepsilon^\infty e^{-\lambda t}T(t)dt, \qquad \lambda > \omega,$$

is the limit in $\mathcal{L}(X)$ of Riemann sums of compact operators since $T(t)$ is compact for all $t > 0$. Therefore $R_\varepsilon(\lambda)$ is compact. But then $R(\lambda)$ is compact due to

$$\lim_{\varepsilon \downarrow 0} \|R_\varepsilon(\lambda) - R(\lambda)\| = 0, \qquad \lambda > \omega.$$

To prove the converse implications, assume now that the map $t \mapsto T(t)$ is continuous from $(0, \infty)$ into $\mathcal{L}(X)$ and that $(\lambda_0 - A)^{-1}$ is compact for some λ_0. Then $(\lambda - A)^{-1}$ is compact for all $\lambda \in \rho(A)$ (Theorem XV.2.3). Fix $\lambda > \omega$ and define

$$S(t)x = \int_0^t e^{-\lambda s}T(s)x\,ds, \qquad t \geq 0.$$

Now $e^{-\lambda s}T(s)$, $0 \leq s < \infty$, is a C_0 semigroup with generator $A - \lambda I$ (see the proof of Theorem 4.5). If we replace $T(t)$ by $e^{-\lambda t}T(t)$ and A by $A - \lambda I$ in Part (iv) of the proof of Theorem 2.2, we obtain that

$$\int_0^t (A - \lambda I)e^{-\lambda s}T(s)x\,ds = e^{-\lambda t}T(t)x - x.$$

Since $A - \lambda I$ is closed, we conclude that

$$S(t)x = (\lambda - A)^{-1}\big(x - e^{-\lambda t}T(t)x\big)$$

is compact for all $t \geq 0$. From the assumption that $t \mapsto T(t)$ is continuous with respect to the operator norm it follows that for any $t_0 > 0$

$$T(t_0) = e^{\lambda t_0}\left(\lim_{h \downarrow 0} \frac{1}{h} \int_{t_0}^{t_0+h} e^{-\lambda s}T(s)ds\right)$$

$$= e^{\lambda t_0}\left\{\lim_{h \downarrow 0} \frac{1}{h}\big(S(t_0 + h) - S(t_0)\big)\right\},$$

with convergence in $\mathcal{L}(X)$. Hence $T(t_0)$ is the limit in $\mathcal{L}(X)$ of compact operators and therefore $T(t_0)$ is compact. $\square$

THEOREM 6.2. *Let $A(X \to X)$ be the generator of a compact semigroup $T(\cdot)$. Then $\sigma(A)$ is either empty or consists of a finite or countable set of eigenvalues of finite type which have no limit point in* $\mathbf{C}$. *Furthermore,*

$$(4) \qquad \{e^{t\mu} \mid \mu \in \sigma(A)\} = \sigma\big(T(t)\big)\backslash\{0\}, \qquad t > 0,$$

and $A\varphi = \mu\varphi$ if and only if $T(t)\varphi = e^{\mu t}\varphi$ for $t \geq 0$.

PROOF. Since A is the generator of a C_0 semigroup, $\rho(A)$ is non-empty. By Theorem 6.1, the operator A has a compact resolvent, and hence the first statement about $\sigma(A)$ is a direct consequence of Theorem XV.2.3.

Assume $A\varphi = \mu\varphi$. Let $A_\lambda = \lambda A(\lambda - A)^{-1}$ be the Yoshida approximant of A. Then $A_\lambda\varphi = \lambda\mu(\lambda - \mu)^{-1}\varphi$, and hence, by Theorems 1.1 and 2.2,

$$\begin{aligned} T(t)\varphi &= \lim_{\lambda\to\infty} e^{tA_\lambda}\varphi = \left\{ \lim_{\lambda\to\infty} \exp\left(t\frac{\lambda\mu}{\lambda - \mu} \right) \right\}\varphi \\ &= e^{t\mu}\varphi, \qquad t \geq 0. \end{aligned}$$

Conversely, assume that $T(t)\varphi = e^{\mu t}\varphi$ for $t \geq 0$. Then

$$\frac{1}{t}\big(T(t)\varphi - \varphi\big) = \left\{ \frac{1}{t}(e^{\mu t} - 1) \right\}\varphi \to \mu\varphi, \qquad t \downarrow 0.$$

Thus $\varphi \in \mathcal{D}(A)$ and $A\varphi = \mu\varphi$.

It remains to prove (4). Take $\mu \in \sigma(A)$. Then μ is an eigenvalue (of finite type) of A. Hence there exists $\varphi \neq 0$ such that $A\varphi = \mu\varphi$. The result of the previous paragraph shows that $e^{t\mu}$ is an eigenvalue of $T(t)$. In particular, $e^{t\mu} \in \sigma\big(T(t)\big)\backslash\{0\}$ for $t > 0$. Next, assume that $\eta \neq 0$ is in $\sigma\big(T(t_0)\big)$ for some $t_0 > 0$. Since $T(t_0)$ is compact and each $T(t)$ commutes with $T(t_0)$, the space $X_0 := \operatorname{Ker}\big(\eta I - T(t_0)\big)$ is a non-zero finite dimensional subspace which is invariant under each $T(t)$. It follows that $T(t)|X_0, 0 \leq t < \infty$, is a C_0 semigroup acting on a finite dimensional space, and hence its generator B is a bounded linear operator with $\mathcal{D}(B) = X_0$. The finite dimensionality of X_0 ensures the existence of an eigenvalue μ of B with corresponding eigenvector $\varphi \in X_0$. Obviously, $\mathcal{D}(B) \subset \mathcal{D}(A)$. Thus $A\varphi = \mu\varphi$ and $\mu \in \sigma(A)$. But then, by the result derived in the second paragraph of this proof, $T(t)\varphi = e^{\mu t}\varphi$ for $t \geq 0$. In particular, $\eta\varphi = e^{\mu t_0}\varphi$, which implies $\eta = e^{\mu t_0}$ (because $\varphi \neq 0$). $\square$

COROLLARY 6.3. *Let $A(H \to H)$ be the generator of a strongly continuous semigroup $T(\cdot)$ on a Hilbert space H. A necessary and sufficient condition that $T(t)$ be compact and selfadjoint for every $t > 0$ is that A is selfadjoint and has a compact resolvent. In this case there exists an orthonormal basis $\varphi_1, \varphi_2, \ldots$ for H consisting of eigenvectors of A with corresponding real eigenvalues $\mu_1, \mu_2, \ldots$ and for each $x \in H$*

$$(5) \qquad T(t)x = \sum_k e^{\mu_k t}\langle x, \varphi_k\rangle\varphi_k, \qquad t \geq 0.$$

PROOF. Suppose that $T(t)$ is compact and selfadjoint for all $t > 0$. Take $\lambda > \omega$ (where ω is as in Lemma 2.1). In the proof of Theorem 6.1 we showed that

$$(6) \qquad (\lambda - A)^{-1} = \int_0^\infty e^{-\lambda t} T(t)\, dt$$

is compact. Moreover, the integral in (6) converges in the operator norm. Since $T(t)$ is selfadjoint for all $t \geq 0$, it follows, by taking adjoints of the Riemann sums converging to $(\lambda - A)^{-1}$, that $(\lambda - A)^{-1}$ is selfadjoint. By Proposition XIV.2.6, this implies that $(\lambda - A)^{-1} = (\lambda - A^*)^{-1}$, because λ is real. Thus $A = A^*$ and A has a compact resolvent.

Next, assume that A is selfadjoint and has a compact resolvent. Choose $\alpha > \omega$ (where ω is as in Lemma 2.1). Then $\alpha - A$ is selfadjoint and has a compact inverse. So we may apply Theorem XVI.5.1 to $\alpha - A$ (instead of A). It follows (make a translation in α) that there exists an orthonormal basis $\{\varphi_1, \varphi_2, \ldots\}$ of H consisting of eigenvectors of A with corresponding real eigenvalues $\mu_1, \mu_2, \ldots$ such that $|\mu_j| \to \infty$ if $\dim H = \infty$ and

$$Ax = \sum_j \mu_j \langle x, \varphi_j \rangle \varphi_j, \qquad x \in \mathcal{D}(A).$$

Since $(\omega, \infty) \subset \rho(A)$, we have $\mu_j \leq \omega$, and hence $\mu_j \to -\infty$ if $\dim H = \infty$. From $A\varphi_j = \mu_j \varphi_j$ it follows (by Theorem 6.2) that $T(t)\varphi_j = e^{\mu_j t}\varphi_j$. Each $x \in H$ has the representation $x = \sum_k \langle x, \varphi_k \rangle \varphi_k$, and therefore

$$T(t)x = \sum_k \langle x, \varphi_k \rangle T(t)\varphi_k = \sum_k e^{\mu_k t} \langle x, \varphi_k \rangle \varphi_k,$$

which proves (5). Since $\mu_j \to -\infty$ if $\dim H = \infty$, the representation (5) shows that $T(t)$ is compact and selfadjoint for each $t > 0$. $\square$

The operator $T(t)$ given by (5) is Hilbert-Schmidt if and only if $\sum_k e^{2\mu_k t} < \infty$. It is a trace class operator if and only if $\sum_k e^{\mu_k t} < \infty$.

We end this chapter with the following example. Let $H = L_2([0, 2\pi])$, and define $A(H \to H)$ by

$$\mathcal{D}(A) = \{f \in H \mid f' \text{ is absolutely continuous}, f'' \in H,$$
$$f(0) = f(2\pi), \ f'(0) = f'(2\pi)\},$$
$$Af = f''.$$

The operator A is densely defined and dissipative since, by partial integration,

$$\langle Af, f \rangle = \int_0^{2\pi} f''(t)\overline{f(t)}\, dt = -\int_0^{2\pi} |f'(t)|^2\, dt \leq 0.$$

An application of Theorem XIV.3.1 shows that $I - A$ is invertible. Hence $1 \in \rho(A)$ and A generates a contraction semigroup $T(\cdot)$ by Theorem 4.2. Now $\varphi_n(t) = (\sqrt{2\pi})^{-1} e^{int}$,

$n = 0, \pm1, \pm2, \ldots$, is an orthonormal basis for H and $A\varphi_n = -n^2\varphi_n$. Hence $T(t)\varphi_n = e^{-n^2t}\varphi_n$ by the arguments given in the second paragraph of the proof of Theorem 6.2. Since

$$\sum_{n=-\infty}^{\infty} e^{-n^2t} < \infty, \qquad t > 0,$$

$T(t)$ is a trace class operator for each $t > 0$. Also, $T(t)$ is selfadjoint, and thus, by Corollary 6.3, the operator A is selfadjoint and has a compact resolvent. The latter statements about A may also be derived from Theorems XIV.3.1 and XVI.1.1.

XIX.7 AN EXAMPLE FROM TRANSPORT THEORY (2)

In this section we return to the half range problem from linear transport theory discussed in Section XIII.9. Here we consider the general case with no restriction on the number of scattering directions, and we shall employ the semigroup theory developed in the present chapter to derive a solution.

Recall from Section XIII.9 that the half range problem concerns an integro-differential equation of the following type:

$$(1) \qquad \mu\frac{\partial\psi}{\partial t}(t,\mu) + \psi(t,\mu) = \int_{-1}^{1} k(\mu,\mu')\psi(t,\mu')d\mu', \qquad -1 \leq \mu \leq 1, \ 0 \leq t < \infty.$$

The scattering function k is a given real-valued and symmetric L_1-function on $[-1,1] \times [-1,1]$. In this section we assume that

$$(2) \qquad k(\mu,\mu') = \sum_{j=0}^{n} a_j p_j(\mu) p_j(\mu'),$$

where p_j is the j-th normalized Legendre polynomial (see [GG], Section I.10) and

$$(3) \qquad -\infty < a_j < 1, \qquad j = 0,1,2,\ldots \ .$$

The problem is to solve (1) under the boundary condition

$$(4) \qquad \psi(0,\mu) = \varphi_+(\mu), \qquad 0 \leq \mu \leq 1,$$

where φ_+ is a given continuous function on $[0,1]$. We shall also require the solution to be bounded in the following sense

$$(5) \qquad \sup_{t\geq 0} \int_{-1}^{1} |\psi(t,\mu)|^2 d\mu < \infty.$$

By writing $\psi(t)(\mu) = \psi(t,\mu)$, we may consider the unknown function ψ as a vector function on $[0,\infty)$ with values in the Hilbert space $L_2([-1,1])$. In this way equation (1) can be written as an operator differential equation

$$(6) \qquad \frac{d}{dt}T\psi(t) + \psi(t) = F\psi(t), \qquad 0 \leq t < \infty,$$

where T and F are the operators on $L_2([-1,1])$ defined by

$$(7) \qquad (Tf)(\mu) = \mu f(\mu), \qquad Ff = \sum_{j=0}^{n} a_j \langle f, p_j \rangle p_j,$$

with $\langle \cdot, \cdot \rangle$ the usual inner product in $L_2([-1,1])$. The derivative in (6) is taken with respect to the norm on $L_2([-1,1])$. Note that in (6) the function $T\psi(\cdot)$ has to be differentiable, which is weaker than requiring the differentiability of ψ.

In the language of the preceding paragraph the boundary condition (4) may be restated as

$$(8) \qquad P_+\psi(0) = \varphi_+,$$

where P_+ is the orthogonal projection from $L_2([-1,1])$ onto $L_2([0,1])$, the closed subspace of $L_2([-1,1])$ consisting of all functions that are zero almost everywhere on $[-1,0]$. The operator T is selfadjoint and the projection P_+ may also be interpreted as the orthogonal projection onto the spectral subspace of T associated with the interval $[0,\infty)$.

The operator F has finite rank and the hypothesis (3) implies that $I - F$ is strictly positive. We shall need the following lemma.

LEMMA 7.1. *Let T and F be as in (7). Then*

$$(9) \qquad \sup_{0 \neq \alpha \in \mathbf{R}} |\alpha|^{1/2} \|(i\alpha - T)^{-1} F\| < \infty.$$

PROOF. Take $\alpha \neq 0$ in $\mathbf{R}$, and put $q_j = (i\alpha - T)^{-1} p_j$. For a suitable positive constant γ_j we have

$$\|q_j\|^2 = \int_{-1}^{1} \frac{1}{\mu^2 + \alpha^2} |p_j(\mu)|^2 d\mu$$

$$\leq \gamma_j \int_{-1}^{1} \frac{1}{\mu^2 + \alpha^2} d\mu = 2\gamma_j \left(\frac{1}{|\alpha|} \arctan \frac{1}{|\alpha|} \right).$$

It follows that

$$\|(i\alpha - T)^{-1} F\| \leq \sum_{j=0}^{n} |a_j| \|q_j\| \leq \frac{C}{|\alpha|^{1/2}},$$

with $C = \sum_{j=0}^{n} |a_j|(\gamma_j \pi)^{1/2}$. $\square$

In what follows we shall not require that T and F are given by (7), but we shall assume that T and F are bounded linear operators acting on an arbitrary complex Hilbert space H and have the following properties:

(i) T is selfadjoint and $\operatorname{Ker} T = \{0\}$,

(ii) F is compact and $I - F$ is strictly positive,

(iii) $\sup\{|\alpha|^{1/2}\|(i\alpha - T)^{-1}F\| \mid 0 \neq \alpha \in \mathbf{R}\} < \infty$.

By Lemma 7.1, the operators T and F in (7) have the properties (i)–(iii). Our aim is to solve the equation:

$$(10) \qquad \begin{cases} (T\psi)'(t) = -\psi(t) + F\psi(t), & 0 \leq t < \infty, \\ P_+\psi(0) = x_+, \end{cases}$$

where P_+ is the orthogonal projection of H onto the spectral subspace of T associated with the interval $[0, \infty)$ and x_+ is a given vector in $\operatorname{Im} P_+$.

We call $\psi\colon [0, \infty) \to H$ a *solution* of (10) if $T\psi$ is continuously differentiable on $[0, \infty)$ and ψ satisfies the two equations in (10). The fact that $T\psi$ is continuously differentiable and the invertibility of $I - F$ imply that a solution of (10) is always a H-valued continuous function.

The first equation in (10) can be rewritten as

$$(11) \qquad (S\psi)'(t) = -\psi(t), \qquad 0 \leq t < \infty,$$

where $S = (I - F)^{-1}T$. As in Section XIII.9 we have to analyse the spectral properties of S. Since $I - F$ is strictly positive, the sesquilinear form

$$(12) \qquad [x, y] = \langle (I - F)x, y \rangle$$

defines an inner product on H which is equivalent with the original inner product $\langle \cdot, \cdot \rangle$ on H, i.e., the norms induced by $[\cdot, \cdot]$ and $\langle \cdot, \cdot \rangle$ are equivalent. The space H endowed with the inner product $[\cdot, \cdot]$ will be denoted by $H^\times$. The operator S is selfadjoint on $H^\times$. Indeed,

$$(13) \qquad \begin{aligned} [Sx, y] &= \langle (I - F)Sx, y \rangle = \langle Tx, y \rangle \\ &= \langle (I - F)^{-1}(I - F)x, Ty \rangle \\ &= \langle (I - F)x, (I - F)^{-1}Ty \rangle = [x, Sy] \end{aligned}$$

for each x and y. It follows that the spectral subspace $H_+^\times$ of S associated with the interval $[0, \infty)$ is well-defined. We write Q_+ for the $[\cdot, \cdot]$-orthogonal projection of $H^\times$ onto $H_+^\times$. Now, consider the operator

$$(14) \qquad S_+ := S|\operatorname{Im} Q_+\colon \operatorname{Im} Q_+ \to \operatorname{Im} Q_+.$$

The operator S_+ is selfadjoint, $\operatorname{Ker} S_+ = \{0\}$ and $\sigma(S_+) \subset [0, \infty)$. It follows that $-S_+^{-1}$ is a densely defined (possibly unbounded) selfadjoint operator,

$$[-S_+^{-1}y, y] \leq 0, \qquad y \in \mathcal{D}(S_+^{-1}),$$

and $(0, \infty) \subset \rho(-S_+^{-1})$. Thus, by Theorem 4.2, the operator $-S_+^{-1}$ is the generator of a contraction semigroup, which we shall denote by $T(t; -S_+^{-1})$.

THEOREM 7.2. *Let T and F have the properties* (i)–(iii). *Then equation* (10) *has a unique bounded solution ψ which is given by*

$$(15) \qquad \psi(t) = T(t; -S_+^{-1})\Pi x_+, \qquad 0 \le t < \infty,$$

where Π is the projection along $\operatorname{Ker} P_+$ onto the spectral subspace of $S = (I - F)^{-1}T$ associated with the interval $[0, \infty]$.

The above theorem is the infinite dimensional analogue of Corollary XIII.9.3. For the proof of Theorem 7.2 we shall not use the method of Section XIII.9 (which was based on equivalence with Wiener-Hopf integral equations), but a more direct method will be employed. To explain the approach followed in the present section, let us return to the finite dimensional case (i.e., let us assume that $H = \mathbf{C}^n$). Rewrite (10) as:

$$(16) \qquad \begin{cases} \psi'(t) = -T^{-1}(I - F)\psi(t), & 0 \le t < \infty, \\ P_+\psi(0) = x_+. \end{cases}$$

Since H is assumed to be $\mathbf{C}^n$, the general solution of the first equation in (16) is given by

$$(17) \qquad \psi(t) = e^{-tT^{-1}(I-F)}y$$

for some $y \in \mathbf{C}^n$. The function $\psi(\cdot)$ has to be a function in $L_2^n([0, \infty))$, which implies that y has to an element from $\operatorname{Im} Q_+$. Also ψ has to fulfil the boundary condition in (16), that is, $y - x_+ \in \operatorname{Ker} P_+$. Thus the vector y in (17) must be chosen in such a way that

$$(18) \qquad y \in \operatorname{Im} Q_+, \qquad y - x_+ \in \operatorname{Ker} P_+.$$

Next, one observes that for each initial value x_+ there exists a unique y satisfying (18) if and only if

$$\mathbf{C}^n = \operatorname{Ker} P_+ \oplus \operatorname{Im} Q_+.$$

Moreover, in that case $y = \Pi x_+$, where Π is the projection along $\operatorname{Ker} P_+$ onto $\operatorname{Im} Q_+$.

The above reasoning presents basically the outline of the proof of Théorem 7.2. The additional difficulties that we meet are due to the fact that T is not boundedly invertible, that is, $0 \in \sigma(T)$.

To prove Theorem 7.2 we have first to establish the existence of the projection Π, i.e., we have to show that

$$H = \operatorname{Ker} P_+ \oplus \operatorname{Im} Q_+,$$

where Q_+ is as in (14). This will be done in the next lemma.

LEMMA 7.3. *The operator*

$$V = (I - P_+)(I - Q_+) + P_+Q_+ \colon H \to H$$

is invertible. In particular, $H = \operatorname{Ker} P_+ \oplus \operatorname{Im} Q_+$.

PROOF. First we show that V is injective. Assume $Vx = 0$. Put $y = Q_+x$. Then $P_+y = P_+Vx = 0$. Hence $y \in \operatorname{Im} Q_+$ and $y \in \operatorname{Ker} P_+$. Recall that the operator S is selfadjoint relative to the inner product $[\cdot,\cdot]$ and $\operatorname{Im} Q_+$ is the spectral subspace of S associated with $[0,\infty)$. It follows (cf. Corollary V.2.4) that $[Sy,y] \geq 0$. On the other hand, $y \in \operatorname{Im} P_+^\perp$ and $\operatorname{Im} P_+$ is the spectral subspace of the selfadjoint operator T associated with $[0,\infty)$. Hence $\langle Ty,y\rangle \leq 0$. But $[Sy,y] = [Ty,y]$ by (13), and therefore $[Sy,y] = 0$. As before, let S_+ be the restriction of S to $\operatorname{Im} Q_+$. Then S_+ is a nonnegative operator, and hence there exists a nonnegative operator B on $\operatorname{Im} Q_+$ such that $B^2 = S_+$ (Theorem V.6.1). From $[By,By] = [Sy,y] = 0$, we conclude that $By = 0$. But then $Sy = S_+y = B^2y = 0$. However, S is injective. Thus $y = 0$, and we have proved that V is injective.

The operator V may be rewritten as

$$(19) \qquad\qquad V = I - (I - 2P_+)(Q_+ - P_+).$$

We shall prove that $Q_+ - P_+$ is a compact operator. When this fact has been established, the invertibility of V follows. Indeed, if $Q_+ - P_+$ is compact, then by (19) the operator V is of the form $I - K$ with K compact. Hence V is a Fredholm operator of index zero (by Corollary XI.4.3), and thus, since $\operatorname{Ker} V = \{0\}$, the operator V must be invertible.

To prove that $Q_+ - P_+$ is a compact operator, choose $\gamma \geq 1$ such that the spectra of T and S are in the open interval $(-\infty,\gamma)$. Let Γ be the oriented boundary of the rectangle with vertices $\pm i$ and $\gamma \pm i$. The orientation of Γ is such that the inner domain is bounded. Take $0 < \varepsilon < 1$, and let Γ_ε be the curve which one obtains from Γ by deleting the points $i\alpha$ with $-\varepsilon < \alpha < \varepsilon$. Put

$$R_\varepsilon = \frac{1}{2\pi i}\int\limits_{\Gamma_\varepsilon} (\lambda - T)^{-1}d\lambda, \qquad R_\varepsilon^\times = \frac{1}{2\pi i}\int\limits_{\Gamma_\varepsilon} (\lambda - S)^{-1}d\lambda.$$

First, we analyse the difference $R_\varepsilon^\times - R_\varepsilon$. Note that

$$(\lambda - S)^{-1} - (\lambda - T)^{-1} = (\lambda - T)^{-1}(S - T)(\lambda - S)^{-1}$$
$$= (\lambda - T)^{-1}FS(\lambda - S)^{-1}.$$

Since F is compact, this implies that $(\lambda - S)^{-1} - (\lambda - T)^{-1}$ is compact for each $\lambda \in \Gamma_\varepsilon$, and hence the difference $R_\varepsilon^\times - R_\varepsilon$ is a compact operator. Next, observe that for $0 \neq \alpha \in \mathbf{R}$

$$(20) \qquad \|S(i\alpha - S)^{-1}\| = \| - I + i\alpha(i\alpha - S)^{-1}\| \leq 2C_1,$$

$$(21) \qquad \|(i\alpha - T)^{-1}F\| \leq \left(\frac{1}{|\alpha|}\right)^{1/2} C_2,$$

where C_1 and C_2 are some constants, not depending on α. To prove (20) one uses Theorem V.2.1 and the fact that S is selfadjoint with respect to the equivalent inner product $[\cdot,\cdot]$; the inequality (21) follows from our hypotheses on the operators T and F (see property (iii)). From (20) and (21) we conclude that

$$\|(i\alpha - S)^{-1} - (i\alpha - T)^{-1}\| \leq 2\left(\frac{1}{|\alpha|}\right)^{1/2} C_1 C_2, \qquad 0 \neq \alpha \in \mathbf{R},$$

and therefore the limit

$$G = \lim_{\varepsilon \downarrow 0}(R_\varepsilon^\times - R_\varepsilon)$$

exists in the operator norm. Since $R_\varepsilon^\times - R_\varepsilon$ is compact for each $\varepsilon > 0$, we also know that G is a compact operator. We shall show that $Q_+ - P_+ = G$.

By using (20) and its analogue with S replaced by T, one sees that the following limits exist in the operator norm:

$$\Omega = \lim_{\varepsilon \downarrow 0} TR_\varepsilon, \qquad \Omega^\times = \lim_{\varepsilon \downarrow 0} SR_\varepsilon^\times.$$

It follows that

$$TG = \lim_{\varepsilon \downarrow 0}\left((I - F)SR_\varepsilon^\times - TR_\varepsilon\right) = (I - F)\Omega^\times - \Omega.$$

Now

$$\Omega = \frac{1}{2\pi i}\int_\Gamma T(\lambda - T)^{-1}d\lambda = \frac{1}{2\pi i}\int_\Gamma \lambda(\lambda - T)^{-1}d\lambda,$$

where the integrals have to be understood as improper integrals (at 0) which converge in the operator norm. From the spectral theory developed in Chapter V we know that $\Omega = TP_+$ (cf. Exercise 32 to Part I). Similarly, $\Omega^\times = SQ_+$, and thus

$$TG = (I - F)SQ_+ - TP_+ = T(Q_+ - P_+).$$

Since T is injective, we obtain $G = Q_+ - P_+$, and thus $Q_+ - P_+$ is compact.

We have now proved that V is invertible. It remains to establish the direct sum decomposition $H = \operatorname{Ker} P_+ \oplus \operatorname{Im} Q_+$. From the definition of V we know that $V(\operatorname{Im} Q_+) \subset \operatorname{Im} P_+$ and $V(\operatorname{Ker} Q_+) \subset \operatorname{Ker} P_+$. Since P_+ and Q_+ are projections and V is invertible, these inclusions cannot be proper. In particular, $V(\operatorname{Im} Q_+) = \operatorname{Im} P_+$. Now, take $y \in H$. Then $z := V^{-1}P_+y \in \operatorname{Im} Q_+$ and

$$P_+y = Vz = (I - P_+)(I - Q_+)z + P_+Q_+z = P_+z.$$

Thus $y - z \in \operatorname{Ker} P_+$, and we have proved that $H = \operatorname{Ker} P_+ + \operatorname{Im} Q_+$. Also $\operatorname{Ker} P_+ \cap \operatorname{Im} Q_+ = \{0\}$, because $\operatorname{Ker} P_+ \cap \operatorname{Im} Q_+ \subset \operatorname{Ker} V$ and the latter space consists of the zero vector only. $\square$

PROOF OF THEOREM 7.2. First we assume that $\psi\colon [0, \infty) \to H$ is a bounded solution of (10) and we prove that ψ admits the representation (15). Recall that S is a bounded selfadjoint operator relative to the inner product $[\cdot,\cdot]$. Choose $\delta > 0$ such that

$$-\frac{1}{2\delta} \leq [Sx,x] \leq \frac{1}{2\delta}, \qquad [x,x] = 1.$$

Our choice of δ implies that $[(\delta S^2 + S)y,y] \leq 0$ for each $y \in \operatorname{Ker} Q_+$. Indeed, let $\{F(\lambda)\}_{\lambda\in\mathbf{R}}$ be the resolution of the identity of the selfadjoint operator S, and take $y \in \operatorname{Ker} Q_+$. Since $\operatorname{Ker} S = \{0\}$, we have $F(0-0) = F(0)$, and hence $Q_+ = I - F(0)$, by Theorem V.5.1. Thus $\operatorname{Ker} Q_+ = \operatorname{Im} F(0)$, and hence

$$[F(\lambda)y,y] = [y,y], \qquad \lambda \geq 0.$$

But then it follows from the spectral theorem for selfadjoint operators that

$$[(\delta S^2 + S)y,y] = \int\limits_{-1/\delta}^{0} (\delta\lambda^2 + \lambda)d[F(\lambda)y,y] \leq 0,$$

because $\delta\lambda^2 + \lambda \leq 0$ for $-\delta^{-1} \leq \lambda \leq 0$.

Now, put $\eta(t) = e^{-t\delta}(I - Q_+)\psi(t)$, $0 \leq t < \infty$, where ψ is the given bounded solution of (10). As ψ satisfies the first equation in (10), formula (11) holds, and hence

$$(22) \qquad \frac{d}{dt}S\eta(t) = -\delta S\eta(t) - \eta(t), \qquad 0 \leq t < \infty.$$

Therefore

$$\frac{d}{dt}[S\eta(t), S\eta(t)] = [(S\eta)'(t), S\eta(t)] + [S\eta(t), (S\eta)'(t)]$$
$$= -2[(\delta S^2 + S)\eta(t), \eta(t)] \geq 0,$$

since $\eta(t) \in \operatorname{Ker} Q_+$ for $t \geq 0$. We conclude that the function $[S\eta(\cdot), S\eta(\cdot)]$ is monotonely increasing. On the other hand for a suitable positive constant C

$$[S\eta(t), S\eta(t)] \leq e^{-2\delta t}C, \qquad 0 \leq t < \infty,$$

because ψ is a bounded solution. Thus $[S\eta(t), S\eta(t)]$ is equal to zero for all $t \geq 0$. Since $\operatorname{Ker} S = \{0\}$, we conclude that $(I - Q_+)\psi(t) = 0$ for $0 \leq t < \infty$.

Put $u(t) = S\psi(t)$, $t \geq 0$, and let

$$S_+ = S|\operatorname{Im} Q_+ : \operatorname{Im} Q_+ \to \operatorname{Im} Q_+.$$

The result of the previous paragraph shows that u is a solution of the following Cauchy problem

$$(23) \qquad \begin{cases} u'(t) = -S_+^{-1}u(t), & 0 \leq t < \infty, \\ u(0) = S_+\psi(0). \end{cases}$$

Note that $u(0) \in \mathcal{D}(-S_+^{-1})$. Since $-S_+^{-1}$ is a generator of a contraction semigroup (see the paragraph preceding Theorem 7.2), we know from Theorem 2.2 that

$$u(t) = T(t; -S_+^{-1})S_+\psi(0), \qquad 0 \le t < \infty,$$

where $T(\cdot\,; -S_+^{-1})$ is the semigroups on $\operatorname{Im} Q_+$ generated by $-S_+^{-1}$. It follows that

$$\psi(t) = S_+^{-1}u(t) = T(t; -S_+^{-1})\psi(0), \qquad 0 \le t < \infty,$$

because of formula (6) in Section 2.

To obtain the representation (15) it remains to prove that $\Pi x_+ = \psi(0)$. From Lemma 7.3 we know that the projection Π is well-defined, and, by definition, $\Pi(I - P_+) = 0$. The function ψ satisfies the second identity in (10). Thus

$$\Pi x_+ = \Pi P_+ \psi(0) = \Pi\psi(0) + \Pi(I - P_+)\psi(0) = \Pi\psi(0).$$

But $\psi(0) \in \operatorname{Im} Q_+$ and $\Pi Q_+ = Q_+$. Therefore, $\psi(0) = \Pi x_+$. This completes the first part of the proof.

It remains to show that any ψ given by (15) is a bounded solution of (10). Thus assume that ψ is given by (15). The values of ψ are vectors in $\operatorname{Im} Q_+ \subset H$. Since the semigroup in (15) is a contraction semigroup, ψ is a H-valued bounded function on $[0, \infty)$. By the strong continuity of the semigroup, ψ is continuous on $[0, \infty)$. Furthermore

$$(24) \qquad P_+\psi(0) = P_+\Pi x_+ = P_+ x_+ - P_+(I - \Pi)x_+ = x_+,$$

and hence ψ satisfies the initial condition in (10). To check the identities in (24), one uses that $x_+ \in \operatorname{Im} P_+$ and $\operatorname{Im}(I - \Pi) = \operatorname{Ker}\Pi = \operatorname{Ker} P_+$. Note that $\Pi x_+ \in \operatorname{Im} Q_+$, and hence

$$T(t; -S_+^{-1})\Pi x_+ = T(t; -S_+^{-1})S_+^{-1}(S_+\Pi x_+).$$

Now $S_+\Pi x_+ \in \mathcal{D}(-S_+^{-1})$. Therefore, by formula (6) in Section 2,

$$\psi(t) = S_+^{-1}u(t), \qquad u(t) = T(t; -S_+^{-1})S_+\Pi x_+.$$

As $S_+\Pi x_+ \in \mathcal{D}(-S_+^{-1})$, the function u is continuously differentiable and its derivative is given by

$$u'(t) = -T(t; -S_+^{-1})\Pi x_+ = -\psi(t), \qquad 0 \le t < \infty.$$

Next, observe that $T\psi(t) = (I - F)S\psi(t) = (I - F)u(t)$ for $t \ge 0$. It follows that $T\psi$ is continuously differentiable and ψ satisfies the first equation in (10). $\square$

The literature on the transport theory and related fields is rich. Here and in Sections XIII.9 and XIII.10 we have only discussed the first basic results.

COMMENTS ON PART IV

Chapter XIV contains standard material dealing with ordinary and partial differential operators in Hilbert space. A more extensive treatment of these topics appear, e.g., in the books Dunford-Schwartz [1], [2], Friedman [1] and Goldberg [1]. The first two sections in Chapter XV extend, in a straightforward manner, the functional calculus and the Riesz projections to unbounded operators. The last two sections in this chapter, which contain a generalization of the Riesz decomposition theorem for a case when the parts of the spectrum of the operator are not disjoint, is based on papers of Bart-Gohberg-Kaashoek [6], [7], [8] (see also Stampfli [1]). Chapter XVI deals with basic properties of unbounded selfadjoint operators and the spectral theorem. The first four sections of Chapter XVII present the standard elements of the Fredholm theory for unbounded operators (cf., Goldberg [1], Kato [1]; also Kaashoek [1]). Section XVII.5, dealing with completeness of eigenvectors and generalized eigenvectors, is concerned with a special case of a theorem from the book Gohberg-Krein [3]. The analysis of the differential operator on the half line in Chapter XVIII has its roots in the Wiener-Hopf theory as developed in the papers of Bart-Gohberg-Kaashoek [2, 3]. The results and the way they are presented seem to be new. Sections 1–6 in Chapter XIX present a first introduction to the theory of strongly continuous semigroups. More about this topic may be found, e.g., in the books Hille-Phillips [1], Davies [1] and Pazy [1]. For further reading about the transport theory see the books of Van der Mee [1], Kaper-Lekkerkerker-Hejtmanek [1], and Greenberg-Van der Mee-Protopopescu [1].

EXERCISES TO PART IV

In the first four exercises $(a_{ij})_{i,j=1}^{\infty}$ is an infinite matrix and $A(\ell_p \to \ell_q)$ is the following operator. The domain $\mathcal{D}(A)$ of A consists of all $v = (v_1, v_2, \ldots) \in \ell_p$ such that

$$(*) \qquad w = \left(\sum_{j=1}^{\infty} a_{1j} v_j, \sum_{j=1}^{\infty} a_{2j} v_j, \ldots \right)$$

is a well-defined sequence in ℓ_q. In particular, for $v \in \mathcal{D}(A)$ the series appearing in $(*)$ are convergent. The action of A is given by $Av = w$. We fix $1 \le p < \infty$, $1 \le q < \infty$.

1. For $i, j = 1, 2, \ldots$ assume that $a_{ij} = 0$ whenever $j \ne i$ and $j \ne i + 1$. Prove that A is a closed densely defined operator and find its conjugate A'. Show that A can be unbounded.

2. Assume that for $i = 1, 2, \ldots$ the sequence $(a_{ij})_{j=1}^{\infty} \in \ell_{p'}$, where $p^{-1} + (p')^{-1} = 1$. Show that A is a closed operator.

3. Let $(a_{ij})_{i,j=1}^{\infty}$ be as in the previous exercise, and assume, in addition, that $(a_{ij})_{i=1}^{\infty} \in \ell_q$, $j = 1, 2, \ldots$. Prove that A is densely defined and find its conjugate A'.

4. Assume that $a_{1j} = j$ and $a_{ij} = 0$ for $i > 1$, where $j = 1, 2, \ldots$. Prove that the operator A is not closable.

5. Suppose $B \in \mathcal{L}(X)$ and $T(X \to X)$ is a closed operator, where X is a Banach space. Show that $\operatorname{Im} B \subset \mathcal{D}(T)$ implies TB is bounded.

6. Let A and K be bounded linear operators acting on a Banach space X. Assume that K is compact and $\operatorname{Im} A \subset \operatorname{Im} K$. Use the result of the previous exercise to show that A is compact.

7. Let $\varphi_1, \varphi_2, \ldots$ be an orthonormal basis of the Hilbert space H, and put $M = \operatorname{span}\{\varphi_n \mid n \ge 1\}$. Take $v \notin M$, and define $A(H \to H)$ by

$$\mathcal{D}(A) = M \oplus \operatorname{span}\{v\},$$

$$A(m + \alpha v) = \alpha v, \qquad m \in M, \ \alpha \in \mathbb{C}.$$

Prove that A is not closable.

8. Given

$$A(t) = \begin{bmatrix} 0 & 1 & 0 & \cdots & 0 \\ 0 & 0 & 1 & \cdots & 0 \\ \vdots & \vdots & \vdots & \ddots & \vdots \\ 0 & 0 & 0 & & 1 \\ -a_0(t) & -a_1(t) & -a_1(t) & \cdots & -a_{n-1}(t) \end{bmatrix}$$

with each $a_k(\cdot)$ integrable on $[a, b]$, let

$$U(t) = I_n + \int_a^t A(s)U(s)ds, \qquad a \le t \le b.$$

Prove that

$$U(t) = \begin{bmatrix} y_1(t) & y_2(t) & \cdots & y_n(t) \\ y_1'(t) & y_2'(t) & \cdots & y_n'(t) \\ \vdots & \vdots & & \vdots \\ y_1^{(n-1)}(t) & y_2^{(n-1)}(t) & \cdots & y_n^{(n-1)}(t) \end{bmatrix}, \qquad a \le t \le b,$$

where $y_k \in \operatorname{Ker} T_{\max,\tau}$ and $y_k^{(j-1)}(a) = \delta_{jk}$, $j, k = 1, \ldots, n$. Here

$$\tau = D^n + \sum_{k=0}^{n-1} a_k(t)D^k$$

and δ_{jk} is the Kronecker delta.

9. Let $\tau = D^2 + a_1(t)D + a_0(t)$, where $a_0(\cdot)$ and $a_1(\cdot)$ are integrable on $[a, b]$, and let $X = L_2([a, b])$. Consider the operator $T(X \to X)$ defined by

$$D(T) = \{g \in \mathcal{D}(T_{\max,\tau}) \mid g(a) = g'(a) = 0\},$$
$$Tg = \tau g.$$

Use Theorem XIV.3.1 and the result of the previous exercise to show that T is invertible and

$$(T^{-1}f)(t) = \int_a^t G(t, s)f(s)ds, \qquad a \le t \le b,$$

with

$$G(t, s) = \frac{u(s)v(t) - u(t)v(s)}{u(s)v'(s) - u'(s)v(s)}, \qquad a \le s \le t \le b,$$

where $u, v \in \operatorname{Ker} T_{\max,\tau}$ and satisfy

$$u(a) = 1, \qquad u'(a) = 0, \qquad v(a) = 0, \qquad v'(a) = 1.$$

10. Let τ and X be as in the previous exercise, and let $T(X \to X)$ be the restriction of $T_{\max,\tau}$ to those $g \in \mathcal{D}(T_{\max,\tau})$ that satisfy $g(b) = g'(b) = 0$. Show that T is invertible and find T^{-1}.

11. Let $X = L_2([1, 3])$, and let τ be the differential expression

$$\tau = D^2 - \frac{1}{t}(t + 2)D + \frac{1}{t^2}(t + 2).$$

Check that $y_1(t) = t$ and $y_2(t) = te^t$ satisfy $\tau y = 0$. Prove that T is invertible and find T^{-1}, where $T(X \to X)$ is defined by

$$\mathcal{D}(T) = \{g \in \mathcal{D}(T_{\max,\tau}) \mid g(1) = g'(1) = 0\},$$
$$Tg = \tau g.$$

12. Given $X = L_2([a,b])$ and $\tau = D^n + \sum_{j=0}^{n-1} a_j(t)D^j$, where each $a_j(\cdot)$ is integrable on $[a,b]$, let $T(X \to X)$ be the restriction of $T_{\max,\tau}$ to those $g \in \mathcal{D}(T_{\max,\tau})$ that satisfy $g^{(j)}(a) = 0$ for $j = 0,\ldots,n-1$. Find the resolvent set $\rho(T)$.

13. Let $X = L_2([a,b])$, and let $\tau = D^n + \sum_{k=0}^{n-1} a_k(t)D^k$, where $a_k \in C^k([a,b])$. Define $T(X \to X)$ by

$$\mathcal{D}(T) = \{g \in \mathcal{D}(T_{\max,\tau}) \mid g^{(k)}(a) = g^{(k)}(b) = 0, \ k = 0,\ldots,n-1\}, \qquad Tg = \tau g.$$

Prove that $T = T_{\min,\tau}$ by showing that T is a closed operator and $T^* = T_{\max,\tau^*}$.

14. Given $H = L_2([0,2\pi])$ and $\tau = D^2 + 1$, find the adjoint of the operator $T(H \to H)$ defined by

$$\mathcal{D}(T) = \{f \in \mathcal{D}(T_{\max,\tau}) \mid f(0) = 0\}, \qquad Tf = \tau f.$$

15. Let $H = L_2([0,1])$, and let τ be a differential expression. Find the adjoint of the operator $T(H \to H)$, where $Tf = \tau f$ for $f \in \mathcal{D}(T)$, in each of the following cases:

(a) $\tau = D^2 + D$,
$$\mathcal{D}(T) = \{f \in \mathcal{D}(T_{\max,\tau}) \mid f(0) = f'(1)\};$$

(b) $\tau = D^3 + 1$
$$\mathcal{D}(T) = \{f \in \mathcal{D}(T_{\max,\tau}) \mid f(0) + f'(0) = f(1) + f'(1), \ f^{(2)}(0) = f^{(2)}(1)\};$$

(c) $\tau = D^2 + tD - t^2$,
$$\mathcal{D}(T) = \{f \in \mathcal{D}(T_{\max,\tau}) \mid f(0) = 0, \ f(1) = f'(0)\}.$$

16. Let X be a Banach space. Suppose $A(X \to X)$ is an operator with $\lambda_0 \in \rho(A)$. Prove that the largest open disc (possibly $\mathbb{C}$) which has its center at λ_0 and does not intersect $\sigma(A)$ has radius ρ^{-1}, where ρ is the spectral radius of $(\lambda_0 - A)^{-1}$.

17. Let $X = C([0,2\pi])$, and let $A(X \to X)$ be defined by

$$\mathcal{D}(A) = \{u \in X \mid u' \in X, \ u(0) = u(2\pi)\}, \qquad Au = -iu'.$$

(a) Show that $\sigma(A)$ consists of the eigenvalues $\lambda_k = k, \ k = 0,\pm 1,\pm 2,\ldots$.

(b) For $\lambda \in \rho(A)$, find a formula for $(\lambda - A)^{-1}v$, where $v \in X$.

(c) Suppose that f is analytic at the integers $k = 0,\pm 1,\pm 2,\ldots,\pm N$ and $f(\lambda)$ is a constant c for $|\lambda| > N$. Prove that

$$(f(A)v)(t) = cv(t) + \sum_{k=-N}^{N} \big(f(k) - c\big)\alpha_k e^{ikt}, \qquad 0 \le t \le 2\pi,$$

where

$$\alpha_k = \frac{1}{2\pi} \int\limits_0^{2\pi} v(s)e^{-iks}\,ds, \qquad k = 0, \pm 1, \ldots, \pm N.$$

18. Let X be a Banach space, and assume that the operator $A(X \to X)$ has a non-empty resolvent set. Suppose $f \in \mathcal{F}_\infty(A)$ has no zeros on $\sigma(A) \cup \{\infty\}$. Prove that $f(A)$ is invertible and find $f(A)^{-1}$.

19. Let $\tau = iD$ and $H = L_2([0,1])$. Define $T(H \to H)$ by

$$\mathcal{D}(T) = \{f \in \mathcal{D}(T_{\max,\tau}) \mid f(0) = f(1) = 0\}, \qquad Tf = if'.$$

Show that T is symmetric. Is T selfadjoint?

20. Let $\tau = iD$ and $H = L_2([0,1])$. Given $\theta \in \mathbf{R}$, define $T_\theta(H \to H)$ by

$$\mathcal{D}(T_\theta) = \{f \in \mathcal{D}(T_{\max,\tau}) \mid f(1) = e^{i\theta} f(0)\}, \qquad T_\theta f = if'.$$

Prove that T_θ is selfadjoint. Show that the operator T in the previous exercise has infinitely many selfadjoint extensions.

21. Let $\tau f = D(pf') + qf$, where $p \in C^2([a,b])$ and $q \in C([a,b])$ are real-valued functions with $p(t) \neq 0$ for all $t \in [a,b]$. Take $H = L_2([a,b])$. In each of the following, $T(H \to H)$ is the restriction of $T_{\max,\tau}$ to those $g \in \mathcal{D}(T_{\max,\tau})$ that satisfy the stated boundary conditions. Determine whether or not the corresponding operator T is selfadjoint.

    (a) $g(a) = 2g(b)$;

    (b) $g(a) = 0$, $g'(b) = 2g(b)$;

    (c) $p(a)g(a) + g'(b) = 0$, $g'(a) = -p(b)g(b)$.

22. Prove that the spectrum of an unbounded selfadjoint operator is unbounded.

23. Let X be a Banach space. Let $T(X \to X)$ and $B(X \to X)$ be linear operators. Assume that T has a non-empty resolvent set and $\mathcal{D}(T) \subset \mathcal{D}(B)$. Prove that B is T-compact if and only if $B(T - \lambda I)^{-1}$ is a compact operator on X for each $\lambda \in \rho(T)$.

24. Let $P(X \to X)$ be a closed operator acting in a Banach space. Assume that $P\mathcal{D}(P) \subset \mathcal{D}(P)$ and $P^2 x = Px$ for all $x \in \mathcal{D}(P)$. Does it follow that P is bounded? (Hint: see the paragraph preceding Theorem I.5.1 in Gohberg-Feldman [1].)

25. Let P be as in the previous exercise. Prove that P is bounded if and only if there exists $\gamma > 0$ such that

$$\|x + y\| \geq \gamma(\|x\| + \|y\|), \qquad x \in \operatorname{Ker} P, \ y \in \operatorname{Im} P.$$

26. Let $A(X \to Y)$ and $B(X \to Y)$ be closed operators acting between Banach spaces X and Y. Suppose that $\operatorname{Im} A \cap \operatorname{Im} B = \{0\}$ and $\operatorname{Im} A + \operatorname{Im} B$ is closed. Prove that $\operatorname{Im} A$ and $\operatorname{Im} B$ are closed.

27. Let X be a Banach space. Suppose that $T(X \to X)$ has a compact resolvent. Prove that for any $B \in \mathcal{L}(X)$ the operator $T + B$ is a Fredholm operator with index zero.

28. Let $T(X \to Y)$ be a closed operator acting between Banach spaces X and Y. Suppose that there exists a $\lambda_0 \in \mathbb{C}$ such that $\lambda_0 - T$ has closed range and $\dim \operatorname{Ker}(\lambda_0 - T) < \infty$. Prove that for any polynomial p the operator $p(T)$ is closed. (Hint: write $p(T) = (\lambda_0 - T)q(T) + cI$ and use induction.)

29. Let C be an unbounded densely defined closed operator with domain and range in the Banach space X. Define $A(X \times X \to X \times X)$ by setting

$$\mathcal{D}(A) = X \times \mathcal{D}(C), \qquad A(x,y) = (Cy, 0).$$

Prove

(a) A is a densely defined closed linear operator;

(b) the abstract Cauchy problem

$$(*) \qquad \begin{cases} u'(t) = Au(t), & 0 \le t < \infty, \\ u(0) = (x_0, y_0), \end{cases}$$

has a unique continuously differentiable solution for each $(x_0, y_0) \in \mathcal{D}(A)$ (find its formula);

(c) A does not generate a strongly continuous semigroup.

How do the above results relate to Theorem XIX.3.1? What can one say about the spectrum of A?

30. Prove that any two C_0-semigroups with the same generator are equal.

31. Let $T(\cdot)$ be a strongly continuous semigroup on the Banach space X with generator A. Define

$$B_\lambda(t)x = \int_0^t e^{\lambda(t-s)} T(s)x\,ds, \qquad t \ge 0.$$

Here $\lambda \in \mathbb{C}$ and $x \in X$. Prove

(a) for every $x \in X$, the vector $B_\lambda(t)x \in \mathcal{D}(A)$ and

$$(\lambda - A)B_\lambda(t)x = e^{\lambda t}x - T(t)x, \qquad t \ge 0;$$

(b) for $x \in \mathcal{D}(A)$,

$$B_\lambda(t)(\lambda - A)x = e^{\lambda t}x - T(t)x, \qquad t \ge 0;$$

(c) the following spectral inclusion holds:

$$\{ e^{t\lambda} \mid \lambda \in \sigma(A) \} \subset \sigma(T(t)), \qquad t \ge 0.$$

30. Let X be the Banach space of all $f \in C([0,1])$ with $f(1) = 0$ endowed with the supremum norm. For $f \in X$ put

$$(T(t)f)(s) = \begin{cases} f(t+s), & t+s \le 1, \\ 0, & t+s > 1. \end{cases}$$

Prove that $T(\cdot)$ is a C_0 semigroup of contractions on X with generator A given by

$$\mathcal{D}(A) = \{f \in X \mid f \in C^1([0,1]),\ f' \in X\}, \qquad Af = f'.$$

Show that $\sigma(A) = \emptyset$, and prove that the inclusion in item (c) of the previous exercise may be strictly proper.

STANDARD REFERENCES TEXTS

For references to standard facts from real and complex analysis, point set topology and functional analysis the following four books are used.

[C]   Conway, J.B., *Functions of One Complex Variable*, Graduate Texts in Mathematics, Springer-Verlag, New York, 1973.

[GG] Gohberg, I., Goldberg, S., *Basic Operator Theory*, Birkhäuser Verlag, Boston, 1981.

[R]   Rudin, W., *Real and Complex Analysis*, McGraw-Hill, New York, 1966.

[W]   Willard, S., *General Topology*, Addison Wesley, Reading, 1970.

BIBLIOGRAPHY

Apostol, C.

[1] On a spectral equivalence of operators, in: *Topics in Operator Theory* (*Constantin Apostol Memorial Issue*), Operator Theory: Advances and Applications, Vol. 32, Birkhäuser Verlag, Basel, 1988, pp. 15–35.

Bart, H., Gohberg, I., Kaashoek, M.A.

[1] *Minimal Factorization of Matrix and Operator Functions*, Operator Theory: Advances and Applications, Vol. 1, Birkhäuser Verlag, Basel, 1979.

[2] Wiener-Hopf integral equations, Toeplitz matrices and linear systems, in: *Toeplitz Centennial*, Operator Theory: Advances and Applications, Vol. 4, Birkhäuser Verlag, Basel, 1982, pp. 85–135.

[3] Convolution equations and linear systems, *Integral Equations and Operator Theory* 5 (1982), 283–340.

[4] The coupling method for solving integral equations, in: *Topics in Operator Theory and Networks, the Rehovot Workshop*, Operator Theory: Advances and Applications, Vol. 12, Birkhäuser Verlag, Basel, 1984, pp. 39–73.

[5] Explicit Wiener-Hopf factorization and realization, in: *Constructive Methods for Wiener-Hopf Factorization*, Operator Theory: Advances and Applications, Vol. 21, Birkhäuser Verlag, Basel, 1986, pp. 235–316.

[6] Fredholm theory of Wiener-Hopf equations in terms of realization of their symbols, *Integral Equations and Operator Theory* 8 (1985), 590–613.

[7] Wiener-Hopf factorization, inverse Fourier transforms and exponentially dichotomous operators, *J. Functional Analysis* 68 (1986), 1–42.

[8] Wiener-Hopf equations with symbols analytic in a strip, in: *Constructive Methods of Wiener-Hopf Factorization* (Gohberg, I. and Kaashoek, M.A., eds.), Operator Theory: Advances and Applications, Vol. 21, Birkhäuser Verlag, Basel, 1986, pp. 39–74.

Bart, H., Kaashoek, M.A., Lay, D.C.

[1] The integral formula for the reduced algebraic multiplicity of meromorphic operator functions, *Proc. Edinburgh Math. Soc.* 21 (1978), 65–72.

Beauzamy, B.

[1] Un operateur sans sous-espace invariant: simplification de l'exemple de P. Enflo, *Integral Equations and Operator Theory* 8 (1985), 314–384.

[2] *Introduction to Operator Theory and Invariant Subspaces*, North-Holland Math. Library 42, Elsevier Science Publ., Amsterdam, 1988.

Calderon, A.P.

[1]   The analytic calculation of the index of elliptic equations, *Proc. Nat. Acad. Sci. USA*
57 (1967), 1193–1194.

Carleman, T.

[1]   Zur Theorie der linearen Integralgleichungen, *Math. Zeitsch.* **9** (1921), 196–217.

Colojoara, L., Foiaş, C.

[1]   *The Theory of Generalized Spectral Operators*, Gordon and Breach, New York, 1968.

Daleckii, Ju.L., Kreĭn, M.G.

[1]   *Stability of Solutions of Differential Equations in Banach Space*, Transl. Math.
Monographs, Vol. 43, Amer. Math. Soc., Providence, R.I., 1974.

Davies, E.B.

[1]   *One-Parameter Semigroups*, London Math. Soc. Monographs, No. 15, Academic
Press, London, 1980.

Dunford, N., Schwartz, J.T.

[1]   *Linear Operators*, Part II: *Spectral Theory (Self Adjoint Operators in Hilbert Space)*,
Wiley-Interscience, New York, 1963.

[2]   *Linear Operators*, Part III: *Spectral Operators*, Wiley-Interscience, New York, 1971.

Enflo, P.H.

[1]   On the invariant subspace problem in Banach spaces, *Acta Mathematica* **158** (1987),
213–313.

Fedosov, B.V.

[1]   Direct proof for the formula for the index of an elliptic system in Euclidean space,
Funct. Anal. Appl. **4** (1970), 339–341.

Fredholm, I.

[1]   Sur une classe d'équations fonctionnelles, *Acta Math.* **27** (1903), 365–390.

Friedman, A.

[1]   *Partial Differential Equations*, Krieger Publ. Co., New York, 1976.

Gel'fand, I.M.

[1]   *Lectures on Linear Algebra*, Interscience Publ., New York, 1961.

Gilbarg, D., Trudinger, N.S.

[1]   *Elliptic Partial Differential Equations of the Second Order* (2nd edition), Springer-Verlag, New York, 1984.

Gohberg, I.C., Fel'dman, I.A.

[1]   *Convolution Equations and Projection Methods for Their Solution*, Transl. Math. Monographs, Vol. 41, Amer. Math. Soc., Providence, R.I., 1974.

Gohberg, I., Kaashoek, M.A.

[1]   Time varying linear systems with boundary conditions and integral operators, I. The transfer operator and its properties, *Integral Equations and Operator Theory* **7** (1984), 325–391.

Gohberg, I., Kaashoek, M.A., Lay, D.C.

[1]   Spectral classification of operators and operator functions, *Bull. Amer. Math. Soc.* **82** (1976), 587–589.

[2]   Equivalence, linearization and decompositions of holomorphic operator functions, *J. Funct. Anal.* **28** (1978), 102–144.

Gohberg, I.C., Kreĭn, M.G.

[1]   The basic propositions on defect numbers, root numbers and indices of linear operators, *Uspekhi Math. Nauk* 12, 2(74) (1957), 43–118 (Russian); English Transl., *Amer. Math. Soc. Transl. (Series 2)* **13** (1960), 185–265.

[2]   Systems of integral equations on a half line with kernels depending on the difference of arguments, *Uspekhi Math. Nauk* 13, 2(80) (1958), 3–72 (Russian); English Transl., *Amer. Math. Soc. Transl. (Series 2)* **14** (1960), 217–287.

[3]   *Introduction to the Theory of Linear Nonselfadjoint Operators*, Transl. Math. Monographs, Vol. 18, Amer. Math. Soc., Providence, R.I., 1969.

Gohberg, I., Krupnik, N.Ya.

[1]   *Einführung in die Theorie der Eindimensionalen Singulären Integraloperatoren*, Mathematische Reihe, Band 63, Birkhäuser Verlag, Basel, 1979.

Gohberg, I., Lancaster, P., Rodman, L.

[1]   *Matrix Polynomials*, Academic Press, New York, 1982.

Gohberg, I.C., Sigal, E.I.

[1]   An operator generalization of the logarithmic residue theorem and the theorem of Rouché, *Mat Sbornik* **84**(1260) (1971), 607–629 (Russian); English Transl., *Math. USSR Sbornik* **13** (1971), 603–625.

Goldberg, S.

[1]   *Unbounded Linear Operators, Theory and Applications*, McGraw-Hill, New York, 1966.

Greenberg, W., Mee, C.V.M. van der, Protopopescu, V.

[1] *Boundary Value Problems in Abstract Kinetic Theory*, Operator Theory: Advances and Applications, Vol. 23, Birkhäuser Verlag, Basel, 1987.

Hilbert, D.

[1] Grundzüge einer allgemeinen Theorie der Linearen Integralgleichungen (Erste Mitteilung), *Nachr. Wiss. Gesell. Göttingen, Math.-Phys. Kl.* (1904), 49–91.

Hille, E., Phillips, R.S.

[1] *Functional Analysis and Semigroups*, Amer. Math. Soc. Colloquim Publ., Vol. XXXI, Amer. Math. Soc., Providence, R.I., 1957.

Horn, A.

[1] On the eigenvalues of a matrix with prescribed singular values, *Proc. Math. Soc.* **5** (1954), 4–7.

Kaashoek, M.A.

[1] Closed linear operators on Banach spaces, *Proc. Acad. Sci. Amsterdam* **A68** (1965), 405–414.

Kaashoek, M.A., Mee, C.V.M. van der, Rodman, L.

[1] Analytic operator functions with compact spectrum, I. Spectral nodes, linearization and equivalence, *Integral Equations and Operator Theory* **4** (1981), 504–547.

Kailath, T.

[1] *Linear Systems*, Prentice-Hall, Inc., Englewood Cliffs, New Jersey, 1980.

Kalman, R.E., Falb, P.L., Arbib, M.A.

[1] *Topics in Mathematical System Theory*, McGraw-Hill, New York, 1969.

Kato, T.

[1] *Perturbation Theory for Linear Operators*, Grundlehren, Band 132, Springer-Verlag, Berlin, 1966.

Kaper, H.G., Lekkerkerker, C.G., Hejtmanek, J.

[1] *Spectral Methods in Linear Transport Theory*, Operator Theory: Advances and Applications, Vol. 5, Birkhäuser Verlag, Basel, 1982.

Keldysh, V.M.

[1] On the eigenvalues and eigenfunctions of certain classes of nonselfadjoint equations, *Dokl. Akad. Nauk SSSR* **77** (1951), 11–14 (Russian); English Transl. in Markus [1].

Knuth, D.E.

[1] *The Art of Computer Programming*, Addison Wesley, Reading, M.A., 1980.

König, H.

[1] *Eigenvalue Distribution of Compact Operators*, Operator Theory: Advances and Applications, Vol. 16, Birkhäuser Verlag, Basel, 1986.

Krupnik, N.Ya.

[1] *Banach Algebras with Symbol and Singular Integral Operators*, Operator Theory: Advances and Applications, Vol. 26, Birkhäuser Verlag, Basel, 1987.

Leiterer, J.

[1] Local and global equivalence of meromorphic operator functions, I, *Math. Nachr.* **83** (1978), 7–29.

[2] Local and global equivalence of meromorphic operator functions, II, *Math. Nachr.* **84** (1978), 145–170.

Lidskii, V.B.

[1] Theorems on the completeness of the system of eigenelements and adjoined elements of operators having a discrete spectrum, *Dokl. Akad. Nauk SSSR* **119** (1958), 1088–1091 (Russian); English Transl., *Amer. Math. Soc. Transl. (Series* 2) **47** (1965), 37–41.

[2] Non-selfadjoint operators with a trace, *Dokl. Akad. Nauk SSSR* **125** (1959), 485–587 (Russian); English Transl., *Amer. Math. Soc. Transl. (Series* 2) **47** (1965), 43–46.

Lorch, E.R.

[1] Return to the self-adjoint transformation, *Acta Sci. Math. Szeged* **12**(B) (1950), 137–144.

[2] *Spectral Theory*, University Texts in the Mathematical Sciences, Oxford University Press, New York, 1962.

Lumer, G., Rosenblum, M.

[1] Linear operator equations, *Proc. Amer. Math. Soc.* **10** (1959), 32–41.

Lyubich, Ju., Macaev, V.I.

[1] On operators with a separable spectrum, *Mat. Sb.* **56** (1962), 433–468 (Russian); English Transl., *Amer. Math. Soc. Transl. (Series* 2) **47** (1965), 89–129.

Macaev, V.I.

[1] A class of completely continuous operators, *Dokl. Akad. Nauk SSSR* **139** (1961), 548–551 (Russian); English Transl., *Soviet Math. Dokl.* **1** (1961), 972–975.

[2] Several theorems on completeness of root subspaces of completely continuous operators, *Dokl. Akad. Nauk SSSR* **155** (1964), 273–276 (Russian), English Transl., *Soviet Math. Dokl.* **5** (1964), 396–399.

Markus, A.S.

[1] *Introduction to Spectral Theory of Polynomial Operator Pencils*, Transl. Math. Monographs, Vol. 71, Amer. Math. Soc., Providence, R.I., 1988.

Markus, A.S., Fel'dman, I.A.

[1] Index of an operator matrix, *Funct. Anal. Appl.* **11** (1977), 149–151.

Markus, A.S., Sigal, E.I.

[1] On the multiplicity of a characteristic value of an analytic operator-function, *Mat. Issled* **5**, No. 3(17) (1970), 129–147 (Russian).

Mee, C.V.M. van der

[1] *Semigroup and Factorization Methods in Transport Theory*, Thesis Vrije Universiteit, Amsterdam, 1981 = Mathematical Centre Tracts **146**, Mathematisch Centrum, Amsterdam, 1981.

Mirsky, L.

[1] Matrices with prescribed characteristic roots and diagonal elements, *J. London Math. Soc.* **33** (1958), 14–21.

Pazy, A.

[1] *Semigroups of Linear Operators and Applications to Partial Differential Equations*, Applied Mathematical Sciences, Vol. 44, Springer-Verlag, New York, 1983.

Pietsch, A.

[1] *Eigenvalues and s-numbers*, Cambridge Studies in Advanced Mathematics, Vol. 13, Cambridge University Press, Cambridge, 1987.

Pólya, G., Szegö, G.

[1] *Aufgaben und Lehrsätze aus der Analysis*, Band I. Grundlehren, Band 19, Springer-Verlag, Berlin, 1964.

Ran, A.C.M., Rodman, L.

[1] A matricial boundary value problem which appears in the transport theory, *J. Math. Anal. Appl.* **130** (1988), 200–222.

Read, C.

[1] Solution to the invariant subspace problem, *Bull. London Math. Soc.* **16** (1984), 337–401.

[2] Solution to the invariant subspace problem on the space ℓ_1, *Bull. London Math. Soc.* **17** (1985), 305–317.

Rickart, C.E.

[1] *Banach Algebras*, R.E. Krieger Publ. Co., New York, 1974.

Riesz, F., Sz.-Nagy, B.

[1] *Leçons d'Analyse Fonctionnelle*, Académie des Sciences de Hongrie, 1955.

Showalter, R.E.

[1] *Hilbert Space Methods for Partial Differential Equations*, Pitman, London, 1977.

Sigal, E.I.

[1] Factor-multiplicity of eigenvalues of meromorphic operator functions, *Mat. Issled* 5, No. 4(18) (1970), 136–152 (Russian).

Stampfli, J.G.

[1] A local spectral theory for operators, IV. Invariant subspaces, *Indiana Univ. Math.* 22 (1972), 159–176.

Stummel, F.

[1] Diskreter konvergenz linearer operatoren, II, *Math. Zeitschr.* 120 (1971), 231–264.

Sz.-Nagy, B.

[1] Perturbations des transformations autoadjointes dans l'espace de Hilbert, *Comment. Math. Helv.* 19 (1947), 347–366.

[2] Perturbations des transformations linéares fermées, *Acta Sci. Math. Szeged* 14 (1951), 125–137.

[3] *Introduction to Real Functions and Orthogonal Expansions*, Oxford Univ. Press, New York, 1965.

Wolf, F.

[1] Operators in Banach space which admit a generalized spectral decomposition, *Nederl. Akad. Wetensch. Indig. Math.* 19 (1957), 302–311.

LIST OF SYMBOLS

$A_\Im$	imaginary part of the operator A, 135, 146
$A_\Re$	real part of the operator A, 146
A^*	(Hilbert space) adjoint of the operator A, 291
A'	conjugate of the operator A, 292
$\overline{A}$	minimal closed linear extension of the operator A, 289
$A\|N$	restriction of A to the subspace of N, 8, 326
$A \geq 0$	non-negative operator A, 82
$AC_n(J)$	a space of functions on the interval J, 295
$A(X \to Y)$	operator A with domain in X and range in Y, 288
$\mathbf{C}$	field of complex numbers, 5
$\mathbf{C}^n$	complex Euclidean n-space
$\mathbf{C}_\infty$	the extended complex plane, 49
$C([a,b])$	the space of complex-valued continuous functions on $[a,b]$, 288
$C^k(\Omega)$	a set of differentiable functions on Ω, 310
$C^\infty(\Omega)$	the set of C^∞-functions on Ω, 310
$C_0^\infty(\Omega)$	the set of C^∞-functions with compact support in Ω, 310
$C^k(\overline{\Omega})$	a set of differentiable functions on $\overline{\Omega}$, 310
$\mathbf{D}$	open unit disk in the complex plane
$\mathcal{D}(A)$	domain of the operator A, 288
D^α	elementary differential operator, 310
$d(A)$	codimension of the range of A, 184
$\partial\Omega$	boundary of the open set Ω, 315
$det\ A$	determinant of an operator matrix A, 194
$\det(I + A)$	determinant of $I + A$ with A a trace class operator, 117
$\dim M$	dimension of the linear space M
E_A	a space spanned by eigenvectors and generalized eigenvectors, 31
$E(t)$	orthogonal projection in a resolution of the identity, 72
$\mathcal{F}(A)$	a family of analytic functions associated with a bounded operator A, 13
$\mathcal{F}_\infty(A)$	a family of analytic functions associated with an unbounded operator A, 323
$G(A)$	graph of the operator A, 289
$H_m(\Omega)$	Sobolev space of order m, 313
$H_m^0(\Omega)$	closure of $C_0^\infty(\Omega)$ in $H_m(\Omega)$, 313
I	identity operator, 5
I_X	identity operator on X, 5
$\Im\lambda$	the imaginary part of the complex number λ
$\operatorname{Im} A$	range (image) of the operator A
$\operatorname{ind}(A)$	index of the operator A, 184
$\mathcal{K}(X)$	the set of compact operators on the Banach space X, 191
$\kappa(\Gamma; 0)$	winding number, 226

$\operatorname{Ker} A$	null space of A
$L_2^m([a,b])$	Lebesgue space of $\mathbf{C}^m$-valued functions, 148
$\mathcal{L}(X)$	Banach space of all bounded linear operators on X, 5
$\mathcal{L}(X,Y)$	Banach space of all bounded linear operators from X into Y, 5
$\lambda_j(A)$	j-th eigenvalue of the compact operator A, 30
$m(\lambda_0, A)$	algebraic multiplicity, 26
$m(\lambda_0, W(\cdot))$	algebraic multiplicity of an operator function, 205
$m(\Gamma, W(\cdot))$	algebraic multiplicity relative to a contour Γ, 205
$n(A)$	dimension of the null space of A, 184
$\nu(A)$	number of non-zero singular values of A, 96
$P_\sigma(A)$	Riesz projection, 9
$\Re\lambda$	real part of the complex number λ
$\mathbf{R}^n$	real Euclidean n-space
$\mathcal{R}_\infty^{m\times m}(\mathbf{R})$	a set of rational $m \times m$ matrix functions, 232
$\operatorname{rank} A$	rank of the operator A
$\rho(A)$	resolvent set of the operator A, 5
$\rho(G,A)$	resolvent set of the pencil $\lambda G - A$, 49
S_1	the trace class operators, 105
S_2	the Hilbert-Schmidt operators, 143
$s_j(A)$	j-th singular value of the operator A, 96, 212
$\sigma(A)$	spectrum of the operator A, 5
$\sigma(G,A)$	spectrum of the operator pencil $\lambda G - A$, 49
$\sigma_{\mathrm{ess}}(A)$	essential spectrum of the operator A, 191, 373
$\mathbf{T}$	unit circle in the complex plane
$T_{\max,\tau,J}$	maximal operator, 295
$T_{\min,\tau,J}$	minimal operator, 300
$T_{R,J}$	a differential operator, 300
$\operatorname{tr} A$	trace of the operator A, 110
W_A	numerical range of A, 167
$^\perp N$	inverse annihilator of N, 292
$S^\perp$	orthogonal complement of S; and also at times, annihilator of S, 292
$(x_1,\ldots,x_n)^T$	transpose of the row vector $(x_1,\ldots,x_n)$
$\langle\cdot,\cdot\rangle$	inner product
$\|A\|_1$	trace class norm of A, 106
$\|A\|_2$	Hilbert-Schmidt norm of A, 143
$\emptyset$	empty set

FSC
www.fsc.org
MIX
Papier aus verantwortungsvollen Quellen
Paper from responsible sources
FSC® C105338